Application of Fracture Mechanics to Cementitious Composites

NATO ASI Series

Advanced Science Institutes Series

A Series presenting the results of activities sponsored by the NATO Science Committee, which aims at the dissemination of advanced scientific and technological knowledge, with a view to strengthening links between scientific communities.

The Series is published by an international board of publishers in conjunction with the NATO Scientific Affairs Division

A	Life Sciences	Plenum Publishing Corporation
B	Physics	London and New York
C	Mathematical and Physical Sciences	D. Reidel Publishing Company Dordrecht and Boston
D	Behavioural and Social Sciences	Martinus Nijhoff Publishers Dordrecht/Boston/Lancaster
E	Applied Sciences	
F	Computer and Systems Sciences	Springer-Verlag Berlin/Heidelberg/New York
G	Ecological Sciences	

Series E: Applied Sciences — No. 94

Application of Fracture Mechanics to Cementitious Composites

edited by

S.P. Shah

Department of Civil Engineering
Northwestern University
Evanston, Illinois
USA

1985 **Martinus Nijhoff Publishers**
Dordrecht / Boston / Lancaster
Published in cooperation with NATO Scientific Affairs Division

Proceedings of the NATO Advanced Research Workshop on Application of Fracture Mechanics to Cementitious Composites, Northwestern University, Evanston, Illinois, USA, September 4-7, 1984

Library of Congress Cataloging in Publication Data

NATO Advanced Research Workshop on "Application of
    Fracture Mechanics to Cementitious Composites"
    Application of fracture mechanics to cementitious
composites.

    (NATO ASI series. Series E, Applied sciences : no. 94)
    "Published in co-operation with NATO Scientific
Affairs Division."
    "Proceedings of the NATO Advanced Research Workshop
on 'Application of Fracture Mechanics to Cementitious
Composites,' Northwestern University, Evanston, Illinois.
U.S.A., September 4-7, 1984"--T.p. verso.
    Includes indexes.
    1. Cement composites--Fracture--Congresses.
2. Fracture mechanics--Congresses. I. Shah, S. P.
(Surenda P.) II. North Atlantic Treaty Organization.
Scientific Affairs Division. III. Title. IV. Series.
TA438.N37 1984   620.1'36   85-10485

ISBN-13: 978-94-010-8764-3   e-ISBN-13: 978-94-009-5121-1
DOI: 10.1007/978-94-009-5121-1

Distributors for the United States and Canada: Kluwer Boston, Inc., 190 Old Derby Street, Hingham, MA 02043, USA

Distributors for the UK and Ireland: Kluwer Academic Publishers, MTP Press Ltd, Falcon House, Queen Square, Lancaster LA1 1RN, UK

Distributors for all other countries: Kluwer Academic Publishers Group, Distribution Center, P.O. Box 322, 3300 AH Dordrecht, The Netherlands

PREFACE

Portland cement concrete is a relatively brittle material. As a result, mechanical behavior of concrete, conventionally reinforced concrete, prestressed concrete, and fiber reinforced concrete is critically influenced by crack propagation. It is, thus, not surprising that attempts are being made to apply the concepts of fracture mechanics to quantify the resistance to cracking in cementious composites.

The field of fracture mechanics originated in the 1920's with A. A. Griffith's work on fracture of brittle materials such as glass. Its most significant applications, however, have been for controlling brittle fracture and fatigue failure of metallic structures such as pressure vessels, airplanes, ships and pipelines. Considerable development has occurred in the last twenty years in modifying Griffith's ideas or in proposing new concepts to account for the ductility typical of metals. As a result of these efforts, standard testing techniques have been available to obtain fracture parameters for metals, and design based on these parameters are included in relevant specifications.

Many attempts have been made, in the last two decades or so, to apply the fracture mechanics concepts to cement, mortar, concrete and reinforced concrete. So far, these attempts have not led to a unique set of material parameters which can quantify the resistance of these cementitious composites to fracture. No standard testing methods and a generally accepted theoretical analysis are established for concrete as they are for metals.

One of the primary reasons for this lack of success is that most of the past work is based on the concept of linear elastic fracture mechanics. However, it is increasingly being realized that because of the large-scale heterogeneity inherent in the microstructure of concrete, strain softening, microcracking and large scale process zone, the classical linear elastic (or the classical elastic-plastic) concepts must be significantly modified to predict crack propagation in concrete. More recently, researchers in many countries are beginning to explore, theoretically, numerically and experimentally these hitherto unexplored aspects of nonlinearity associated with crack growth in cementitious composites as well as in ceramics and rocks.

The recently increased understanding and awareness of unusual aspects of crack growth has resulted, for example, from: optically observing crack growth in double torsion and double cantilever types of specimens; electron microscopy observations of crack growth in compact tension specimens; use of infrared spectroscopy and optical interference microscopy to study process zone; development of finite element programs to include the nonlinear process zone in structural modeling; theoretical analysis which includes the tensile strain softening in the process zone in front of the crack-tip; application and extension of continuum damage theory; analysis of stochastic aspects of crack growth; and better understanding of mixed mode fracture criteria.

As a result of this increased understanding of fracture processes and better numerical and theoretical modeling, it is also beginning to be recognized that use of nonlinear fracture mechanics can be advantageous in rational analysis of the behavior of concrete structures. Situations where fracture mechanics can be a useful tool include: impact and impulsive loading, dynamic shear fracture, some aspects of bond between reinforcement and concrete, and predictions of deflections and ductility.

The primary purpose of this workshop was to bring various researchers together in order to (1) delineate problems and prospects of applying and modifying fracture mechanics concepts to cementitious composites; (2) to synthesize various theoretical, numerical, analytical, experimental and microstructural aspects of crack propagation, and (3) to focus attention on yet unexplored but critical areas of research.

This proceedings volume contains the final contributions of lecturers and reporters.

I hope that the efforts of all who have contributed to this book will produce lasting and worthwhile results.

January 1985
Evanston, Illinois   Surendra P. Shah

CONTENTS

ACKNOWLEDGMENT

This workshop was one of a series of NATO Advanced Research Workshops and was sponsored by Materials Science Program of NATO Scientific Affairs Division [Program Director; Dr. Craig Sinclair]. Additional financial support was provided by the U.S. Army Research Office, Metallurgy and Materials Science Division [Director: Dr. George Mayer; Program Manager, Dr. John Bailey]. Assistance and encouragement of Dr. Robert D. Carnahan, Senior Vice President of Science and Technology, U.S. Gypsum Company and Dr. Jack Summerfield, Director, Administrative Services, U.S. Gypsum Company, were critical to the success of the workshop.

Advice and assistance received from Dr. Bruno Boley, Dean of The Technological Institute; Mrs. Lillian Warren, his Administrative Secretary; Dr. Raymond Krizek, Chairman, Department of Civil Engineering, Northwestern University; and from the members of the Organizing Committee are deeply appreciated.

Thanks are due to Mary D. Hill for her tireless, efficient and expert work as Workshop Secretary; Vellore S. Gopalaratnam, as Workshop Coordinator, and to Reji John and Yeou-Shang Jenq for their editorial assistance.

The editor would like to acknowledge continuous support provided by Mrs. Henny Hoogervorst, Acquisition Editor, Martinus Nijhoff Publisher.

SECTION I

ADVANCES IN NONLINEAR FRACTURE MECHANICS

APPLICATION OF FRACTURE MECHANICS
TO CEMENTITIOUS COMPOSITES
NATO-ARW - September 4-7, 1984
Northwestern University, U.S.A.
S. P. Shah, Editor

NON-LINEAR RESPONSE OF CONCRETE: INTERACTION OF SIZE, LOADING
STEP AND MATERIAL PROPERTY

G. C. Sih

Institute of Fracture and Solid Mechanics
Lehigh University
Bethlehem, Pennsylvania 18015 USA

1.   INTRODUCTION

Materials that may appear to be homogeneous to the naked eye
are, in fact, highly inhomogeneous when viewed through a microscope. To a design engineer, geometric irregularities alone are
not sufficient for defining the strength of material. The load
carrying capacities of structural members as a function of their
corresponding elongations are necessary. Such information can be
obtained by subjecting specimens with simple geometries to tensile
or compressive loads at controlled loading rates. The resulting
data plotted in terms of uniaxial stress and strain can be linear
or nonlinear depending on the combined effect of specimen size and
geometry, loading rate and material type. This procedure presents
fundamental difficulties in design when the stress and strain response becomes nonlinear. Lacking in particular is a knowledge of
how uniaxial test data could be used under service conditions
where the stress states are multiaxial.

The classical approach of identifying macroparameters such as
stress, strain, etc., with the failure of continuum elements in
combined stress states has not been successful, especially when
the deformation process is irreversible. It is now common knowledge that irreversibility is the cause of material damage due to
permanent deformation and slow crack growth (1-3). This process
is inherently loadtime history dependent which implies that *material damage is a function of load increment*. In other words, the
way with which damage is accumulated depends on each load step and
cannot be assumed arbitrarily. This fundamental character of the
physical process applies to all materials and concrete is no exception.

3

At the engineering scale level, the mechanical behavior of concrete is highly nonlinear. This presents difficulties to the designers who must anticipate structural components behavior for situations other than those tested. A major objective of nonlinear failure analysis is, therefore, to seek a theory such that the nonlinear data could be interpreted linearly via a damage parameter. The simple scheme of interpolation can then be used to unravel the complex interaction of design parameters. The crack extension resistance curves or simply R-curves in fracture mechanics are intended to linearize 1) the data associated with failure by yielding and fracture. Local and global failure must therefore be distinguished by a single criterion 2) and should be addressed covering the full range of failure behavior from plastic collapse to brittle fracture. End results should be expressed in terms of the *limiting values of the load and geometry*. Too often, they are camouflaged in the theory in terms of damage parameters.

One major difference between metal alloy and concrete perhaps lies in collecting reliable uniaxial data. No great difficulties are encountered in achieving a homogeneous stress distribution in a metal specimen under tension until necking begins. The same procedure, however, becomes problematic when a concrete or metal specimen is loaded in compression. Because of the end constraints due to friction, the mid-portion of the specimen tends to bulge and results in a nonhomogeneous stress field. The idea of pre-cracking concrete test specimens has thus attracted the attention of many recent investigators (7). A popular configuration is that of the three-point bend specimen with an edge crack. Data on the resistance of concrete to initiate rapid fracture can thus be collected by making use of the concept of linear elastic fracture mechanics (LEFM). Values of K_{1c} have been reported and studied by varying aggregate sizes, volume fractions of the mixture, moisture conditions, etc. As a rule, however, concrete fracture is always preceded by slow crack growth (8-10). This intervening stage of material damage is significant and cannot be treated by LEFM.

---

1. Those models based on crack opening displacement (COD) (4) and the J-integral (5) are of limited use as the resulting R-curves are still highly nonlinear and provides no predictive capability. A more specific example is given in (6) for the polycarbonate material whose stress and strain response is similar to that of concrete with a softening behavior.

2. Two or more independent failure criterion are frequently used to describe slow crack growth and the onset of rapid fracture. Arbitrariness in the model is then introduced at the expense of predictive capability.

The present investigation addresses the incremental damage of
concrete by slow crack growth prior to rapid fracture or separation of the material. Damage accumulation will be estimated by
application of the strain energy density theory in conjunction
with the pseudo-linear elastic (12) and incremental plasticity
stress analysis. The former assumes elastic unloading and latter
plastic unloading. Both models account for the accumulation of
permanent damage as the load steps are increased. Unloading characteristics of the material are shown to have a pronounced effect
on the history of damage even though the same failure criterion
was employed. R-curves for different specimen thickness, loading
steps and material types are obtained and appear as straight lines
in a S (strain energy density factor) versus a (crack length) plot.
The results clearly show that the strain energy density criterion
controls the qualitative feature of the R-curves with straight
lines preserved in both models. Changes in material unloading or
the way with which damage accumulates affect only the slopes of
the straight lines. Both the pseudo-linear and incremental plasticity models are considered to be restricted in application.
Discussion on their limitations 3) is beyond the scope of this
work and will be reserved for future communications.

2. DAMAGE ACCUMULATION CRITERION: STRAIN ENERGY DENSITY

Failure is known to initiate locally leading to eventual global instability in solids. Within the framework of continuum mechanics, the material building blocks consisting of finite size
elements 4) are assumed to be *homogeneous* with known mechanical
properties. Inhomogeneity in stress or energy state, however, occurs within the system as the state of affairs do vary from element to element for each load increment. This inhomogeneity is
best reflected through the fluctuation of energy stored in a unit
volume of material element known simply as the strain energy density 5):

---

3. Depending on the material type, the pseudo-linear analysis
tends to underestimate energy dissipated by permanent deformation
and incremental plasticity has the opposite influence. A more realistic model of material damage that preserves the degree of homogeneity from the uniaxial data to different locations of a multiaxial stress state is being developed at Lehigh University.

4. The size of these elements must be sufficiently large such
that irregularities in material microstructure would have a small
influence on the global behavior of the continuum.

5. In this work, the term "strain energy density" refers to
the energy in a volume element that is available for dissipation
as material is damaged.

$$\frac{dW}{dV} = \int_0^{\varepsilon_{ij}} \sigma_{ij} d\varepsilon_{ij} \qquad (1)$$

where σ_{ij} and ε_{ij} stand for the stress and strain components. Equation (1), of course, applies to any unique relation of σ_{ij} and ε_{ij}. Each material is assumed to possess a critical value of dW/dV or $(dW/dV)_c$ that can be obtained from the area under the true stress and true strain curve at fracture (13) regardless of whether the response is linear or nonlinear. Aside from the most idealized situation where fracture occurs suddenly with little or no other forms of energy dissipation, material elements are damaged gradually prior to total separation. This gradual damage is usually referred to as yielding. Let the energy dissipated during this process be denoted by $(dW/dV)_p$. Hence, the available release energy at failure is

$$\left(\frac{dW}{dV}\right)_c^* = \left(\frac{dW}{dV}\right)_c - \left(\frac{dW}{dV}\right)_p \qquad (2)$$

The way with which material elements undergo gradual damage is assumed to occur incrementally and follow the relation

$$\left(\frac{dW}{dV}\right)_c \text{ or } \left(\frac{dW}{dV}\right)_c^* = \frac{S_1}{r_1} = \frac{S_2}{r_2} = --- = \frac{S_j}{r_j} = --- = \text{const.} \qquad (3)$$

In equation (3), S_j are the strain energy density factors that correspond to the growth increments 6) r_j ($j = 1,2,---,n$), Figure 1. The "constant" is known from uniaxial test data and denotes a given damage state. For instance, S_j/r_j may terminate at S_c/r_c corresponding to global instability by rapid crack propagation or at S_a/r_a representing crack arrest. The rates of energy dissipation per unit volume are therefore described by incremental values of S_j and thus uniquely define the various failure modes including plastic collapse and brittle fracture as the two extremes. This will be discussed more specifically in connection with the subsequent numerical examples.

---

6. The first increment of growth r_1 is assumed to be larger than the radius of the core region, say r_0, which serves as a limiting distance of the continuum mechanics theory.

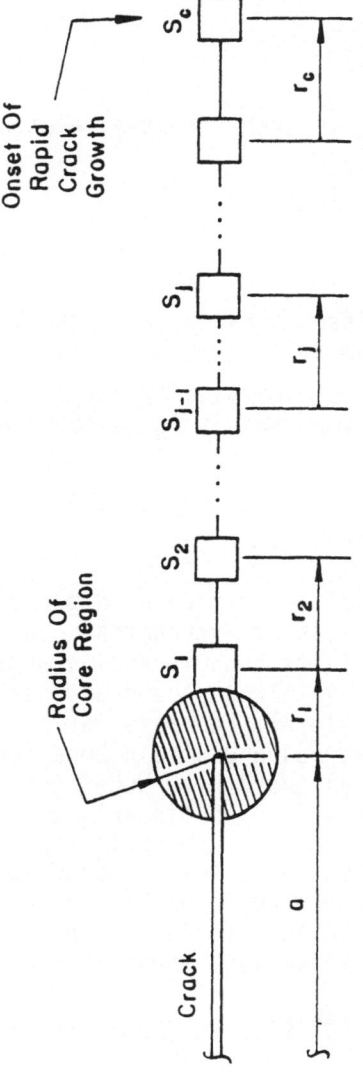

Figure 1. Incremental crack growth up to global instability.

The LEFM approach is simply a limiting case of the more general assumption stated by equation (3). Should fracture occur suddenly with no warning, then all energy is released at incipient fracture and the ligament that triggers this unstable process is

$$r_c = \frac{S_c}{(dW/dV)_c} \tag{4}$$

The critical value S_c is then directly related to the ASTM fracture toughness value K_{1c}:

$$S_c = \frac{(1+\nu)(1-2\nu)}{2\pi E} K_{1c}^2 \tag{5}$$

in which ν is the Poisson's ratio and E the Young's modulus. The measurement of K_{1c} and S_c must necessarily involve scatter as the amount of overload 7) to initiate fracture instability differs from specimen to specimen and is not accounted for in equation (5).

3. THREE-POINT BEND CONCRETE SPECIMEN: YIELDING AND SLOW CRACK GROWTH

Concrete is generally regarded as a two-phase composite system consisting of a framework of aggregates held together by cement paste. The effects of its nonhomogeneity on failure initiation and by crack growth have already been discussed in detail (8). At the macroscopic scale level, concrete may be assumed to be quasi-homogeneous and its stress and strain behavior follows those described in Figures 2(a) and 2(b). Under gradual compression, the stress increases with strain nonlinearly up to the point of fracture, Figure 2(a). Permanent damage is manifested by the fact that loading and unloading follow different paths. When the specimens are deformed by controlling the strain rate, the material continues to carry load and stretch after the peak stress. This feature is known as "softening" shown in Figure 2(b).

3.1 Specimen Size, Material Type and Controlled Deflection

The three-point bend specimen with an initial crack length of $a_0 = 5$ cm in Figure 3 is depicted for investigation. The specimen length h and depth t are kept, respectively, at 60 cm and

---

7. Since the critical condition or load σ_c corresponds to metastable equilibrium, it is by definition that a load larger than σ_c must be applied to trigger fracture.

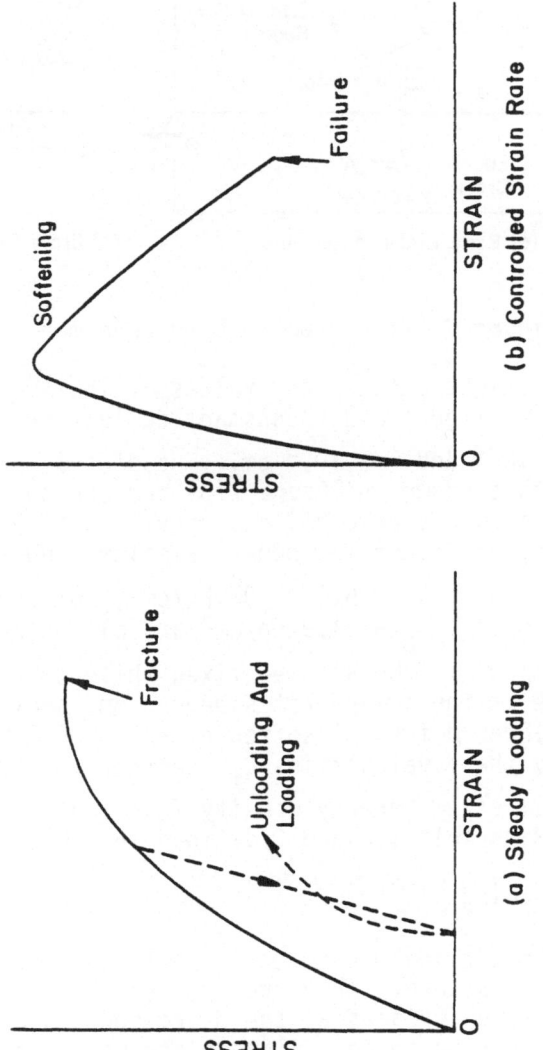

Figure 2. Stress and strain response of concrete.

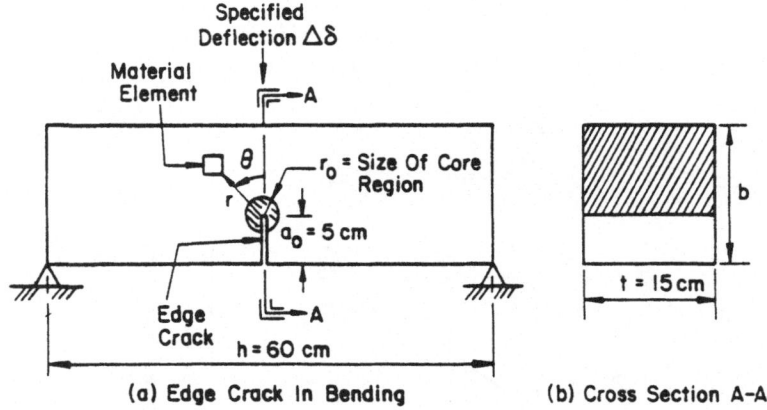

Figure 3. Three-point bend specimen.

15 cm while the height b takes the values of 15, 45, 222 and 310 cm. Load is controlled through constant deflection steps $\Delta\delta$ = 4 x 10^{-3}, 2 x 10^{-3} and 1 x 10^{-3} cm. The stress and strain behavior follows that shown in Figure 2(b) and can be approximated with sufficient accuracy by a bilinear curve as illustrated in Figures 4. Since all concretes behave similarly up to the point u, the Young's modulus E = 36.5 x 10^4 kg/cm^2, Poisson's ratio ν = 0.1, ultimate stress σ_u = 31.9 kg/cm^2 and ultimate strain ε_u = 0.87 x 10^{-4} cm/cm will be assumed fixed while the softening character can change as the damage rates depend on specimen size and loading step. Three different softening rates will be analyzed and specified by the final strains ε_f together with their corresponding critical strain energy density functions. They will be referred to as Material A, B and C defined in Table 1.

3.2 Elastic Unloading

Two of the basic failure modes in continuum mechanics are yielding and fracture. They are physically associated with the creation of free surfaces at the microscopic and macroscopic scale level. Referring to the bilinear stress and strain relation in Figure 4(a), permanent damage occurs when the strain surpasses the point u, say at p. The material then experiences a reduction in modulus from E to E* as the slope of the line op is smaller than that of ou. The term "elastic unloading" is used to describe this assumed behavior of shifting the path of reversibility from

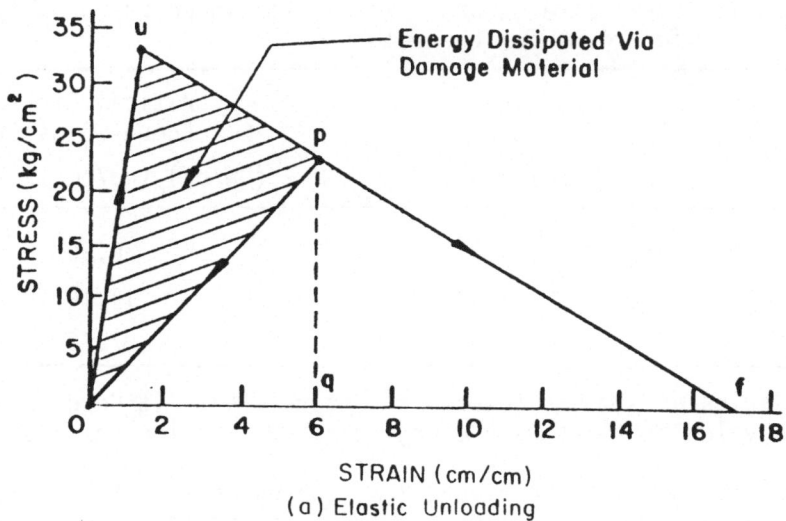

(a) Elastic Unloading

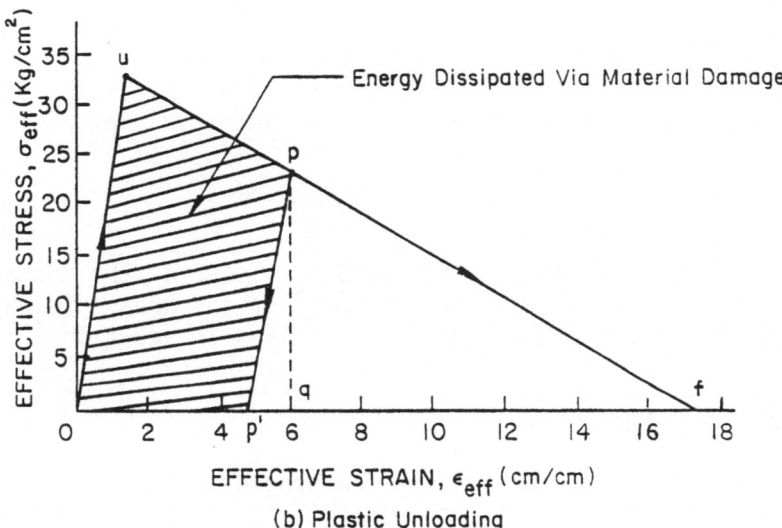

(b) Plastic Unloading

Figure 4. Unloading and bilinear behavior of concrete.

ou to op. Following the model 8) developed in (12), E* is discretized to 25 increments:

$$E^*(n) = \frac{26-n}{25} E, \quad n = 1, 2, ---, 25 \tag{6}$$

---

8. This quasi-linear damage model was first applied to analyze the progressive failure of metal alloys (12) by yielding and crack growth and later applied to concrete (8,10,14).

Table 1. Final strains and critical strain energy density functions for concrete

Material Type	Final Strain ε_f (cm/cm $\times$ 10^{-3})	Critical Strain Energy Density $(dW/dV)_c$ (kg/cm^2 $\times$ 10^{-3})
A	16.0	26.90
B	8.0	14.14
C	4.0	7.70

which varies as p moves towards f, Figure 4(a). It depends on the current value of the strain energy density function, dW/dV.

The region shaded in Figure 4(a) corresponds to the energy dissipated as a result of permanent damage and it can be calculated from

$$\left(\frac{dW}{dV}\right)_p = \frac{1}{2}\left(\sigma_u\varepsilon - \sigma\varepsilon_u\right) \tag{7}$$

in which σ and ε are the uniaxial stress and strain at p given by

$$\sigma = E^*\varepsilon = \frac{\sigma_u(\varepsilon_u+\varepsilon_f)}{\varepsilon_f+(\sigma_u/E^*)} \tag{8}$$

The available energy density for sudden release at the instant p is not the total area dW/dV but $(dW/dV)^*$ or the area opfq which excludes the dissipated energy density given in equation (7), i.e.,

$$\left(\frac{dW}{dV}\right)^* = \frac{1}{2}\left(\sigma_u\varepsilon_u + \sigma_u\varepsilon_f - \sigma_u\varepsilon + \sigma\varepsilon_u\right) \tag{9}$$

Accounted for in equation (9) are the progressive damage of elements local to the crack tip prior to total collapse of the system.

3.3 Plastic Unloading

When the material is damaged, it suffers permanent deformation or strain. This can be evidenced upon unloading of the specimen. For most of the engineering materials including concrete, unloading occurs along a path such as pp' in Figure 4(b) that is approximately parallel to the original loading path ou. The unrecoverable energy density $(dW/dV)_p$ at p is now oupp' and is considerably larger than that assumed in the model of elastic unloading. The available energy density $(dW/dV)^*$ is obviously reduced accordingly. A major assumption of the theory of plasticity is that the uniaxial stress and strain curve coincides with the effective stress σ_{eff} and effective strain ε_{eff} curve as shown in Figure 4(b). The on-set of plastic flow is thus determined by comparing 9)

$$\sigma_{eff} = \frac{1}{\sqrt{2}} \sqrt{(\sigma_1-\sigma_2)^2+(\sigma_2-\sigma_3)^2+(\sigma_3-\sigma_1)^2} \tag{10}$$

with the uniaxial yield strength of the material. Hence, only damage due to distortion is included. The energy density associated with volume change or dilatation 10) can be significant in regions local to the crack tip and is neglected in equation (10). This is one of the main reasons why plasticity theory is known to be inadequate for analyzing failure by yielding and crack growth. Nevertheless, it is useful to gain insight into the approximate nature of the plasticity theory that tends to overexaggerate on failure by distortion while the quasi-linear damage model under-estimates energy dissipation.

4. CRACK GROWTH RESISTANCE CURVES

As mentioned earlier, crack growth resistance curves are useful in that the nonlinear load versus deflection data can be related to the damage parameters. To this end, the finite element method is applied to calculate the strain energy density function ahead of the slowly moving edge crack in Figure 3 as the load is increased incrementally. The results pertaining to the quasi-linear damage model in (14) are also shown subsequently by dotted curves. They are presented for the purpose of comparison with those obtained from the incremental theory of plasticity.

4.1 Specimen Size Effect

Figure 5 displays the variations of the strain energy density factor S with crack growth $\Delta a = a-a_0$ for Material C with a constant $\Delta\delta/b$ ratio of 2.6×10^{-4}. Specimen size is altered by taking b = 15, 45 and 222 cm. The critical strain energy density factor S_c

---

9. The expression for ε_{eff} is similar to that for σ_{eff} except that the principal stresses σ_j (j = 1,2,3) are replaced by the principal strains ε_j (j = 1,2,3).

10. The energy dissipated by volume change and shape change are inherently coupled and unseparable in the nonlinear case. It is fundamentally unsound to arbitrarily assume any relation involving some proportion of dilatation and distortion energy density.

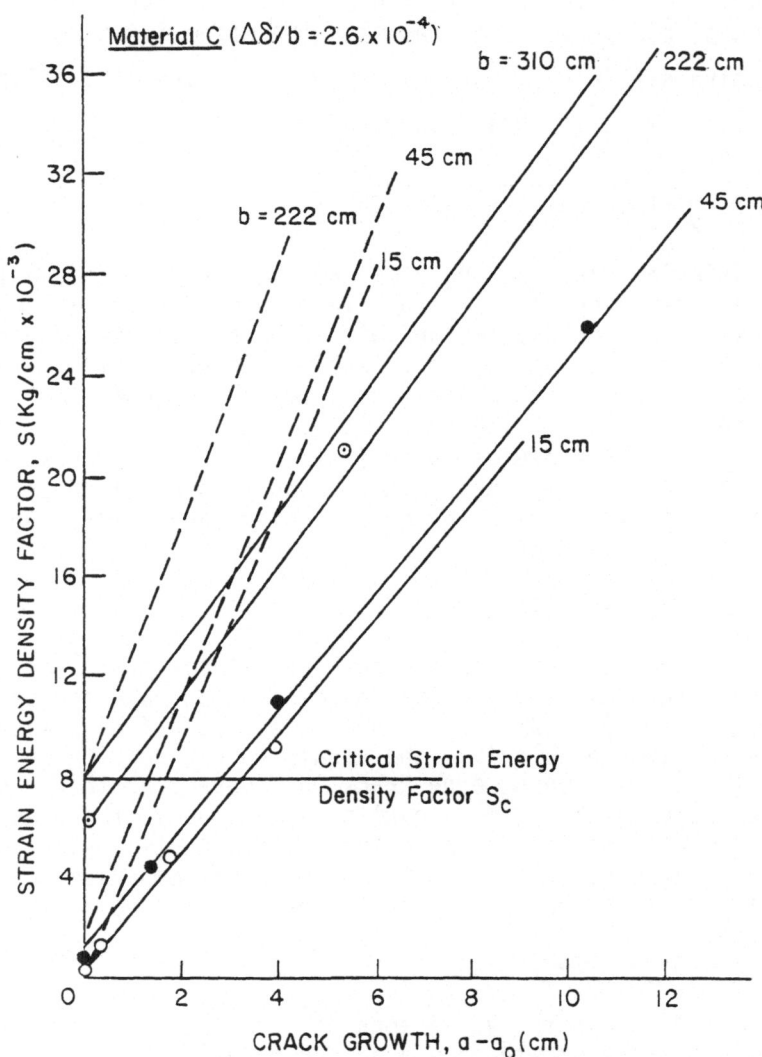

Figure 5.   Strain energy density factor as a function of crack growth for Material C, $\Delta\delta/b = 2.6 \times 10^{-4}$ and different specimen size.

for Material C in Table 1 is 8.0×10^{-3} kg/cm which corresponds to a K_{1c} value of 143.36 kg/cm$^{3/2}$. Those crack lengths that correspond to the onset of rapid fracture can be determined from the intersections of the S versus Δa curves with the S_c = const. line.

Slow crack growth is seen to be enhanced as b is decreased, a frequently observed phenomenon in testing. Complete brittle fracture for the quasi-linear model is reached when b = 222 cm while the plasticity theory yields a larger value of b equal to 310 cm. The

15

plasticity model also predicts considerably more slow crack growth. This is a consequence of parallel unloading where more energy is dissipated by plastic deformation. The specimen would behave more softly and less work would be required for each subsequent crack growth. A comparison of the load versus deflection variations is given in Figure 6. As a result of dS/da = const., interpolation may thus be used to obtain the relation between b and Δa other than those displayed in Figure 5. Numerical values of S and Δa can be found in Table 2. The straight line feature of the S versus Δa behavior checked well with the experimental data on polycarbonate (6).

4.2 Change in Loading Step

The change in loading step Δδ is exhibited in Figure 7. Three sets of S versus a curves with data given in Table 3 will be dis-

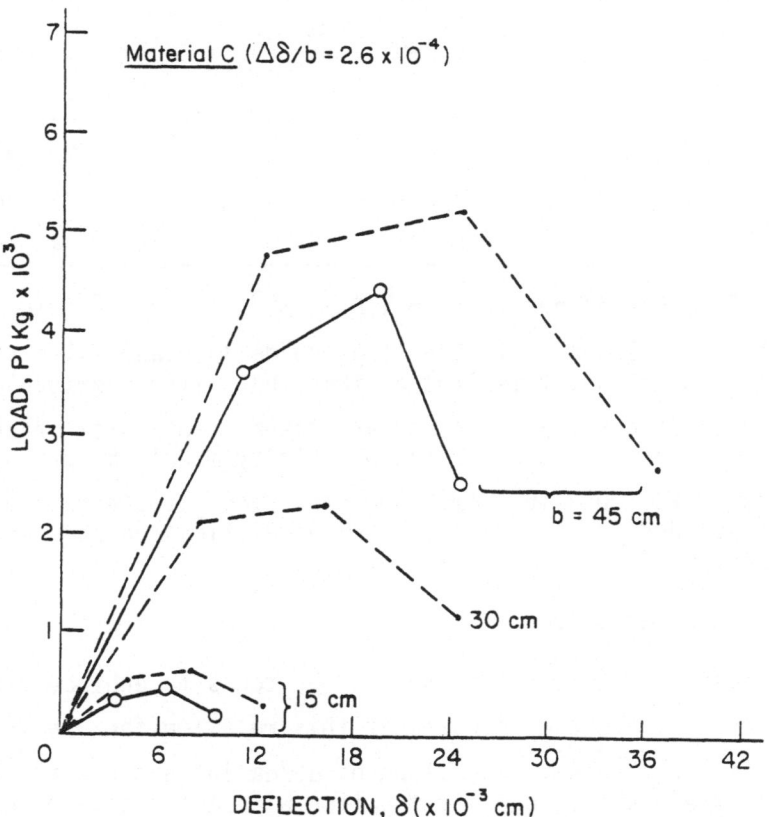

Figure 6.   Load versus deflection for Material C, $\Delta\delta/b = 2.6 \times 10^{-4}$ and different specimen size.

Table 2. Crack growth data for Material C, $\Delta\delta/b = 2.6 \times 10^{-4}$ and different specimen size

Dimension b (cm)	Crack Growth Δa (cm)	Strain Energy Density Factor S (kg/cm x 10^{-3})
15	0.000	0.441
	0.336	1.346
	1.654	5.272
	3.877	8.874
45	0.000	0.821
	1.118	4.470
	3.821	10.814
	10.246	25.699
222	0.000	6.126
	5.248	20.991
	23.658	73.640
	54.737	124.319

cussed. They correspond to $\Delta\delta = 1.0 \times 10^{-3}$, 2.0×10^{-3} and 4.0×10^{-3} cm. In general, the slope dS/da increases with $\Delta\delta$. For a given S_c value, this implies that the critical crack length decreases if larger loading steps are taken. More slow crack growth is again predicted for the plasticity model except when $\Delta\delta$ is increased to 4.0×10^{-3} cm. However, since strain rate effects are not included in the analysis, the predictions on change in $\Delta\delta$ are subject to further examination.

4.3 Toughness Variation

Increase in fracture toughness or S_c tends to enhance slow crack growth. Figure 8 illustrates this relation for b = 15 cm and $\Delta\delta = 4.0 \times 10^{-3}$ cm. The slope of dS/da increases with the critical strain energy density function with Material A being the largest. By holding the specimen size and loading step constant, the energy input remains the same for both the quasi-linear and plasticity model. Since elastic unloading underestimates the dissipation energy, more energy would be available to prolong slow crack growth. This is the reason why the slopes of the dotted

Table 3. Crack growth data for Material C, b = 15 cm and
different loading steps

Displacement δ (cm x 10^{-3})	Crack Growth Δa (cm)	Strain Energy Density Factor S (kg/cm x 10^{-3})
$\Delta\delta = 1.0$ x 10^{-3} cm		
1	0.106	0.423
2	0.164	0.654
3	0.167	0.667
4	0.205	0.821
5	0.228	1.106
6	0.230	1.117
7	0.349	1.398
8	0.350	1.399
$\Delta\delta = 2.0$ x 10^{-3} cm		
2	0.111	0.442
4	0.162	0.651
6	0.273	1.091
8	0.293	1.172
10	0.403	1.613
12	0.593	2.377
14	0.749	2.997
16	1.053	4.211
$\Delta\delta = 4.0$ x 10^{-3} cm		
4	0.313	1.252
8	0.497	1.987
12	1.131	4.522
16	2.087	8.349

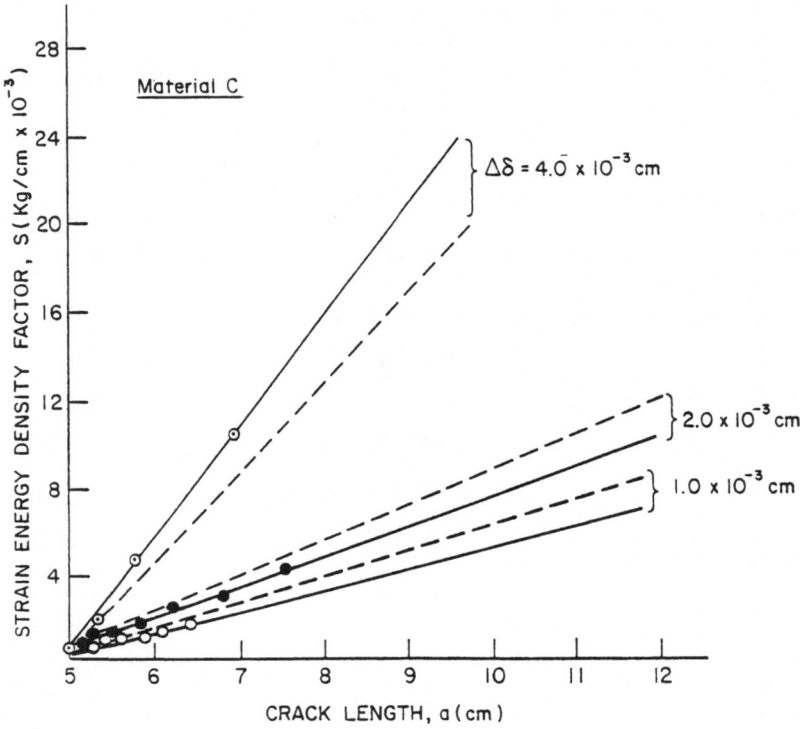

Figure 7. Strain energy density factor as a function of crack length for Material C, b = 15 cm and different deflection step $\Delta\delta$.

lines are less than those for the solid lines. Refer to Table 4 for details. The data for Material C has already been given in Table 2.

5. CONCLUDING REMARKS

Based on the uniaxial stress and strain data for concrete, the structural behavior of the three-point bend specimen with a crack is predicted by assuming elastic and plastic unloading. While the qualitative feature of the R-curves in the strain energy density factor versus crack growth plot is unchanged, the difference in energy dissipation of the two models clearly affects the predictions quantitatively. This has been illustrated for the entire range of failure modes from the very small specimens that tend to fail by plastic collapse to the very large ones that are more vulnerable to brittle fracture. Most significant is that the R-curves are *straight lines* which permit the designers to use interpolation and extract information on allowable load and component size for situations other than those tested.

Table 4. Crack growth data for b = 15 cm, $\Delta\delta = 4.0 \times 10^{-3}$ cm and different material

Material Type	Displacement δ (cm × 10^{-3})	Crack Growth Δa (cm)	Strain Energy Density Factor S (kg/cm × 10^{-3})
A	4	0.046	1.123
	8	0.079	1.921
	12	0.108	2.617
	16	0.165	4.012
	20	0.216	5.233
	24	0.298	7.218
	28	0.392	9.477
B	4	0.106	1.198
	8	0.142	1.605
	12	0.231	2.611
	16	0.317	3.587
	20	0.496	5.611
	24	0.773	8.733
	28	1.161	13.116

The size of structural members is known to influence the failure mode and hence the load carrying capacity. This can be best illustrated by a plot of the allowable load as a function of the parameter b in the three-point bend specimen. The results can be normalized to the maximum load $P_{max}^{(2)}$ at incipient fracture as defined by ASTM (15) through K_{1c}:

$$P_{max}^{(2)} = \frac{tb^{3/2}}{hf(a_0/b)} K_{1c} \tag{11}$$

in which $f(a_0/b)$ is given by

$$f(\frac{a_0}{b}) = 2.9 \left(\frac{a_0}{b}\right)^{1/2} - 4.6 \left(\frac{a_0}{b}\right)^{3/2} + 21.8 \left(\frac{a_0}{b}\right)^{5/2}$$
$$- 37.6 \left(\frac{a_0}{b}\right)^{7/2} + 38.7 \left(\frac{a_0}{b}\right)^{9/2} \tag{12}$$

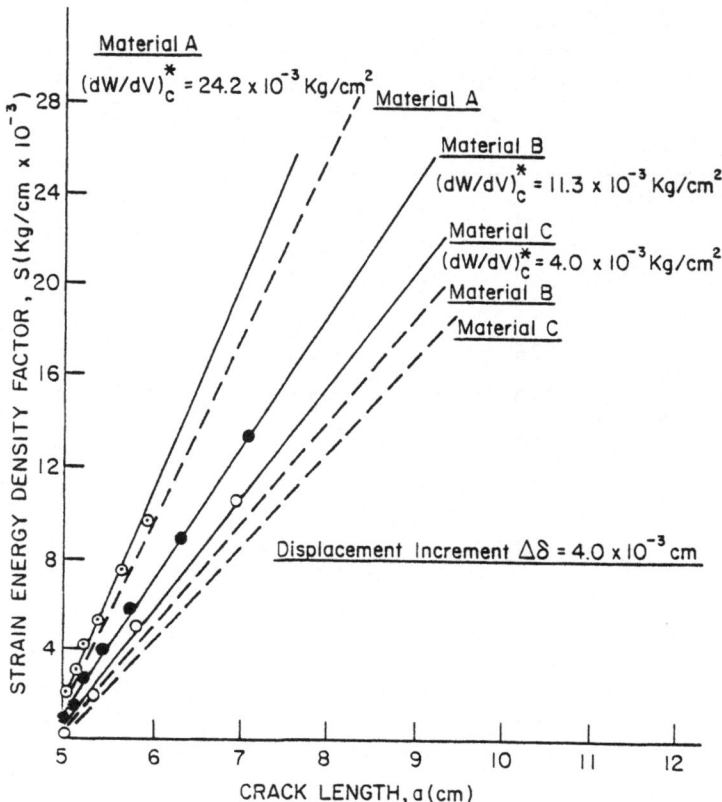

Figure 8.   Strain energy density factor as a function of crack
length for b = 15 cm, $\Delta\delta = 4 \times 10^{-3}$ cm and different
softening rates.

Figure 9 plots the variations of load percentage with b such that
100% is identified with brittle fracture. The other extreme limit
load $P_{max}^{(3)}$ corresponding to plastic collapse can be estimated from

$$P_{max}^{(3)} = \frac{2}{3} \frac{\sigma_u t(b-a_0)^2}{h} \tag{13}$$

where $b-a_0$ is the ligament size at collapse. The loads $P_{max}^{(1)}$ and
$P_{max}^{(4)}$ refer, respectively, to the cases of elastic and plastic unloading. These results are summarized in Figure 9. For a given
specimen size, limit analysis overestimates the failure load up to
$b \leq 120$ cm serving as an upper bound. The elastic-plastic damage
model gives the lower bound failure loads. All these models contain approximations that may not conform with reality. In partic-

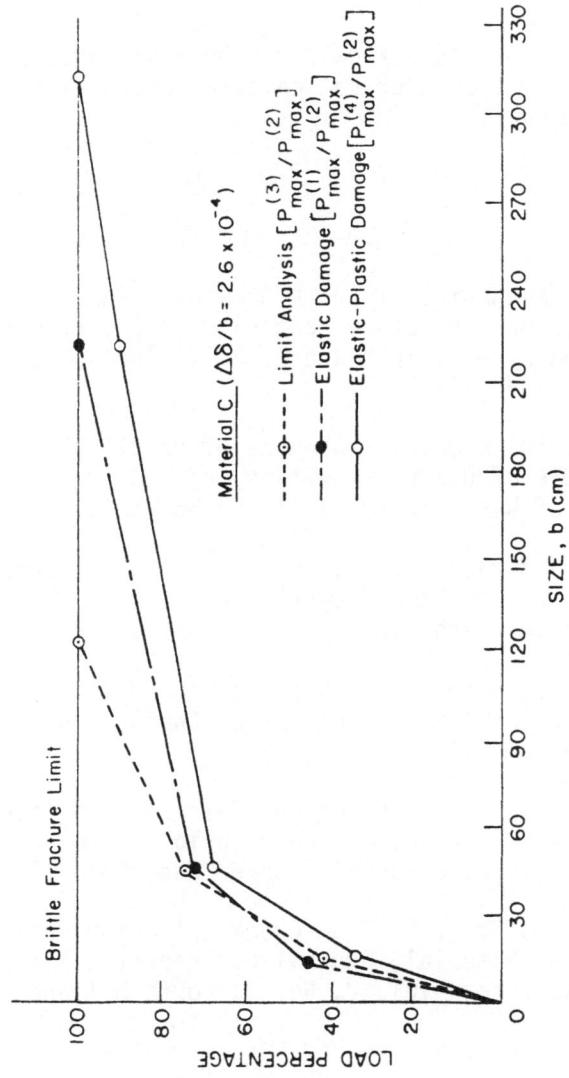

Figure 9. Load percentage versus specimen height for Material C,
$\Delta\delta/b = 2.6 \times 10^{-4}$ and different models.

ular, the strain effects are not included and they can differ from element to element in regions close to the crack tip. The association of this nonhomogeneity in the stress strain behavior with uniaxial test data must be understood and requires a change in the basic philosophy of the continuum mechanics approach with regard to the way of generating the constitutive relations.

ACKNOWLEDGEMENT

The author wishes to acknowledge the assistance provided by Mr. D. Y. Tzou for the numerical calculations on the R-curves for the plasticity model.

REFERENCES

1. Sih, G. C., Mechanics of Crack Growth: Geometrical Size Effect in Fracture, Fracture Mechanics in Engineering Application, edited by G. C. Sih and S. R. Valluri, Sijthoff and Noordhoff (1979) 3-29.

2. Sih, G. C., The Mechanics Aspects of Ductile Fracture, Continuum Models of Discrete Systems, edited by J. W. Provan, University of Waterloo Press (1977) 361-386.

3. Sih, G. C. and Madenci, E., Crack Growth Resistance Characterized by the Strain Energy Density Function, International Journal of Fracture Mechanics, Vol. 18, No. 6 (1983) 1159-1171.

4. Ballarini, R., Shah, S. P. and Keer, L. M., Crack Growth in Cement Based Composites, Journal of Engineering Fracture Mechanics (to appear).

5. Bernstein, H. L., A Study of the J-Integral Method Polycarbonate, AFWAL-TR-82-4080, Air Force Wright Aeronautical Laboratories, Wright-Patterson Air Force Base, Ohio (August 1982).

6. Sih, G. C. and Tzou, D. Y., Crack Extension Resistance of Polycarbonate Material, Journal of Theoretical and Applied Fracture Mechanics, Vol. 2, No. 3, North Holland (in press).

7. Mindess, S., The Cracking and Fracture of Concrete: An Annotated Bibliography, 1928-1980, Materials Research Series, Report No. 2, I.S.S.N. 0228-4251, The University of British Columbia, Vancouver (1981).

8. Sih, G. C., Mechanics of Material Damage in Concrete, Fracture Mechanics of Concrete: Material Evaluation and Testing, edited by A. Carpinteri and A. R. Ingraffea, Martinus Nijhoff Publishers, The Hague (1984) 1-29.

9. Ingraffea, A. R. and Saouma, V., Numerical Modeling of Discrete Crack Propagation in Reinforced and Plain Concrete, Fracture Mechanics of Concrete: Structural Application and Numerical Application, edited by G. C. Sih and A. DiTommaso, Martinus Nijhoff Publishers, The Hague (1984).

10. Carpinteri, A., Scale Effects in Fracture of Plain and Reinforced Concrete Structures, Fracture Mechanics of Concrete: Structural Application and Numerical Application, edited by G. C. Sih and A. DiTommaso, Martinus Nijhoff Publishers, The Hague (1984).

11. Gdoutos, E. E., Mixed Mode Crack Extension, Engineering Application of Fracture Mechanics, Vol. II, Martinus Nijhoff Publishers, The Hague (1984).

12. Sih, G. C. and Matic, P., A Pseudo-Linear Analysis of Yielding and Crack Growth: Strain Energy Density Criterion, Defects, Fracture and Fatigue, edited by G. C. Sih and J. W. Provan, Martinus Nijhoff Publishers, The Hague (1983) 223-232.

13. Gillemot, L. F., Criterion of Crack Initiation and Spreading, International Journal of Engineering Fracture Mechanics, Vol. 8 (1976) 239-253.

14. Carpinteri, A. and Sih, G. C., Damage Accumulation and Crack Growth in Bilinear Materials with Softening: Application of Strain Energy Density Theory, Journal of Theoretical and Applied Fracture Mechanics, Vol. 2, No. 2, North Holland (in press).

15. Plane Strain Crack Toughness Testing of High Strength Metallic Materials, edited by W. F. Brown, Jr. and J. E. Srawley, ASTM Special Technical Publication No. 410 (1966).

APPLICATION OF FRACTURE MECHANICS
TO CEMENTITIOUS COMPOSITES
NATO-ARW - September 4-7, 1984
Northwestern University, U.S.A.
S. P. Shah, Editor

FRACTURE PROCESS ZONE OF CONCRETE

A. S. Kobayashi,* N. M. Hawkins,** D. B. Barker***
and B. M. Liaw*

University of Washington
*Department of Mechanical Engineering
**Department of Civil Engineering
Seattle, WA 98195
***On leave from University of Maryland at the time
of this study

ABSTRACT

Double displacement compliance and replica techniques were
used to determine the macro and total crack lengths, respectively
in nineteen crack-line wedge-loaded, double cantilever beam (CLWL-
DCB) mode I concrete specimens. Details of the fracture process
zone, which lies between the macro and hairline crack tips, were
determined through numerical experiments involving a finite element model of the CLWL-DCB concrete specimens. With a critical
crack tip opening displacement (CTOD) at the macrocrack tip as a
subcritical crack growth criterion, this fracture process zone was
then incorporated into a finite element model of CLWL-DCB concrete
specimen which was then driven in its propagation mode to reproduce some of the subcritical crack growth experiments. This
finite element program was also used in its propagation mode to
replicate mixed-mode fracture experiments involving diagonal
tension fractures of modified CLWL-DCB concrete specimens.

INTRODUCTION

Fracture research in concrete can be divided into two cat-
egories, linear elastic [1-6] and nonlinear [7-11] fracture
mechanics. Those who advocate linear elastic fracture mechanics
(LEFM) use visual and compliance techniques to establish the
required crack dimensions from which the stress intensity factors

are computed. The importance of this computation is underscored in Reference [4] which attributes the unjustified demise of LEFM in concrete fracture studies to the erroneous stress intensity factor used in Reference [12]. While the advances [13,14] in finite element method have reduced if not eliminated problems in computational accuracy, few advances in the crack detection techniques in concrete have been made during this period of controversies.

In the following sections, the authors will present crack length measurements which were obtained by a new crack detection technique. These crack measurements are then used interactively to drive a finite element program in its generation mode which in turn yielded the aggregate interlocking forces ahead of the macrocrack tip.

MEASURED CRACK LENGTHS

The fracture specimens used in a series of experiments [15] conducted by the authors and their colleagues are the crack-line-wedge-loaded, double-cantilever beam (CLWL-DCB) sepcimens described in ASTM E 561-80, Standard Recommended Practice for R-Curve Determination. The specimen and loading mechanism which is shown schematically in Figure 1, is a rigid loading system where stable crack growth can be controlled through the applied wedge displacement. The specimens used in the mode I experiments are shown in Figure 2. This type of specimen has been extensively characterized so that crack growth can be determined by measuring crack opening displacements at two locations V_1 and V_2 in Figure 2, along the crack line.

Two different sized specimens were used in the study. The normal specimen dimensions were 2 in. (51mm) in thickness and W = 15 in. (381mm) with the starter notch length of a_{mo} = 5.25 in. (133mm) and a_{mo}/W = 0.35. Another group of specimens were 50% larger with a thickness of 3 in. (76mm) and W = 22.5 in. (572mm) with a_{mo}/W = 0.35. A miniature 5,000 lb. (22.5 kN) load cell was embedded in the split-pin loading fixture for monitoring crack line load. The two crack line displacements, V_1 and V_2, were monitored with clip gages. The three transducer outputs were recorded on a digital data acquisition system for later plotting and analysis.

Great care was exercised in the construction of the loading frame and during the experiments to eliminate any out-of-plane bending, which would introduce errors in crack-line displacement measurements, of the sepcimen. A double displacement compliance relation was used to determine the effective elastic crack length in the specimen. The crack opening displacements were an order of

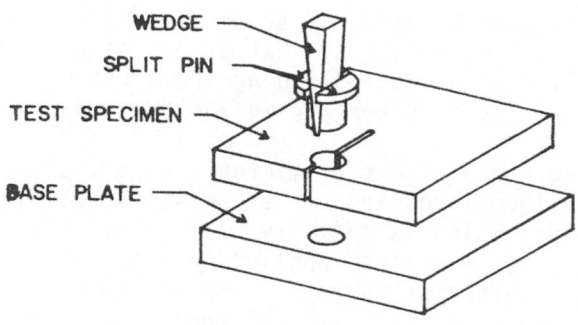

Figure 1.   Crack-Line Wedge-Loading Assembly

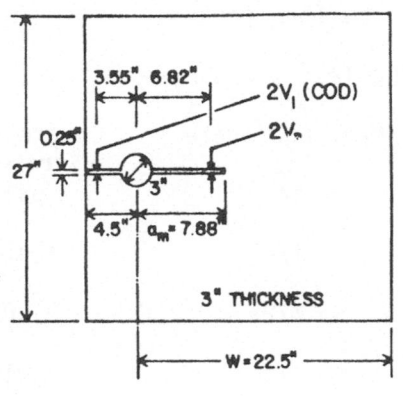

(a) LARGE SPECIMEN

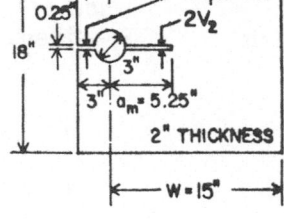

(b) SMALL SPECIMEN

Figure 2.   CLWL DCB Specimens

magnitude smaller than those measured in typical metal CLWL-DCB specimens and thus errors in crack opening displacement had to be minimized. This high resolution desired in crack opening displacement measurement meant that even small extraneous displacements caused by specimen bending had to be eliminated. Complete elimination is extremely difficult. Extraneous displacements were detected by a small loading and unloading cycle where the out-of-plane bending appears as a hysteresis in the $2V_1 - 2V_2$ plot.

The crack length and extent of microcracking was measured by a replicating technique involving use of an acetylcellulose film. The technique, which is commonly used in electron microscopy and fractography, was first applied by Kobayashi and Fourney [16] for microcrack detection in rock. For our test series, one surface of the fracture specimen was polished with a carborundum stone while the concrete was still green and replicas of the polished surface were taken at specified load intervals. The replicating films, when inspected with oblique illumination under a low power microscope, i.e., 20-140 power, clearly showed the extent of surface cracking. The replicating technique consistently followed surface cracks 1/2 to 1 inch (13-25mm) further in length than that determined by viewing the surface directly under 10x magnification.

The test series consisted of four groups with four specimens per group. The group 2 specimens were tested twice. So that, in all, nineteen CLWL-DCB specimens were tested. Table 1 summarizes the major variable changes between the four groups.

Table 1 - CLWL-DCB Test Groups

Group	No. of Specimens	Size W in.	Max. Aggregate Size in.	Compressive Strength psi
1	4	15	1/4	4,470
2	4	15	1/2	3,900
2'	3	15	1/2	5,010
3	4	15	1/4	9,070
4	4	22.5	1/2	3,470

The concrete and aggregate for all the specimens was carefully isolated and stored before casting and thus the same concrete ingredients were used throughout the test series. The cement was a type III, high early strength Portland cement, and the sand and gravel were obtained locally. There were a total of five castings for the four groups with various mix proportions and aggregate grading. All the specimens in each test group were cast from the same mix and cured in plastic bags before testing. Average curing time for Groups 1 through 4 and Group 2' was five and ten weeks respectively. Test cylinders were also cast with the specimens and compressive strength tests conducted on the same day as completion of the mode I test. The testing for each group of specimens was completed within a period of a maximum of five days.

As mentioned earlier, great care was exercised in placing the specimens in the loading frame to minimize any out-of-plane bending. Once the specimen alignment was confirmed, loading commenced at a rate of about 1/60 inch/minute (420 m./minute) at the load line. Typically, loading was stopped five times during the test to take a replica of the specimen surface. Even with the relative slow rate of loading, it was evident by watching the displacement transducer output that cracking continued for some time after loading was stopped. Thus, sufficient time, typically 5 - 10 minutes, had to be allowed for the displacement and load outputs to stabilize before each replica was taken.

Once a replica had been taken, a small unload-reload cycle was applied to the specimen in order to obtain an unloading-reloading compliance curve. Due to machine constraints, unloading was conducted under manual machine control at a rate as close as possible to the loading rate.

For several specimens a portion of the specimen containing the crack was excised after carefully removing the specimen from the loading fixture. This smaller piece was then sectioned along the midplane of the specimen. The crack length seen was then compared with that recorded by the replicating film. Total crack lengths were identical within the researchers' resolution capabilities for determining crack lengths on the sectioned surfaces. The mid-plane crack was, however, less straight than the surface crack, and followed a more random path that weaved around aggregate particles.

A typical double displacement compliance plot from one of the tests is shown in Figure 3. The crack opening displacement, $2V_1$, nearest the mouth is the abscissa and the crack opening displacement, $2V_2$, closest to the crack tip is the ordinate. Figure 4 is the corresponding compliance plot for the crack opening displacement, $2V_1$. The crack line load was measured with the miniature load cell inside the split pin. The slope of the compliance plot is related to the crack length. From these two compliance plots, two crack lengths can be calculated at each point where a replica was taken. The first length can be calculated from the slope of a secant drawn from the origin to the specific point on the compli-

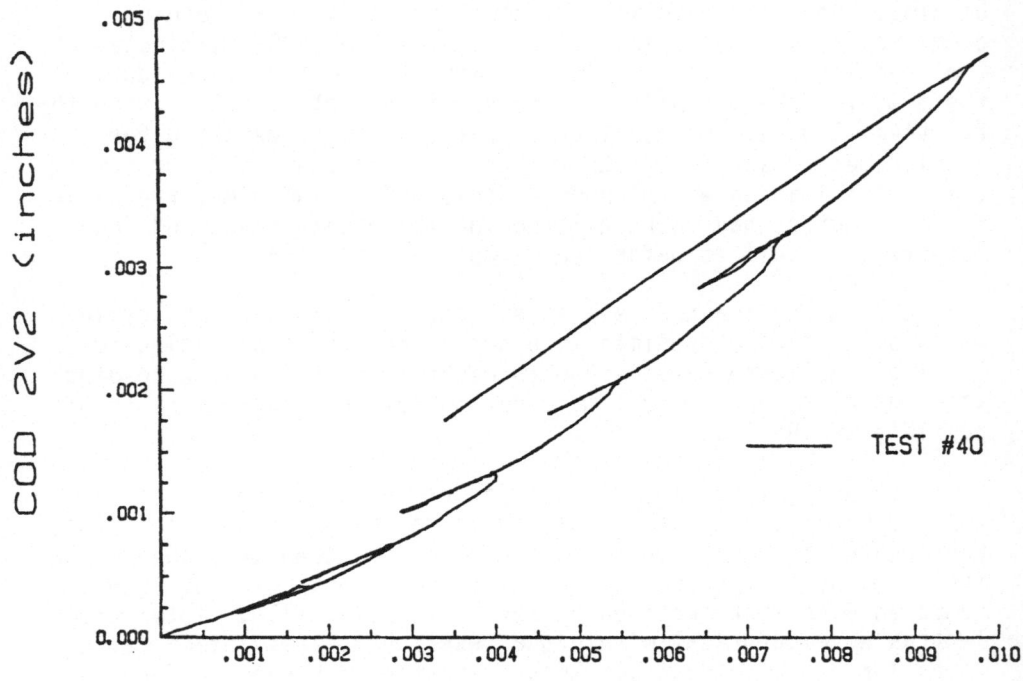

Figure 3. Displacement Record - Large Specimen.

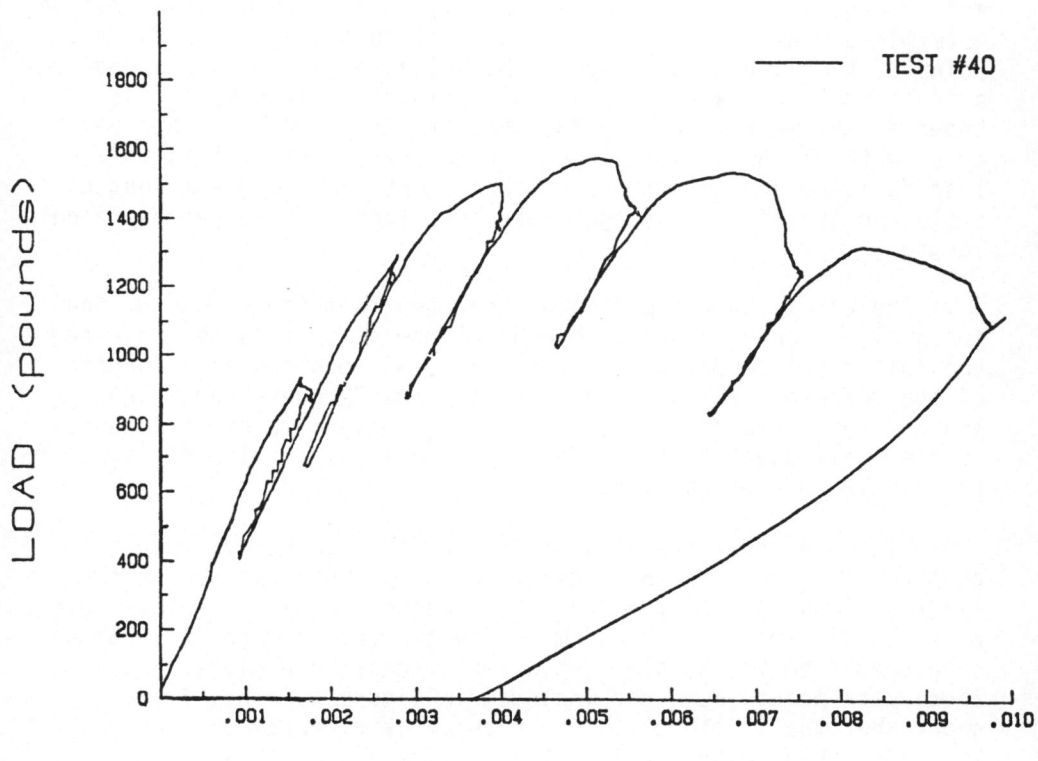

Figure 4. Load vs Displacement Relation - Large Specimen

ance curve. This slope corresponds to an effective crack length a. That effective crack length is the elastic equivalent crack length that includes the effect of the crack-tip nonlinear region. In the terms common for metals testing, the effective crack length is the physical crack size augmented by the effects of crack-tip plastic deformation. For the specimens reported here, the effective crack length is the equivalent length of an elastic traction free crack that includes compensation for the non-linear process zone region at its tip caused by microcracking and aggregate interlock. A second "effective" crack length can be determined from the slope of the partial unloading curve for any point of the loading envelope. The unloading slope corresponds to a traction free crack. In metals testing this length, a_m, is known as the macrocrack length, that is, the length of the open crack without the plastic zone. In concrete testing this un-loading slope should correspond to a traction free crack length excluding the effects of aggregate interlock and the process zone length.

The effective crack length, computed from the slopes of the secant line for both the double displacement compliance curve and the load displacement compliance curve, was plotted as a function of the measured hairline crack length, the "process zone" length, and the total crack length. There was no significant difference in the small scatter of the data about a straight line of best fit for the two different plots.

Figure 5 shows all data points for the macrocrack length computed from the double displacement compliance plots. Crack lengths computed from the load-displacement compliance plots were consistently smaller and had more scatter than the crack lengths predicted from the double displacement compliance plots, thus these data points were not included in Figure 5. This figure shows that the length of the macrocrack is directly proportional to the crack as measured from the replica of the total crack length.

Figure 5 also indicates that the fracture process zone or the region of microcracking and aggregate interlock continues to grow in size as the crack grows. With the larger sized specimens, this region of microcracking and aggregate interlock was measured in excess of 4.5 in. (114mm) with no sign of reaching a constant value.

FRACTURE PROCESS ZONE

The sensitivity of the replica technique resulted in observed lengths for the fracture process zone where aggregate interlocking prevails which were in excess of those reported previously

[8,11]. In theory, such a fracture process zone is akin to that used in fracture mechanics [17,18] in metals. More importantly, however, it was found that the relationship between the lengths of the macrocrack and the fracture process zone was reproducible for different specimens and different observers. The data points for the average curve in Figure 5 were used to determine the characteristics of the crack opening resistance within that zone.

Initially, the generation mode [19] of finite element analysis was used to conduct a series of numerical experiments which mimicked actual experiments. Specifically, attempts were made to match the applied wedge load versus crack opening displacement relation for the CLWL-DCB specimens by an iterative procedure involving a postulated crack opening resistance for the experimentally determined fracture process zone. The objective was to extract the otherwise unobtainable information by hybrid experimental-numerical analysis. These numerical experiments resulted in an improved model of the strip damage zone that incorporates recent findings for other materials. The propagation mode [19] of finite element analysis was then used to extend the crack and the crack extension histories obtained both numerically and experimentally were compared.

The crack propagation criterion, imposed on this analysis for stable crack growth was the CTOD criterion [21,22] used by Wecharatana and Shah [11]. Extensive trial-and-error study showed that the measured lengths of the macrocrack and the fracture process zone could be duplicated with the stress distribution for the fracture process zone shown in Figure 6. The unloading zone, which trails the fracture process zone and in which f'_t is fully developed, is modeled after Ref. [20] and is a special case of the zone suggested in Ref. [11]. The length of the unloading zone was chosen such that the CTOD's at the macrocrack tip and at the beginning of the unloading zone were 0.0015 and 0.0005 inch, respectively. The average value of these CTOD's coincides with the critical CTOD used by Wecharatana and Shah [11] their constant f'_t model.

Figure 7 shows a typical wedge load versus COD relation. It also illustrates the computational sequences in the propagation analysis using the finite element model with the fracture process zone of Figure 7, for a small CLWL-DCB specimen. With increasing applied wedge displacement, the crack tip of the fracture process zone is advanced one nodal distance when the crack tip element reaches the tensile strength of the concrete. Kinematically admissible equilibrium positions, which also satisfy the two prescribed CTOD's, are then sought through a sequence of iterative processes. As shown in Figure 7, equilibrium is achieved at wedge loads lower than those predicted prior to crack extension. The final wedge loads are in good agreement with those observed

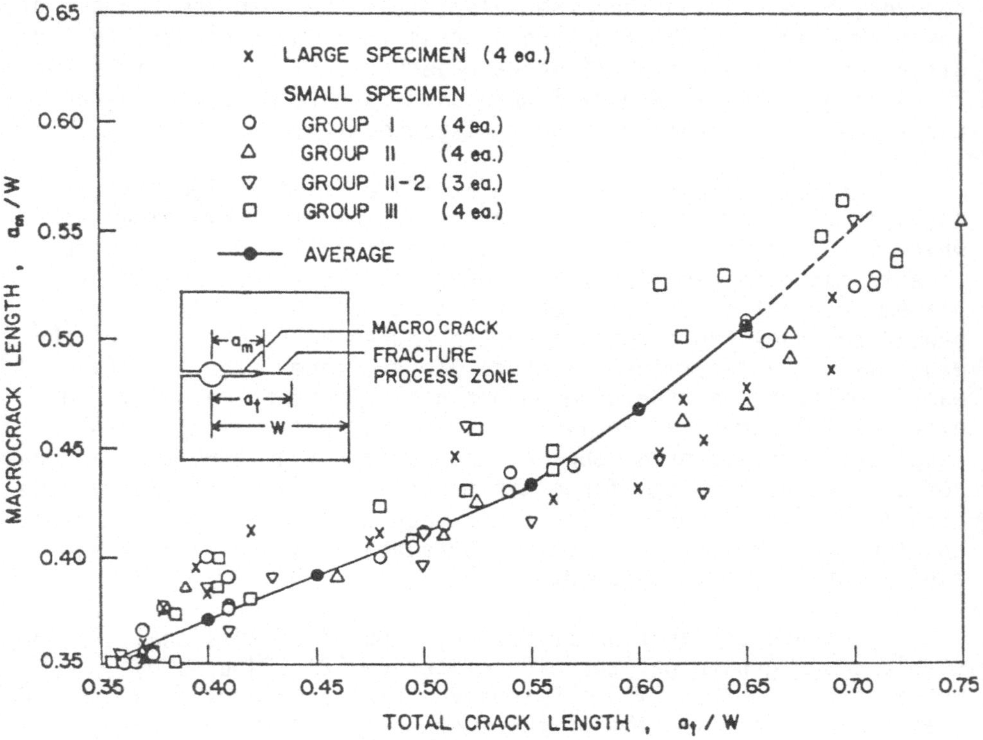

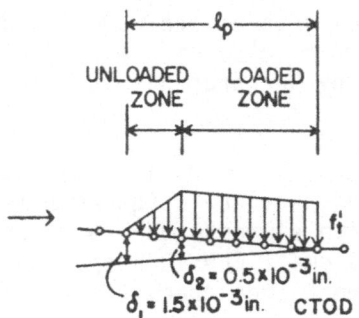

Figure 5. Total Crack Length vs Macrocrack Length

Figure 6.　Fracture Process Zone

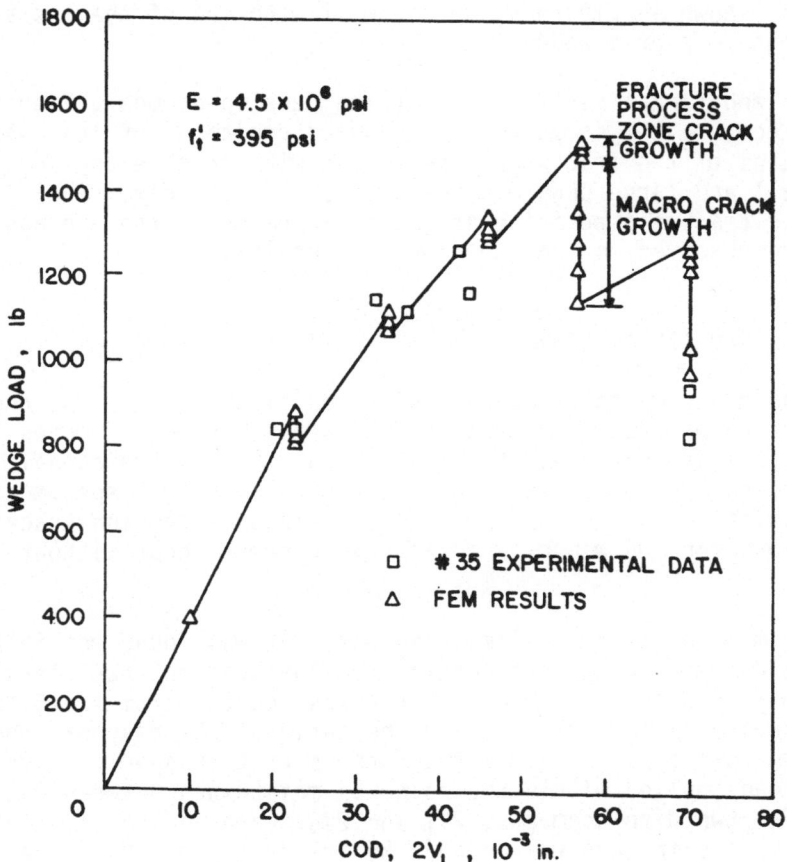

Figure 7. Computation Steps for Equilibrium Crack

experimentally. Figure 8 shows the agreement between the test data and the numerical results for two additional experiments where good agreement was again obtained.

Figure 9 shows the wedge load versus COD results for the four large CLWL-DCB specimens. There is a larger scatter in the experimental results than for the small CLWL-DCB specimens. For comparison, the Wecharatana-Shah model was incorporated in the finite element model and a propagation analysis was examined. The results, shown in Figure 9, agreed well with one of the tests with the larger measured wedge load.

As additional verification of the developed model, Figures 10 and 11 compare the computed and measured positions of the hairline crack tips of the fracture process zone with crack extensions for the small and large CLWL-DCB specimens, respectively. The computed crack tip position is generally close to the average position observed in the experimental results.

DIAGONAL TENSION FRACTURE

The experimental procedure with modifications and the concept of fracture process zone were used to analyze combined modes I and II fracture in concrete. Figure 12 shows the added diagonal compression loading imposed on the original CLWL-DCB specimens. This modified CLWL-DCB specimen models diagonal tension cracking conditions for the web of a reinforced concrete beam without shear.

From a series of preliminary tests, it was found possible to produce diagonal tension fracture from the original cast-in-situ crack provided that the crack length was reduced from the 5.25 in. used previously to 4.25 in.; and the ratio of the diagonal compression load applied to the two corners of this specimen, to the wedge load applied within the hole was maintained between 0.5 and 2.0. The two displacements, $2V_1$ and $2V_2$, were monitored. In addition, a slip gage was added 1.55 in. from gage $2V_2$. The diagonal load D was applied by a jack acting within a self reacting frame and its force monitored by a load cell placed at the reaction point. The load from that diagonal force was distributed to the specimen through a steel block that projected around the corner of the specimen and 1 inch down either side.

The tests conducted to date are summarized in Table 2. The maximum aggregate size has been kept constant at 1/2 in. The ratio of the diagonal force to the wedge force has been held constant at ratios between about 0.5 and 0.7 until a predetermined diagonal force has been reached at which time the diagonal force has been kept constant and only the wedge force altered. The

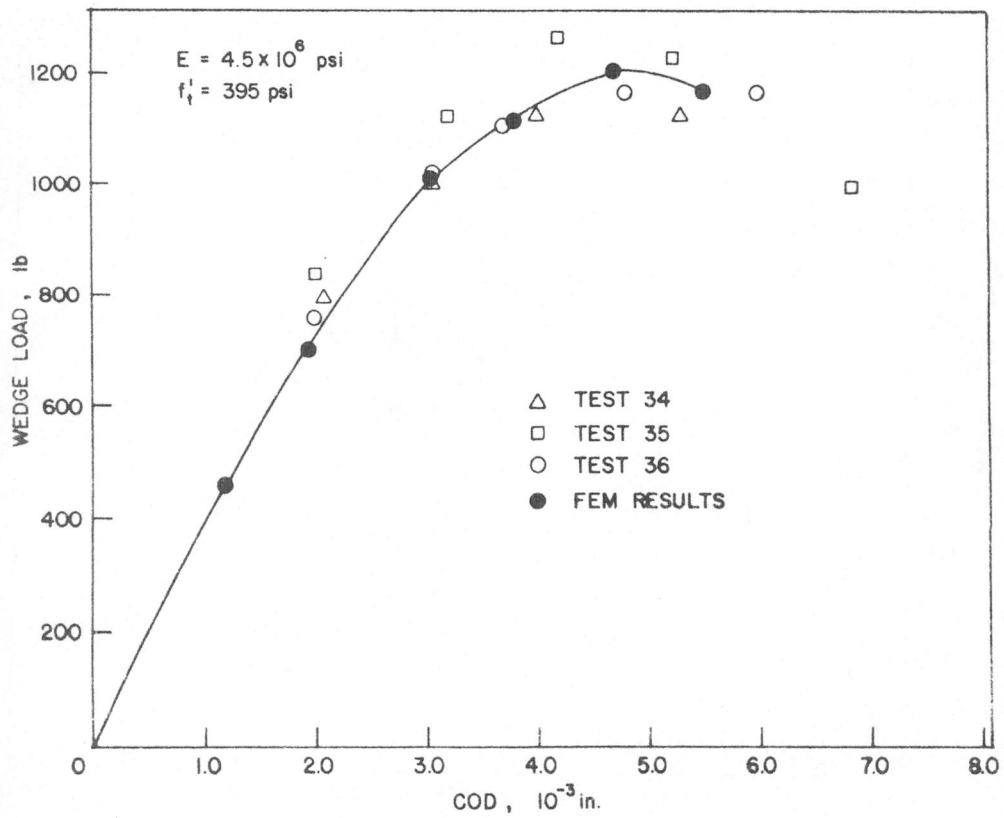

Figure 8. Wedge Load vs COD Curves. Experimental and
Analytical Results

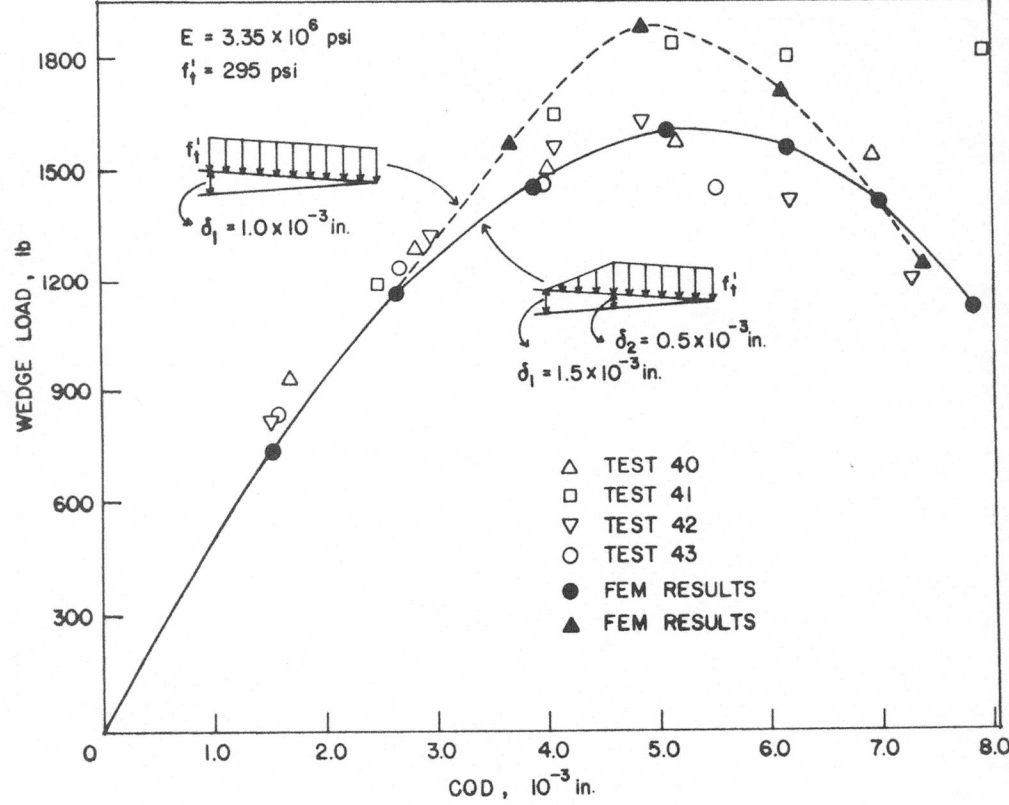

Figure 9. Wedge Load vs COD Curves. Experimental and
Analytical Results

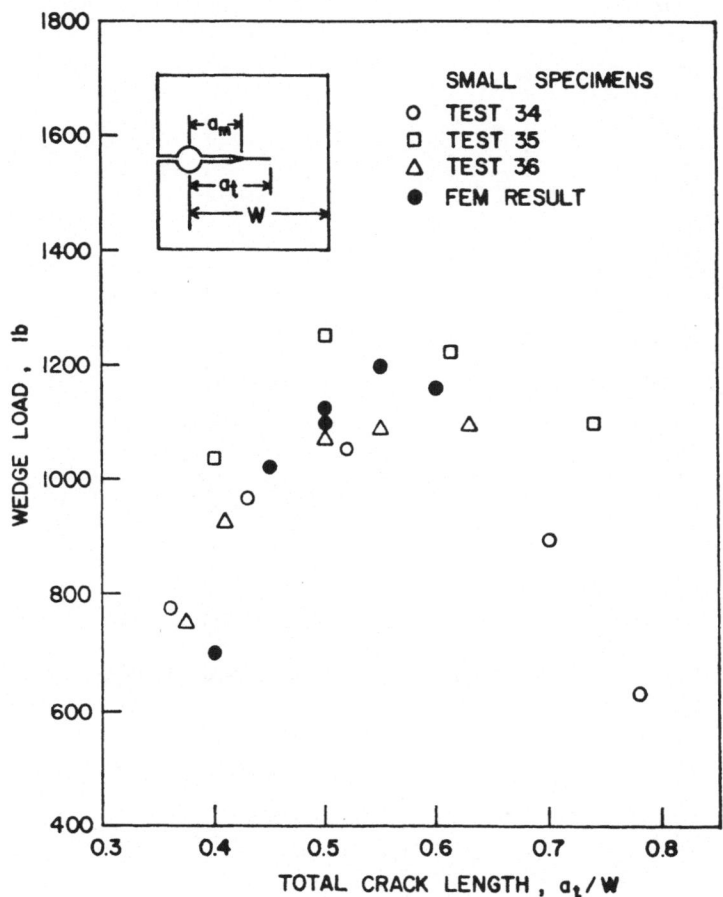

Figure 10. Wedge Load vs Total Crack Length (a$_t$/W)

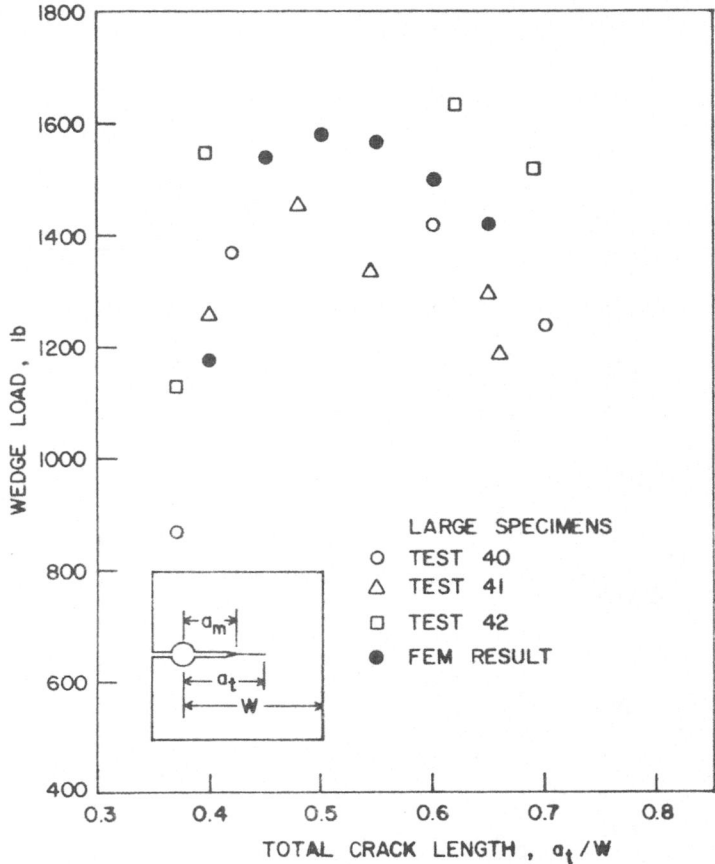

Figure 11. Wedge Load vs Total Crack Length (a_t/W)

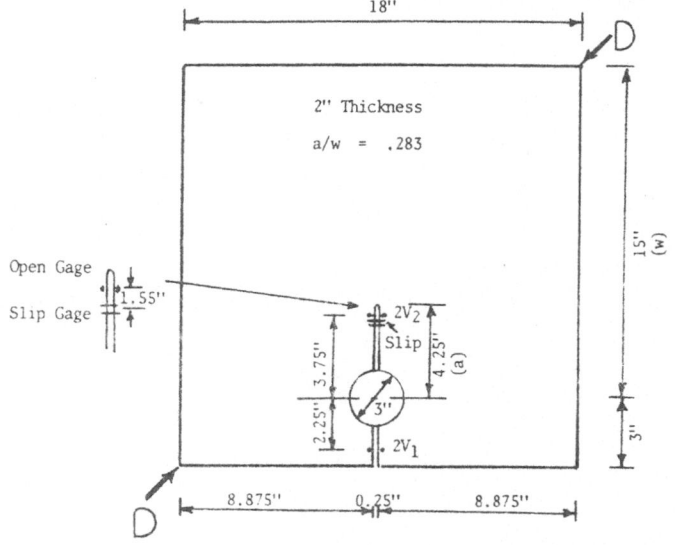

Figure 12. CLWL-DCB Specimen with Diagonal Compression Loading

specimens have had concrete strengths of about 3,500 psi.

A typical crack propagation result is shown in Figure 13. The crack grew in what appeared to be the principal stress direction. Again, a replicating film was used to measure the extent of the micro-crack. However, that operation was found to be more difficult than for the Mode I loading because the crack tended to close while readings were being taken.

Table 2 - Diagonal Tension Fracture Tests

Specimen	Maximum Agg. Size	Force Ratio	Preset Diag. Force	Concrete Compress. Strength
Mixed 01	.50	.53	790	3,500
Mixed 02	.50	.62	810	3,500
Mixed 03	.50	.71	810	3,500
Mixed 04	.50	.62	1,050	3,500
Mixed 11	.25	.53	820	3,300
Mixed 12	.25	.62	820	3,300
Mixed 21	.50	.53	820	3,770
Mixed 22	.50	.62	850	3,770
Mixed 31	.50	.53	600	3,770
Mixed 32	.50	.62	600	3,770
Mixed 33	.50	.71	600	3,770

$$*\text{Force Ratio} = \frac{\text{Diagonal Load}}{\text{Crack-line Load}}$$

Unfortunately, while the wedge loading was displacement controlled, the diagonal load was force controlled and thus the resultant stress redistribution effects with time probably contributed to crack closing. In the result shown here, the crack did not effectively start to extend until a wedge load of 988 lbs. (Stage R_2), and the maximum wedge load of 1,423 lbs., was achieved at Stage R_5 when the crack was only about 2 inches long and barely visible with the naked eye.

Figure 14 shows the wedge and diagonal compression loads vs COD relations. The maximum wedge force was achieved at a crack mouth opening displacement of approximately 3×10^{-3} in. after

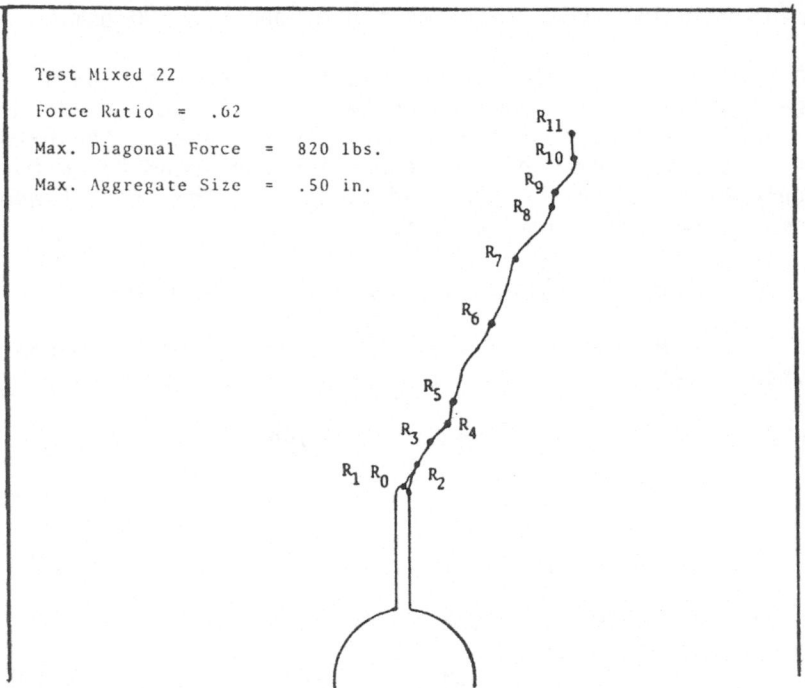

Test Mixed 22

Force Ratio = .62

Max. Diagonal Force = 820 lbs.

Max. Aggregate Size = .50 in.

Figure 13.   Diagonal Tension Crack Extension

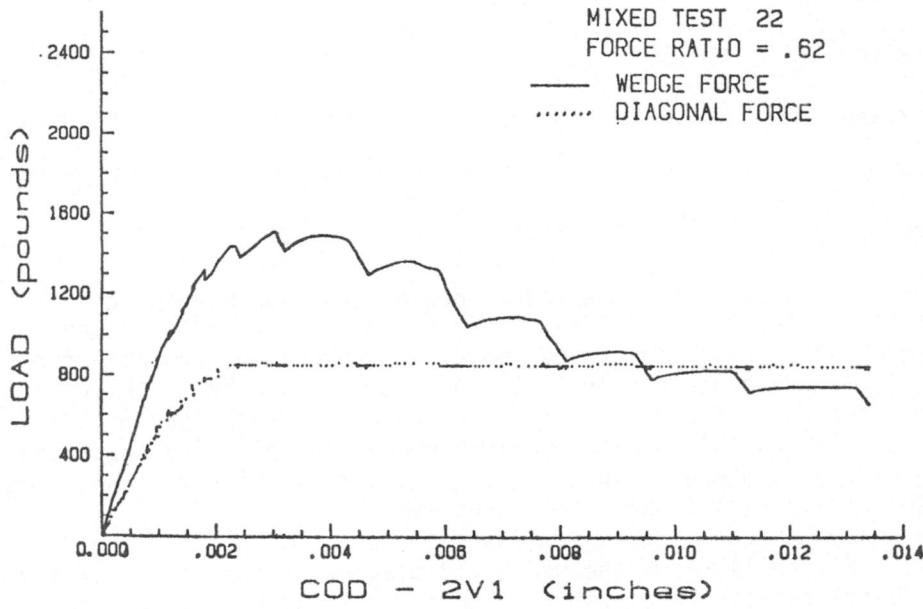

Figure 14.   Loads vs Crack Opening Displacement

which the wedge load decreased with increasing displacements until the wedge load dropped to the value of the diagonal load.

Figure 15 shows the relation between the crack mouth opening displacement and the crack tip opening displacement. Figure 16 shows the relation between the crack tip opening displacement and the slip. The slip does not begin until $2V_2$ exceeds about 0.5 x 10^{-3} in. Then it increases almost linearly with $2V_2$ at a rate of about one quarter of $2V_2$.

In spite of relatively wide variations in the ratio of the diagonal to wedge force and in the maximum diagonal force, the direction of the crack propagation was almost constant for different specimens as shown in Figure 17. That direction was at about an angle of 30° to the axis of the cast-in-place crack. A finite element analysis in the generation mode was thus conducted with a constant angle of 30° for the diagonal tension crack with the fracture process zone shown in Figure 6. For this brittle material, it was also assumed that the mixed mode crack tip stresses did not alter the postulated fracture process zone.

The mesh used for the finite element analysis is shown in Figure 18. The specimen analyzed in detail to date is the specimen "Mixed 22" which had a force ratio of 0.62 and a maximum diagonal force of 820 lbs. First, it was observed that up until the limiting diagonal force was reached, the direction of the crack agreed well with the maximum compressive stress direction. Measured and predicted loads and displacements versus the measured crack length are shown in Figure 19. There is reasonable agreement between measured and predicted results with the large deviations being for the slips, which are predicted to be greater than those measured, and the wedge load which was found to increase faster, and to fall more slowly with increasing crack lengths than predicted. It is believed those discrepancies are due primarily to friction effects from the rollers applying the loads to the loading wedge, and investigations are underway to determine the magnitude of those effects. Through a numerical experiment, shear effects behind the advancing crack tip, calculated in accordance with models developed by several researchers, were found to have a negligible effect on the relationships for predicted values in Figure 19.

CONCLUSIONS

A fracture process zone with the variable crack opening resistance and prescribed CTOD's, has been used successfully to predict stable crack growth in CLWL-DCB concrete specimens which were loaded under mode I and mixed modes I and II.

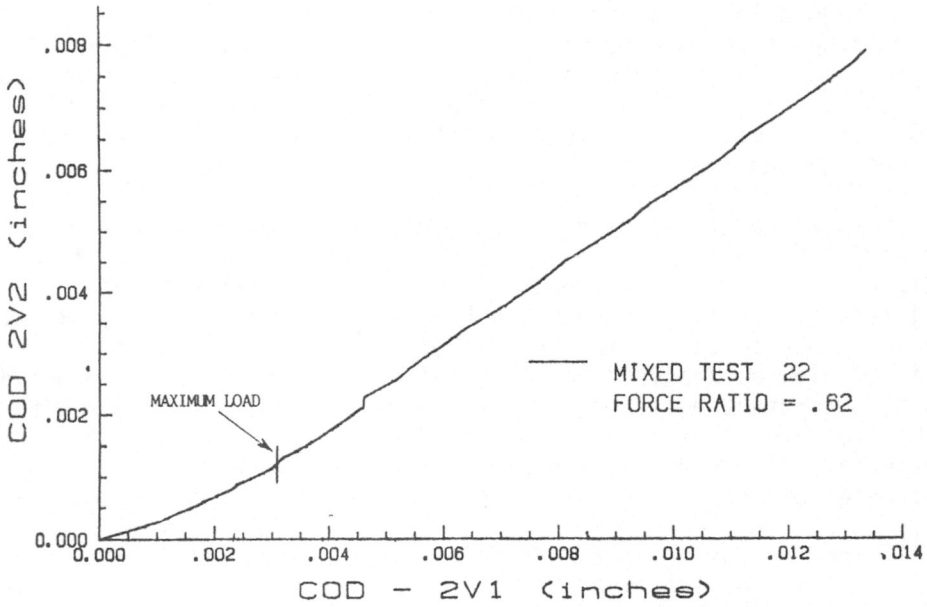

Figure 15. Crack End and Crack Mouth Opening Displacement

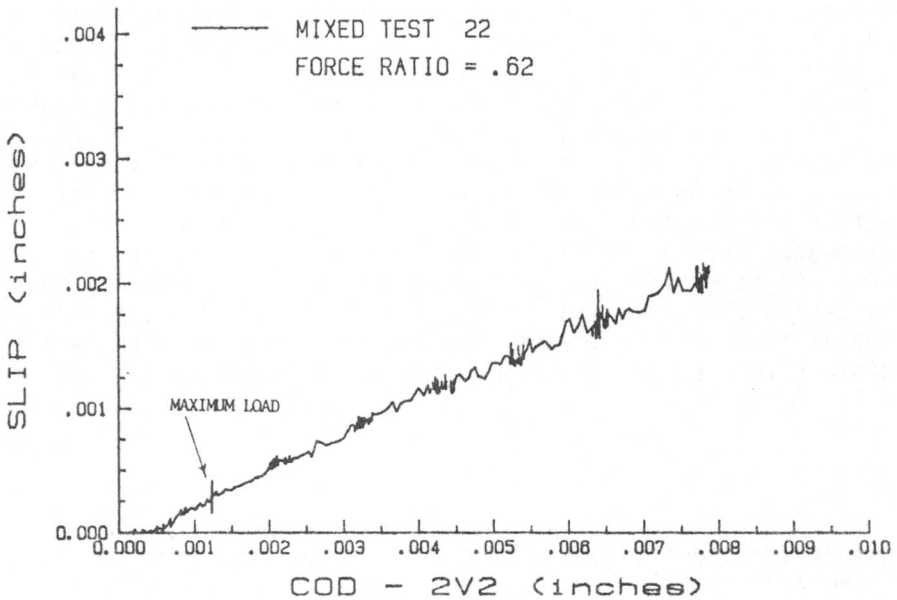

Figure 16. Slip and Crack Mouth Opening Relationship

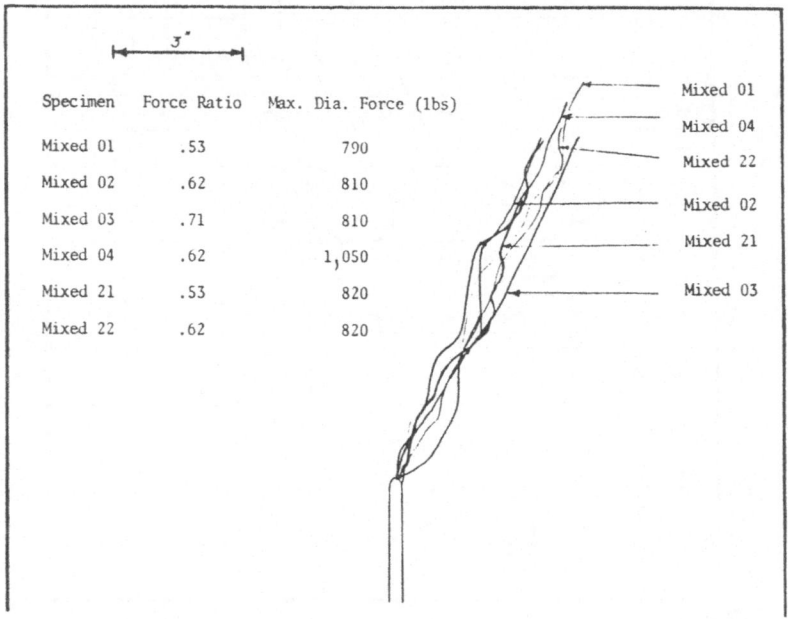

Specimen	Force Ratio	Max. Dia. Force (lbs)
Mixed 01	.53	790
Mixed 02	.62	810
Mixed 03	.71	810
Mixed 04	.62	1,050
Mixed 21	.53	820
Mixed 22	.62	820

Figure 17. Comparison of Diagonal Tension Cracks

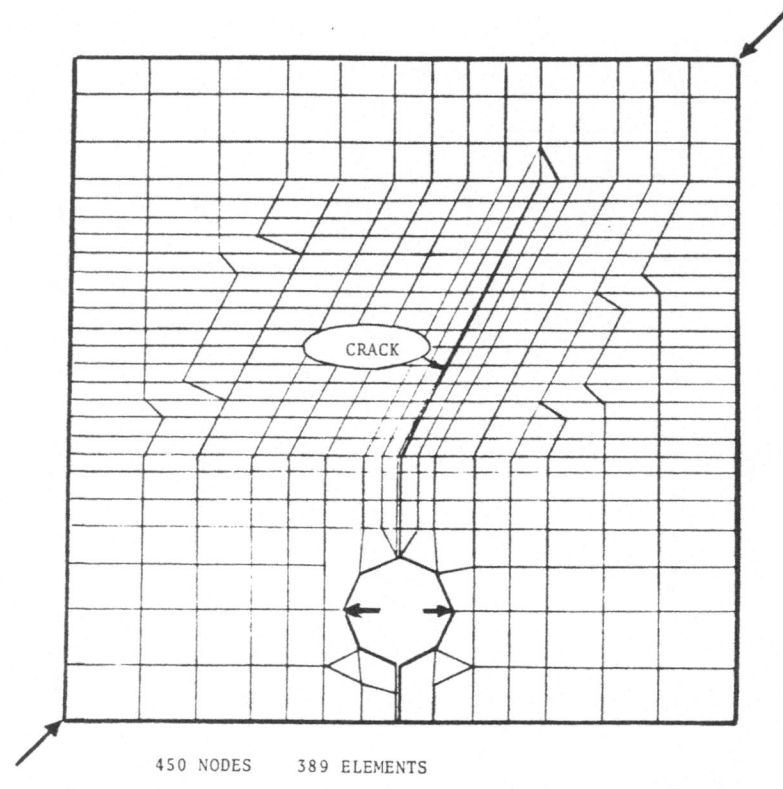

450 NODES   389 ELEMENTS

Figure 18. Finite Element Breakdown of Diagonal Tension
Specimen

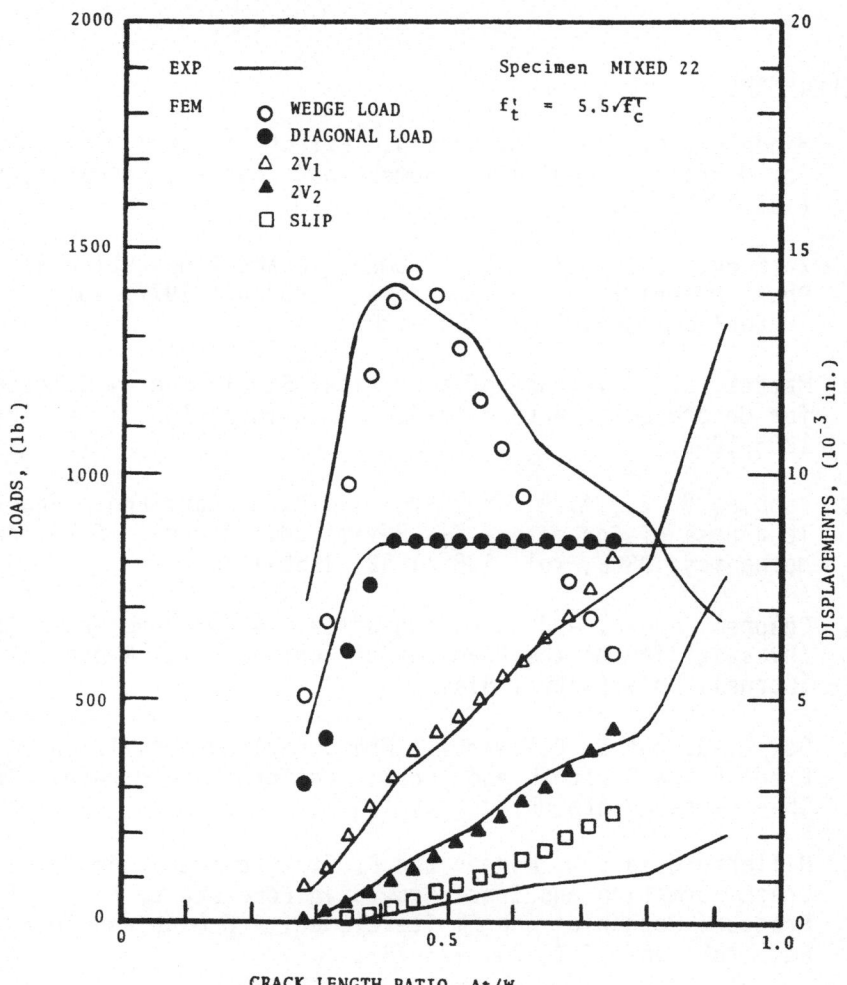

Figure 19. Comparison of Measured and Predicted Crack Lengths, Loads and Displacements

48

ACKNOWLEDGEMENT

This research was funded by the National Science Foundation under Grant No. CEE-8117029 for which the Program Manager is Dr. M. P. Gauss, NSF.

REFERENCES

1. Mindness, S. and J. S. Nadeau. Effect of Notch Width on K_{IC} for Mortar and Concrete. Cement and Concrete Research, vol. 6 (1976) 529-534.

2. Zaitsev, J. W. and F. H. Wittman. Crack Propagation in a Two-Phase Material Such as Concrete. Fracture 1977, vol. 3 (Waterloo, ICF4, 1977) 1197-1203.

3. Mazars, J. Existence of a Critical Strain Energy Release Rate for Concrete. Fracture 1977, vol. 3 (Waterloo, TCF4, 1977) 1205-1209.

4. Saouma, V. E., A. R. Ingraffea and D. M. Catalano. Fracture Toughness of Concrete - K_{IC} Revisited. Journal of Engineering Mechanics, ASCE, vol. 108 (1982) 1152-1166.

5. Chappell, J. F. and A. R. Ingraffea. A Fracture Mechanics Investigation of the Cracking of Fontana Dam. Report 81-7 (Cornell University 1981).

6. Go, C.-G. and S. E. Swartz. Fracture Toughness Techniques to Predict Crack Growth and Tensile Failure in Concrete. Report 154 (Kansas State University 1983).

7. Hillerborg, A., M. Modeer and P. E. Petersson. Analysis of Crack Formation and Crack Growth in Concrete by Means of Fracture Mechanics and Finite Element. Cement and Concrete Research, vol. 6 (1976) 773-781.

8. Petersson, P. E. Fracture Energy of Concrete: Method of Determination. Cement and Concrete Research, vol. 10 (1980) 79-89, and Fracture Energy of Concrete: Practical Performance and Experimental Results. Cement and Concrete Research, vol. 1, (1980) 91-101.

9. Bazant, Z. P. and S. S. Kim. Plastic-Fracturing Theory for Concrete. Journal of Engineering Mechanics, ASCE, vol. 105 (1979) 407-428.

10. Bazant, Z. P. and B. H. Oh. Crack Band Theory for Fracture of Concrete. Materieuax et Constructions, vol. 16 (1982) 155-177.

11. Wecharatana, M. and S. P. Shah. Prediction of Nonlinear Fracture Process Zone in Concrete. Journal of Engineering Mechanics, ASCE, vol. 109 (1983) 1231-1246.

12. Kesler, C., D. Naus and J. Lott. Fracture Mechanics - Its Applicability to Concrete. Proceedings of International Conference on Mechanical Behavior Materials, vol. 4 (1972) 113-124.

13. Scordellis, A. C. Finite Element Analysis of Reinforced Concrete Structures. Proceedings of Specialty Conference on Finite Element Method in Civil Engineering (Montreal 1972) 71-114.

14. Saouma, V. E. Interactive Finite Element Analysis of Rein-forced Concrete: A Fracture Mechanics Approach. Report 81-5 (Cornell University 1981).

15. Baker, D. B., N. M. Hawkins, F.-L. Jeang, K.-Z. Cho and A. S. Kobayashi. Experimental Investigation of Concrete Fracture in a CLWL Specimen. To be submitted.

16. Kobayashi, T. and W. L. Fourney. Experimental Characteriza-tion of the Development of Micro-Crack Process Zone at a Crack Tip on Rock under Load. 19th US National Symposium on Rock Mechanics (May 1979).

17. Dugdale, D. S. Yielding of Steel Sheets Containing Slits. J. of Mechanics, Physics and Solids. vol 8 (1960) 100-104.

18. Barrenblatt, G. J. The Mathematical Theory of Equilibrium Crack in the Brittle Fracture. Advances in Applied Mechanics. vol. 7 (1962) 55-125.

19. Kobayashi, A. S. Dynamic Fracture Analysis by Dynamic Finite Element Method - Generation and Propagation Analyses. Non-linear and Dynamic Fracture Mechanics. ed. by N. Perrone and S. N. Atluri, ASME-AMD-vol. 35 (1979) 19-36.

20. Lee, O. S. and A. S. Kobayashi. Crack Tip Plasticity of a Tearing Crack. Submitted for publication in Fracture Mechanics (16) ASTM STP.

21. Kanninen, M. F., E. F. Rybicky, R. B. Stonesifer, D. Broek, A. R. Rosenfield, C. W. Marschall and G. T. Hahn. Elastic-Plastic Fracture Mechanics for Two-Dimensional Stable Crack Growth and Instability Problems. Elastic-Plastic Fracture. ed. by J. D. Landes, J. A. Begley and G. A. Clarke, ASTM STP 668 (1979) 121-150.

22. Shih, C. F., H. G. deLorenzi and W. R. Andrews. Studies on Crack Initiation and Stable Crack Growth. Elastic-Plastic Fracture, ed. by J. D. Landes, J. A. Begley and G. A. Clarke, ASTM STP 668 (1979) 65-120.

APPLICATION OF FRACTURE MECHANICS
TO CEMENTITIOUS COMPOSITES
NATO-ARW - September 4-7, 1984
Northwestern Univeristy, U.S.A.
S. P. Shah, Editor

NONLINEAR ANALYSIS FOR MIXED MODE FRACTURE

R. Ballarini, S.P. Shah and L.M. Keer

Department of Civil Engineering
Northwestern University
Evanston, Illinois 60201

INTRODUCTION

When fracture occurs in materials such as concrete, rock,
mortar and some glassy polymers, stable crack growth often
precedes unstable crack propagation. This paper will address
the problem of characterizing slow crack growth. In the first
section a description of the process zone (the cause of the
slow crack growth) will be presented, along with a literature
review of work performed to understand the effect of this non-
linear zone. Recent models used to theoretically and experimentally predict the size and effect of the process zone will
be explained in detail in the second section. Having
explained the importance of including the process zone in
theoretical models, a model will be presented to predict Mixed
Mode fracture in materials which exhibit slow crack growth.
This model is an extension of the model developed by one of
the authors [Shah] to predict Mode-I fracture. Understanding
Mixed Mode crack propagation is necessary to solve many structural engineering problems, and an example of such a problem
will be given in the last two sections. This example is the
problem of predicting the tensile capacity of short anchor
bolts embedded in concrete. In future communications the
predictions of the model will be compared to experimental
results. Some preliminary experimental results are presented
in the last section.

THE PROCESS ZONE

Linear elastic fracture mechanics (LEFM) has been applied successfully to predict crack propagation in materials such as glass and high strength steels. In these materials plastic flow in the region surrounding a crack tip is relatively small, and only one physical quantity associated with the fracture process is needed to characterize crack propagation. This parameter is often the critical stress intensity factor (K_{IC}), a measure of the critical state of stress at a crack tip at the onset of unstable crack propagation. In cement based composites (concrete, mortar), geomaterials and some polymeric materials a zone surrounds a crack tip within which a significant amount of microcracking and/or other inelastic deformations occur. Furthermore, because cracks in these materials can be tortuous, mechanical interlock arises from aggregate or fiber bridging (Figure 1). The zone within which the mechanical interlock and microcracking occur is referred to as the process zone for rock and ceramic materials [1,2]. In some glassy polymers this inelastic zone is called crazing [3,4,5].

Another example in which mechanical interlock may occur is at the cracking of the cancellous bone/PMMA interface, which may be a failure mechanism for artificial joints. Mechanical modeling using this failure mechanism has been initiated by Mak, et al. [6], Mak and Keer [7], Keer and Meade [8], and Clech, et al. [9]. In these papers the cracked interface is assumed to behave as a no-slip crack and in [8] a closing pressure is assumed in the form of a non-linear spring over the entire crack surface.

Attempts have been made to theoretically and experimentally predict the length of the non-linear process zone. The most commonly used theoretical model is the Dugdale model. In this model the inelastic displacement in the process zone (termed plastic zone) is represented by a closing pressure taken equal to the yield stress of the material acting on a very narrow strip ahead of the traction free crack tip. The length of the strip is calculated by assuming that the stress intensity factor due to the yield stress cancels out the stress intensity factor due to the applied far-field stresses.

For materials such as concrete and rock which exhibit limited deformation capacity it is doubtful that the assumption of zero singularity is valid. Ballarini, Shah and Keer [10] have shown that for brittle cement based composites the energy term associated with the stress intensity factor is comparable to the inelastic energy consumed in the process zone and thus cannot be ignored. In addition, for concrete and rock the closing pressure is not constant but depends on the crack opening displacements [11,12,13,14]. Because of this phenomenon determining fracture parameters is reduced to solving a problem with complicated coupling terms. For Mode-I the problem has been approached by using a non-linear integral equation method [10], a smeared crack approach [15], finite element analysis [16] and several fictitious crack models [16,17].

Several experimental techniques have been employed to determine the size of the process zone for various materials. Cedolin, et al. [18] used high sensitivity moire inter-ferometry to obtain a precise description of the fracture process zone in concrete specimens subjected to tensile loading. Knab, et al. [19] used fluorescent thin sections to observe the microstructural details of the fracture zone both near the surface and in the interior of mortar specimens. Other techniques such as scanning electron microscope [2], acoustic emission [2,20], ultrasonic probing [21] and X-Ray diffraction [1] have also been employed to determine the size of the process zone.

FRACTURE MODELS

Kaplan [22] was the first to apply LEFM concepts to the study of fracture in concrete and mortar. He tested pre-notched beams in three-point and four-point bending, and both beam size and initial notch length were varied in this study. He observed that for the same beam size, the critical energy release rate (a material property in LEFM) values remained constant for different notch lengths. These values increased with increasing beam size, however. Kaplan pointed out that these discrepancies were a result of slow crack growth prior to instability. Kaplan's study was the first to indicate that the effect of slow crack growth prior to brittle fracture should be included in theoretical models used to predict crack growth in cement-based composites. More recently, models have been proposed which include the effect of the process zone. Hillerborg [17] modeled the process zone as an extension of the actual crack subjected to a closing pressure which depends on the crack opening displacements. In this model the stress intensity factor at the tip of the fictitious crack is elim-inated, and it is assumed that the crack will grow in the case of Mode-I fracture, when the stress at the tip of the ficti-tious crack reaches the tensile strength of the material. Catalano and Ingraffea [16] developed a finite element model in which the mechanical interlock along the crack is modeled as a closing pressure that resists crack opening. Their model differs from Hillerborg's model in that the stress singularity at the crack tip is not eliminated and no process zone ahead of the visible crack is considered. Wecharatana and Shah [11,13,14] proposed a model similar to that of Catalano and Ingraffea, but instead of considering the process zone to be just mechanical interlock along the visible crack, they also included the zone ahead of the visible crack in which micro-cracking and other inelastic deformations occur. Ballarini, Shah and Keer [10] improved the model proposed in [13] by employing a more accurate analytical technique to calculate physical quantities such as crack opening displacements and stress intensity factors. In their model the process zone is modeled as an extension of the stress free crack subjected to a closing pressure that depends on the crack face displace-ments. Crack propagation is assumed to occur when the crack opening displacement at the tip of the stress free crack reaches a critical value. To predict results, the elasto-statics problem shown in Figure 2 was solved using a Green's function approach together with integral transform techniques. Stress intensity factors and crack opening displacements were obtained by numerically solving a non-linear singular integral equation. The model was used to predict results for concrete specimens and fiber reinforced mortar specimens in three and four point bending. Figures 3 and 4 are comparisons between predicted and observed values of crack mouth opening. In Figure 3 the model used is denoted by Model 2. Model 1 is a model which does not include the effect of the process zone ahead of the visible crack. Some investigators [17] have defined the fracture energy as the area under the descending portion of the uniaxial stress displacement curve. For the concrete specimens for which results are given in Figure 3 and Table 4 the area is .225 lb-in/in^2. As shown in Table 1 the energy associated with the stress intensity factor is com-parable with the fracture toughness mentioned. Therefore it is doubtful whether Dugdale type modeling is applicable to materials such as concrete.

Kobayashi, et al. [23] incorporated the model described above into a finite element model of CLWL-DCB concrete spec-imens which was then driven to reproduce experiments. Through the use of a replica technique they measured the length of the process zone and compared the results with the results obtained using their numerical procedure. Because of the sensitivity of the replica technique their observed lengths were in excess of those reported by other investigators [13, 24, 25]. Their numerical model was used to predict stable crack growth not only for Mode-I but also Mixed Mode loading.

Sih [26] approached the problem of predicting the non-
linear response of concrete beams in three point bending by
applying the strain energy density criterion. He argued that
models based on the J-integral and critical C.O.D. resulted in
resistance curves that were highly non-linear. By assuming
elastic and plastic unloading and the application of critical
strain energy density criterion Sih obtained resistance curves
in the form of strain energy density as functions of crack
length for various material properties and beam geometries.
He argues that the significance of his model is that the R-
curves are straight lines which permit designers to use interpolation and extract information for situations other than
those tested. This model was not, however, verified experimentally. It would be interesting to see how well
experimental results can be predicted from this model.

CRACK OPENING AND CRACK SLIDING RESISTANCE MODEL FOR MIXED MODE FRACTURE

Experiments with concrete, rock and fiber-reinforced
concrete specimens subjected to uniaxial tension in a
relatively stiff testing system have shown that the post-peak
displacements are essentially a result of the opening of a
single crack [27,28,29]. If the microcracking occurring prior
to the peak load can be ignored, the uniaxial stress-
displacement relationship in the post-peak region can be
assumed to be an approximate function for the crack opening
resistance in the process zone. The models in [8, 9, 10, 11,
13, 14, 15, 16, 17] replaced the process zone by this kind of
function. These models were used to characterize Mode-I
fracture. To model Mixed Mode Fracture it is necessary to
include a resistance to shear loading, and this will be done
by assuming a function for crack sliding resistance in the
process zone. Analytical models for cracking of concrete
subject to shear which include aggregate interlock have been
proposed (see, e.g. [30]).

TENSILE CAPACITY OF SHORT ANCHOR BOLTS (An example of Mixed Mode fracture)

There are many design situations where the pull-out
failure of rigid anchors in brittle (tension-weak) materials
is a critical consideration. Anchor bolts are often used as
connections in concrete structure, roof bolts in rock tunnels
and tie backs in rocks. In each situation failure may occur
as a result of the bolts pulling out of the matrix. Therefore, one of the design considerations is the pull-out

capacity of the bolts. Another situation where pull-out failure arises is in the so called pull-out test. This is a non-destructive test which is used to measure the in-situ strength of concrete in early ages (Figure 5). This test involves inserting a circular steel disc in fresh concrete, pulling out at desired ages, and correlating the pull-out load with the compressive strength of the concrete. As a final example, connections and fasteners involving reinforced plastics are often made by inserting relatively rigid steel rods or anchors in a plastic matrix [31].

When a relatively rigid steel anchor is pulled out from a brittle material like concrete, a cone of fractured matrix material often pulls out with the anchor. The failure process during pull-out is complex and likely includes (Figure 6) initiation of cracking probably under Mixed Mode conditions, propagation of cracks probably primarily under Mode-I, and then friction and geometrical interlocking along the cracked surfaces. The current design procedures often make various simplifying assumptions such as primarily compression failure, plasticity type shear failure or principal tensile failure along the surface of the cone. Some exploratory research conducted by the authors showed that these assumptions may be valid only for a certain restricted geometry of the anchor bolts and the support conditions but may be otherwise erroneous.

Klinger and Mendonca [32] performed a literature review on the tensile capacity of short anchor bolts. They compared results of available tensile tests with predictions of six design procedures currently available for computing nominal capacity of short anchor bolts and welded studs loaded mono-tonically in tension. It was shown that for those tests governed by concrete rather than steel failure, these methods are significantly unconservative (Figure 7) and show consider-able scatter. This work suggests that more accurate procedures are necessary to safely predict the ultimate capacity of anchor bolts.

Ottosen [33] analyzed the Lok-test by means of an axi-symmetric finite element analysis. This analysis followed the progression of circumferential and radial cracking by means of an iterative smeared cracking procedure. His analysis showed that circumferential cracks begin at the disc edge and propa-gate towards the reaction ring. In addition, he observed that large compressive stresses run from the disc in a rather narrow band towards the support. He postulates that it is these com-pressive struts which constitute the load carrying mechanism.

From this he concludes that the failure in pull-out tests is by crushing of concrete and not by cracking of concrete.

Stone and Carino [34] performed large-scale pull-out tests in an effort to provide experimental data. Their experiments showed that three distinct phases occur prior to ultimate failure of a pull-out test. These phases are marked by discontinuities in the load-strain histories of embedment gages placed along the failure surface (Figure 8). They concluded that large tensile stresses result in the start of circumferential cracking at a load equal to 33% of the ultimate, and that shear failure of the matrix and degradation of aggregate interlock occurs at 80% of ultimate load. This study suggests that since the failure mechanisms are complex, it is doubtful that for all the geometries the ultimate load is directly related to just the compressive strength.

PRELIMINARY EXPERIMENTAL RESULTS ON ANCHOR PULL-OUT

Plane stress tests were conducted using mortar as a matrix material. The geometry of the specimen and the loading conditions are shown in Figure 6. The variables were (1) the depth of the anchor embedment and (2) the support reaction distance. During the test, the length and inclination of the cracks were microscopically monitored. In addition, the load-slip relationship was continually measured (see Figures 6A and 6B).

These preliminary results showed that different failure mechanisms exist in a pull-out test. These mechanisms are governed by the geometry of the test. As shown in Figure 9 for short reaction distances cracking initiates at approximately 50% of the ultimate load. At this load cracking was observed through the microscope and the load-slip curve becomes non-linear. These results are similar to those reported in [34]. (The discrepancy between 33% in [34] and 50% here is probably due to 3-D effects and differences in matrix properties).

For this configuration an increase in load is needed for the cracks to continue propagating. For longer reaction distances, on the other hand, cracking starts at ultimate load (Figure 10) and the load drops as cracks propagate. In both cases the cracks tend to propagate from the sharp edge of the steel anchor towards the support reactions. Figure 11 shows that the relationship between pull-out load and independently measured compressive strength is also a function of geometry. Tests were performed at three ages (1, 2 and 4 days). For

short reaction distances the percent increase in pull-out load
(using the first day results as a reference) is the same as
the percent increase in compressive strength. For the longer
distances, on the other hand, the compressive strength
increases at a faster rate than the pull-out load. It is
interesting to note that in the limiting case of when the
supports are not there the problem becomes similar to the
short anchor bolt problem; these results might explain why
some of the design procedures that are currently available do
not give satisfactory predictions of bolt capacity. That is,
for this geometry it seems that the results are not related to
the compressive strength in the same manner as they are for
the short reaction distance case. A better understanding of
what is being measured in this test is needed, and it is
expected that the theoretical model will explain the different
failure mechanisms.

THEORETICAL APPROACH

The problem to be studied is the two dimensional anchor
pull-out. A first attempt at solution to such a problem was
made by Miller and Keer [35], who used a complex variable
approach to quantify the cracking that might develop at the
tip of an anchor being pulled vertically in an infinite
elastic medium. This solution did not model the exact
situation for two reasons:

1. The loaded surface of the half-space was approximated
   by concentrated loads in an infinite space.

2. The cracks that branched from the anchor were assumed
   to be straight.

Within this context the solution appeared to be surprisingly
accurate with regard to modeling the experimentally observed
3-D phenomena. In this paper the 2-D case will be modeled by
incorporating the free surface effect correctly so that crack
growth near the surface loads can be correctly assessed (see
Figure 12).

The problem of an anchor embedded in a half-space was
solved to obtain the response of the structure before cracking
initiates. The problem is formulated in terms of the complex
potentials of Muskhelishvili [36]. The stresses and displacements can be expressed in terms of the analytic functions

Φ and Ψ as:

$$\sigma_{xx} + \sigma_{yy} = 4 \text{ Real } [\Phi(z)] \tag{1}$$

$$\sigma_{yy} - i\sigma_{xy} = \Phi(z) + \overline{\Phi(z)} + z \overline{\Phi'(z)} + \overline{\Psi(z)} \tag{2}$$

$$2\mu\left(\frac{\partial u}{\partial x} + i \frac{\partial v}{\partial x}\right) = \kappa\Phi(z) - \overline{\Phi(z)} - \overline{z\Phi'(z)} - \overline{\Psi(z)} \tag{3}$$

where $i = \sqrt{-1}$, $z = x + iy$, μ is the shear modulus, $\kappa = 3-4\nu$ for plane strain and $\kappa = (3 - \nu)/(1 + \nu)$ for plane stress, ν being Poisson's ratio. Primes denote differentiation with respect to z, and bars imply complex conjugation. The anchor is modeled by a rigid unbonded plate in an elastic half-space loaded vertically. The support reactions are modeled by concentrated forces applied on the surface. The boundary conditions along the anchor are:

$$2\mu\left(\frac{\partial u}{\partial x} + i \frac{\partial v}{\partial x}\right)^+ = 0 \tag{4}$$

$$(\sigma_{yy} - i\sigma_{xy})^- = 0 \tag{5}$$

where the subscripts + and - refer to the upper and lower surfaces of the plate, respectively.

By using the Green's functions for a dislocation in a half-space and a point force in a half-space the problem is reduced (omitting the details which will be presented in a future communication) to the following set of coupled singular integral equations for the unknown dislocation and point force densities:

$$\int_{-c}^{c} \alpha(\xi) \left\{ \frac{\kappa - 1}{\xi - x} + K_1(x,\xi) \right\} d\xi$$

$$+ \int_{-c}^{c} \overline{\alpha(\xi)} K_2(x,\xi) d\xi + \pi i(\kappa + 1) \alpha(x)$$

$$\tag{6}$$

$$+ \int_{-c}^{c} \beta(\xi) \left\{ \frac{-2}{\xi - x} + K_3(x,\xi) \right\} d\xi$$

$$+ \int_{-c}^{c} \overline{\beta(\xi)} K_4(x,\xi) d\xi + f_1(x) = 0 \quad -c \leq x \leq c$$

$$\int_{-c}^{c} \alpha(\xi) \left\{ \frac{-2\kappa}{\xi - x} + K_5(x,\xi) \right\} d\xi + \int_{-c}^{c} \overline{\alpha(\xi)} K_6(x,\xi) d\xi$$

$$+ \int_{-c}^{c} \beta(\xi) \left\{ \frac{1-\kappa}{\xi - x} + K_7(x,\xi) \right\} d\xi + \int_{-c}^{c} \overline{\beta(\xi)} K_8(x,\xi) d\xi \tag{7}$$

$$- \pi i (\kappa + 1) \beta(x) + f_2(x) = 0 \qquad\qquad -c \leq x \leq c$$

$$\int_{-c}^{c} \alpha(\xi) d\xi = \frac{P}{2\pi i(\kappa + 1)} \tag{8}$$

$$\int_{-c}^{c} \beta(\xi) d\xi = 0 \tag{9}$$

The quantities $\alpha(\xi)$ and $\beta(\xi)$ are defined as:

$$\alpha(\xi) = \frac{-\partial}{\partial \xi} \frac{(F_x + iF_y)}{2\pi(\kappa + 1)} \tag{10}$$

$$\beta(\xi) = \frac{\mu e^{i\theta}}{\pi i (\kappa + 1)} \frac{\partial}{\partial \xi} \left\{ [u_r] + i [v_\theta] \right\} \tag{11}$$

where F_x and F_y are the x and y components of the point force and $[u_r]$ and $[v_\theta]$ are the r and θ components of the displacement jump $(K_1 \ldots K_8, f_1$ and f_2 will be presented in a future communication).

Equations 6-9 were solved numerically and the results were used to calculate physical quantities.

Figures 13-16 are contour plots of maximum and minimum principal stresses. The results presented are all for the case h = 2c. It can be seen that the location of the support reactions has a significant influence on the stress fields. In Figures 13 and 14 the supports are not present, and in Figures 15 and 16 they are. In all cases very high stresses are present near the tips of the anchor. The maximum principal stresses decay faster in the case when the support reactions are present than for the case when they are not. This is a result of the high compressive stresses produced by the concentrated forces. The minimum principal stresses are also highly influenced by the support reactions. For the case when the support reactions are not present, very high compressive stresses are present at the tips of the anchor. These stresses decay rapidly away from the edges of the anchor. When the support reactions are present, on the other hand, very high compressive stresses exist in a region extending from the tips of the anchor to the supports. These results suggest that cracks will grow from the anchor to the supports. This phenomenon has been observed experimentally (Figure 6 and 6A).

Table 2 illustrates how the compliance changes as the embedment and support distances change. It can be seen that compliance increases as (1) the depth of the embedment decreases and (2) as the support distance increases. These results indicate that different failure mechanisms might be observed in an experiment where the geometry of the test changes. This has been shown to be true.

In future communications the results of the problem which includes cracking will be presented, along with predictions of the mathematical model. These predictions will be compared with experimental results.

ACKNOWLEDGMENT

The research reported here was supported by a grant from the Air Force Office of Scientific Research (AFOSR-820243); Program Manager, Lt. Col. Lawrence D. Hokanson. Leon M. Keer is grateful for support from the National Science Foundation, grant MEA 8117106.

REFERENCES

1. Friedman, M., Handin, J. and Alani, G., "Fracture Surface Energy of Rocks", International Journal of Rock Mechanics and Mineral Sciences, Vol. 9, 1972, pp. 757-766.

2. Hoagland, R.G., Hahn, G.T. and Rosenfield, A.R., "Influence of Microstructure on Fracture Propagation in Rock", Rock Mechanics, Vol. 5, 1973, pp. 77-106.

3. Williams, J.G., "Visco-Elastic and Thermal Effects on Crack Growth in PMMA", International Journal of Fracture Mechanics, Vol. 8, No. 4, December 1972, pp. 393-401.

4. Weidmann, G.W., and Doll, W., "Some Results of Optical Interface Measurements of Critical Displacements at a Crack Tip", International Journal of Fracture Mechanics, Vol. 14, 1978, pp. R189-R193.

5. Chudnovsky, A., Palley, I. and Baer, E., "Thermodynamics of the Quasiequilibrial Growth of Crazes", Journal of Material Science, Vol. 16, 1981, pp. 35-44.

6. Mak, A.F., Keer, L.M., Chen, S.H. and Lewis, J.L., "A No-Slip Interface Crack", Journal Applied Mechanics, Vol. 47, 1980, pp. 342-350.

7. Mak, A.F. and Keer, L.M., "A No-Slip Edge Crack on a Bimaterial Interface", Journal Applied Mechanics, Vol. 47, 1980, pp. 816-820.

8. Keer, L.M. and Meade, K.P., "A note on a No-Slip Crack", Journal Applied Mechanics, Vol. 49, 1982, pp. 454-455.

9. CLech, J.P., Keer, L.M. and Lewis, J.L., "A Crack Model of a Bone Cement Interface", Journal of Biomechanical Engineering, in press.

10. Ballarini, R., Shah, S.P. and Keer, L.M., "Crack Growth in Cement Based Composites", Engineering Fracture Mechanics, in press.

11. Wecharatana, M. and Shah, S.P., "Slow Crack Growth in Cement Composites", A.S.C.E. Journal of the Structural Division, Vol. 108, No. ST6, June 1982, pp. 1400-1413.

12. Labuz, J.F., Shah, S.P. and Dowding, C.H., "Experimental Analysis of Crack Propagation in Granite", International Journal of Rock Mechanics and Mining Sciences", in press.

13. Wecharatana, M. and Shah, S.P., "Predictions of Non-Linear Fracture Process Zone in Concrete", A.S.C.E. Journal EMD, Vol. 109, No. 5, October 1983, pp. 1231-1246.

14. Wecharatana, M. and Shah, S.P., "A Model for Predicting Fracture Resistance of Fiber Reinforced Concrete", Cement and Concrete Research, Vol. 13, 1983, pp. 819-829.

15. Bazant, Z.P. and Oh, B.H., "Crack Band Theory for Fracture of Concrete", Materials and Structures, Vol. 16, No. 93, May-June, 1983, pp. 155-178.

16. Catalano, D.M. and Ingraffea, A.R., "Concrete Fracture: A Linear Elastic Fracture Mechanics Approach", Report No. 82-1, Dept. of Structural Engineering, Cornell University, November 1982.

17. Hillerborg, A., Modeer, M. and Peterson, P.E., "Analysis of Crack Formation and Crack Growth in Concrete by Means of Fracture Mechanics and Finite Elements", Cement and Concrete Research, Vol. 6, 1976, pp. 773-782.

18. Cedolin, L., Dei Poli, S. and Iori, I., "Experimental Determination of the Fracture Process Zone in Concrete", Cement and Concrete Research, to Appear.

19. Knab, L.I., Walker, H.N., Clifton, J.R., and Fuller, E.R., Jr., "Fluorescent Thin Sections to Observe the Fracture Zone in Mortar", Cement and Concrete Research, to Appear.

20. Swan, G. and Alm, O., "Sub-Critical Crack Growth in Stripa Granite: Direct Observations", 23rd Symposium of Rock Mechanics, 1982, pp. 542-550.

21. Swanson, P.L. and Spetzler, H., "Ultrasonic Probing of the Fracture Process Zone in Rock Using Surface Waves", Proceedings of the 25th U.S. National Symposium on Rock Mechanics, Held at Northwestern University, Evanston, Illinois, June 25-27, 1984.

22. Kaplan, M.F. "Crack Propagation and the Fracture of Concrete", Journal A.C.I., Vol. 58, 1961, pp. 591-610.

23. Kobayashy, A.S., Hawkins, N.M., Barker D.B., and Liaw, B.M., "Fracture Process Zone of Concrete", Presented at Application of Fracture Mechanics to Cementitious Composites, NATO-ARW, Sept. 4-7, 1984, Northwestern University, U.S.A., S.P. Shah, Editor.

24. Petersson, P.E., "Fracture Energy of Concrete: Method of Determination", Cement and Concrete Research, Vol. 10, 1980, pp. 78-89.

25. Petersson, P.E., "Fracture Energy of Concrete: Practical Performance and Experimental Results", Cement and Concrete Research, Vol. 1, 1980, pp. 91-101.

26. Sih, G.C., "Non-Linear Response of Concrete: Interaction of Size, Loading Step and Material Property", Presented at Application of Fracture Mechanics to Cementitious Composites NATO-ARW, Sept. 4-7, 1984, Northwestern University, U.S.A., S.P. Shah, Editor.

27. Shah, S.P., Stroeven, P., Daluisen, D. and Van Stekelenberg, P., "Complete Stress-Strain Curves for Steel Fiber Reinforced Concrete in Uniaxial Tension and Compression", Proceedings, International Symposium, Rilem-ACI-ASTM, Sheffield, Sept. 1978, pp. 339-408.

28. Petersson, P.E., "Fracture Mechanical Calculations and Tests for Fibre-Reinforced Cementitious Materials", Advances in Cement-Matrix Composites, Proceedings, Symposium Materials Research Society, Annual Meeting, 1980, pp. 95-106.

29. Gopalaratnam, V.S. and Shah, S.P., "Softening Response of Plain Concrete in Direct Tension", Journal A.C.I., in press.

30. Bazant, Z.P. and Gambarova, P.G., "Crack Shear in Concrete: Crack Band Microplane Model", Report, Center for Concrete and Geomaterials, Northwestern University, 1983.

31. Gordon, J.E., Structures, Da Capo Press, 1978, pp. 138-141.

32. Klinger, R.E. and Mendonca, J.A., "Tensile Capacity of Short Anchor Bolts and Welded Studs: A Literature Review", Journal A.C.I., Vol. 79, No. 4, July-August 1982, pp. 270-279.

33. Ottosen, N.S., "Non-Linear Finite Element Analysis of a Pull-Out Test", A.S.C.E. Journal of the Structural Division, Vol. 107, No. 4, April 1981, pp. 591-603.

34. Stone, W.C. and Carino, N.J., "Deformation and Failure in Large-Scale Pull-Out Tests", Journal A.C.I., Vol. 80, No. 6, November-December 1983, pp. 501-513.

35. Miller, G.R. and Keer, L.M., "An Approximate Analytical Model of Anchor Pull-Out Test", ASME Journal of Applied Mechanics, Vol. 104, 1982, pp. 768-772.

36. Muskhelishvili, N.I., Some Basic Problems in the Theory of Elasticity, Noordhoff, Leyden, The Netherlands, 1954.

Stress-Free Crack Length (inches)	Process Zone Length (inches)	Load P (lbs.)	K_{IP}	$K_{I\hat{\sigma}}$ (lb./in.$^{3/2}$)	K_{IT}	$G = \dfrac{K_{IT}^2(1-\nu^2)}{E}$ (lb.-in./in^2)	$CMOD_P$ (inches)	$CMOD_T$ (inches)
3.00	1.20	4440	1440	-497	943	.19	.00244	.00230
3.66	1.14	3960	1468	-467	1001	.22	.00281	.00268
4.37	1.03	3600	1532	-417	1115	.27	.00331	.00318
5.14	.86	3240	1609	-368	1241	.33	.00389	.00378
5.81	.79	2880	1686	-327	1359	.40	.00457	.00447
6.51	.69	2520	1763	-288	1475	.47	.00541	.00531
7.24	.56	2280	1855	-252	1603	.55	.00644	.00635
7.85	.55	1800	1950	-222	1728	.64	.00773	.00764
8.60	.40	1560	2221	-152	2069	.92	.0102	.0101
9.37	.23	1200	2378	-114	2264	1.1	.0125	.0125

(Fracture Energy is .225 lb-in/in^2)

Table 1 -- Energy associated with stress intensity factor (Ref. [10])

$\dfrac{h}{2c}$	$\dfrac{d}{2c}$	$\dfrac{\mu\Delta}{p}$
.75		.368
1.00		.332
1.25		.309
∞		.210
.75	.75	.186
.75	1.50	.320
1.00	1.00	.208
1.00	1.50	.272
1.25	1.25	.215

Table 2 - Compliance as Function of Geometry

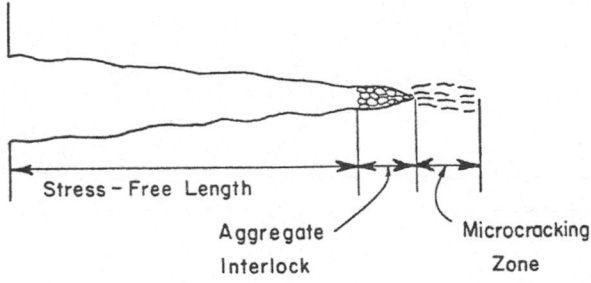

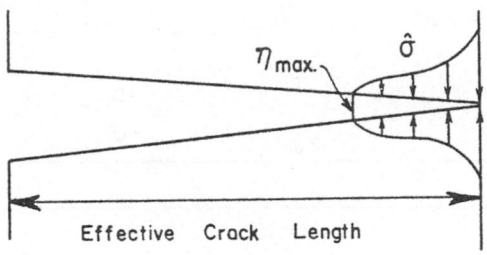

Figure 1 - Model for Process Zone

68

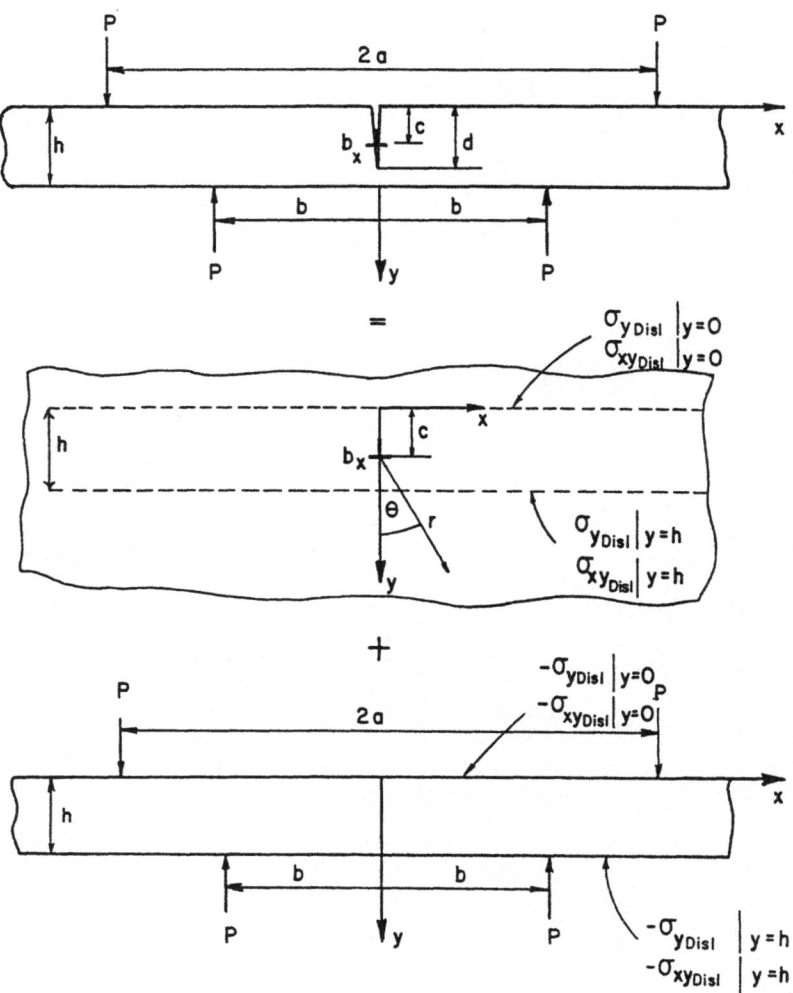

Figure 2 – Superposition Scheme

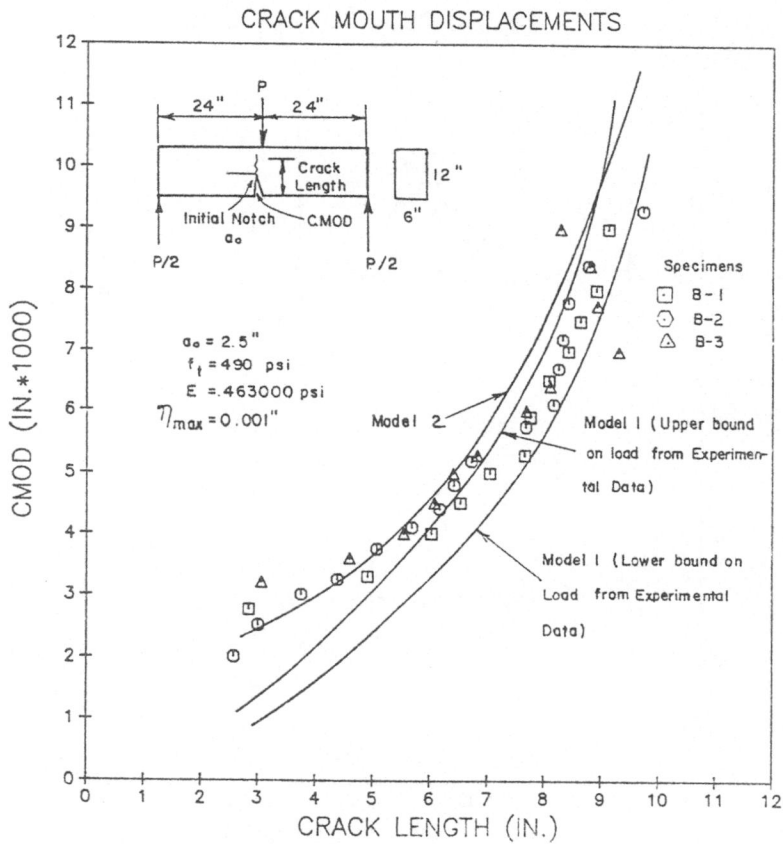

Figure 3 – Comparison of Analytical & Experimental Results
for Concrete (Ref. [10]).

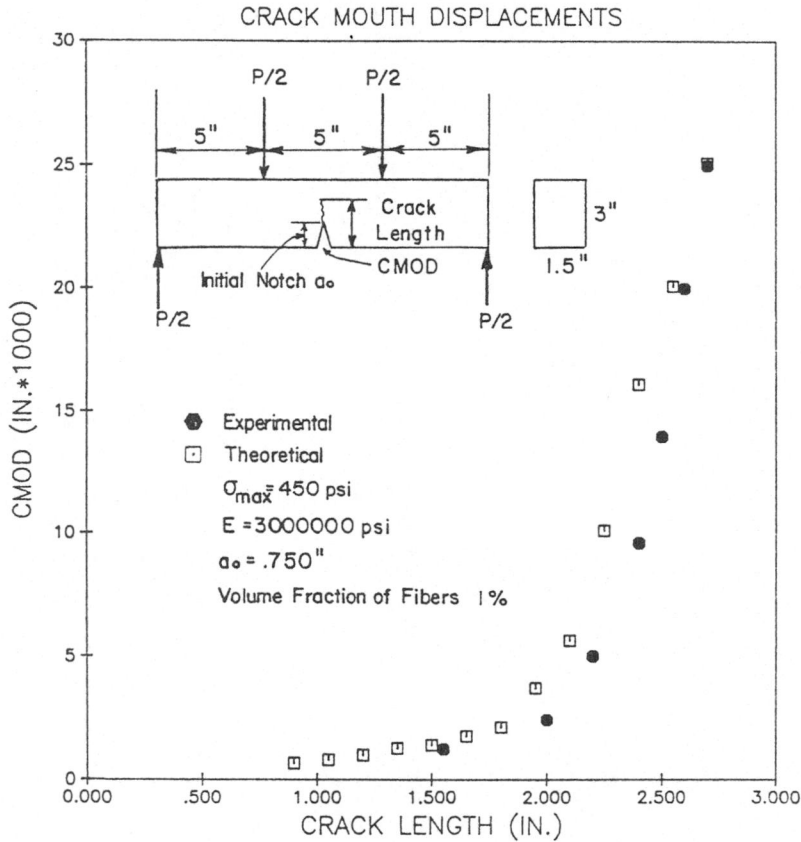

Figure 4 – Comparison of Analytical and Experimental Results
for Fiber-Reinforced Mortar (Ref. [10]).

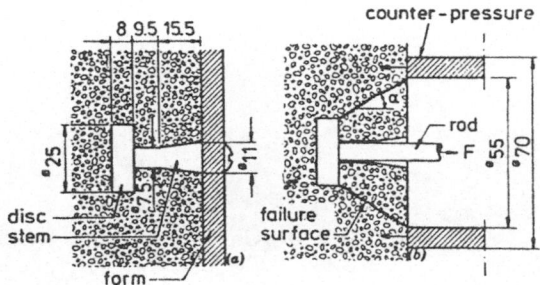

Figure 5 - Lok-Test (All Dimensions in Millimeters Ref. [33]).

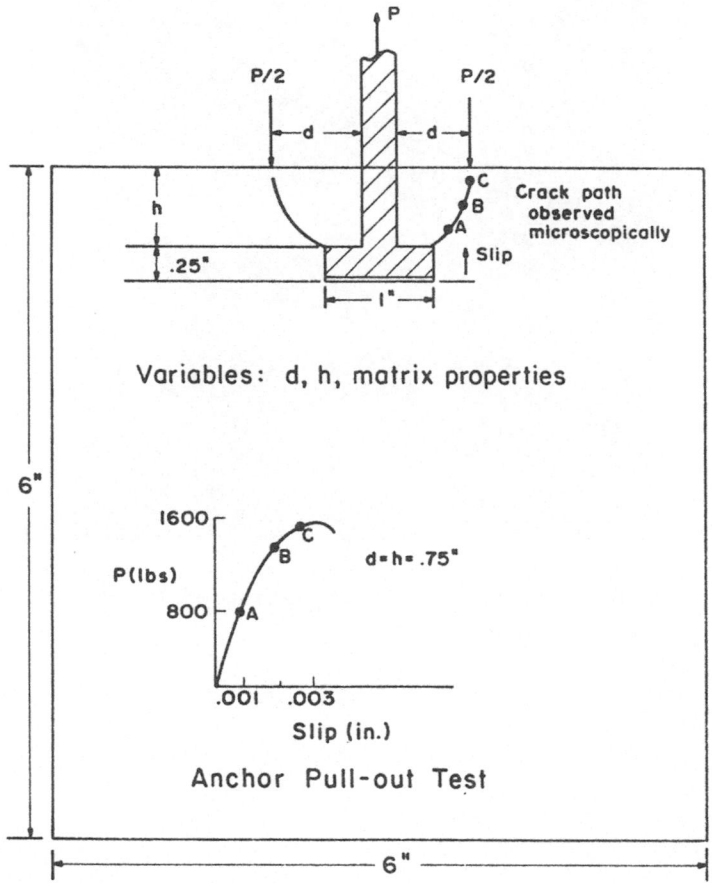

Figure 6 - Configuration of 2-D Pull-Out Test

Figure 6A -- Microscopic Observation of Crack Tip
Locations

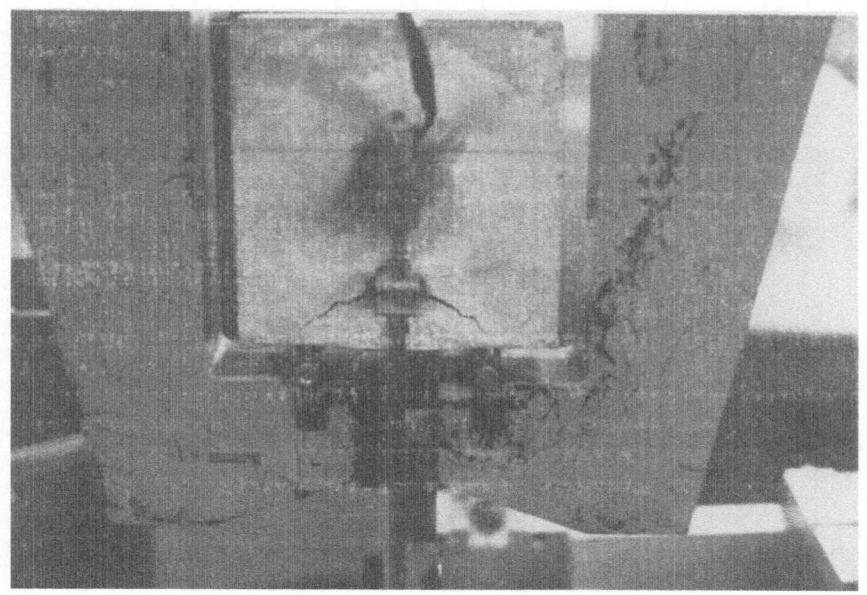

Figure 6B – Typical Pull-Out Failure

74

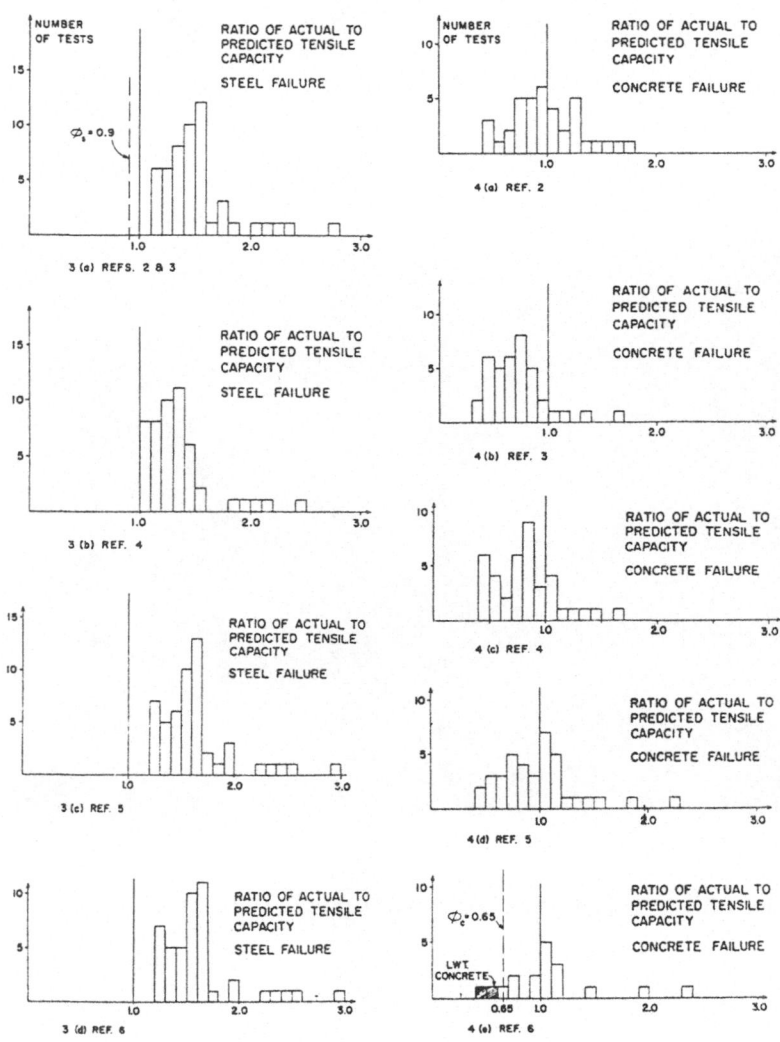

Figure 7 - Predictions of Currently Available Design
Procedures (Ref. [32]).

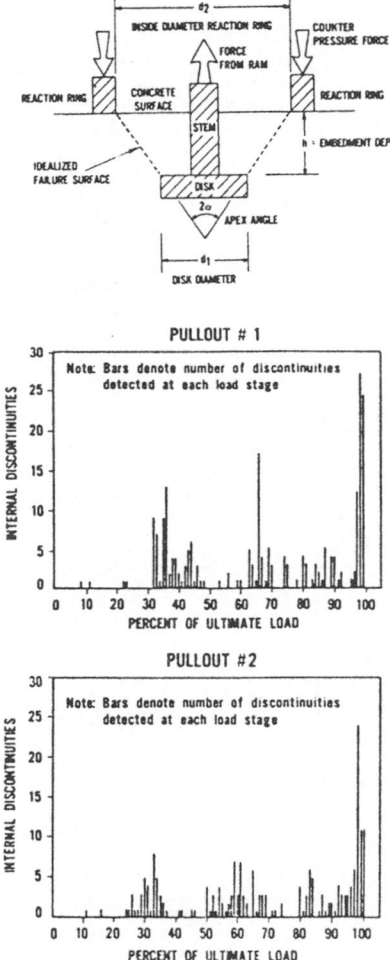

Figure 8 - Discontinuities in Load-Strain Histories
in Large Scale Pull-Out Tests (Ref. [34])

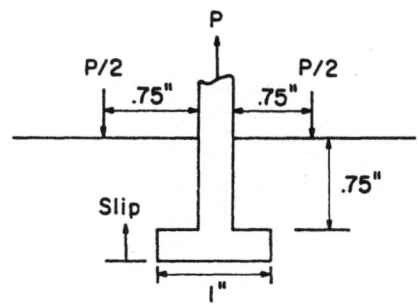

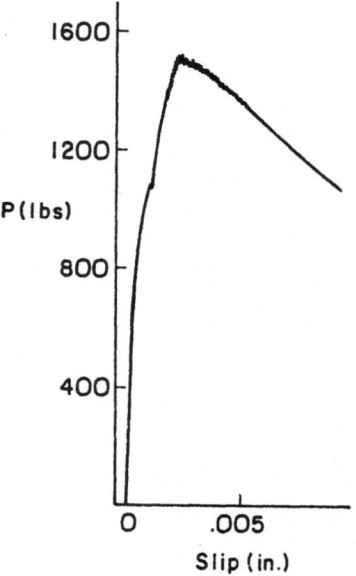

Figure 9 - Load-Slip (Configuration 1)

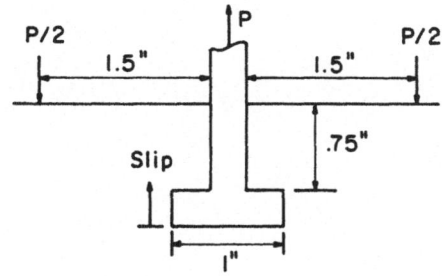

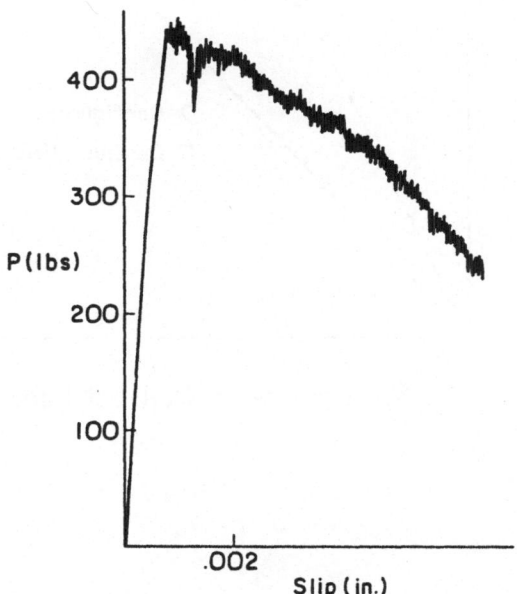

Figure 10 -- Load-Slip (Configuration 2)

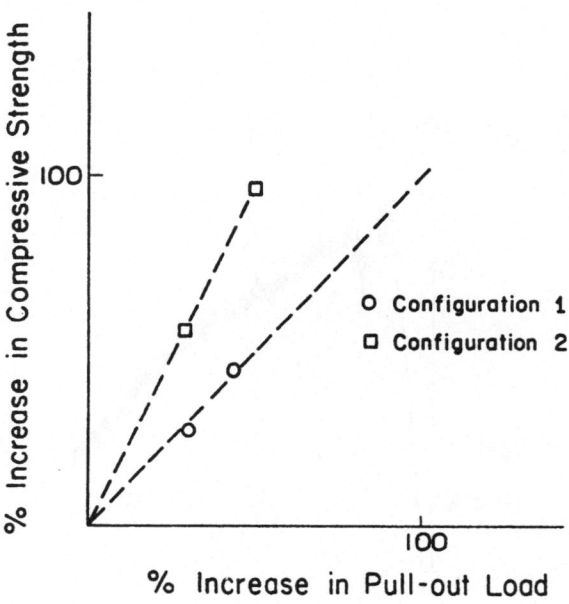

Figure 11 - Relationship Between Pull-Out Load and
Compressive Strength

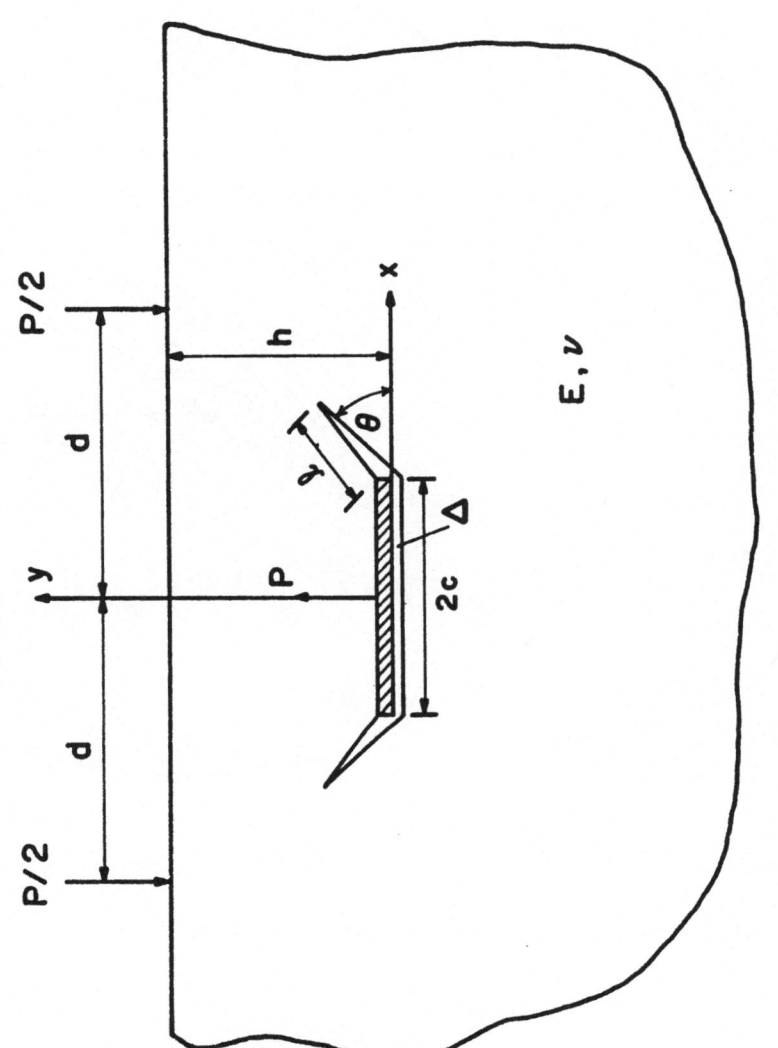

Figure 12 - Mathematical Model

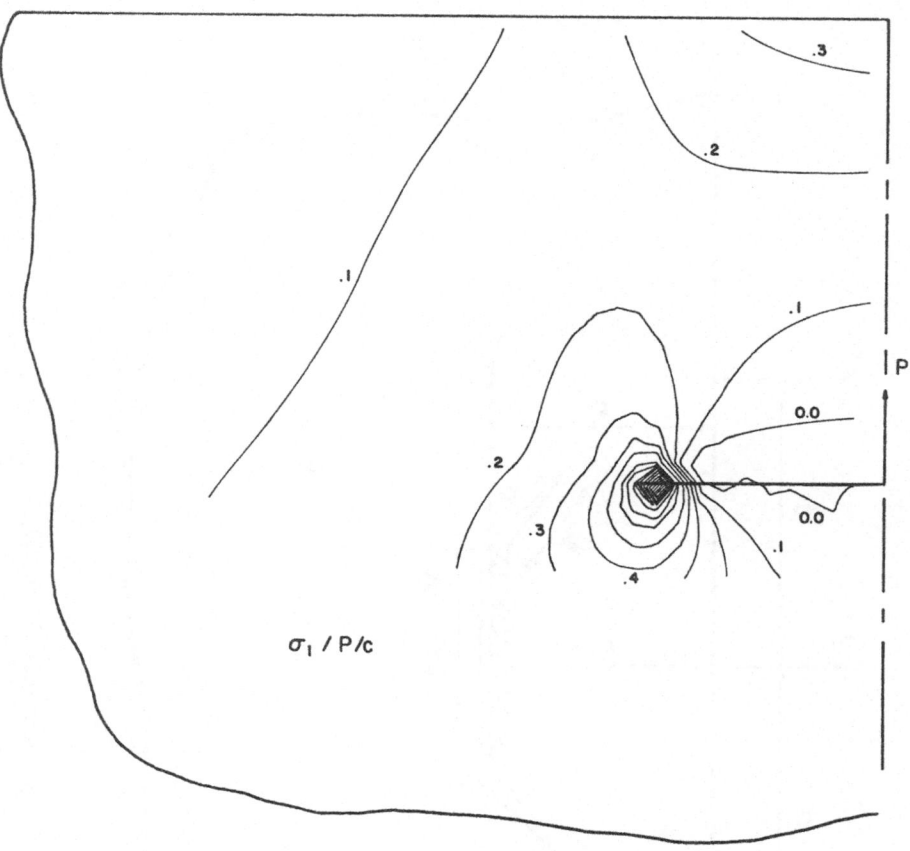

Figure 13 — Maximum Principal Stresses (h = 2c)

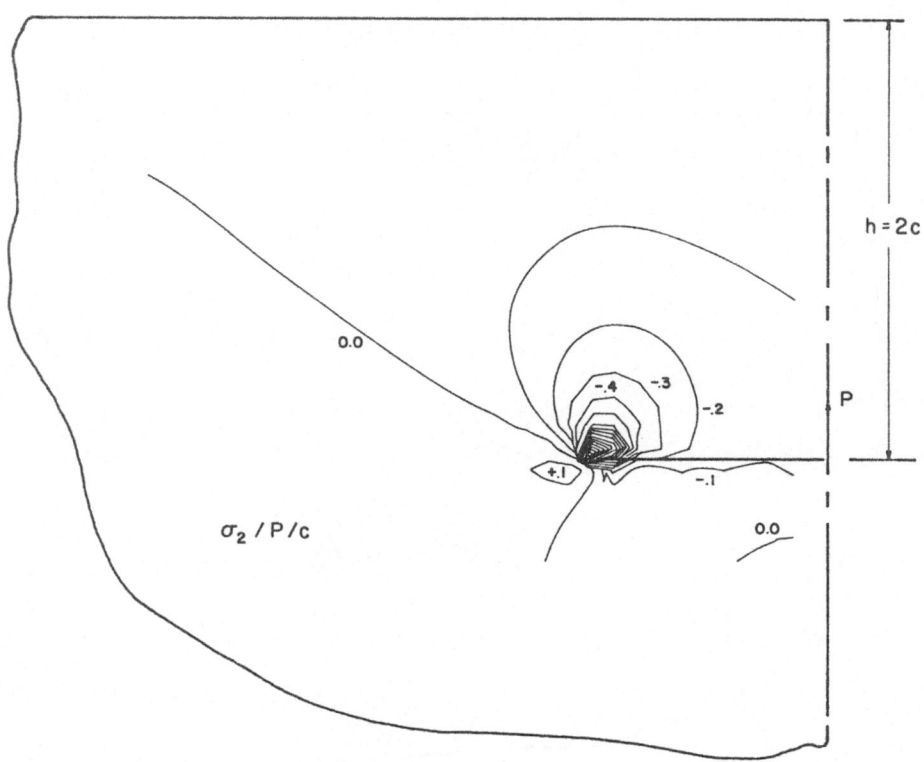

Figure 14 – Minimum Principal Stresses (h = 2c)

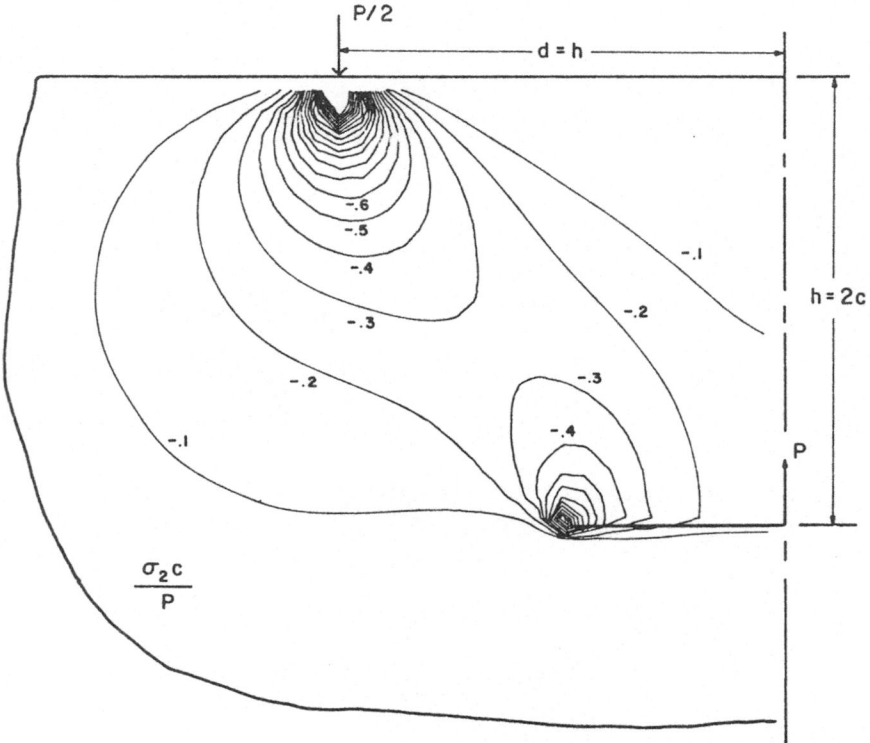

Figure 15 – Maximum Principal Stresses (h = d = 2c)

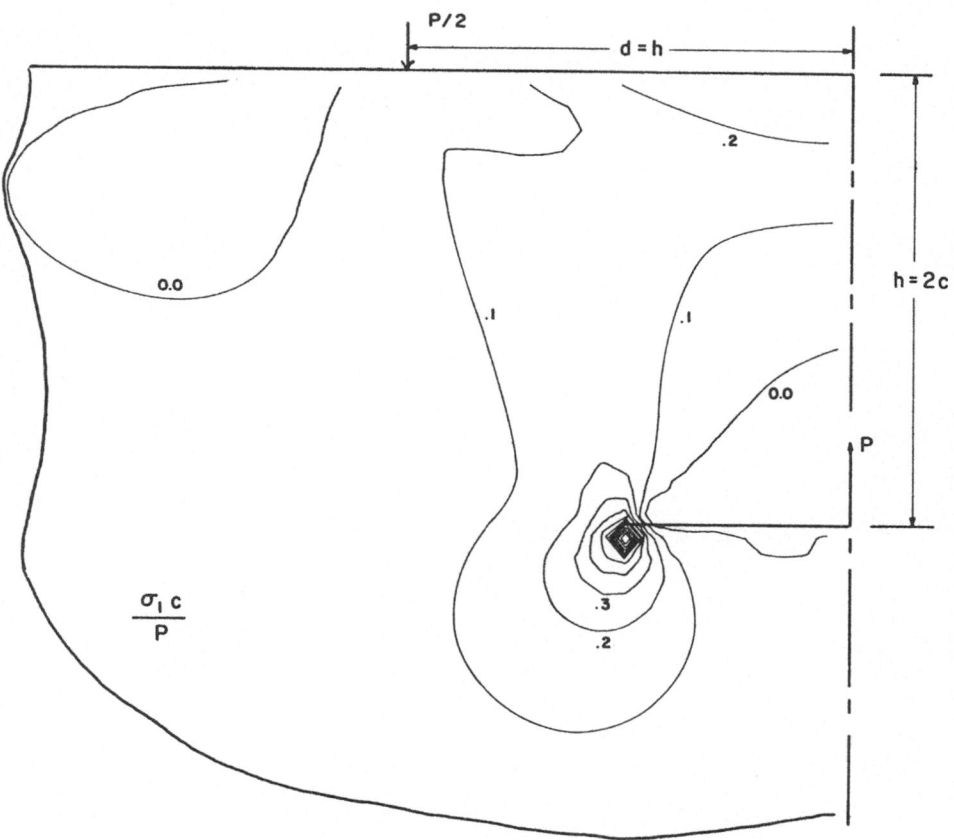

Figure 16 – Minimum Principal Stresses (h = d = 2c)

SECTION II

FRACTURE PROCESSES IN CEMENT COMPOSITES : EXPERIMENTAL OBSERVATIONS

**APPLICATION OF FRACTURE MECHANICS
TO CEMENTITIOUS COMPOSITES**
NATO-ARW - September 4-7, 1984
Northwestern Univeristy, U.S.A.
S. P. Shah, Editor

ON THE CRACKING IN CONCRETE AND FIBER-REINFORCED CEMENTS

Sidney Diamond and Arnon Bentur[*]
School of Civil Engineering, Purdue University
West Lafayette, IN 47907, U.S.A.

ABSTRACT

The results of an extensive series of observations on details of cracking in pastes, mortars, concretes, and steel- and glass-fiber reinforced cement composites are reported and interpreted. The specimens were specially-prepared materials loaded while under observation in a scanning electron microscope equipped with a system permitting observation of saturated specimens maintained in a wet environment. The results bear on the adequacy of current fracture mechanics approaches in modeling these systems, and certain areas are suggested where modifications of current ideas and experimental practices may be in order.

1.   INTRODUCTION

Cracking processes in cement composites (mortars, concretes, and fiber-reinforced cements) are important influences on mechanical behavior and on durability in aggressive environments. The unusual stress-strain curve ordinarily displayed by concrete is usually explained by the development of bond and then matrix cracks at successively higher stress levels (1). The flexural and tensile behavior of fiber-reinforced cement composites is explainable in terms of crack arrest by the fibers (2), and subsequent bridging over the opened cracks by the fibers, prior to their eventual pull-out (3). In reinforced concrete structures the control of cracking is an important parameter in minimizing corrosion of the steel (4).

---

[*]Permanent Address: Faculty of Civil Engineering, Technion - Israel Institute of Technology, Haifa, Israel.

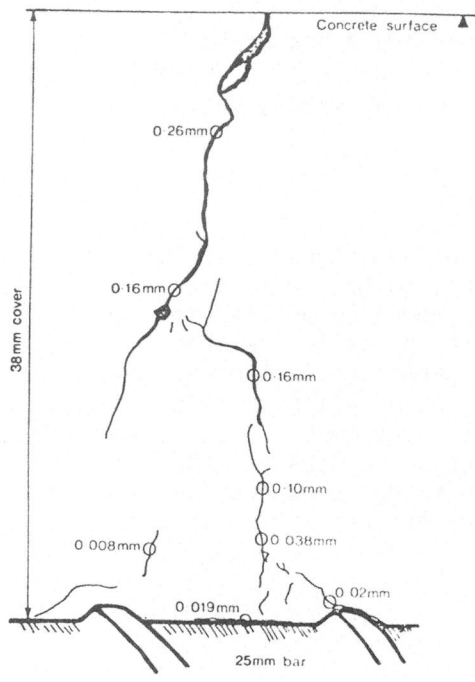

Fig. 1 - Cracking patterns in reinforced concrete (5) observed by optical microscony. The crack extends from the surface of the concrete cover.

Although cracks in cement composites are usually modeled as straight, parallel-sided discontinuities, the actual geometry of real cracks in concrete can be quite complex, as shown for example in Fig. 1. The complexity of the cracking process itself can be deduced from the results of many fracture mechanics studies: inevitably it is concluded that fracture cannot be adequately described by the concepts of linear elastic fracture mechanics, which are ordinarily successful in describing cracking processes in most brittle materials.

In attempting to model fracture in concrete, most workers find it necessary to invoke the presence of some sort of "process zone" ahead of the advancing crack tip. This zone is assumed to be partly damaged, but capable of transmitting some stress laterally across its area. Although the behavior of such a damaged process zone has been characterized numerically in various analytical models, the actual nature of the physical processes that are supposed to be taking place is not usually specified. Terms such as "microcracking" and "aggregate interlocking" have been used, but without much documentation.

There have been a number of attempts by optical microscopy (6) and by x-ray methods (7) to study the cracking processes. These are ordinarily applied to two distinct types of testing situations: axial loading tests designed to determine stress-strain curves, and tests of notched specimens designed to determine fracture mechanics parameters. Unfortunately, the optical microscopy method only detects cracks that appear on the surface being examined. The x-ray methods permit identification and mapping of interior cracks, but only in thin slices of concrete, of the order of 1 cm or less in thickness.

Despite these limitations, such studies have revealed something of the complexity of the cracking processes of concrete under axial loading. They have indicated that bond cracks between the aggregate pieces and the mortar matrix are developed first, ordinarily at applied stresses of the order of 30% of failure stress. Matrix (mortar) cracks are detected only at somewhat higher applied stress levels. Crack instability occurs as a general phenomenon only when bond and matrix cracks intersect and become continuous (1).

Quantitative information presented in such studies is usually confined to tallies of total crack length. In axial loading studies attempts are made to estimate separately the lengths of bond and matrix cracks at the various loading stages. In fracture mechanics tests, the position of the tip of the crack and the length of the main crack induced at the several stages of slow crack growth are usually reported. Little information is

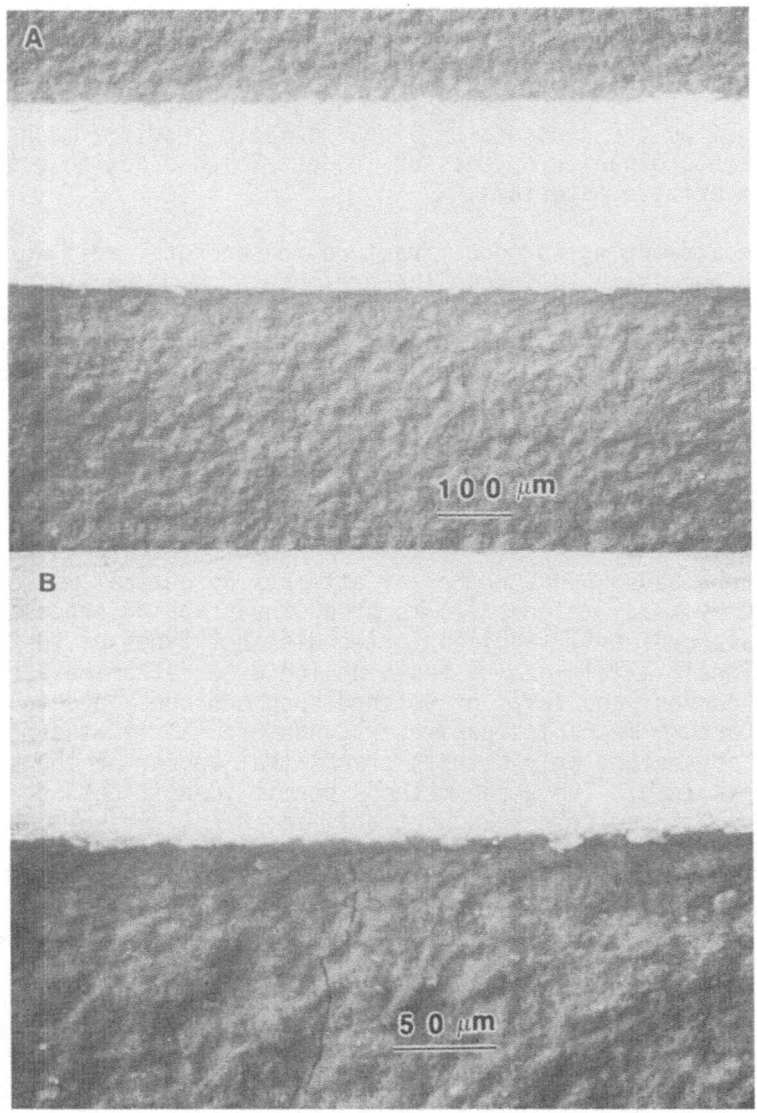

Fig. 2 - Observations of cracks in steel fiber reinforced cement by
SEM:
  (a) Lower magnification (100x) in which no crack is
      observed,
  (b) Higher magnification (300x) of the same field
      indicating the presence of a crack.

ordinarily provided with respect to crack widths, or the geometry of the developing crack pattern. These characteristics are quite important, especially when considering the possibility of occurrence of damaged zones ahead of the crack tip itself.

To some extent this lack of documentation of details other than crack length reflects the tendency of theorists to prefer to view the crack for analytical purposes as a single straight discontinuity. This tendency is perhaps reinforced by the very real limitations of optical microscopy with respect to revealing the finer features of the crack system, where much of the complexity resides.

Some specific reference to these limitations may be appropriate at this point. At a magnification of ca. 100x, for example, an optical microscope can resolve details at best only down to the order of five to ten μm; the depth of focus is smaller than about 2 μm. Cracks narrower than about 5 μm cannot be seen at all; those slightly wider can be observed clearly only if the surface has been polished so that its local roughness is less than about 2 μm. While such polishing is often done prior to examing concrete in petrographic studies, especially in connection with durability problems, it is rarely done in fracture studies. The use of dye penetrants to impregnate the cracks and improve detectability is helpful, but sometimes misleading, since such penetrants impregnate pores and other voids as well as cracks. Practically speaking, the effective resolution of most optical microscopy techniques that have been used in studies of cracking in concrete is probably not much better than perhaps 20 μm.

These limitations strongly influence the interpretation of the results of investigations of cracking and internal damage under applied loads. For example, recent studies (8,9) have indicated that compressive stress-strain curves of concretes and even pastes show non-linearity at low strains, accompanied by irreversible damage to the structural capacity of the specimen. No cracks are usually reported to occur as a result of loading in the low load range, but it is quite possible that cracks are produced, but are simply too narrow to detect by the optical microscopy methods ordinarily used.

An instructive illustration of the absolute need for high resolution in the detection of the extent of cracking in cement composites is presented in Fig. 2. The first micrograph, Fig. 2a, shows a scanning electron microscope (SEM) field of a loaded steel fiber reinforced cement specimen taken at a magnification of 100x. No crack is observable, even under careful study of the original micrograph. However, as seen in Fig. 2b, at a higher magnification (300x) the same area is seen to have a fine but definite crack about 0.3 μm in width. This is more than an order of magnitude, and perhaps close to two orders of magnitude narrower than the finest crack that could have been observed by the usual optical microscopy techniques. As an illustration, the narrowest cracks detected in Fig. 1 were 20 μm in width.

In fracture mechanics testing, optical microscopy is used to establish the position of the tip of an advancing crack in the stable (slow) crack growth stages of crack propagation. The validity of the results depend on the validity of the location of the crack tip. This in turn naturally depends on the resolution of the instrument used to locate it. One must view the validity of most published results with reservation. For example, on viewing Fig. 2a, one might readily conclude that the induced crack has not reached the steel fiber. Examining the same field at the higher effective resolution of Fig. 2b leads to an obviously different conclusion.

In view of the limitations of optical microscopy, it has been recognized for some years that such studies need higher resolution than optical (or x-ray) methods can provide. The SEM, with its high resolution and remarkable depth of field is obviously superior for the purpose. However, the SEM ordinarily requires specimens to be dried and coated in preliminary preparation, and to be exposed to high vacuum of the order of 10^{-5} torr in the instrument. Under such conditions the cracking induced by loading will be influenced by, and likely confounded with, cracks produced by the shrinkage in consequence of drying and evacuation. Thus there is ordinarily an element of uncertainty in the interpretations of cracking patterns observed in the SEM.

A number of publications report SEM studies of cracking in various cement composites (10 - 13). In two of these (11, 12) the specimens have been observed while being loaded directly in the SEM. However, in none of these studies was drying and exposure to high vacuum avoided; in consequence, the significance of the results is somewhat limited.

In a system used by Mindess and Diamond (14-16) it became possible to avoid these difficulties and study the cracking processes of cement composites loaded within the SEM, but not exposed to previous drying, and with only limited exposure to evaporation within the instrument. It was possible to remove the sample from underwater storage and mount them immediately in the loading device in the SEM, with observation and loading starting almost immediately thereafter. The specimens remain sensibly saturated under examination. The surface loses its sheen, but measured moisture content loss is only 1 and 2 percent in the upper mm or so of the specimen and practically none in the body; thus the testing and observation are effectively carried out under "saturated surface-dry" conditions. A discussion of some of the technical details of the operating system used and illustrations of some typical results has recently been presented at the 6th annual meeting of the International Cement Microscopy Society (17); some of the details will be provided in the next section of this report.

Most of the experimental work using this device has been carried out to study cracking induced in a standard compact tension device ordinarily used in studying fracture mechanics properties of metals. In operation a wedge-loading assembly is used to produce a crack from a precast notch. A separate device for compressive loading has also been used. Initially the systems were applied to the study of cracking in cement pastes and mortars (15,16,18,19); more recently we have investigated concretes with small aggregates (up to 7 mm), and the various types of fiber-reinforced cement composites (20 - 22). Taken collectively, these studies provide information on the cracking characteristics of the whole range of cement composites. The present report represents an attempt to review and interpret the results of this entire program of investigation as one connected whole.

2. SEM SYSTEM, SPECIMEN HANDLING, AND OPERATING PARAMETERS

2.1 Sem system

The scanning electron microscope used in these studies is an ISI Super III-A unit, which is equipped with a wide and capacious specimen chamber easily accessible through a full-length door hinge. Provision for feed-through of a large number of electrical contacts from outside the instrument makes it easy to set up devices within the specimen chamber and control them electrically from external controls.

The SEM is equipped with a Robinson backscatter detector which provides excellent resolution, but more important, permits differential pumping of the specimen chamber, which can be maintained at low vacuum levels of the order of 0.5 torr, instead of the 10 torr necessary for operation of the electron column. It has been possible to operate in a "wet environmental" mode, with air being first drawn through successive water-filled bubbler chambers to become water saturated, before being pumped at reduced pressure through the specimen chamber. While resolution is inevitably degraded by such operation, satisfactory resolution for imaging cracks in cement systems can be obtained, and useful magnifications of up to 1000x secured. When water-

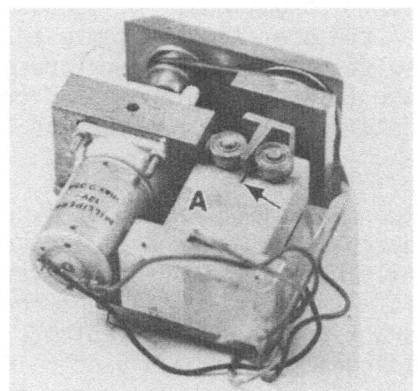

Fig. 3 - Photograph of compact tension apparatus with specimen
("A") mounted in position for loading. Note the precast
notch immediately beneath the loading wedge (arrow).

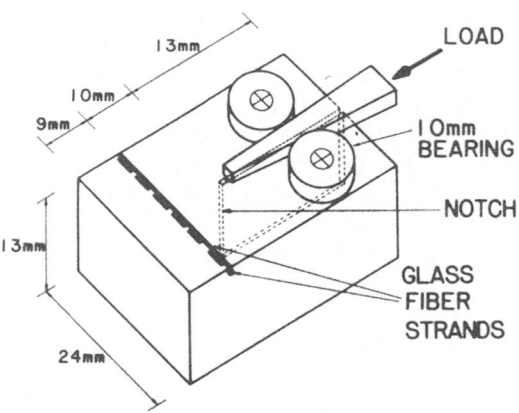

Fig. 4 - Description of specimen and loading arrangement in tests
of specimens reinforced with glass fibers.

saturated specimens are examined some surface drying may take place, but the water loss is negligible; nevertheless, the possibility of slight shrinkage cracking occurring cannot be entirely excluded, especially with pastes under extended observation.

2.2 Specimen Configuration and Experimental Details

In nearly all of the work to be described here, specimens were prepared using Type I portland cement. Mixing was ordinarily carried out in a special mixing device designed to accomplish mixing under vacuum so as to preclude incorporation of large air voids; some systems were mixed in a standard Hobart mixer of the type specified in ASTM C 305. Specimens were typically cast in special molds of the desired configuration, demolded at 1 day, and cured under limewater for several weeks prior to testing.

The specimen size used is naturally limited by the size of the sample chamber in the SEM and the range of motion permitted by the stage controls. Most of the testing has been done with compact tension specimens, 32 mm long, 24 mm wide, and 13 mm thick. The precast notch from which the crack is produced is 13 mm long, and runs through the full thickness of the specimen.

In operation the specimens are wedge-loaded under external control while under continuous SEM observation and continuous monitoring by a load cell connected to an external recorder. A photograph of the device is provided in Fig. 3.

In special studies of certain fiber-reinforced systems a modified loading pattern was invoked, in which the wedge loading was applied only to the exposed face of the specimen, where single fibers or strands had been prepositioned (Fig. 4). Actually the fiber was prepositioned slightly below the as-cast surface of the specimen, to avoid laitance effects; the overlying material was removed by fine grinding prior to loading in the SEM. In developing this loading arrangement it was taken into account that only the interaction of the propagating crack with the fiber at the upper surface of the specimen can be observed. A plane stress loading condition on a very thin specimen only slightly thicker than the glass fiber strand would be ideal, but such a specimen would be extremely difficult to cast and manipulate. The more practical alternative chosen here was to apply the wedge load only at the exposed face of the specimen, where the fiber is prepositioned. The tensile field is greatest at the top of the specimen and diminishes with depth. While this arrangement is by no means ideal it has been shown in preliminary testing to be effective, reproducible, and well suited to the qualitative study contemplated.

The device used for testing in compression is shown in Fig 5. The specimens tested here are small, slender bars 20 mm long, 6 mm wide, and 6 mm thick, and the loading is normal to the bar axis. A tensile failure is induced along the bar axis due to the Poisson effect. Partly because of end restraint, the crack tends to be stable in about the central two-thirds of the length of the specimen.

We have found by experience that proper preparation of the surface to be examined is an important prerequisite for successful observation of the cracks that develop. All microscopy inherently is a process of imaging two-dimensional surfaces at enlarged magnifications. Details on the surface need to be readily interpretable, and ideally the surface should faithfully represent the interior bulk of the specimen. We prepare such surfaces by wet grinding on a lapping wheel, using successively finer grit sizes, and finishing with a No. 900 or No. 1200 fine polishing grit. Sufficient material is removed so that the as-cast surface and any laitance is completely discarded.

The smooth surface so produced is appropriate for observation of cement pastes and fiber-reinforced cements, but in studying mortars and concretes one needs to image and distinguish the outlines of the sand and coarse aggregate particles. Siliceous grains should provide some elemental contrast with the calcium-rich paste matrix through differential production of backscattered electrons, but the amount of such contrast is generally insufficient, and it is entirely lacking with limestone aggregates. To enhance our ability to see the boundaries of the sand and aggregate grains, we ordinarily use a final roughing step after fine polishing the surface. A brief lapping with a relatively coarse grit preferentially erodes the paste, adding topographic contrast to whatever elemental contrast might exist. Unfortunately, soft aggregate particles may erode as rapidly as the paste matrix, leading to inability to establish the desired topographic contrast.

2.3 Operating Parameters

When the wet environmental mode of operation is used, the electron column is plugged with a removable plug having an aperture large enough to pass the electron beam, but small enough so that molecular gas transport is quite slow. Thus the column can be maintained at high vacuum by the main diffusion pump of the instrument, while the specimen chamber is separately pumped with moist air streaming through it. Resolution is sufficient so that cracks as narrow as 0.1 μm can readily be detected. For elucidating still finer details, the fractured portions of certain specimens may be examined at very high magnifications.

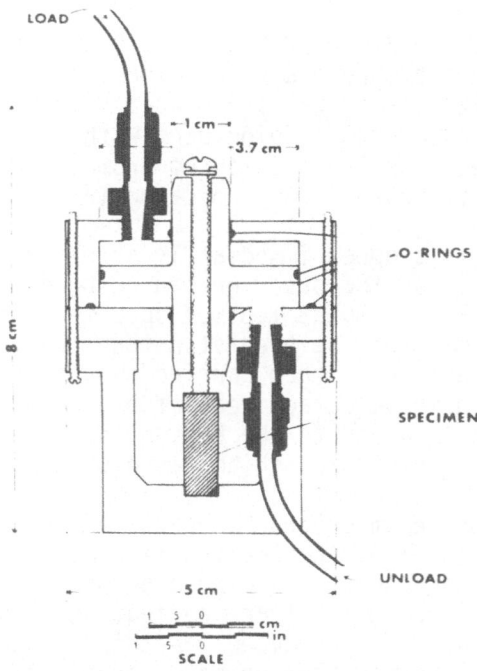

Fig. 5 - Detailed drawing of compression device and specimen. Loading and unloading ducts are connected to external nitrogen gas cylinder.

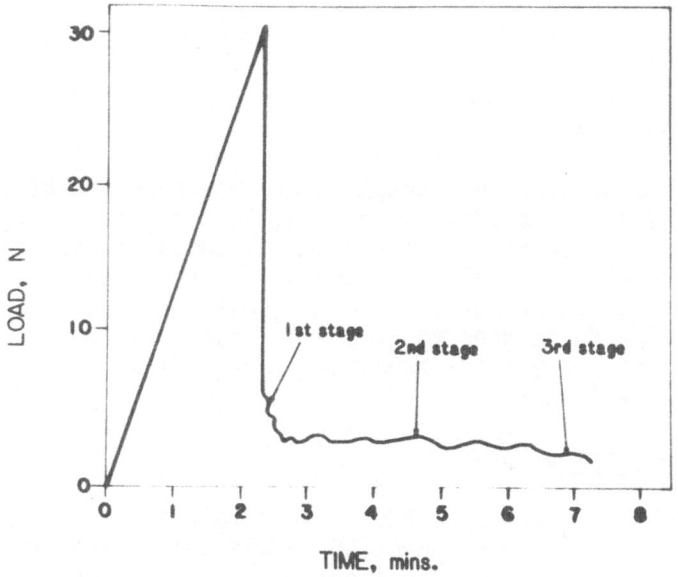

Fig. 6 - Typical load-time curve.

In the usual operating protocol the specimen is first examined for any evidence of pre-cracking. The load is then applied and increased until the crack is produced. This is always a sudden and clear-cut event. Coincident with the appearance of a crack on the screen the load monitored from the output of the load cell falls off sharply, as indicated in Fig. 6.

The loading is stopped immediately after the crack appears, that is, as quickly as the observer can respond. At this stage the observed crack in compact tension specimens ordinarily extends forward about 10 mm from the precast notch. Detailed observation of the cracking pattern is then carried out, first at low magnification, and then in selected areas at higher magnification. Special attention is given to the region that appears to contain the crack tip, or tips if branching has occurred.

After the desired micrographs have been obtained at this initial stage of loading, the loading is usually resumed, resulting in widening and progressive extension of the crack or cracks. This is stopped after an arbitrary period, and a second-stage round of documentation is produced. A third stage may also be examined, or the load may then be advanced continuously until the main crack extends all the way across the specimen and the specimen separates into two pieces. When desired, one of these can be remounted so as to expose the actual failure surface; this surface can then be examined either in the wet environmental mode, or the specimen can be dried and coated for subsequent conventional examination at high magnification.

3. OBSERVATIONS

3.1 Introduction

In this section we attempt to provide representative illustrations and observations relating to the details of cracking in the various specific types of cement composite systems examined. While there are similarities among all of the patterns developed, there are also important differences that have practical consequences and these need to be pointed out specifically. In all the micrographs presented here the overall direction of crack propagation is upward.

3.2 Cement Pastes

The crack ordinarily induced in cement pastes at "first cracking" has a width of about 30 μm where it leaves the precast notch in compact tension specimens, and about the same value near the center of the axial tension crack in compressively-loaded specimens. In the compact tension specimens, the width progressively diminishes in the forward direction, and it may be about 3 μm wide about 10 mm forward of the notch. Thus the "aspect ratio" of the initially-formed crack is of the order to 70 or so.

The crack assumes what appears to be a generally straight path at low magnification, but higher magnification reveals that the actual path is composed of an assemblage of short, zig-zag straight segments deviating from the overall direction of propagation (Fig. 7). The individual straight segments are of the order of 60 μm in length, but vary somewhat in this regard.

As the tip of the crack is approached, the path becomes less regular. In addition to being narrower, the crack tends to become visibly discontinuous in several places, as is indicated in Fig. 8. These places are often where there are shifts in the mutual orientation of adjacent straight segments. The apparent discontinuity at the surface may be local, and the underlying crack within the specimen may well be continuous.

A feature sometimes, but not frequently, observed in cement pastes is branching close to the area of the crack tip itself. An illustration is provided in Fig. 9. Usually only two branches are produced, with one of them ordinarily being wider than the other. On further loading, invariably only one of the branches is activated and increases in width; the other maintains its original width and does not extend visibly. This effect is illustrated in Fig. 9b. The active branch is not necessarily the branch that was wider initially; often it is not. The fact that only one of two branches is activated on additional loading seems to hold good wherever branching is found.

It is extremely difficult to trace a crack to its tip and be absolutely certain that the tip has in fact been located; the crack seems to narrow and become indefinite as its width is reduced to that of the individual particle of hydrated cement paste. The field surrounding a typical crack tip is shown in Fig. 10. As far as can be established, the termination occurs at the intersection of the faint crack with the approximately 60 μm residual unhydrated cement grain at "A" in the micrograph. The width of the crack here is no more than a few tenths of a μm. There is some branching in the region near the termination, faint branching being visible in the original micrograph above and to the left of "B", to the left of "C", and above and to the left of "D". The extent to which these features will be visible in the printed version of this publication is uncertain.

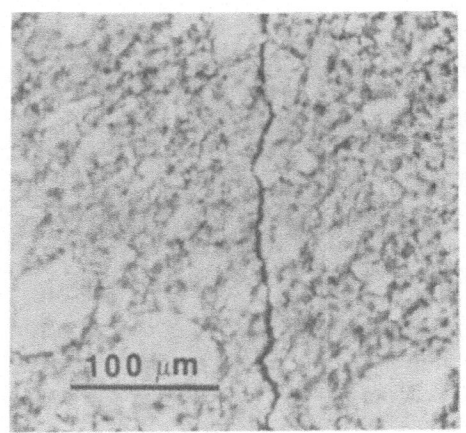

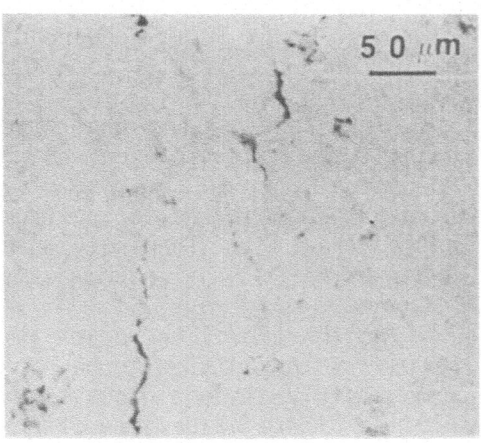

Fig. 7 - Typical portion of a
loading crack induced
in cement paste (16).

Fig. 8 - Illustration of apparent discontinuity
in loading crack in
paste (18).

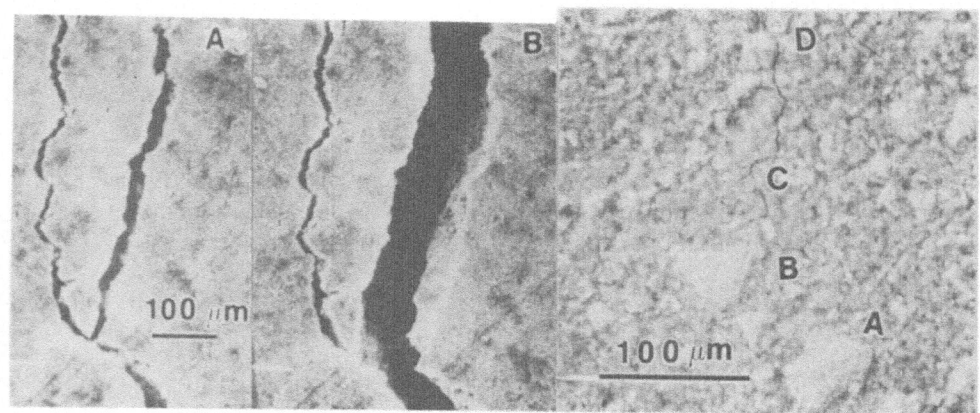

Fig. 9 - Detail of crack extension effects at point
of branching showing that
only one branch is "active"
in crack extension:
a, (left) is the initial
crack, b, (right) is the
crack after extension by
additional wedge dis-
placement (18).

Fig. 10 - Region of crack tip
in cement paste.
Actual apparent
termination is at "A",
"B", "C" and "D" mark
locations of branching
(16)

3.3 Mortars

The initial cracks induced by loading compact tension specimens of mortars show lengths and widths similar to those reported previously for cement paste specimens. However, one might expect somewhat more tortuosity in the overall pattern, and branching is much more common. An illustration is provided in Fig. 11. The crack forks at "A" and two parallel branches are produced. The left branch terminates at "B"; the right branch continues on.

The tortuous character of the crack path is obvious even at this low magnification of 65x.

Other features commonly observed with mortar specimens are surface discontinuities similar to that evident in Fig. 8; and occasional multiple cracking, an illustration of which is provided in Fig. 12. Multiple cracking tends to occur where the crack leaves the perimeter of a sand grain, and also around air voids.

The interactions of propagating cracks with sand grains is of particular interest in these studies. In air-dried specimens which we have examined in the past, shrinkage-induced bond cracks can be observed around some of the grains prior to any loading. An illustration is provided in Fig. 13. In such air-dried specimens, when load cracks are produced they tend to run along segments of the cement-sand grain interface, as is shown in the bottom of Fig. 12a; the process results in a clean separation or "debonding" between the sand grain and the surrounding paste, as seen in Fig. 14. This feature is likely conditioned by incipient pre-existing perimeter cracks induced by shrinkage stresses (Fig. 13).

Occasionally, but rarely, a loading crack is observed to run through, rather than around, a sand grain. This may cause the sand grain to shatter, as can be seen in Fig. 15.

In air-dried specimens, if a loading crack happens to pass between two closely-spaced grains, there seems to be some probability of partial debonding along the perimeters of both of the adjacent grains, as seen in Fig. 16a. In such areas, one of the two debonding cracks is usually dominant, the other being smaller or more localized (Fig. 16b).

The overall characteristics of loading cracks, including such features as tortuosity, branching, multiple crack occurrences, etc. are similar in mortars tested after previous air drying and mortars tested in the wet mode without previous drying. However, the interactions of the cracks with sand grains are somewhat different for the mortars that have been continuously wet. Here

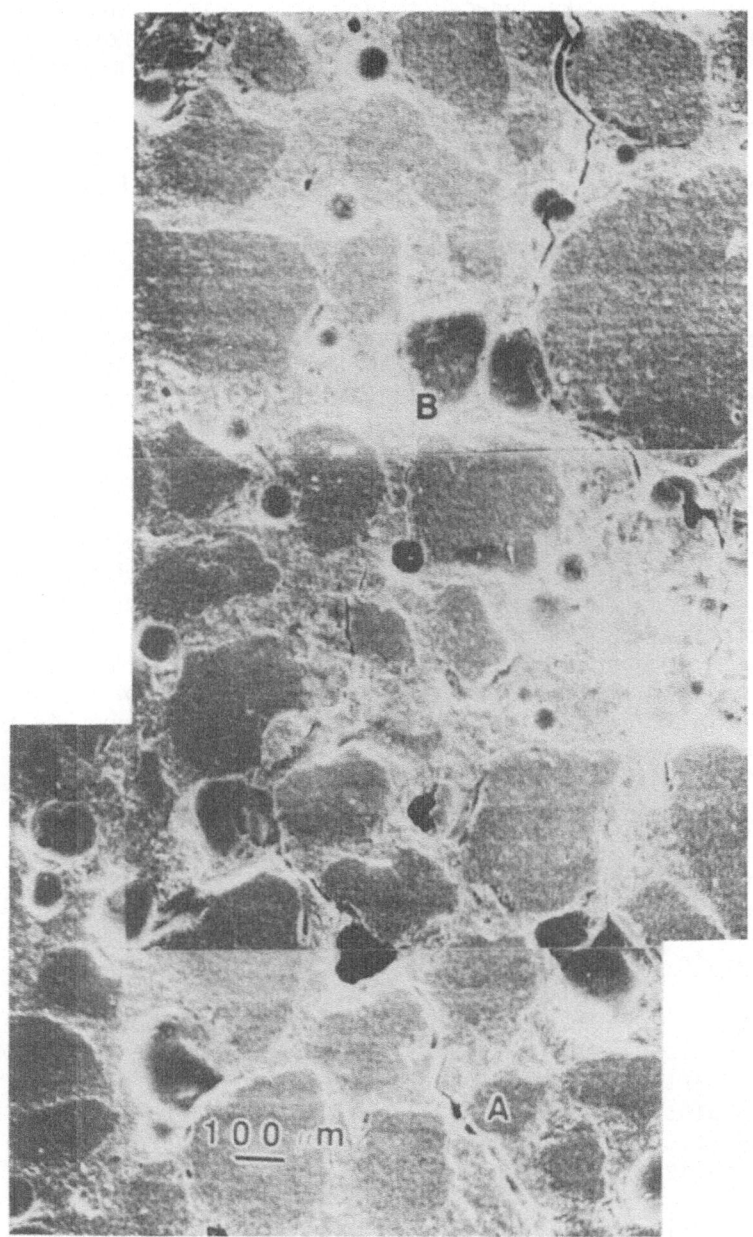

Fig. 11 - Typical portion of a loading crack induced in air dry
         mortar (19).

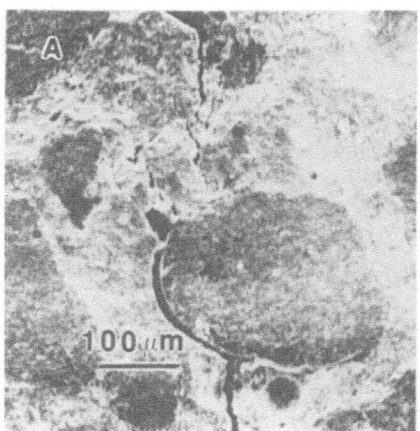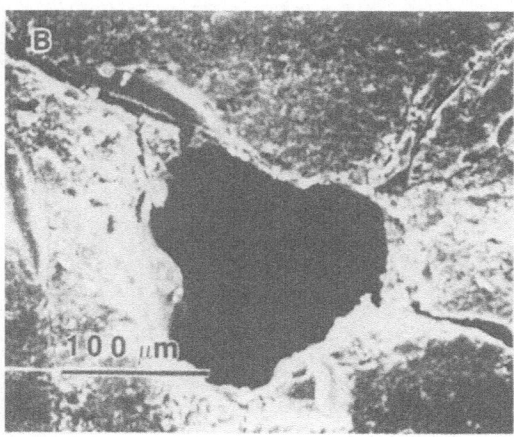

Fig. 12 - Debonding and multiple cracking in air dry mortar,
(a) around sand grain, (b) around air void (19).

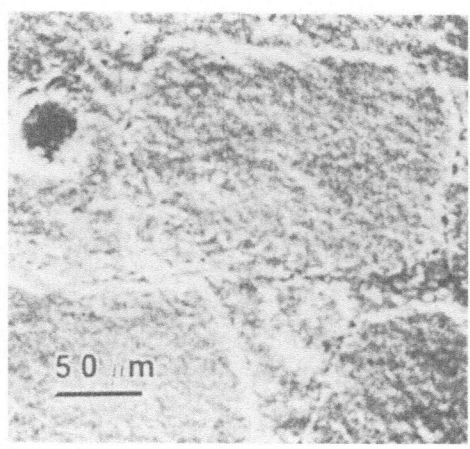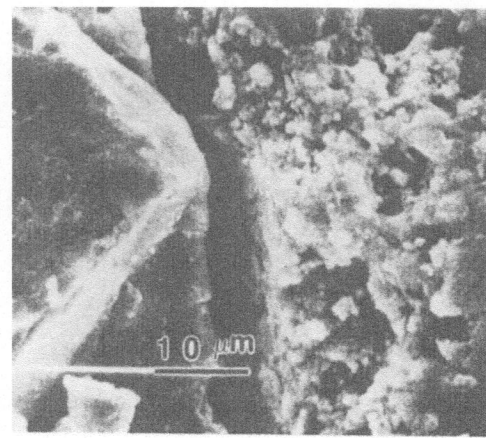

Fig. 13 - Bond crack prior
to loading, in air
dry mortar (19).

Fig. 14 - Bond crack in air
dry mortar (19).

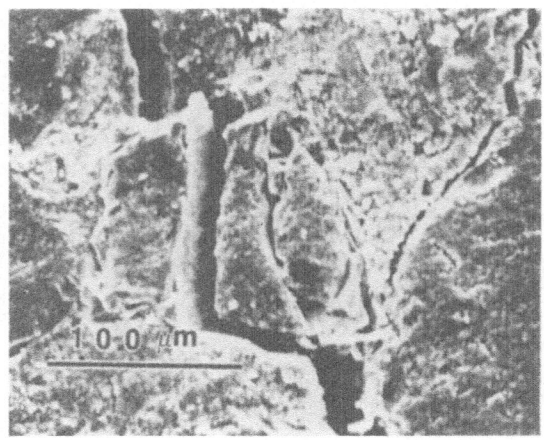

Fig. 15 - A shattered sand particle in loaded, air dry mortar (19).

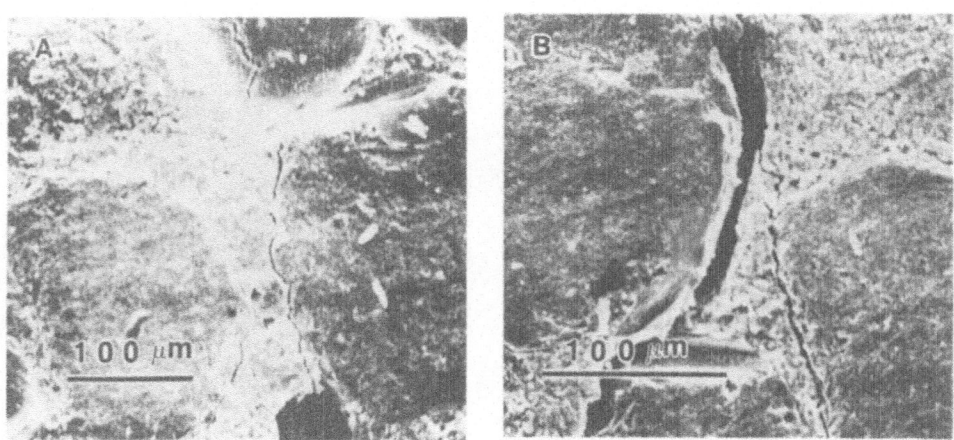

Fig. 16 - Separate bond cracks around adjacent sand particles in air dry mortar (19).

no debonding cracks, incipient or otherwise, are observed before
loading. The loading crack takes a meandering course to avoid
the sand grains, but it rarely runs along the actual sand grain -
paste interface. An idea of the path usually pursued can be
gotten from Fig. 17.

A view of the actual surface of fracture produced in a wet
mortar that was loaded until the crack caused its separation into
two halves is shown in Fig. 18. The original polished surface of
the specimen appears as the straight line at the upper left corner
of the micrograph. The undulatory character of the path taken by
the crack through the bulk of the specimen is apparent, as is the
fact that very little, if any, actual sand grain surface has been
exposed. There is always paste adhering over the mounds that
mark the positions of the sand grains. This is clearly in consequence of the fact that the crack runs not along the actual
sand grain - cement paste interface, but through the paste at
some distance from the sand grain itself.

In the limited compressive testing of mortars that we have
done, always on continuously wet specimens, the results have been
quite similar in local detail to those described above. Compressive testing results ordinarily in the formation of only a
single main crack, which appears suddenly when the compressive
stress applied approaches the strength of the mortars; roughly
40 MPa in the specimens tested. The crack is perhaps 30 μm wide
in the center, and narrows toward either end; it typically runs
around sand grains, but in the paste between them rather than at
the interfaces themselves.

3.4 Concretes

Only a limited number of concrete specimens have been tested;
in consequence, the results reported here, while of great interest,
are somewhat preliminary in nature.

The concrete specimens tested were prepared with coarse
aggregate pieces up to a maximum size of 7 mm; the use of larger
sized aggregate was found to preclude making specimens of the
limited size that could be tested. The aggregates were strong
crushed quartzite grains of excellent shape and quality.

It is obvious that the small size of the specimen compared
to the size of the aggregate precludes any interpretation of such
specimens with respect to fracture mechanics parameters; our
purpose here was not to attempt to do so, but simply to study
the geometry of the crack system as it developed.

Fig. 17 - Load crack in wet mortar. The crack path in the paste between the sand particles is tortuous and when it approaches the grain it runs along its surface but not at the actual interface (16).

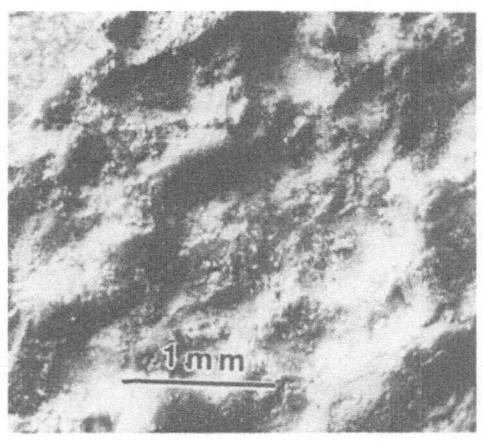

Fig. 18 - Fragment of fractured mortar specimen, remounted to show the undulatory crack-created surface. The bumps are sand grains with a layer of paste separating them from the new surface created by the crack (15).

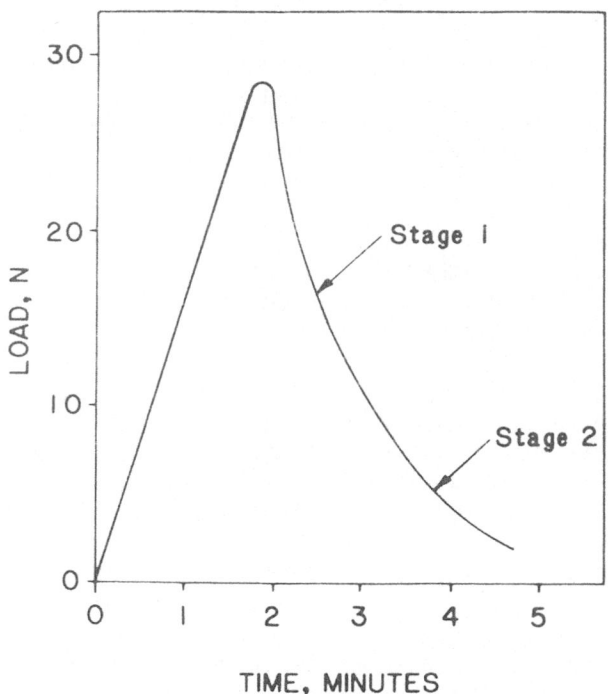

Fig. 19 - Load-time curve of concrete specimen.

The general characteristics of the crack produced in the compact tension testing of this concrete were found to be consistent in many respects with those reported for concretes tested in larger specimens and examined using optical microscopy. The overall tortuosity of the crack path is very high, since the cracks must propagate around densely-spaced aggregate pieces as well as sand grains.

In testing concrete, the correspondence between the first visual appearance of the crack on the SEM monitor and the peak of the load-cell output plot was maintained as rigorously as with pastes and mortars. However, with concrete the post-peak load did not drop nearly as far. Thus, as indicated in Fig. 19, the specimen will continue to carry 50 percent or more of its peak load when the movement of the loading wedge has stopped. At this stage the crack has propagated forward 5 to 10 mm from the precast notch, depending on the particular location of the aggregate pieces with respect to the notch. It should be recalled that the aggregate pieces are up to 7 mm in size.

In these specimens it is practically impossible to locate a point that might reliably be defined as the tip of the crack. The zone where the crack appears to terminate is usually located between two adjacent aggregate pieces. Such zones are characterized by many small branched cracks, which are much less than 1 μm in width.

A number of interesting observations have been made with respect to phenomena observed in these zones on resumption of loading after the initial crack has been produced and studied. Unfortunately, the available contrast between aggregate pieces and surrounding pastes in these concrete specimens is quite limited, and it may be difficult to see some of the features described. However, by careful comparison of the image on the SEM monitor with the original micrographs, the boundaries of the aggregate pieces and sand grains have been resolved, and the positions of the cracks accurately delineated. Drawings are provided to indicate the position of the cracks and the adjacent aggregate and sand particles in some of the illustrations discussed.

The zone surrounding the apparent crack tip after the initial load crack had been produced, and after it had been advanced (i.e. first and second loading stages) are shown in Figs. 20a through 20c. The details observed at the first loading stage will be discussed before proceeding to a discussion of the changes documented in Fig. 20c.

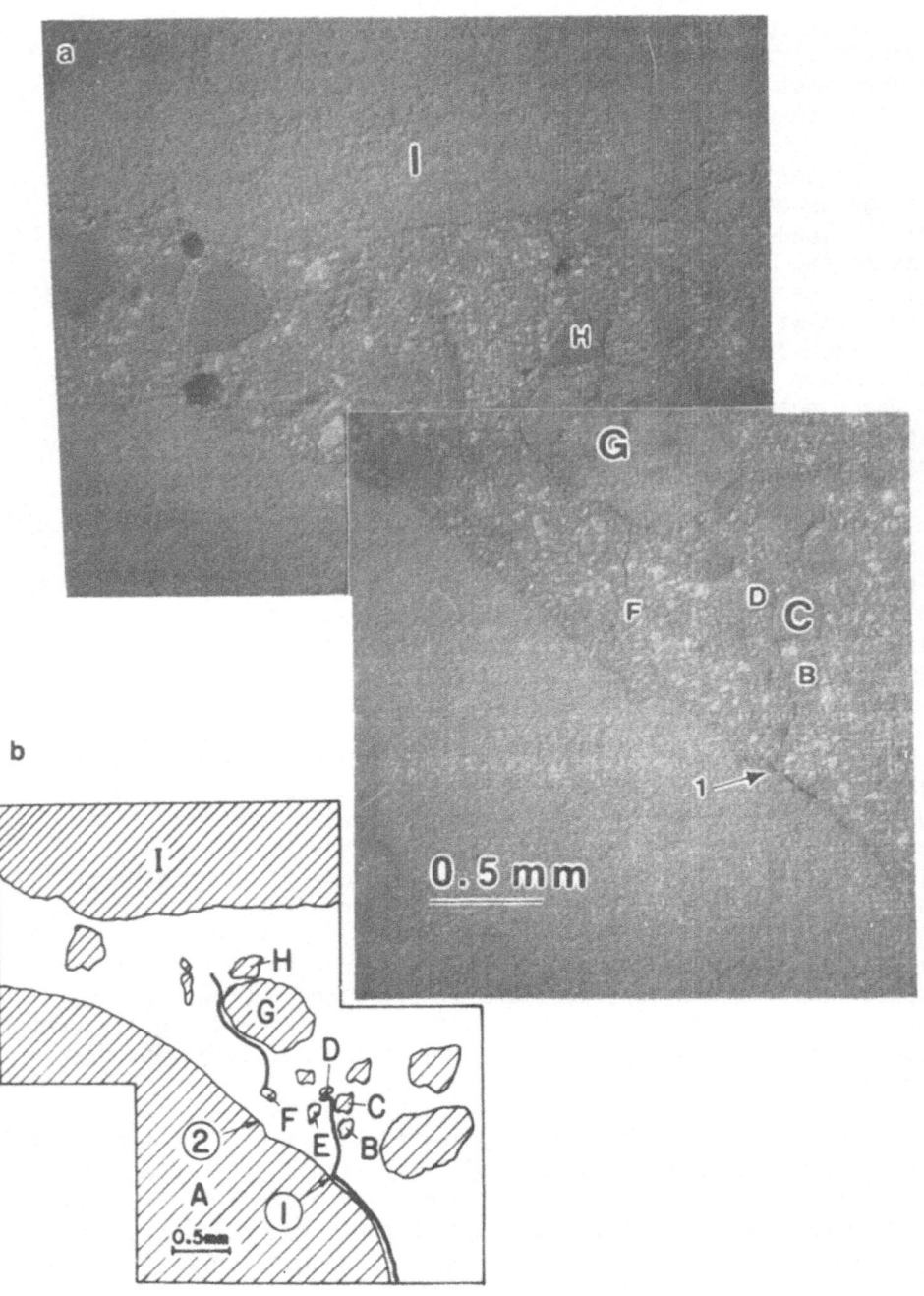

Fig. 20a, b - Zone of crack tip in concrete at the first stage of loading.

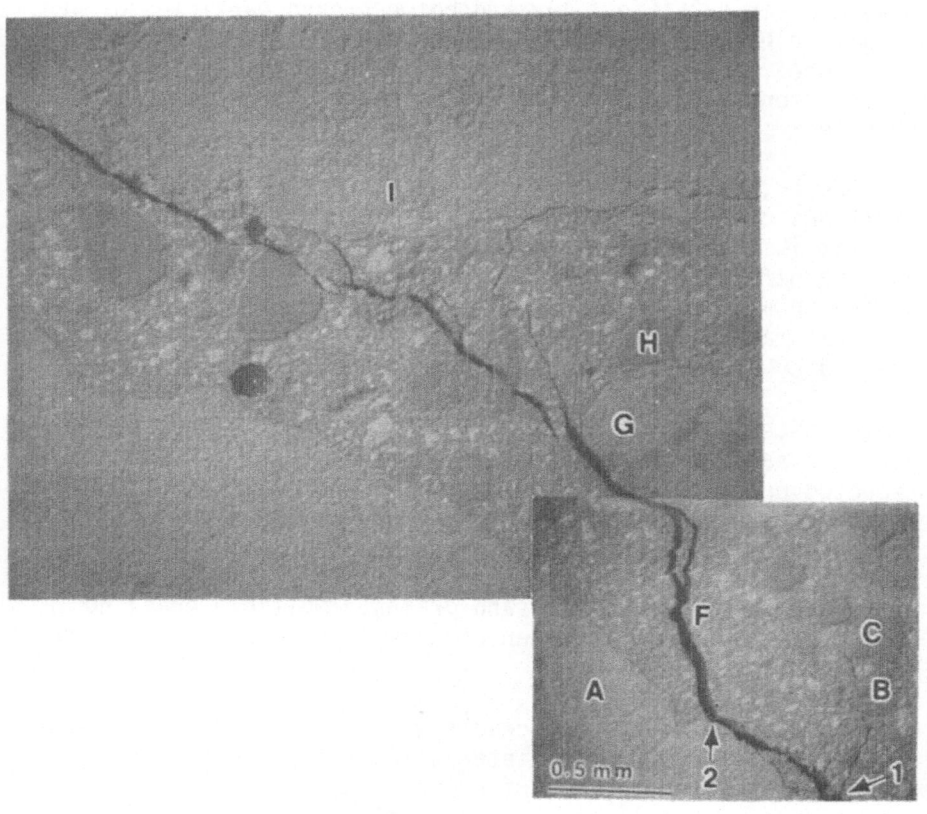

Fig. 20c - The same zone as in a, at the second stage of
loading after the crack propagated beyond it.

In its advance from the precast notch, the original crack ran around aggregate piece "A", and after passing this aggregate piece veered off into the mortar matrix. It apparently terminated within the mortar between aggregate pieces "A" and "I" (Figs. 20a and 20b). At the low magnification provided in these figures, 35x, it appears that the zone encompassing the tip of the crack contains two discontinuous cracks, one terminating near sand particle "D", the other starting at sand particle "F" and terminating just past the large sand particle "G". Both cracks are about 5 μm in width.

Examination of the same area at higher magnification, 100x, indicates the existence of a complex network of narrower cracks, less than 1 μm in width, in the zone between aggregate piece "A" and sand grain "F" (Fig. 21), and between sand particle "G" and aggregate piece "I" (Fig. 22). These finer cracks are tortuous and branched, and run through cement paste matrix material as well as around sand grains; many seem to be discontinuous and to terminate in the paste.

However, at still higher magnification, 1000x, there are indications of the existence of still another finer network of paste matrix cracks, of width less than 0.3 μm. For example, the higher magnification view of the center of the field in Fig. 22 provided in Fig. 23 indicates that the two apparently discontinuous cracks marked "a" and "b" in Fig. 22b are actually interconnected by an array of even narrower cracks.

The interactions of these cracks with concrete aggregate pieces and sand grains present some familiar features and some unexpected ones. The interactions with sand grains in concrete seem to be generally similar to those previously described in mortars. Effects like multiple cracking around sand grains are frequently observed (Fig. 24). Many of these multiple cracks are in a direction radial to the sand grains. Multiple cracking seems to be more common between adjacent sand grains than around isolated ones.

The interactions of the cracks with coarse aggregate pieces are somewhat different. One interesting feature occasionally observed was the penetration of cracks into the aggregate pieces for some distance, this effect taking place even in the first loading stage. Instances of this are visible in Figs 21 and 22.

The effects of reloading the specimen show up in Fig. 20c. The tortuous character of the advancing crack and its many branches are evident in this micrograph. The branches seem to be extensions of the pre-existing main crack. For example, in Fig. 20a

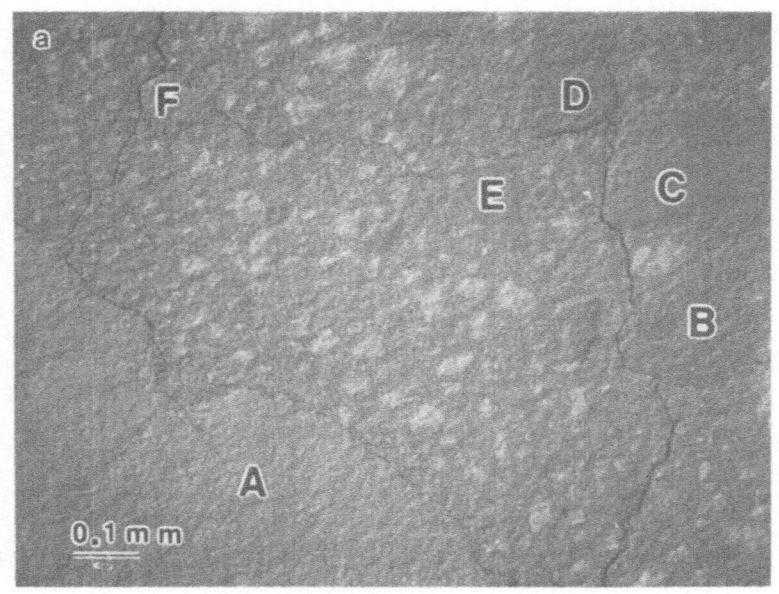

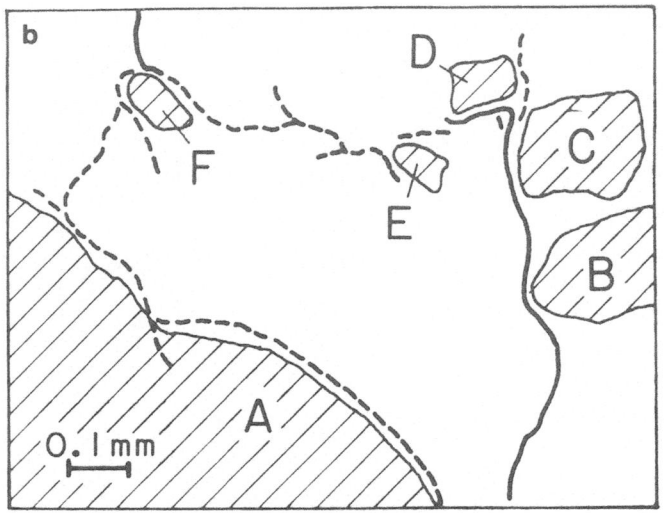

------ cracks of width less than 1.0 μm

——— cracks wider than 1.0 μm

Fig. 21 - Higher magnifications (100x) of the field between
aggregates A and F. (Fig. 20a, b) at the first stage
of loading, showing network of cracks of width of 1 μm
and less between the aggregates and along their surfaces.

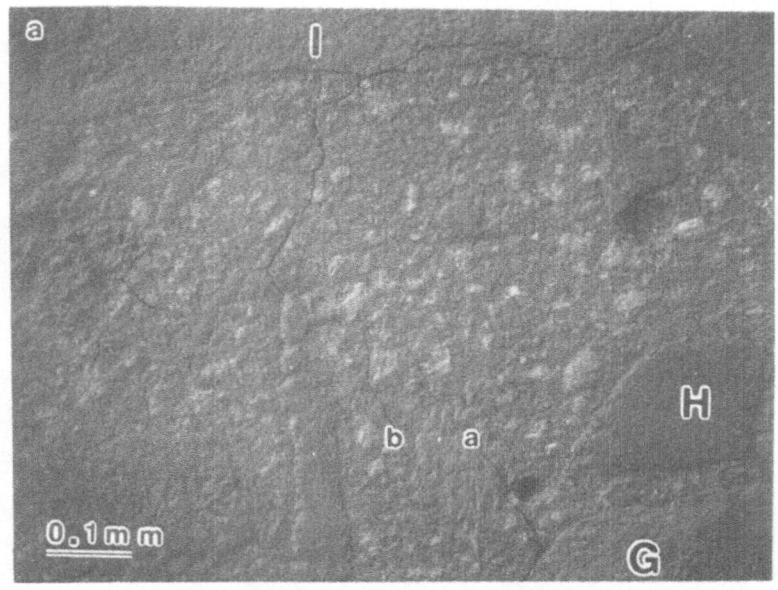

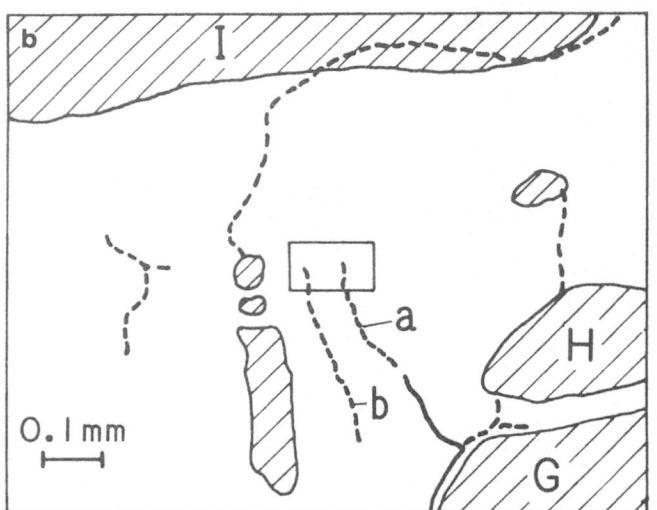

----- cracks of width less than 1.0 μm

——— cracks wider than 1.0 μm

Fig. 22 - Higher magnification (100x) of the field between ag-
         grates G and I (Fig. 20a, b) at the first stage of
         loading, showing network of cracks of width of 1 μm
         and less between the aggregates and along their surfaces.

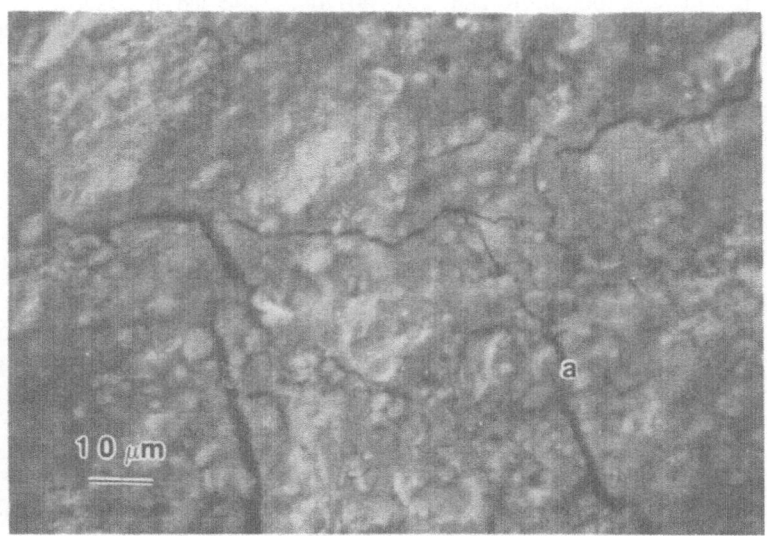

Fig. 23 - Higher magnification (1000x) of the center field of
Fig. 22 showing cracks smaller than 0.3 μm which seem
to run in the paste between two wider cracks (a and b)
that can also be observed in Fig. 22.

Fig. 24 - Multiple radial cracks between adjacent sand grains
in concrete.

and b, it can be seen that at point 1 the main crack path seems
to be leaving aggregate piece "A" and running toward sand grain
"B". The area can be seen at higher magnification in Fig. 25.
On further loading, this section of the crack was not activated,
and it remained as a branch on a new main crack that propagated
further along the boundary of aggregate piece "A" before taking
off at point 2.

A similar effect was observed around the large sand grain
"G". At the first stage one can see a network of cracks propagating from this grain toward aggregate piece "I", on a path approximately normal to the surface of the aggregate piece (Fig. 22).
On reloading, these cracks remained inactive, and a new main
crack started off from sand grain "G" along a different path at
about 45 degrees to the perimeter of the aggregate piece.

It is worth commenting on the overall crack path originally
observed after the first loading, and comparing it to that developed after the loading was resumed. On first loading the crack
tended to propagate in the direction forward from the precast
notch, which is roughly perpendicular to aggregate piece "I"
which lies in its path and forms an obstacle to its further
advance. After the second stage loading the general crack path
changes its orientation so as to make an angle of approximately
45 degrees with the notch, so that it can pass around aggregate
piece "I" without crossing it. This sequence of events suggests
that the aggregate surface may be acting as a crack arrestor in
the first stage, causing the crack to stop in front of it. When
loading is resumed, the halted crack remains inactive, and a new
path around the aggregate piece is eventually activated.

3.5 Fiber Reinforced Cements

The geometry of cracking observed when an ongoing load-
induced crack intersects a reinforcing fiber was specifically
investigated in a series of studies, (20 - 22) and turns out to
be quite complex. Four distinct modes of behavior have been
described, as shown schematically in Fig. 26. Type I is simple,
undeviated passage of the crack through, or rather around the
reinforcement, as illustrated in Fig. 27 with a steel fiber.
This is quite rare, less than 10% of the steel and glass fiber
specimens examined showing this behavior.

The Type II pattern of Figure 26, which shows two branches
re-uniting posterior to the reinforcement with no overall lateral
displacement, was even more rare, being observed only a few times
and only with glass fiber reinforced specimens. An illustration
is provided as Fig. 28.

Fig. 25 - Deviation of the main crack at point 1 (Fig. 20b)
from the aggregate surface into the mortar matrix,
observed at higher magnification (700x).

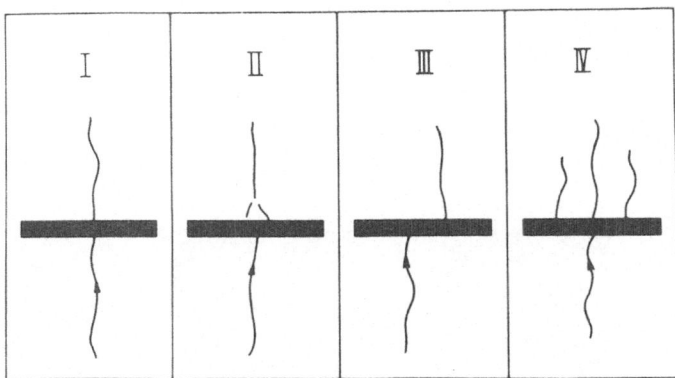

Fig. 26 - The four cracking patterns observed at the intersection
of a propagating crack and a fiber placed normal to
its path (20).

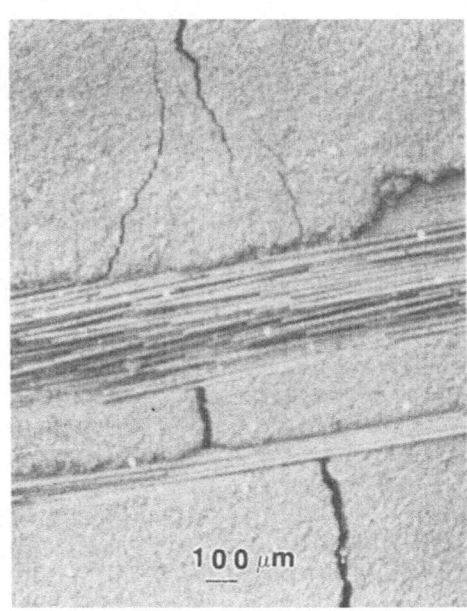

Fig. 27 - Type I crack in steel   Fig. 28 - Type II crack in glass
       fiber reinforced cement   fiber strand rein-
       (21).   forced cement (20).

Crossing events in which the advancing crack is shifted laterally some distance along the fiber were much more common. Simple shifting, designated as Type III, was observed in more than 50 percent of the events observed with glass fiber reinforcement, and about 30% of those observed with steel fiber reinforcement. An illustration is provided as Fig. 29. In this mode the advancing crack seems to be temporarily blocked at the fiber and then shifts laterally between 0.5 and 2.5 mm before resuming its forward progress.

The Type IV pattern, which involves branching into multiple post-fiber cracks was observed about half the time with steel fibers and somewhat less commonly with glass fibers. This mode of interaction involves branching either into two to four branches (Fig. 30), or into a large number of finer cracks of varying widths (Fig. 31). The post-fiber multiple cracks seemed to run parallel to each other in the roughly forward direction. Some of them tended to merge into a secondary post-fiber main crack that continued to propagate forward; others appeared to terminate individually a short distance from the fiber, as seen in Fig. 31b.

Specimens were also examined with arrays of parallel fibers across the prospective crack path. These showed a mix of event types similar to those occurring with single-fiber specimens. In a typical example shown in Fig. 32, a single Type I interaction is observed (Fiber 3); the other interactions were of the more complex Types III and IV. At the low magnification of Fig. 32a it appears that the crack terminates at the fifth steel fiber, but as seen in Fig. 32b at higher magnification (200x), there is multiple cracking beyond the last fiber.

What might happen when an advancing crack intersects a fiber at less than a right angle is seen in Fig. 33, where 10 mm-long steel fibers are seen randomly oriented on the horizontal plane of the specimen. The crack extends from the tip of the brass loading wedge at "A", which just covers up the end of the pre-cast notch. A Type IV event is shown with Fiber 1, where the crack crosses at nearly a right angle. Where the intersection is with an inclined fiber, such as Fiber 4, the crack tends to turn and run parallel to the fiber for at least some distance along its length.

In examining the details of the individual intersections in Figs. 27 to 33, one can observe many apparent discontinuities between the crack anterior and posterior to the fiber. This is obviously an artifact of the experimental arrangement, and continuity is expected if one could observe the region underneath the fibers in the underlying cement paste. A technique was

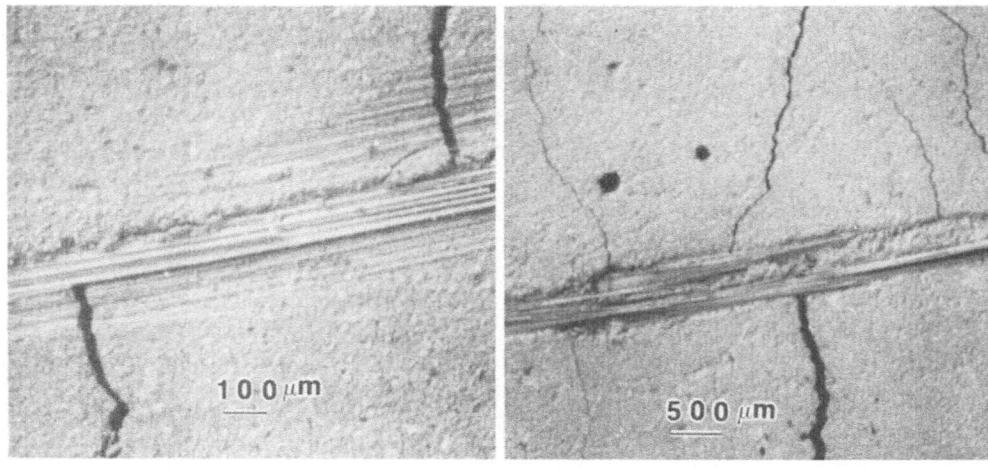

Fig. 29 - Type III crack in glass fiber strand reinforced cement (20).

Fig. 30 - Type IV crack in glass fiber strand reinforced cement showing separation into 4 branches (20).

Fig. 31 - Type IV cracks in steel fiber reinforced cement showing branching into many narrower cracks that merge into a main crack, and also discontinuous microcracks.
   (a) General appearance of the fiber and the crack in front and behind it,
   (b) Higher magnification of the post fiber cracks, showing also discontinuous microcracks (21).

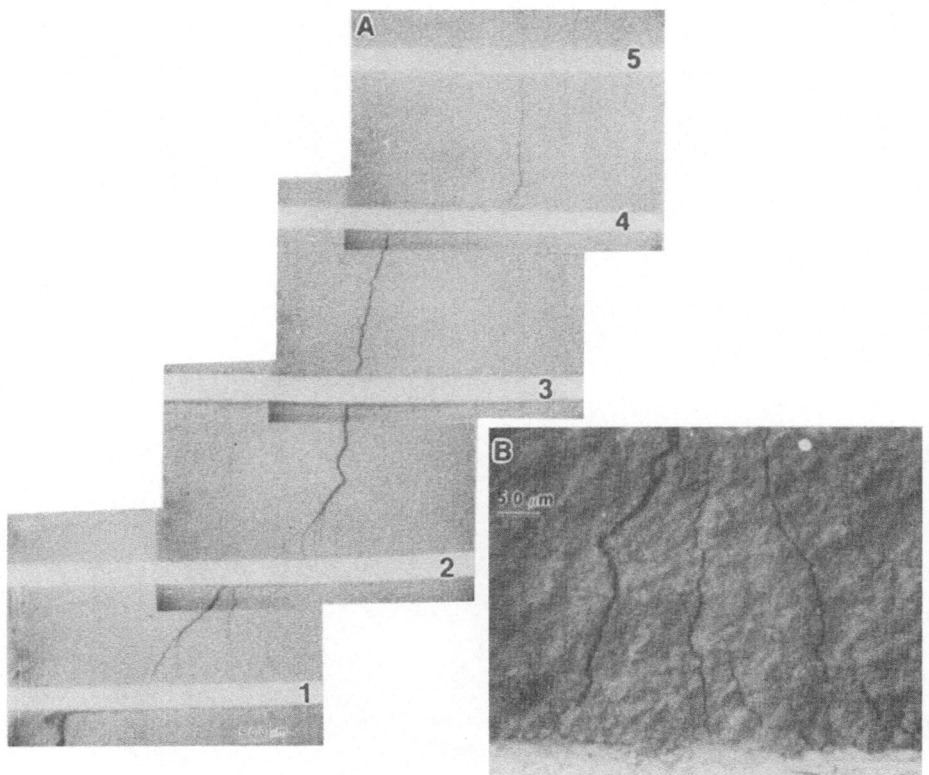

Fig. 32 - (a) Collage of five micrographs showing crack induced
in a compact tension specimen intersecting with five
parallel steel fibers oriented normal to the notch.

(b) Higher magnification (200x) of the field after
fiber 5, showing that the crack extended beyond it,
exhibiting multiple cracking (22).

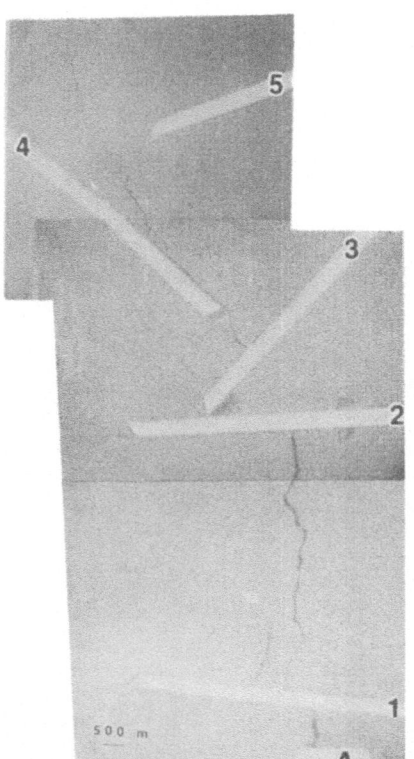

Fig. 33 - Collage of three micrographs showing crack
induced in a compact
tension specimen inter-
secting five steel
fibers at varying
orientations (22).

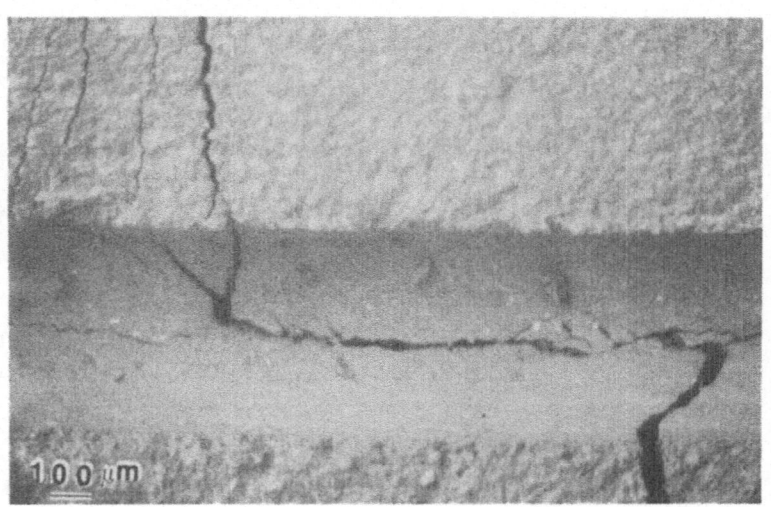

Fig. 34 - Complex cracking at the cement paste surface
immediately underneath the steel fiber in specimens
showing Type IV cracking pattern (21).

developed to accomplish this, by peeling off individual steel
fibers in previously loaded and cracked specimens, so as to expose
the underlying grooves. The specimens had to be removed from the
SEM to accomplish this, but they were then immediately returned
to the wet system for further observation.

Fig. 34 shows a typical result. The crack path in the
groove under the fiber, marking the lateral shift, is clearly
continuous. The crack then turned again and resumed a path
parallel to its original one. The branching shown at the left
side of the groove where the second turn occurred is commonly
observed.

3.5.1 Debonding and "Pseudo-debonding". One of the little
understood but important aspects of the interaction of cracks
with reinforcing fibers is the question of exactly what happens
at the fiber-cement paste interface as the crack reaches this
zone. The mechanical behavior of the composite is in part
conditioned by the response at this interface. In consequence,
we have attempted to document the effects taking place here with
both steel and glass fiber composites. While there are some
similarities, the responses in the two types of systems are
sufficiently different that they need to be discussed separately.

3.5.2 Steel fiber reinforced cements. A significant amount of
true interfacial debonding was found to occur at the steel fiber
interface when the oncoming crack first intersects a steel fiber.
Illustrations are provided in Fig. 35. Note that the bright
steel surface is exposed by the laterally-spreading crack in
both instances, confirming that the crack is actually running
through the true interface.

On the other hand, it was found to be actually more likely
with steel fibers to find that instead of true debonding, a sort
of "pseudo-debonding" event occurred as the crack approached the
fiber. Such an interaction is shown in Fig. 36. Here the main
crack, travelling upward in the micrograph, seems to have been
arrested in the matrix just ahead of the steel fiber; it then
splits into two branches, each of which turned 90 degrees in
opposite directions, and proceeded to run parallel to the steel
fiber, but perhaps 20 μm away from it. This is not true debonding, since the separation is entirely within the matrix and no
debonding effect is felt at the actual steel-cement paste interface, but it may have a generally similar effect.

It should be emphasized that both debonding and pseudo-debonding are local effects; the distance along or parallel to the
fiber that is usually affected may not extend more than a few

hundred μm laterally from the point of intersection of the original crack with the fiber. Beyond this distance the interface or the surrounding paste ordinarily seem to be left intact.

It is interesting to note that both true debonding and pseudo-debonding were observed not only with steel fibers positioned to be perpendicular to the oncoming crack, but also with fibers at lesser inclinations. Here also, pseudo-debonding events were more frequent than true debonding events.

3.5.3 Glass fiber reinforced cement. The behavior at the interface of glass fiber reinforced cement systems cannot be discussed quite as simply as that for steel fiber reinforced cements because of the complexity of the reinforcement system itself. In the glass system the actual reinforcing unit is not a single fiber but a lens-shaped strand, which is a bundle of about 200 fine glass filaments, each about 10 μm in diameter. The interactions of the crack with the outside filaments of the strand may be quite different from those with interior filaments.

True debonding was occasionally observed at the interface between outside filaments and the surrounding paste matrix. However to study this kind of interaction most conveniently, the glass fiber must be exposed on the surface being examined, i.e. the removal of overlying material must be continued until the fiber is exposed. However, glass, being a brittle material, is subject to mechanical damage from the polishing grit particles, and it is uncertain whether the interaction of a subsequent crack with an exposed, possible damaged filament is not changed in some way from what would have occurred in the interior.

To get around this problem, specimens were designed with two strands positioned one above the other. The upper strand was designed to serve as a "marker" strand; in preparing the polished surface, polishing was terminated when it was encountered. The specimen was then loaded, and in examining results we concentrated on areas where the passage of the crack exposed the underlying strand, i.e. the strand that was not exposed to mechanical damage by the surface polishing action.

As it well known to those familiar with glass fiber cement composites, the mechanical properties tend to change with time. Initially ductile composites tend to become embrittled by prolonged exposure to hot, wet environments. The micro-mechanical responses observed when loading compact tension specimens were quite different for ductile composites than for those that had lost their ductility.

Fig. 37 provides a typical view of an underlying strand where the main crack had passed through and the loading had continued sufficiently to widen the crack and permit easy observation. True debonding is evident on all of the filaments exposed. All of the glass filaments bridge over the full width of the crack and present clean surfaces in the exposed areas; indeed, there are practically no hydration product particles visible between the filaments. On additional wedge movement, cracks such as this widen even further and expose additional glass filament surface which is also clean and smooth. Thus it appears that any resistance to lateral displacement and eventual pull-out offered by the glass is entirely frictional in nature.

This situation is in great contrast with what is observed in similar specimens that were embrittled by aging treatments and had lost their ductility. An illustration is provided in Fig. 38. Here passage and subsequent widening of the crack resulted in a clean parting of all of the filaments after only a little bit of pull out; thus the bridging action could not be sustained in this material.

A number of observations made on these specimens provided clues as to the cause of the brittle behavior. In contrast to the almost empty space observed between the filaments in the ductile composite, here the spaces are found to be almost filled with cement hydration products. An illustration is provided in Fig. 39. These products, including many areas that are primarily calcium hydroxide, appear to effectively cement the individual filaments together. In some of the specimens examined there was clear evidence of local flexural failure of such cemented or partly-cemented filaments, occurring where the crack was shifted laterally at its intersection with the strand, i.e. in Type III or Type IV responses. An illustration is provided in Fig. 40. It appears that the cemented-together filaments have lost their flexibility; in consequence, their ability to accommodate the large deflections imposed by the widening of the crack was reduced. In this particular strand, some of the filaments marked "A" in the figure, have remained free of hydration products, and show flexible behavior sufficient to accommodate the imposed displacements without parting. Stucke and Majumdar (10) have suggested that flexural failure of filaments that are tightly bound by hydration products may be an important cause of aging embrittlement of glass fiber cement composites; observations such as these seem to support such a view.

124

Fig. 35 - Debonding at the steel fiber-cement paste matrix, at the actual interface (22).

Fig. 36 - Arrest of crack in the matrix in front of a steel fiber, and its continued propagation in a different course in the matrix, parallel to the fiber at about 20 μm away from the steel surface (20).

Fig. 37 - Pulled out glass filaments bridging over a load crack in glass fiber strand reinforced cement prepared with Cem FIL-2 glass fiber strand, cured for 14 days in lime water.

Fig. 38 - Broken glass filaments observed at the load induced crack in glass fiber strand reinforced cement, prepared with E glass fiber strand, after accelerated aging (20).

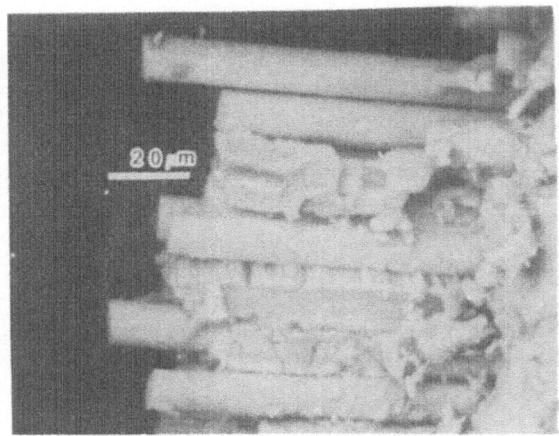

Fig. 39 - Broken glass filaments observed at the load induced
crack in glass fiber reinforced cement prepared with
Cem FIL-1 glass fiber strand, after accelerated aging.
The spaces between the filaments are filled with
hydration products.

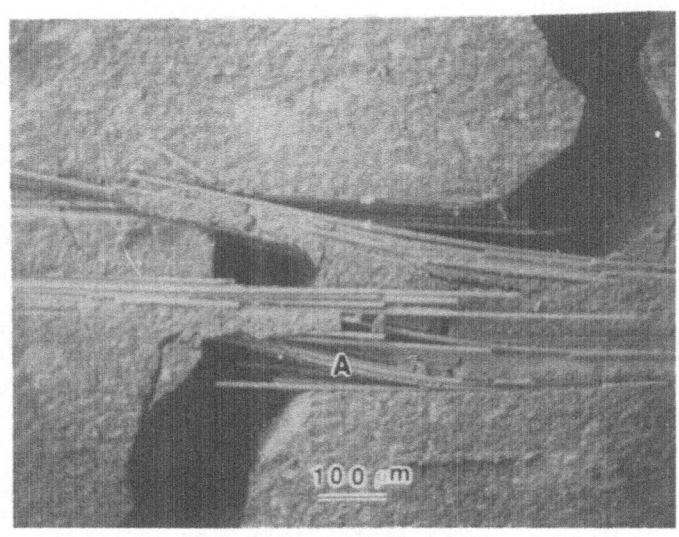

Fig. 40 - Failure of Cem FIL-1 glass fiber strand after aging
in water at 20^{0}C for one year. The shift in the crack
path resulted in bending of the strand.

4. DISCUSSION

It should be now be evident that the cracking processes that occur in cement-based materials are not only complex, but are strongly influenced by the nature of the specific material, i.e. cement paste, mortar, concrete, steel fiber reinforced cement, or glass fiber reinforced cement. Because of the complexity of the processes that seem to be occurring in the different systems, it is not surprising that analytical models based on relatively simple assumptions have their shortcomings.

In the following section we will attempt to point out certain relevant features of each type of system.

4.1 Mortars and Concretes

The geometric features of the cracking patterns observed in mortars and concretes are clearly more complex than is ordinarily assumed by existing theories. The crack path is never straight, even in the limiting case of hardened cement pastes. The zig-zagging of segments around the general path of crack extension observed in these SEM studies is essentially the same as that reported by Higgins and Bailey (23), who used high resolution optical microscopy. The individual segments of crack extension are roughly of the same order of length as that of the larger residual unhydrated cement grains, that is, several tens of μm. The tip of a stable crack in pastes is usually arrested by one such grain, and a limited region of branching can usually be observed around it.

The complexity of the crack system increases considerably when sand grains, and especially when coarse aggregate pieces are present. At least six distinct microstructural elements associated with irregularities in the crack path can be identified:

1. "Micro-level" tortuosity, i.e. the occurrence of connected short segments of varying local orientation, in those portions of the crack which run in the paste matrix between aggregate particles. This is entirely similar to the tortuosity observed in paste specimens.

2. "Macro-level" tortuosity, resulting from the fact that the crack is forced to propagate around the aggregate particles. The deviations from the general direction are thus of the order of size of the sand or coarse aggregate particles. When sufficiently wide, they can be observed by optical microscopy.

These deviations may occur at sites of local debonding around the perimeter of certain aggregate pieces, but are not confined to such sites.

    3. Branching from the main crack. Branching is seldom observed in pastes except near an arrested tip, but it is fairly common in mortars and even more so in concretes. In concrete the branches can be relatively long, often extending over distances commensurate with the size of an aggregate piece. The width of branch cracks can vary from 50 μm or more down to as little as 0.1 μm; thus many of the narrower branches and extensions would not be detectable by optical microscopy.

    4. Multiple cracking, i.e. subdivision into three or more branches at a single locus. Such crack subdivision sometimes occurs in the paste between two adjacent sand grains, as the main crack passes through, but it is not confined to such locations.

    5. Local cracking, consisting of sets of short cracks radial to the perimeter of individual sand grains, possibly associated with prior differential shrinkage of the paste around the grains.

    6. Apparent discontinuities of the main crack in the plane of the specimen being observed. This occurs only occasionally in pastes, but more commonly in mortars and concretes.

    Most of these irregularities presumably arise from the effects of disturbances lying in the crack path. The deviations, branching, and multiple cracking all occur during the rapid stage of crack propagation, and it is not possible to follow the actual sequence of events in the absence of a high-speed camera. Unfortunately, scanning microscopes, being raster instruments, are inherently unsuitable for this purpose. Alford (24) recently reported interesting results of such dynamic studies of cracking events using a high-speed camera in connection with an optical microscopy system.

    In the absence of high-speed photographic capability in the SEM, useful information can still be obtained if observations are carried out at near the tip of the stable crack after the initial loading has been completed, and subsequently in the same area after the crack has been extended by further loading. We have done this and believe that the insight so obtained is helpful, although naturally, difficulties remain.

    For one thing, as stated previously, it is always difficult to establish a point on the stable crack that can be positively identified as the crack tip. In mortars, and especially in concretes, the crack becomes almost undefinable as it nears its tip. It is inherently narrow, tortuous, and seemingly discontinuous in this area, and what is worse, it seems often to branch into still narrower microcracks of the order of a few tenths of a μm in width. This is smaller than the size of many hydrated cement (C-S-H) particles that constitute the finer elements of the matrix through which the crack is passing. These finest "tip region" microcracks seem often to run toward a neighboring aggregate grain.

When loading is resumed to extend a previously-stable crack in concrete, it can usually be seen that the crack changes course so as to bypass any aggregate piece or sand grain lying ahead of it. In so doing, the most advanced segment of the crack usually becomes an inactive branch. It seems that the stable crack is ordinarily arrested at or near an inclusion, and the fact that considerable microcracking usually is developed in the region of the end of the arrested crack, suggests the occurrence of strong local energy absorbing processes. Further progress of the crack apparently requires an alternative path, i.e. the crack prefers to propagate from some point where the resistance is less. Thus it seems to typically resume its advance not from the crack tip, i.e. the furthest extension of the stable crack, but from some point further behind. This leaves the segment leading to the original tip as an inactive branch.

Superficially, these observations sound like they might relate to the "process zone" or "damaged zone" discussed in theoretical fracture mechanics treatment of mortars and concretes. There are certainly similarities. However, in non-linear fracture mechanics modeling, it is often implied that a boundary can be drawn between the main crack, assumed to be straight and unbranched, and the process zone ahead of the visible crack tip. The present observations suggest that such a boundary can not be identified; there is no straight, unbranched "main crack" region, and the full length of the crack is tortuous, frequently branched, and always irregular. Furthermore, the regions of fine multiple cracking described above are found near the ends of branch cracks as well as at the tip of the furthest-advanced crack segment.

In the light of these observations, it seems to us that one should be quite cautious with criteria conventionally used to define the cracking process. For example, the crack opening displacement (COD) criterion, used with considerable success with metals, does not appear to be geometrically appropriate for cement composite systems.

Having introduced consideration of local energy absorption in connection with the potential for further crack propagation from the arrested tip, we should also remark on the overall implications with respect to energy considerations of the general pattern of cracking that has been observed. It is obvious that a considerable amount of fracture energy must be dissipated in creating the large amount of branch and multiple cracking that is seen to occur in concrete. Even in the absence of branching, the development of a tortuous crack path itself requires extra energy. These energy-dissipative processes need to be taken specifically into account in developing fracture models for concretes, and perhaps for mortars as well.

It appears that perhaps some of these processes are over-looked in existing fracture mechanics modeling, in part at least, because of certain aspects of the experimental methods applied to the study of fracture in concrete. In addition to inadequate resolution in the methods used to examine the progress of the crack, some specific shortcomings that might be pointed out in the light of the present results include the following:

(1) Fracture mechanics parameters are often calculated on the basis of the length of the sawn or precast notch, without observing and adjusting for the slow stable crack growth that has preceded the onset of fast cracking. At the critical stage the tip of the crack has presumably moved well away from the notch, and the extended segment is tortuous and branched, rather than straight and parallel sided.

(2) Observations of crack propagation are often carried out on a surface that has not been ground. The as-cast surface layer is rich in paste and lacking in aggregate, and the trace of the crack observed in it is likely to be quite different from that of the crack in the underlying concrete. Many of the tortuous geometric complexities are likely to be smeared out and not observed on the paste-rich surface.

(3) In many testing arrangements such as the commonly-used double cantilever and double torsion systems, the specimen is cast with a double groove in order to force the crack to propagate in a strictly determined straight path. As a result, the tortuos-ity and branching which ordinarily occurs in concrete is to some extent artificially eliminated.

4.2 Fiber Reinforced Cement Composites

The successful application of fiber reinforced cement com-posites is dependent on their post-cracking performance; thus the

details of cracking induced in these materials is of particular
importance.

Most models assume that the initial crack produced by loading runs straight across the matrix and past the fibers, and
that as this crack widens the fibers bridge across it. The post-
cracking behavior of the composite is then usually assumed to be
conditioned by the resistance offered by the fibers to further
widening of the crack and pull out of the fibers. Pull out
parameters are thought to be quantifiable from the results of
separate pull out tests.

Observations presented here suggest that the cracking processes in fiber reinforced cement composites are much more complex
than usually assumed. Important features relevant to the modeling
of such composites include the following:

(1) Geometry of the cracking pattern. Cracks induced from
a precast notch toward fibers placed across their paths do not
typically continue in a straight path across the fibers, as is
usually assumed; rather, nearly always they undergo lateral displacement for a considerable length along the fibers, and the
process is often accompanied by branching and the formation of
microcracks as well. Cracks intersecting inclined fibers also
change direction so as to run parallel to the fibers, but appear
to do this without much branching or other difficulty, and may
simply go around the fibers and not cross them at all. In the
former situation, the energy dissipated in the shifting and
branching of the crack before it crosses the fiber may contribute
to the toughness of the composite. This response needs to be
taken into account when modeling the system.

(2) Pull-out processes. The processes of pull-out, as a
secondary stage in the cracking of fiber reinforced cement
composites, are also more complex than is usually assumed, and
are not effectively duplicated in simple pull-out testing arrangements. The present observations indicate that in pulling out
after the crack has passed and widened, there is a region of
debonding developed, often extending between the two ligaments
of the passing crack, when the posterior ligament is offset
with respect to the anterior one. However the debonding observed
is not symmetrical with respect to the fiber axis. A short
length of debonding is often seen adjacent to the anterior ligament on the anterior side; this terminates, and is succeeded by
a segment of debonding on the posterior side of the fiber, adjacent to the other ligment. This asymmetry suggests that there
is local bending of the fiber in addition to pull out taking
place. Stucke and Majumdar (10) considered such bending in

their discussion of the embrittlement of GRC. Evidence such as that indicated in Fig. 40 of the present work suggests that indeed such bending may be the immediate cause of failure of glass filaments in a glass fiber strand that has lost its flexibility as the result of the deposition of hydration products within the strand.

(3) Crack arrest and debonding. As indicated earlier, a crack propagating toward a fiber is often arrested several tens of µm ahead of the fiber, and changes its course so as to run parallel to the fiber for a considerable distance. As a result local "pseudo-debonding" occurs, "pseudo" in the sense that the debonding takes place not at the interface but through the matrix parallel to it. This behavior is at variance with the true debonding ordinarily assumed in models of the system. It provides a likely explanation of why certain surface treatments to steel fibers that improve pull-out resistance in pull-out tests do not necessarily improve the mechanical properties of the composite itself.

(4) Fiber orientation. As mentioned previously, cracks approaching inclined fibers usually change direction so as to run parallel to them for a considerable distance, sometimes to the end of the fiber, but sometimes eventually turning again to cross the fiber and resume forward progress. The behavior is reminiscent of cross-slip or double cross-slip of dislocations circumventing tangles or other obstacles to free glide in ductile metals. The existence of this mechanism suggests that inclined fibers may not contribute much to the post-cracking load carrying capacity of the composite. This obviously needs to be investigated further as a specific function of angle of incidence, and perhaps taken into account quantitatively in calculating the orientation efficiency factors for fibers in models of fiber-reinforced systems.

4.3 Interface Effects and the Cook-Gordon Crack Arrest Mechanism

The interface zone surrounding any reinforcement in a cement paste composite system is substantially different from the "bulk" paste further away from the foreign inclusion; it has a more complex structure, the details of which seem to play a large part in the various fracture processes.

To begin with, cement composites exposed to drying out tend to develop shrinkage cracks along aggregate surfaces prior to loading. Even in the absence of exposure to drying, chemical shrinkage effects can induce weakness at the actual interface. These weakened interfaces constitute preferred paths for the local passage of loading cracks. Where this occurs, clean separation between the aggregate surface and the paste matrix, such as was indicated in Fig. 14 of the present paper, is observed.

However, in the absence of drying, loading cracks that run along the actual interface between the paste and the reinforcement are not frequent. This finding appears to be a consequence of the structure of the interfacial zone itself, which is quite different from the usual structure of the "bulk" cement paste. Some discussion of this structure is required in order to explain the occurrence of the pseudo-debonding that has been observed.

Usually it is found that the interfacial zone consists of (a) a thin (1 or 2 μm thick) duplex film in actual contact with the reinforcement, (b) outside of this, a zone of perhaps 10 to 30 μm thickness, which in reasonably well hydrated systems is largely occupied by relatively massive calcium hydroxide crystals, with occasional interruptions of more porous regions, and outside of this, (c) a highly porous layer parallel to the interface. A schematic representation of this structure is shown in Fig. 41. The "Pseudo-debonding" cracks noted especially in steel fiber reinforced composites apparently run along the outer porous layer parallel to the reinforcement.

An interesting hypothesis that is corollary to this concept can be inferred from the discussion of the effects of weak interfaces in composites provided by Cook and Gordon (25). These authors suggested a specific crack arrest mechanism that may explain some of the mechanical features observed in the present study.

This mechanism is presented schematically in Fig. 42. Fig. 42a shown the expected stress field ahead of a propagating elliptical crack. In Fig. 42b a diagram is provided of the crack arrest that is presumed to occur when the oncoming crack intersects a weak interface. The tensile stress field moving parallel to the crack, diagrammed as σ_x in Fig. 42a, has its maximum value at a location ahead of the crack rather than at its tip. Cook and Gordon suggested that such a stress field interacts with a weak interface to produce cracking along the interface before the propagating crack tip has reached it (Fig. 42b (2). When the crack tip subsequently reaches the interface it is blunted by the crack parallel to the fiber that is already in place (Fig. 42b (3)). As a result, the stress concentration will be reduced, and the crack will continue to propagate in the path of least resistance parallel to the fiber, rather than crossing it.

This pattern is quite similar to what seems to be occurring

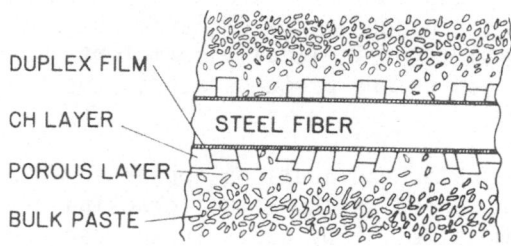

Fig. 41 - Schematic description of the microstructure of steel
fiber-cement paste interface.

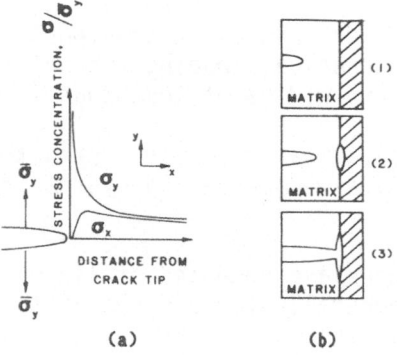

Fig. 42 - Cook-Gordon crack arrest mechanism (25).

      (a) Stress concentration in front of an elliptical
          crack,

      (b) Sequence of events taking place in the propagating
          of a crack in the matrix toward an inclusion with
          a weak interface.

especially in steel fiber cement composites, except that the weak zone is not at the interface but rather in the porous layer previously described that occurs a few tens of μm away from it. To complete the picture, it appears that propagation parallel to the fiber continues until the crack intercepts one of the occasional porous regions in the CH layer where the CH deposition has been incomplete or is locally deficient, where it presumably again changes course and proceeds across the fiber in the forward direction.

These suggested explanations appear to account for most of the features observed in the actual cracking patterns of fiber reinforced composites. Obviously independent confirmation by others is required before they can be said to have been fully established, but we believe that they will turn out to be essentially correct.

5. CONCLUSIONS

A number of specific conclusions are drawn from this work. Because of the large volume of material presented, it seems appropriate to subdivide the conclusions into groups relating to specific types of cement composite systems.

(a) Cement pastes

1. Stable cracks induced in previously undried portland cement pastes by mechanical loading are initially about 30 μm in width but taper to widths of less than 1 μm near their tips.

2. Cracks of similar dimensions result from purely tensile loadings applied by a wedge-loading system or from compressive loadings.

3. These cracks are composed of linked short segments zig-zagging around a generally forward propagation direction, and are locally parallel-sided.

4. Only restricted branching occurs, mostly near the tip of the stable crack.

5. The width near the tip is only a few tenths of a μm, and the actual tip is difficult to locate with precision. The crack usually appears to terminate at or near a large unhydrated cement grain.

(b) Mortars

6.   Cracks of similar dimensions to those induced in pastes
are induced with mortars, but the details are influenced by
whether the specimens have been subjected to drying.

7.   The cracks are more tortuous, and more branched than are
the cracks produced in cement paste.

8.   In previously-dried specimens, cracks run in part through
the actual interface between sand grains and paste, but cracks
in undried specimens almost invariably avoid the interface although they may run parallel to it for some distance.

(c) Concretes

As indicated earlier, only a relatively few concrete specimens have been examined. The conclusions concerning concrete
listed below should therefore be regarded as preliminary in nature.

9.   The initial crack induced in concrete specimens is of
similar dimensions to those in mortars or pastes, but cracks in
concrete follow a much more tortuous path.

10. Cracks induced in concrete show much more branching, and
multiple subdivision into three or more branches occurs in places.

11. The tip of the crack is no easier to distinguish in
concrete than in simpler systems, although here it often appears
to be associated with a large sand grain or an aggregate piece.
An extensive system of fine branches occurs near the terminal
section of a stable crack, and a local system of still-finer
multiple cracks appears in the immediate vicinity of the tip itself.

12. When loading is resumed, the extension of a previously
stable crack occurs not from the original tip, but from an anterior locus; the segment leading to the original tip is not
ordinarily re-activated.

13. While the subdivision and branching seen to occur near
the tip zone in the concretes examined, is in some respects
reminiscent of what is expected in a "process zone", there is no
physical distinction corresponding to separate lengths of
"straight, open crack" behind a crack tip and "process zone
microcracking" ahead of a crack tip. Indeed, straight segments
longer than several tens of μm do not occur.

14. Measurements of fracture mechanics parameters in concrete as usually carried out are not only deficient in resolution, but incorporate certain features tending to minimize recognition of the full extent of the tortuosity and branching of cracks in concrete. Crack opening displacement concepts and measurements in particular, do not conform with the geometry of cracking in concrete, and must be regarded as entirely pro-forma.

(d)   Steel Fiber Reinforced Cement Composites

15. A classification of four distinct modes of interaction observed between an advancing load crack and a fiber positioned across its path has been devised.

16. The observed occurrences of cracks running straight across a fiber, as assumed to occur in most analytical treatments, are relatively rare.

17. Most cracks appear to be arrested by encountering a layer of dense calcium hydroxide crystals just ahead of the fiber, and are displaced laterally several mm, before turning again and proceeding forward across the fiber.

18. This lateral shifting takes place in a relatively porous zone several tens of μm from the actual interface; it is not true debonding at the interface but "pseudo-debonding".

19. This process of "pseudo-debonding" apparently explains the lack of effectiveness of surface treatments designed to strengthen the actual steel fiber-paste interfacial bond.

20. Considerable branching and microcracking often accompany the lateral shifting of the advancing crack, and some true debonding may take place between places where parallel secondary cracks run past the fiber.

21. Fibers positioned at less than right angles to the advancing crack seem to cause the crack to turn and run parallel to the fiber without much branching or microcracking being produced; the crack may continue along the inclined path so as to completely bypass the fiber.

22. The Cook-Gordon crack arrest mechanism proposed to explain lateral debonding of cracks along the reinforcement in composites characterized by weak interfaces can be modified so as to to explain many of the features observed in these systems.

(e) Glass Fiber Reinforced Cement Composites

23.   Lateral shifting also occurs frequently with glass fiber cement composites, which develop crack patterns somewhat similar to those seen in steel fiber composites.

24.   In glass fiber systems, when the crack is widened subsequent to first cracking, true debonding often takes place prior to pull-out.

25.   In our experiments, evidences of local bending are observed at this stage, and the effective resistance to pull-out is obviously affected by this. It is difficult to see how such bending can be avoided in the post-cracking behavior of glass fiber reinforced composites, especially at the high deformations sometimes desired in service.

26.   Ductile composites, where spaces between the filaments within the strands remain open, are able to support such bending stresses without local failure, and pull-out of the filaments occurs. Embrittled composites have these spaces filled with hydration products to the extent that the flexibility of the individual filaments is lost and fracture across the filaments occurs after little or no pull-out.

6.   ACKNOWLEDGEMENTS

Although much of the material reported herein is new, the present report is in part a summation and interpretation of a number of individual investigations carried out over a period of some six years. We are particularly indebted to our colleague, Professor S. Mindess of the University of British Columbia, who is responsible for the design and construction of the testing devices used, and who has participated extensively and effectively in most of the earlier work. A separate general lecture by Professor Mindess appears in the Proceedings of this Workshop.

Thanks also are due to Mrs. Janet Lovell for her superb maintenance of the scanning electron microscope and her assistance with various aspects of specimen preparation and related activities.

This paper is a contribution from the School of Civil Engineering, Purdue University. We acknowledge with thanks the contributions of the Joint Highway Research Project, Purdue University and the Indiana State Department of Highways, toward the support and maintenance of the Scanning Electron Microscopy facility.

138

7. REFERENCES

1. Glucklich, J. (1965), "The Effect of Microcracking on the Time Dependent Deformation and the Long Term Strength of Concrete," in A. E. Brooks and K. Newman (Eds.), Int. Conf. The Structure of Concrete, London, Cement and Concrete Association, pp. 176-189.

2. Romualdi, J.P. and Mandel, J.A. (1964), "Tensile Strength of Concrete Affected by Uniformly Distributed and Closely Spaced Short Lengths of Wire Reinforcement," American Concrete Institute Journal, Proceedings, Vol. 60, No. 6, pp. 657-670.

3. Aveston, J. and Kelly, A. (1973), "Theory of Multiple Fracture of Fibrous Composites," Journal Materials Science, Vol. 8, No. 3, pp. 352-362.

4. ACI Committee 224, (1980), "Control of Cracking in Concrete Structures," Concrete International, Design and Construction, Vol. 2, No. 10, pp. 35-76.

5. Beeby, A.W. (1979), "Concrete in the Oceans - Cracking and Corrosion", Technical Report No. 2, CIRIA/EG, Cement and Concrete Association, Department of Energy, United Kingdom.

6. Slate, F.D., (1983), "Microscopic Observation of Cracks in Concrete, with Emphasis on Techniques Developed and Used at Cornell University", in Wittman F. H. (Ed.) "Fracture Mechanics of Concrete," Elsevier pp. 75-84.

7. Slate, F.D. (1983), "X-Ray Technique for Studying Cracks in Concrete, with Emphasis on Methods Developed and Used at Cornell University," ibid, pp. 85-94.

8. Spooner, D.C. (1972), "The Stress-Strain Relationship for Hardened Cement Pastes in Compression," Magazine of Concrete Research, Vol. 24, No. 79, pp. 85-92.

9. Spooner, D.C. and Dougill, J.W. (1975), "A Quantitative Assessment of Damage Sustained in Concrete During Compressive Loading," Magazine of Concrete Research, Vol. 27, No. 92, pp. 151-160.

10. Stucke, M.S., and Majumdar, A.J. (1976), "Microstructure of Glass Reinforced Cement Composites," Journal of Materials Science, Vol. 11, No. 6, pp. 1019-1030.

11. Derucher, K.N. (1978), "Application of the Scanning Electron Microscope to Fracture Studies of Concrete," Building and Environment, Vol. 13, No. 2, pp. 135-141.

12. Tait, R.B. and Bohm, H. (1980), "In Situ Scanning Electron Microscope Observations of Double Torsion Fracture of Concrete," Proceedings Electron Microscopy Society of South Africa, Vol. 10, pp. 17-18.

13. Darwin, D. and Attiogbe, E.K., (1983), "Load Induced Cracks in Cement Paste," in Proceedings Fourth Engineering Mechanics Division Specialty Conference, ASCE, Purdue University.

14. Mindess, S. and Diamond, S. (1980), "A Preliminary Study of Crack Propagation in Mortar," Cement and Concrete Research, Vol. 10, No. 4, pp. 509-519.

15. Diamond, S., Mindess, S. and Lovell, J. (1983), "Use of a Robinson Backscatter Detector and 'Wet Cell' for Examination of Wet Cement Paste and Mortar Specimens Under Load," Cement and Concrete Research, Vol. 13, No. 1, pp. 107-113.

16. Mindess, S. and Diamond, S. (1982), "A Device for Direct Observation of Cracking of Cement Paste or Mortar under Compressive Loading within a Scanning Electron Microscope," Cement and Concrete Research, Vol. 12, No. 5, pp. 569-576.

17. Diamond, S., Mindess, S., Bentur, A. and Lovell, J. (1984), "Development and Applications of Devices to Study Cracking with the SEM," Proceedings 6th Annual Cement Microscopy Conference, Albuquerque, N.M. USA

18. Diamond, S. and Mindess, S., (1980), "Scanning Electron Microscopic Observations of Cracking in Portland Cement Paste," in Proceedings of the Seventh International Congress on the Chemistry of Cement, Paris, Vol. III, pp. VI-114 - VI-119.

19. Mindess, S. and Diamond, S. (1982), "The Cracking and Fracture of Mortar," Materials and Structures, Vol. 15, No. 86, pp. 107-113.

20. Bentur, A. and Diamond, S. (1984), "Fracture of Glass Fiber Reinforced Cement," Cement and Concrete Research, Vol 14, No. 1, pp. 31-42.

21. Bentur, A. and Diamond, S., "Crack Patterns in Steel Fiber Reinforced Cement Paste," accepted for publication, Materials and Structure (Paris).

22. Bentur, A., Diamond, S. and Mindess, S., "Cracking Processes in Steel Fiber Reinforced Cement Paste," in preparation.

23. Higgins, D.D. and Bailey, J.E. (1976), "A Microstructural Investigation of the Failure Behavior of Cement Paste," in Proceedings Conference "Hydraulic Cement Paste: Their Structure and Properties", Cement and Concrete Associations, pp. 283-296.

24. Alford, N.N. (1982), "Dynamic Considerations of Fracture in Mortars" Materials Science and Engineering, Vol. 56, No. 3, pp. 279-287.

25. Cook, J. and Gordon, J.E., (1964), "A Mechanism for the Control of Crack Propagation in All Brittle Systems," Proceedings Royal Society, Vol. 282A, pp. 508-520.

**APPLICATION OF FRACTURE MECHANICS
TO CEMENTITIOUS COMPOSITES**
NATO-ARW - September 4-7, 1984
Northwestern Univeristy, U.S.A.
S. P. Shah, Editor

FRACTURE AND DAMAGE MECHANICS OF CONCRETE

Dominique FRANCOIS

Ecole Centrale des Arts et Manufactures - Grande Voie des
Vignes F 92290 - CHATENAY MALABRY

ABSTRACT

Under stress microcracks propagate in the cement paste and arrest when reaching the aggregates. A stable distribution of microcracks results, absorbing a large amount of energy compared with
the fracture toughness of the cement paste.

At the tip of a macrocrack a damaged microcracked zone develops. An exact solution of the problem was given in mode III,
while the approximate solution in mode I shows that in concrete
this zone is quite wide.

Very large specimens are thus needed to perform a valid K_{IC}
measurement on concrete. Some experiments using a DCB specimen are
described.

On smaller specimens the damaged zone spreads across the whole
section. The behavior can be explained by damage mechanics, the
damage parameter being identified with tensile or bending tests in
displacement control. The damage threshold can be represented by a
statistical variable following Weibull's statistics. There is a
good correlation between the evolution of the damage parameter and
the acoustic emission.

The irreversible strain of concrete is the result of the
release of internal stresses as elements break.

Finally, the damage theory can help to predict the behavior of
concrete in fatigue.

INTRODUCTION

The root of fracture mechanics lies in the works of Griffith who studied glass, a brittle material. It was later extended to metallic constructions. It is surprising that there is practically no application in the realm of concrete which does not display any plastic deformation and which fractures in a brittle manner in tension. This must be due in part to the traditional way in which the concrete structures are designed neglecting completely the tensile resistance and taking into account only the parts that work in compression. However Fig. 1 illustrates a more fundamental difficulty. It shows various results of fracture toughness measurements on concrete taken from the literature. They appear to increase with the specimen size, and they are difficult to extrapolate to large constructions. This is connected with the micromechanisms of the fracture of concrete, leading to the formation of a large number of small stable microcracks. This process absorbs much energy within a large damaged zone surrounding the main crack. In order to maintain its autonomy large size specimens are needed. The aim of the present paper is to explain why such behavior is observed, to show how valid K_{IC} measurements were carried out, and finally how damage mechanics can lead to a more practical way to deal with the fracture of concrete structures.

MICROCRACKING OF CONCRETE (Fig. 2)

Concrete contains numerous microcracks (1, 2); under stress they can extend. Their propagation is easy in the cement paste which has a low fracture toughness $2\gamma_M$. The corresponding stress is given by:

$$\sigma_M = (2E_M\gamma_M/\pi C_M)^{1/2}$$

where E_M is the Young's modulus of the cement paste and C_M the size of the crack.

When such a crack meets an aggregate, according to Wittman the fracture energy is increased and the crack is stopped, if the corresponding stress σ_A given by:

$$\sigma_A = [2 E_A\gamma_A/\pi(C_M + \Delta C)]^{1/2}$$

is larger than σ_M. In this last expression E_A is the Young's modulus and $2\gamma_A$ the fracture toughness of the aggregate, while CM + ΔC is the crack length when it reaches the aggregate.

This model over simplifies the actual situation, but explains in essence why the cracks which otherwise would propagate in a catastrophic manner are stablized by the aggregates. A modification leading to the same conclusion considers that they do not cross the aggregates, but that they follow the interfaces. They then behave rather like a notch whose tip radius is equal to the radius R of the aggregates, and whose length is equal to d + 2R, d being the distance between aggregates. Their volume fraction f is thus $(4\pi/3)$ $(R/d)^3$. The propagation will now continue if the local stress at the tip of the notch, i.e. at the side of the inclusion, reaches the fracture stress σ_e of the cement paste. As

$$\sigma_{local} = \sigma[1 + 2 (4 + d/2R)^{1/2}]$$

The fracture stress of the concrete would be given by:

$$\sigma_R = \sigma_C \{1 + 2 [1 + (\pi+\sigma_f)^{1/3}]^{1/2}\}$$

Now σ_C is approximately related to γ_M and E_M by:

$$\sigma_C \approx (2 E_M\gamma_M / \pi C_0)^{1/2}$$

where C_0 is the effective maximum crack length of the pores in the cement paste. The ratio σ_R/σ_M is thus given by:

$$\sigma_R/\sigma_M = (C_M/C_0)^{1/2} \{ 1 + 2 [1 + (\pi + \sigma_f)^{1/3}]^{1/2}\}$$

C_M/C_0 for a given size of aggregates can be considered as being proportional to d/C_0, that is to say to $R/C_0 f^{1/3}$, σ_R/σ_M is thus an increasing function of the volume fraction f of the aggregates.

The most questionable simplification of these models, and of the related computer simulation experiments, is the two dimensional representation of the concrete structure. A crack does not meet an aggregate along its whole front. It would be more reasonable to assume that it must overcome an average fracture toughness equal to

$(1 - f)2\gamma_M + f2\gamma_A$ if the crack crosses the aggregates and may be

$(1 - f)2\gamma_M + f.2\gamma_M.\dfrac{R}{C_0}$ if it goes along the interfaces.

In any case the microcracks which are initially present in the cement paste, mostly at the interfaces of the aggregates, extend under stress, until they meet the next aggregates where they are stabilized. Each microcrack extension will absorb an energy roughly equal to $2\gamma_M.\dfrac{\pi d^2}{4}$. The fracture stress would be:

$$\sigma_R \approx [\frac{4E\gamma}{\pi d^m} (1 - f + f\ R/C_0)]^{1/2} \approx \sigma C^{f^{2/3}}$$

THE DAMAGED ZONE

Bui and Ehrlacher (3) considered how a damaged zone (also called a process zone) could be created at the tip of a crack in a brittle material. Bazant (4) introduced a similar approach. Bui and Ehrlacher assumed that the material was elastic below some fracture stress σ_R , and that when it was reached sudden fracture occured. They studied the propagation of a damaged zone Z whose boundary was ∂Z (Fig. 3). It is similar to the propagation of a notch, except that along this boundary ∂Z the flux of mass is conserved and there is a positive jump of specific entropy. Energy is dissipated along ∂Z . In the damaged zone the stress is zero. The problem can be solved analytically in mode III, for a small scale damage zone in steady state. Small scale damage means that both stress and strain decrease at infinity as $r^{-1/2}$, r being the distance to the tip. Under quasi static loading, the boundary ∂Z has the shape of a cusped cycloid. The thickness of the damaged zone is such that $2h_0 = (K_{III}/\tau_R)^2$ (Fig. 2) where τ_R is the shear fracture stress. The usual relation between the energy release rate G_{III} and K_{III} holds:

$$G_{III} = \frac{K_{III}^2}{\mu}$$

μ being the shear modulus and fracture occurs when $G = G_{IIIC}$ the fracture toughness of the material.

This rigorous analysis can be extended in mode I: in that case the boundary of the damaged zone can no longer be found analytically. A finite element analysis shows that it is very similar to the cusped cycloid of mode III, except that it is more blunted. The thickness of the damaged zone is such that:

$$2\ h = 2(K_I/\sigma_R)^2$$

In concrete the fracture energy G_{IC} must be large because much energy is dissipated in the extension of the microcracks. If $K_{IC} = \sqrt{EG_{IC}/(1 - v^2)} \simeq 2$ MPa$\sqrt{m}$ and $\sigma_R \approx 4$ MPa , h reaches 0.25 m. The damaged zone in concrete must be very large and extremely big specimens are needed in order to maintain the conditions of small scale damage.

MEASUREMENT OF THE FRACTURE TOUGHNESS K_{IC} OF CONCRETE

At Laboratoire Central des Ponts et Chaussees, very large DCB specimens are being tested to measure the fracture toughness K_{IC} of concrete (5, 6, 7, 16). They are schematically shown on Fig. 4. They include a central thinner part and two prestressed wings in order to prevent the deviation of the crack propagation. They are loaded with an hydraulic ram inserted inside the notch at the lower end. It is activated by a servohydraulic system in displacement control.

During the test the opening of the crack is continuously recorded (Fig. 5). Partial unloadings at regular intervals allow to measure the compliance and thus to deduce the effective crack length. The compliance function was found by a finite elements computation. This gave also the stress intensity factor K_I and the evolution of the fracture toughness K_{IC} can thus be plotted as the crack propagates. Figure 6 shows results obtained on a concrete whose composition is given in Table I. It can be seen that K_{IC} fluctuates between 2 and 2.3 MPa $\sqrt{m}$. It is believed that this behavior is due to local heterogeneities of the concrete.

Figure 7 shows how much higher is the fracture toughness of fiber reinforced concrete and conversely, the low fracture toughness of low density aggregates concrete.

DAMAGE MECHANICS OF CONCRETE

Obviously routine K_{IC} measurements of concrete are out of the question owing to the very large size of the specimens. At the other extreme, instead of propagation of a pseudocrack, the loading of a small specimen develops a large number of microcracks in a damaged zone which spreads over the whole section. This is the realm of damage mechanics (Fig. 8) (8, 9, 10). On the damaged section of the material acts an effective stress given as a function of the applied stress by:

$$\sigma_{eff} = \sigma/(1 - D)$$

D being the damage parameter.

Most generally it is a tensor. It is taken as a scalar for the sake of simplification. The essential hypothesis of the theory is to replace the stress by the effective stress in the constitutive equation of the material. The problem is now to relate D to the deformation. Mazars (11) identified the damage using tensile tests carried out by Terrien (LMS Ecole Polytechnique) or his own bending tests (Fig. 9). They were all carried out under displacement control in order to record the post peak behavior. Below a threshold ε_{Do} there is no microcracking and $D = 0$. When the strain becomes larger than D_0, the unloading slope becomes smaller and smaller than the loading one, because the damage lowers the effective modulus. Neglecting the irreversible deformation, the damage parameter is written

$$D = 1 - \varepsilon_{Do}(1 - A)/\varepsilon - A/\exp [B(\varepsilon - \varepsilon_{Do})]$$

The concrete is characterized by 3 empirical constants A, B and the threshold ε_{Do}. It was found that they had the following values for the concrete of Table I.

$$a = 0.8 \quad B = 2 \times 10^4 \quad \varepsilon_{Do} = 0.6 \times 10^{-4}$$

TABLE I
COMPOSITION OF THE CONCRETE

	weight	volume (dm^3)
gravel 4/12 mm	1 103 kg	419.4
sand 0/5 mm	706 kg	266.4
cement HP	353 kg	113.14
water	201 kg	201
Total	2 363 kg	1 000

at 28 days mean specific mass.   2.303
          mean compression resistance 44.4 MPa
          mean Young's modulus   33,266 MPa

However it is well known that the resistance of concrete is statistically distributed and that the mean values are different in tension and in bending. This behavior is of course related to the statistical distribution of the initial microcracks and is well explained by Weibull's weakest link theory.

Mazars (12) introduced this distribution in the damage theory leading to a statistical distribution of the thresholds ε_{Do}. The most probable value of ε_{Do} is shown to be given by

$$\varepsilon_{Do} = (\omega_0/\int_V g^m \, dv)^{1/m}$$

where w_0 and m are Weibull's parameters and g is a function giving the spacial distribution of the strain in the stressed volume V. For the concrete tested m was found to be equal to 5 and w_0 to 8.66×10^{-19}.

That this damage theory is not a pure mathematical tool was clearly demonstrated by experiments using acoustic emission (13). Beams were stressed in bending while the acoustic emission was recorded using two transducers at each end. It was thus possible not only to obtain the rate of events but also to locate them along the beam. Figure 10 shows a comparison between the number of counts of acoustic emission along the beam and the computed damage. This last result was obtained by a finite elements method where the stress was replaced by the effective stress (11, 12). Both evolutions are quite similar and the same was found for different types of loadings.

THE ROLE OF INTERNAL STRESSES

In his modelization Mazars neglected the irreversible strain which is observed after unloading. It can be attributed to internal stresses following a very simple example given by Bui (14) (Fig. 11). A three bars structure is initially in a state such that the middle one is in tension while the other two are in compression. The material is elastic and brittle. When the three bars are loaded in tension, the middle one breaks first and this results in a sudden elongation of the other two and in a lowering of the compliance. Upon unloading a residual elongation is observed corresponding to the release of the initial compression. This can be generalized and the post peak tensile behavior can be reproduced by a set of elements containing an initial distribution of internal stresses.

DAMAGE IN FATIGUE

The results of fatigue tests on concrete are relatively scarce. It is possible, as for other materials, to draw an S. N. curve giving the life as a function of the stress amplitude. However the results display a large scatter. It was found that a much better fit was obtained when the life was plotted as a function of the irreversible strain which is accumulated at each cycle. Specimens were tested in cyclic bending while the elongation of the most highly stressed region of the beam was continuously recorded using a strain gage. A progressive deformation was observed whose amplitude per cycle was an increasing function of the applied load.

This behavior can be accounted for if it is assumed that at

each cycle some constant damage is added and that it is proportional to the strain. In a fatigue tensile test the evolution of the strain would be given by $\varepsilon = \sigma_{eff}/E$ where ε and σ_{eff} stand for the maximum strain and effective stress respectively. If the rate of damage accumulation per cycle is $\Delta D/\Delta N = K \varepsilon$, K being a constant of proportionality, the effective stress after N cycles is given by:

$$\sigma_{eff} = \sigma/(1 - K\varepsilon N)$$

and the strain evolution follows $\varepsilon(1 - K/\varepsilon N) = \sigma/E$.

A parabolic behavior is obtained. The strain per cycle $d\varepsilon/dN$ increases slowly and becomes infinite when ε reaches $2\sigma/E$, which would correspond to an instability. At this point $N = N_R = E/4K\sigma$.

This kind of calculation is easily transposed to a beam in bending. In that case the rate of damage is assumed to be zero for negative strains. This model predicts a behavior which is quite similar to the oberved strain evolution: for a large number of cycles it increases almost linearly. There is a fast acceleration before fracture. The coefficient of proportionality between the rate of damage per cycle and the strain amplitude is a function of the R ratio, i. e. the ratio of the minimum to the maximum load. Figure 12 compares the calculated and the measured strain evolution and Figure 13 shows a comparison between the predicted and the experimental life.

This model needs to be checked against more experimental datas, and remains crude. However it gives encouraging predictions.

CONCLUSIONS

The fracture of the concrete involves the development of a large number of small microcracks which absorb a high energy compared with the fracture toughness of the cement paste. In a large enough specimen they concentrate in a pseudo macrocrack whose width is equal to $2(K_I/\sigma_R)^2$, σ_R being the fracture stress in tension. The fracture toughness for a normal concrete fluctuates between 2 and 2.3 MPa$\sqrt{m}$.

When the specimen is not large enough, the damaged zone

spreads. The fracture can be studied by damage mechanics, the damage parameter being identified as a function of a strain with tensile or bending tests in displacement control. The damage begins above a threshold which is statistically distributed. It shows a good correlation with acoustic emission.

The irreversible strain of concrete results from the progressive release of the internal stresses as the elements break.

The behavior of concrete in fatigue can be predicted by an accumulation of damage, the rate of which is proportional to the strain amplitude and depends upon the R ratio.

REFERENCES

1. Wittmann, F. H., "Structure and Fracture Mechanics of Concrete in Mecanique de la rupture", Conseil International de la langue francaise - 103, rue de Lille - PARIS (1983) 193.

2. Wittmann, F. H., "Mechanisms and Mechanics of Fracture of Concrete in Advances in Fracture Research", D. Francois, Pergamon, Oxford (1982) 4, 1467.

3. Bui, H. D. and Ehrlacher, A., "Propagation of Damage in Elastic and Plastic Solids in Advances in Fracture Research", D. Francois Pergamon, Oxford, (1982), 2, 533.

4. Bazant, Z. P., "Application of Fracture Mechanics Concept in Structural Analysis of Concrete and Geomaterials in Mecanique de la rupture. Conseil International de la langue francaise, 103, rue de Lille - Paris - (1983) 233.

5. Sok, C, Baron, J. and Francois, D., "Mecanique de la rupture appliquee au beton hydraulique", Cement and Concrete Research (1979) 9, 641.

6. Benkirane, M., "Propagation d'une fissure dans le beton precontraint, Thesis, Universite de Technologie de Compeigne, (1982).

7. Chuy, S, Benkirane, M. E., Baron, J. and Francois, D., "Crack propagation in prestressed concrete; interaction with reinforcement", Advances in Fracture Mechanics, D. Francoix Pergamon, Oxford (1982), 4, 1507-1514.

8. Kachanov, L. N., Akad, Nauk, S.S.R., Ord. Tekh. Nauk., (1958) 8, 26-31.

9. Rabotnov, Y. M., "Creep Fracture ", Proc. XII Int. Cong. Appl. Mech., 1968, Stanford, Springer, (1969).

10. Lemaitre, J. and Mazars, J. "Application de la theorie de l'endommagement au comportement non lineaire et a la rupture du beton de structures, Annales ITBTP, (1982) 401, 249.

11. Mazars, J. "Mechanical Damage and Fracture of Concrete Structures", Advances in Fracture Mechanics, D. Francois Pergamon, Oxford, (1982), 4, 1499.

12. Mazars, J., "Probabilistic Aspects of Mechanical Damage in Concrete Structures", Int. Conf. on Fracture Mechanics, Technology applied to Material Evaluation and Structural Design, Melbourne (1982).

13. Walter, D., "Characterisation physique des degradation liees aux phenome nes de tension. Thesis. Universite de Technologie de Compiegne (1982).

14. Bui, H. D., "Rupture et endommagement: Application aux materiaux fragiles in Mecanique de lat rupture", Conseil International de la langue francaise, 103 rue de Lille - Paris, (1983), 35.

15. Alliche, A., "Comportement en fatigue de la pate de ciment durcie", Thesis, Universite de Technologie de Compiegne (1982).

16. Benkirane, M. E. and Acker, P., "Apports de la mecanique de la rupture dans le comportement des betons in Mecanique de la rupture, Conseil International de la langue francaise, 103 rue de Lille - Paris - (1983) 177.

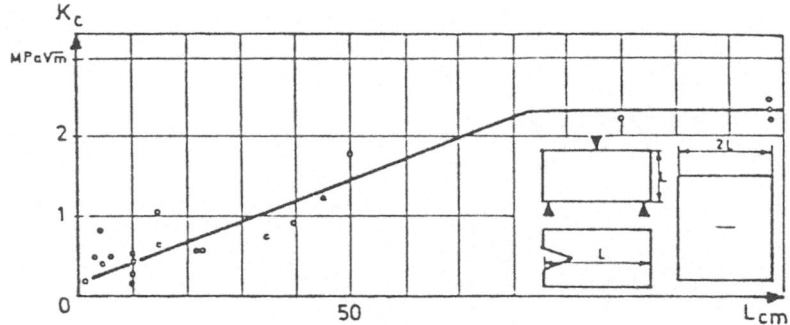

Fig. 1 - Various results of the fracture toughness K_C taken from the literature as a function of the size of the specimens.

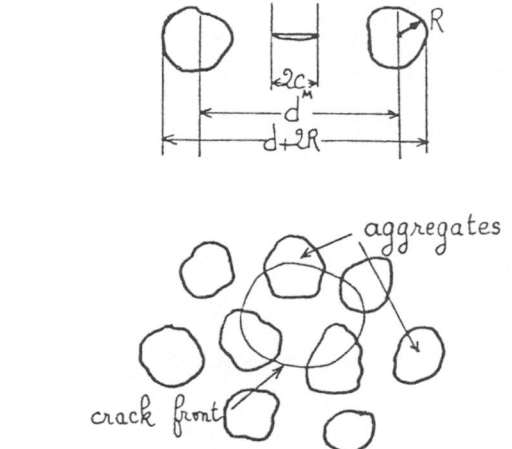

Fig. 2 - Microcracking of concrete

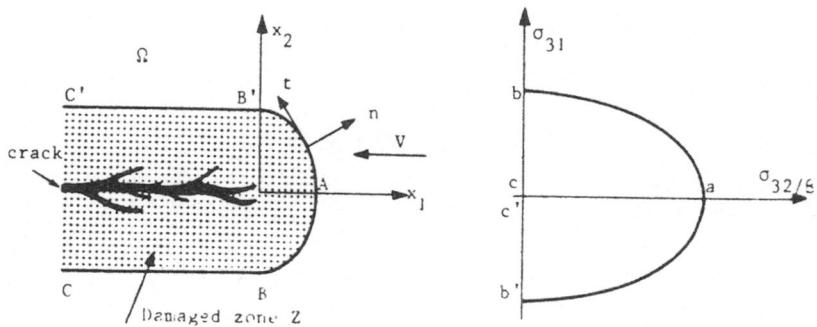

Fig. 3 - The damaged zone in mode III according to Brei and Ehrlacher (3).

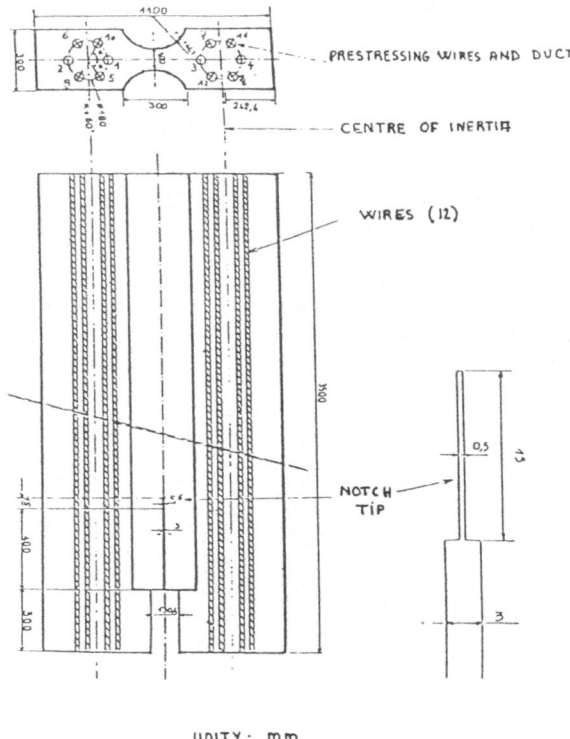

unity: mm

Fig. 4 - DCB specimen to measure K_{IC} in concrete

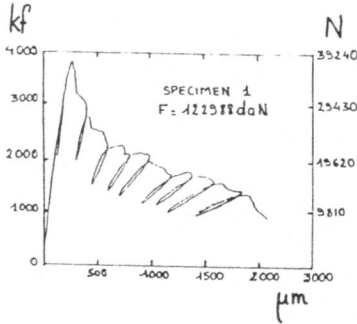

Fig. 5 - Recording of the load versus the crack
opening in a DCB specimen test.

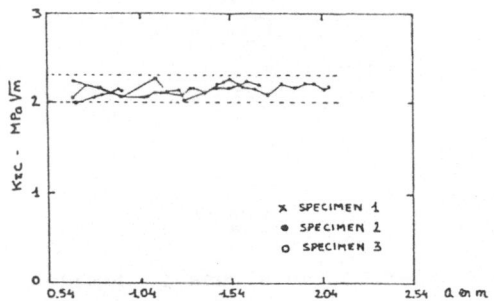

Fig. 6 - Results of the measurements of K_{IC} as a function of the crack propagation in a DCB specimen

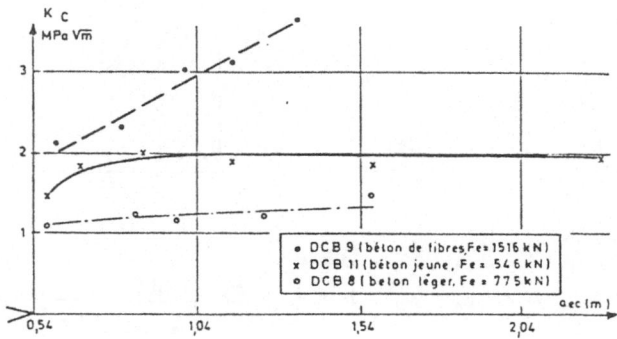

Fig. 7 - Fracture toughness of fiber reinforced concrete

$$\sigma_{eff} = \frac{\sigma}{1 - D}$$

$$\sigma_{eff} = \varepsilon E_o (1 - D)$$

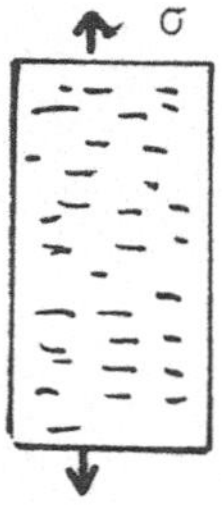

Fig. 8 - The effective stress in damage mechanics

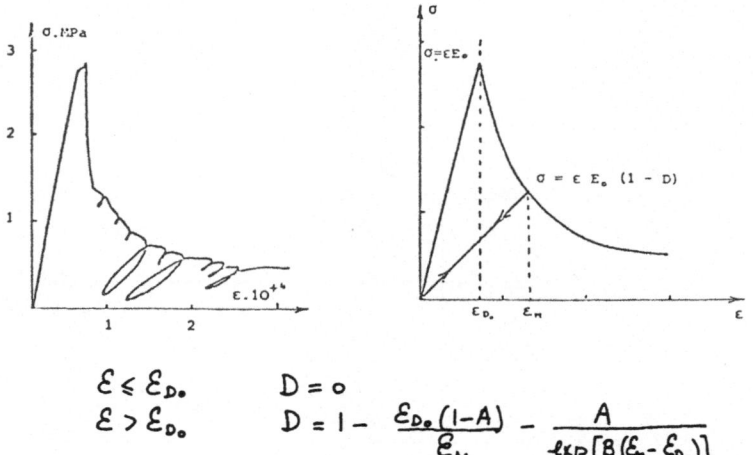

$$\varepsilon \leqslant \varepsilon_{D_o} \qquad D = 0$$

$$\varepsilon > \varepsilon_{D_o} \qquad D = 1 - \frac{\varepsilon_{D_o}(1-A)}{\varepsilon_M} - \frac{A}{exp\left[B(\varepsilon_n - \varepsilon_{D_o})\right]}$$

Fig. 9 - Tensile tests on concrete (Terrien) and identification of the damage parameter (Mazars).

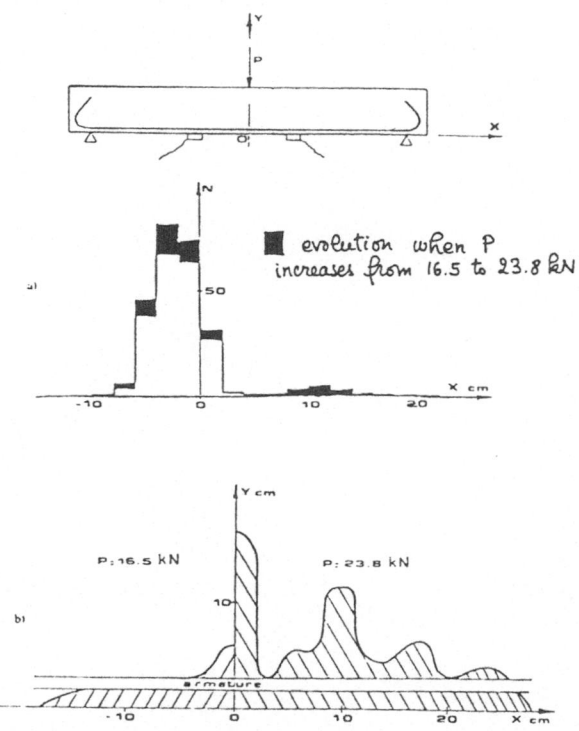

Fig. 10 - Comparison of the evolution of the damage parameter D and of the acoustic emission along the length of a beam in bending.

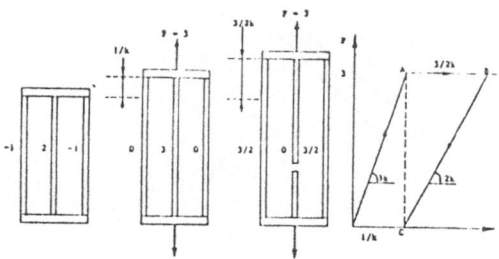

Fig. 11 - The three bars model of Bui to explain the irreversible strain of concrete by the release of internal stresses.

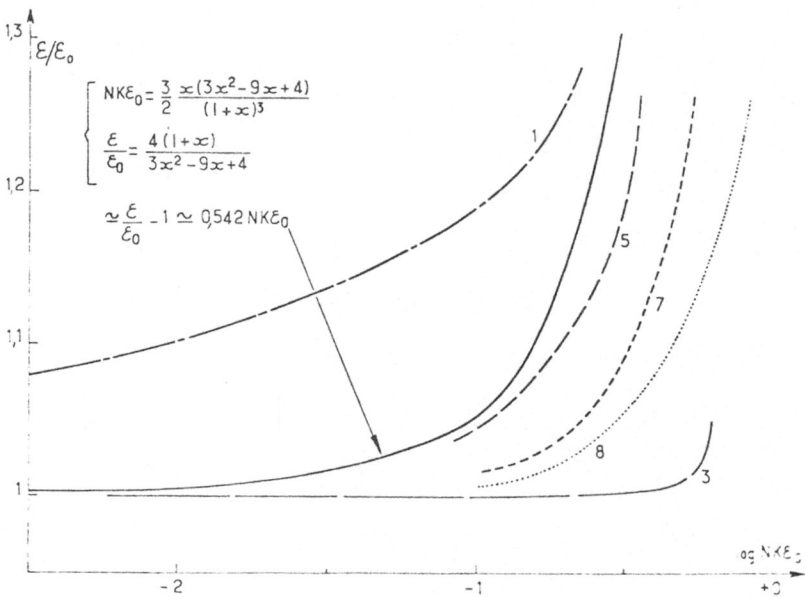

Fig. 12 - Comparison of predicted and measured evolution of the strain of beams in fatigue bending.

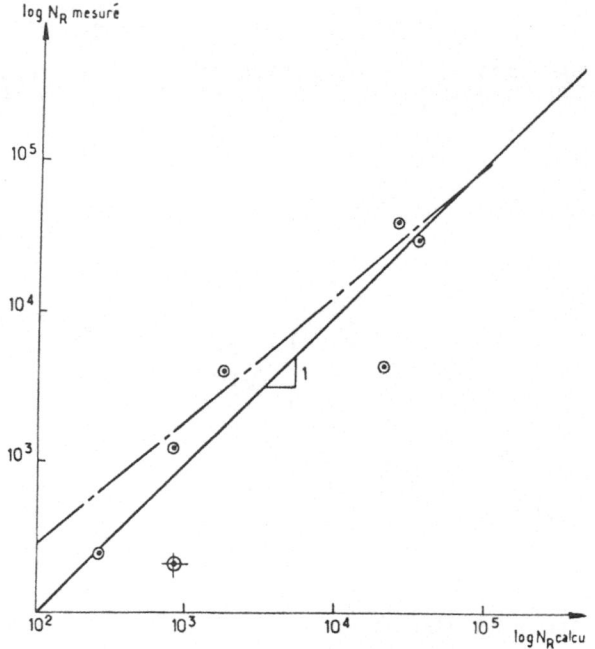

Fig. 13 - Comparison of the predicted and the measured fatigue lifes.

**APPLICATION OF FRACTURE MECHANICS
TO CEMENTITIOUS COMPOSITES**
NATO-ARW - September 4-7, 1984
Northwestern University, U.S.A.
S. P. Shah, Editor

FRACTURE PROCESSES IN FIBRE REINFORCED CEMENT SHEETS

by A J Majumdar and P L Walton

Department of the Environment, Building Research Establishment,
Building Research Station, Garston, Watford, Herts, England.

SUMMARY

In cement composites containing fibres the matrix fails
first but subsequent crack growth is suppressed to a certain
extent by the fibres. If sufficient fibres are present, multiple
cracking of the matrix occurs and the composite finally fails by
fibre fracture. These materials are so markedly non-linear that
even the elasto-plastic methods of fracture analysis, such as
R-curves, are probably not applicable to them in principle. Some
success has been claimed for these methods in explaining the
fracture processes in an intermediate type of fibre cement
composite exemplified by asbestos cement.

When the fibres are reactive towards cements, as is the case
with glass fibres, fracture processes in the composite change
with time in some environments. Failure processes in these
materials when used in components are complex and do not often
lend themselves to a quantitative assessment.

The fracture processes in fibre reinforced cement sheets are
reviewed in this paper with particular reference to experimental
observations.

1.0 INTRODUCTION

Fibre reinforced cement sheets in the form of asbestos
reinforced cement and autoclaved calcium silicate products have

been used in the construction industry for many years. During
the last decade several new fibre cement composites have appeared
on the scene containing, either singly or jointly, fibres made
from glass, cellulose, polyolefins, polyvinyl alcohol,
polyacrylonitrile etc. Some of these composites have been
developed to replace asbestos-based products that are considered
to be hazardous to health. Others, for example glass fibre
reinforced cement (grc) have found a much wider application. The
potential of relatively more expensive fibres such as carbon or
polyamide as reinforcement for cement has also been assessed in
the laboratory.

Much work has already been done to elucidate the basic
fracture mechanisms in fibre cement composites and theories have
been developed for predicting composite properties. Some aspects
of this ongoing research are discussed in this paper with
particular emphasis on experimental observations.

2.0 THEORETICAL CONSIDERATIONS

2.1 General Remarks

For reasons of simplicity it is customary to treat fibre
cement composites as two-phase mixtures consisting of fibre and
matrix. In practical composites variables used in fabrication
play a very important role in controlling the properties of the
products. For fibres the principal fabrication variables are
their volume fraction, aspect ratio, form, orientation and if
bundles are used, their degree of filamentisation. Cement
properties are governed by the fineness of the cement powder, the
water/cement ratio used in fabrication and curing methods
employed. If external additions such as sand pulverized fuel
ash (pfa) or slag or admixtures are used, these also influence
the properties of the matrix in a significant way. The
interaction between fibre and matrix is fundamentally dependent
on the development of a bond joining them and this is influenced
by fabrication methods. Practical composites, therefore, tend to
show a wide variation in properties and confirmation of theory by
experiment nearly always poses a difficult problem.

For many years Wittmann(1) has advocated the use of
hierarchical models applicable to micro, meso and macro levels of
concrete structure in studying the fracture processes in concrete.
The same concepts are relevant for fibre reinforced cement
materials although the extent of interaction in this case may be
different from that seen in concrete.

In practical fibre cement composites the failure strain of
the fibre is always much higher than that of the matrix and the

matrix fails before the fibre. It is, therefore, reasonable to consider the fracture of the cement matrix first.

2.2 The Cement Matrix

In fibre cement composites the matrix is very rarely neat cement paste; a proportion of sand or similar material is nearly always present to reduce the drying shrinkage of the paste. This increases the heterogeneity of the matrix, and provides innumerable weak interfaces that constitute a primary source of crack formation in the matrix.

The cement paste itself is multiphase; its microstructure and phase constitution evolve with time if hydration is allowed to continue. In the case of hydration of alite, the most abundant phase in Portland cement, a quantitative picture of this evolution is emerging[2] but similar knowledge about the more complex cement paste is rather incomplete at present. Nevertheless, it is generally agreed that fully hydrated and hardened Portland cement pastes (hcp) contain approximately 70% by volume of the near amorphous calcium silicate hydrates (C-S-H) and 20% of $Ca(OH)_2$, the rest consisting of minor phases[3].

The cement matrix is usually highly porous, porosity exceeding 25% being not uncommon. A part of this porosity is microstructurally associated with the gel-like C-S-H phase, the size of the pores being very small, a few nanometres only. The very large pores, a millimetre or more in dimensions are introduced during fabrication. The pore size distribution in cements has an important bearing on their response to stress. In addition, the cement matrix contains microcracks that may form due to thermal and moisture movements or chemical effects.

When cement is hydrated and cured in the usual way, the development of porosity and microcracking in the material cannot be prevented and it responds in a non-linear manner when stress is applied showing a small permanent deformation when the load is removed. When sand or other aggregates are added to the matrix the non-linearity shown by the material increases significantly. Slow crack growth rather than catastrophic failure becomes a notable feature in the response of such materials to stress as evidenced in the case of mortar, for example, by its low fatigue endurance limit compared to its static fracture stress.

The literature on the application of linear elastic fracture mechanics (LEFM) to cement and concrete is extensive; Mindess[4] and Swamy[5] have reviewed all but the most recent work in this area. There is a wide variation in the values of measured LEFM fracture parameters such as the critical strain energy release rate G_{1c} or the critical stress intensity factor K_{1c}

even for hcp and although some of these variations arise from differences in fabrication and curing of specimens, other factors more directly related to experiments, for example the size and geometry of the of the specimen, or the test methods employed have been shown to influence the results in a significant way. If some of these limitations are removed it still remains an open question whether the measured K_{1c} or G_{1c} values would be more reproducible and hence could be used in design.

Much interest has been shown in the last two or three years in the properties of the very strong macrodefect free (MDF) cement developed by Birchall and co-workers(6,7). Initially it was suggested that a LEFM approach explained the strength properties of the material satisfactorily but more recently Kendall et al(7) have proposed that a modified Griffith relationship may be more appropriate. In this relationship, the tensile strength, σ, is given by:

$$\sigma = \left[\frac{E_o R_o (1-p)^3 \exp(-kp)}{\pi c} \right]^{\frac{1}{2}} \tag{1}$$

p is the volume fraction of pores of very small size (gel pores) and 2c is the length of 'crack-like' pores that are a few millimetres long and may amount to only 10% of the total porosity of cements; E_o and R_o are the Young's modulus and fracture energy respectively of cement at zero porosity; k is an empirical constant. The importance of the size, shape and distribution of pores in cement and concrete in modelling their fracture behaviour has also been stressed by Wittmann(1).

There are difficulties in treating hcp as a Griffith solid; for example Higgins and Bailey(8a) observed that hcp does not appear to be notch-sensitive for shallow notches. They and others have also noted that fracture paths in hcp are often tortuous and show crack branching on a significant scale. Energy expended in these processes cannot be accounted for in the LEFM treatment of fracture, and Eden and Bailey(8b) suggest that the enhanced strength of MDF cements may be due to microstructural improvements rather than a reduction in flaw size.

As far as fracture processes are concerned there is little doubt that even at small applied stresses the stress intensity in front of some of the pre-existing flaws in hcp can rupture the rather weak cohesive bonds. It appears that in young cement pastes fracture paths may preferentially go through the C-S-H, and large $Ca(OH)_2$ crystals may act as crack arresters(9); but in hcp, and mortar the evidence is less certain. Tabor(10) has suggested that in hcp the bond between hexagonal plates of

$Ca(OH)_2$ may be the weakest link. For partially hydrated pastes and mortar the interface between the C-S-H or $Ca(OH)_2$ and sand or clinker grains provides another source of weakness and debonding at these interfaces will constitute an important feature of the fracture processes involved. Detailed microstructural work using advanced electron microscope techniques such as that of Mindess and Diamond[11] will be required to throw further light on this subject.

2.3 Fibre Cement Composites

When fibres are incorporated in the cement matrix they modify its cracking behaviour and if sufficient numbers of fibres are present the matrix is reinforced in various ways. Fracture processes that lead to eventual failure of these materials include in the main debonding, slipping and pull-out of fibres, and fibre fracture in addition to the cracking of the matrix already described. Each of these processes is complex in practical composites but much has been learnt in the past few years from simple models.

Fracture processes in materials have to be rationalised, as far as possible, with their stress-strain behaviour under different forms of loading. In this respect most attention has been paid to the behaviour of cement composites in tension. Broadly speaking, three different types of tensile stress-strain diagrams are obtained with these materials, as shown in Fig 1[12]. The difference between the theoretical curves A and B arises because in the former case there are not enough fibres in the composite to support the load after the matrix cracks at its ultimate failure strain ε_{mu}. In this case the composite fails by single fracture. The curve B is typical of composites that show multiple fracture of the matrix. The condition for this to occur is, for continuous aligned composites

$$\sigma_{fu} V_f > \sigma_{mu} V_m + \sigma'_f V_f \tag{2}$$

where σ_{fu} is the ultimate strength of the fibre and V_f its volume fraction, σ'_f is the stress on the fibre when the matrix fails, V_m is the volume fraction of the matrix and σ_{mu} its tensile strength. The fibre volume fraction that marks the transition from single to multiple fracture is the critical fibre volume $V_{f,crit}$. Curve C, representative of asbestos cement, is a special case.

Romualdi and Batson[13] were the first to suggest that the actual cracking stress of cement and concrete is enhanced by incorporating fine fibres. They regarded the fibres as limiting the cracks in the matrix by exerting forces that oppose displacements in the material ahead of the crack. It was also assumed that no slippage takes place at the fibre/matrix interface. The

162

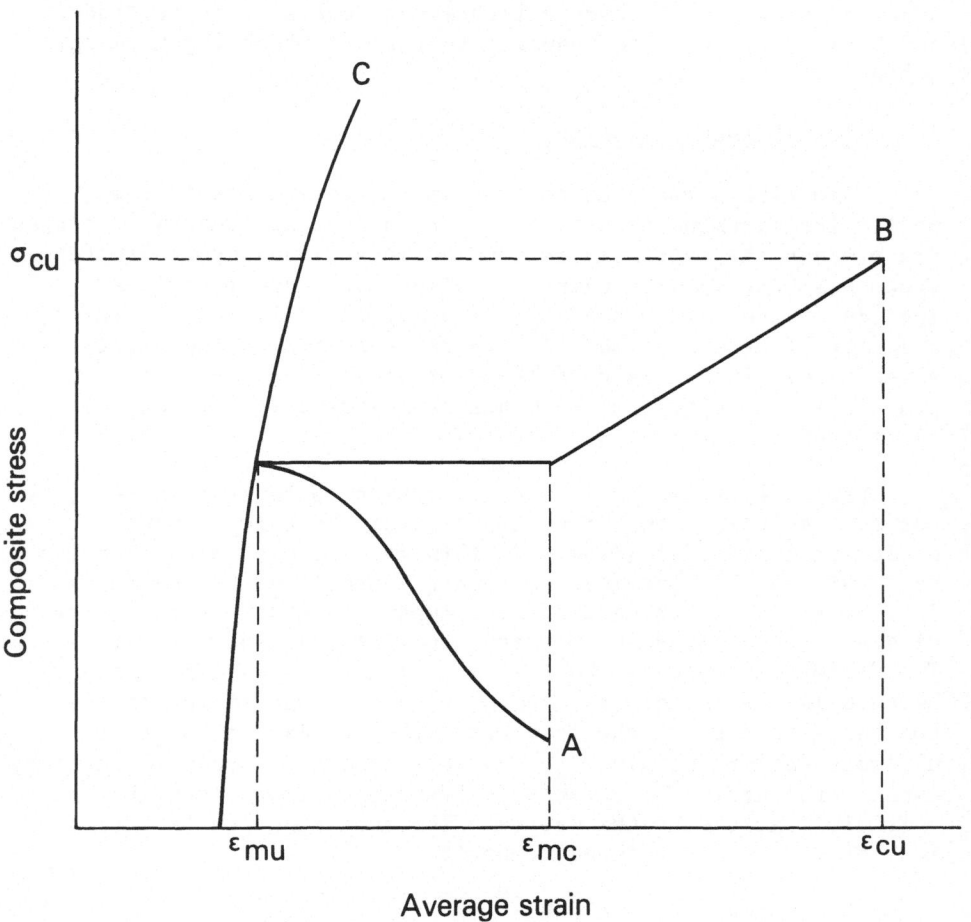

Figure 1   Theoretical tensile stress-strain curves for different
types of fibre cement composites (after Hannant(12)).

increase in the tensile cracking stress of the cement matrix
predicted by Romualdi and Batson and by the very similar elastic
theory of Aveston and Kelly(14) is not realised in practice
because the interfacial stresses developed soon break the bond
between the fibre and the matrix.

The non-elastic theory of Aveston, Cooper and Kelly(15), the
so-called ACK model predicts a much smaller increase in the first

cracking strain of the composite. The cracking strain is
calculated from the equation

$$
\varepsilon_{muc} = \left[\frac{12\tau \, \gamma_m \, E_f \, V_f^2}{E_c \, E_m^2 \, r \, V_m} \right]^{1/3}
\tag{3}
$$

where the subscripts f, m and c refer to the fibre, matrix and
composite respectively and r is the radius of the fibre; τ is
the interfacial shear strength and γ_m is the fracture surface
work in forming a crack in the matrix. It has been pointed out
recently by Korczynskyj et al(16) that the above relationship
does not take into account energy requirements for crack growth,
the analysis being concerned with two conditions only: the
uncracked condition and the situation in which a crack has
propagated completely across the specimen. If the value of ε_{muc}
predicted from Equation 3 is less than the failure strain of
the unreinforced matrix, the fibres are assumed not to influence
the matrix properties in any way.

An alternative first-crack model has been proposed by
Korczynskyj et al(16) for brittle-matrix composites based on the
Griffith energy criterion. A pre-existing matrix crack in the
composite is assumed to be surrounded by an elliptical zone of
relaxation and the strain distribution within this zone is
calculated with the help of a computer assuming a frictional bond
between the fibre and the matrix. It is shown that crack growth
is modified by the presence of fibres and that the matrix
cracking strain increases continuously with V_f, its magnitude
depending on various fibre and matrix parameters. The analysis
suggests that under certain circumstances even at appreciable
composite strains long cracks may remain stable in matrices with
low work of fracture (eg cement) because of extra energy
absorption at the fibre/matrix interface and by the fibre.

A similar but slightly different model has been proposed by
Hughes(17) and some results from his calculations showing the
effect of fibre on matrix crack growth are given in a recent paper
by Hannant et al(18). Unlike Korczynskyj et al, Hughes does not
assume a constant size for the relaxation zone. It is shown that
the effect of fibre addition is to produce a closing pressure on
existing cracks – an idea implicit in the crack arrest theory of
Romualdi and Batson(13). This model also predicts a continuous
increase of matrix failure strain with V_f (see Fig 2) and this
does not depend on whether the $V_{f,crit}$ is exceeded.

The new fracture models mentioned above predict that in
cement composites significant stresses can develop in the fibre
even before the matrix cracks propagate rapidly. According to

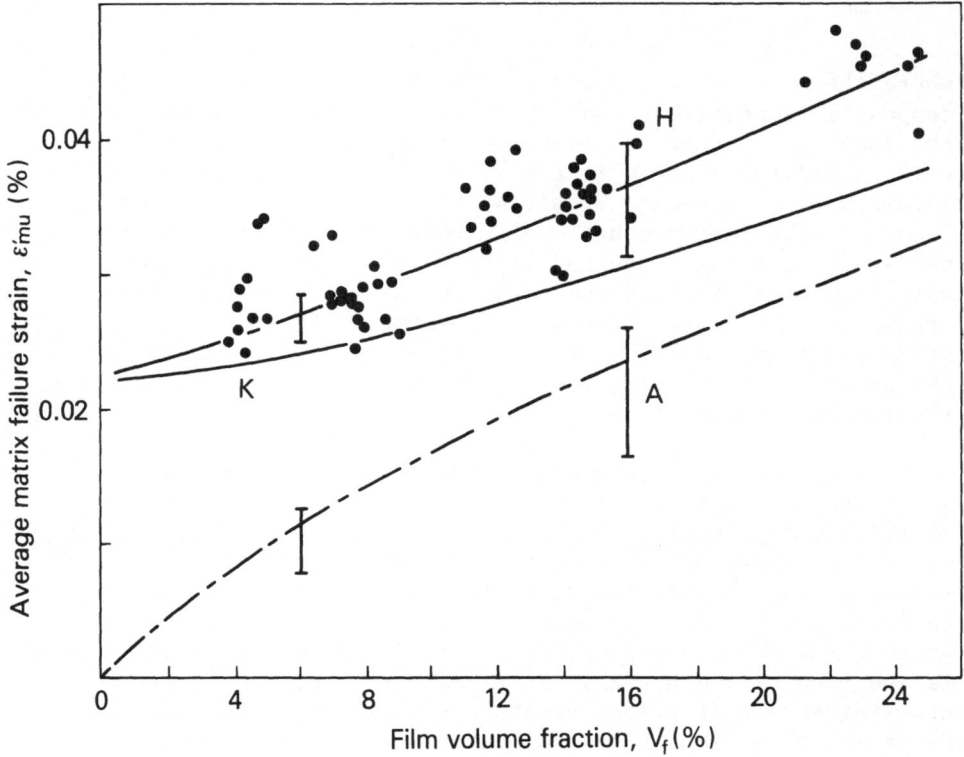

Figure 2   Relation between average matrix cracking strain and film
volume fraction of a composite containing fibrillated
polypropylene. H Hughes(17), K Korczynskyj et al(16),
A Aveston et al(15) (after Hannant et al(18)).

Hannant et al(18), the maximum stress that is produced in the
fibre in this situation is given by the relationship

$$\sigma_{f,max} > (2\tau \; B' \; E_f/r)^{\frac{1}{2}} \qquad\qquad (4)$$

where B' is the opening of the crack due to differential slip
between fibre and matrix. Assuming a value of B' = 1μm, the
maximum stresses in glass fibres in cement approach 300 MPa which
is an order of magnitude greater than the value of $E_f\varepsilon_{mu}$ used in
the law of mixtures. If glass fibre strengths are reduced very
greatly as a result of corrosion by cement and they fall below

300 MPa or thereabouts, it is thus theoretically possible for the fibres to fracture before the matrix cracks.

Fracture processes after matrix cracking have been described by Aveston et al(15,19). For composites containing more than the critical volume of fibres once the enhanced matrix cracking strain is exceeded further cracking will ensue and the ACK model predicts that the matrix will be traversed by a set of parallel cracks with an average spacing of 1.364 x' where

$$x' = \frac{\sigma_{mu} V_m}{V_f} \frac{r}{2\tau}$$ (5)

The total strain at the limit of multiple cracking is given by $\varepsilon_{mc} = \varepsilon_{mu} (1 + 0.659\alpha)$ where $\alpha = E_m V_m/E_f V_f$. On further loading, the fibres stretch and slip relative to the blocks of the cracked matrix which now takes no further part in sharing the increased load. The tangent modulus at this stage is given by $E_f V_f$. The composite fails at a stress

$$\sigma_{cu} = \sigma_{fu} V_f (1 - \frac{\ell_c}{2\ell})$$ (6)

where ℓ_c is the critical fibre length, and at a strain

$$\varepsilon_{cu} = \varepsilon_{fu} \left(1 - \frac{\ell_c}{2\ell}\right) - 0.341\alpha \, \varepsilon_{mu}$$ (7)

The calculated stress strain curves for aligned carbon fibre cement composites are shown in Fig 3. For composites containing random discontinuous fibres expressions such as equation 6 or 7 are modified by the inclusion of an efficiency factor for fibre orientation, several suggested values of which are to be found in the literature.

For composites containing a planar distribution of random short fibres Laws et al(20) have proposed that their response to stress can also be calculated by considering (a) the contribution of the fibres that 'hold' (elastic response) and (b) that of the fibres that slip (frictional response). This is illustrated in Fig 4. After the maximum stress has been reached there is still a post-failure stress capacity which is largely frictional in nature, arising as fibres that have slipped are pulled out of the matrix and one crack opens up. Two bond parameters are needed to describe the fracture processes - τ_s the strength of the bond that opposes slipping and τ_d the frictional bond operating over the fibres that are slipping.

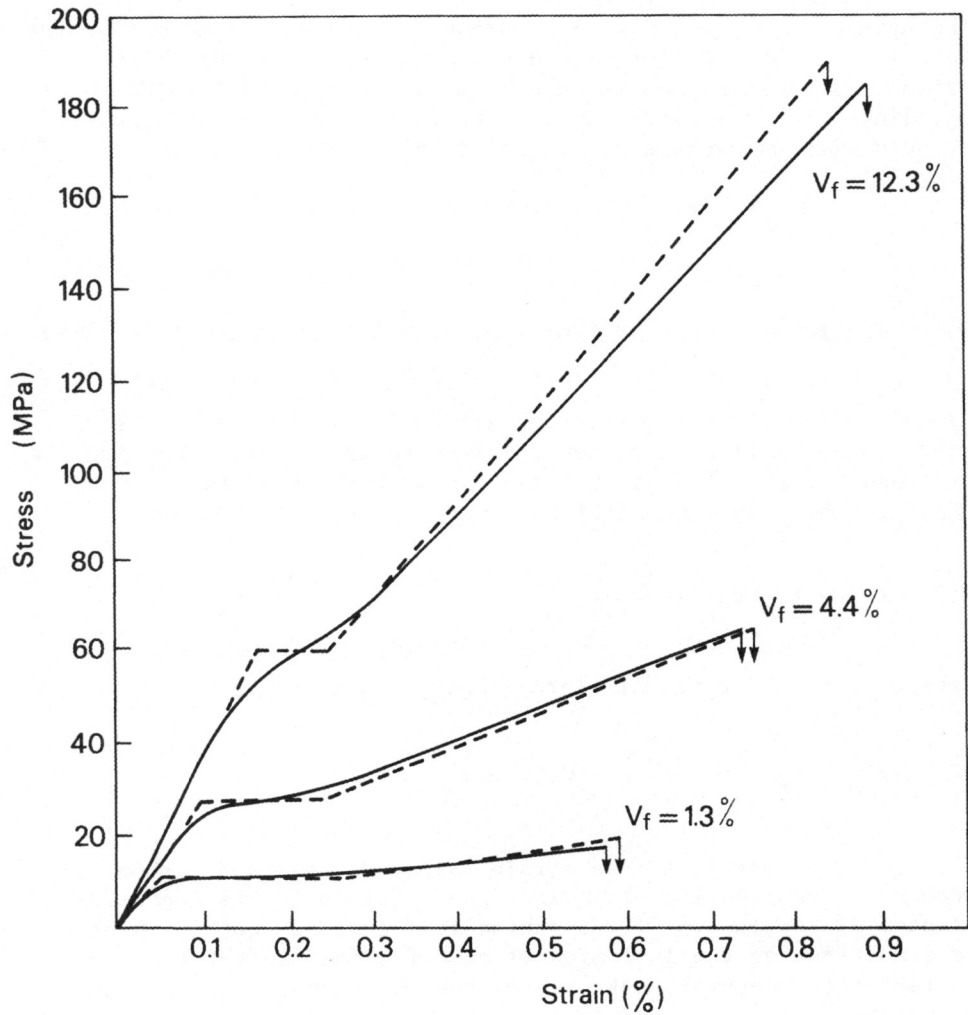

Figure 3   Tensile stress–strain diagrams for continuous carbon-
fibre reinforced cement.   Full lines are the
experimental curves, broken lines are the curves
predicted by the ACK theory (after Aveston et al(19)).

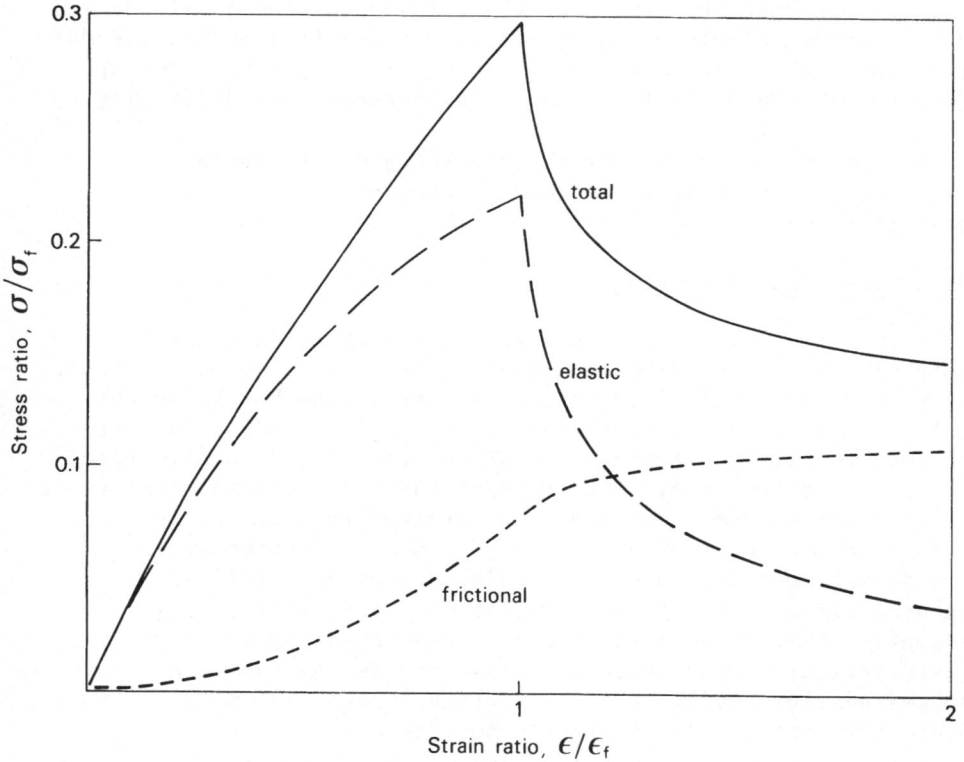

Figure 4   The stress-strain response of a fibrous mat.

3.0 EXPERIMENTAL OBSERVATIONS

From the experimentalist's point of view there are great difficulties in verifying the details of the fracture models described in the previous section. Cement is a highly variable material and its properties are very sensitive to slight changes in humidity and temperature. The values of its fracture parameters such as failure strain or fracture surface energy that apply to a fibre reinforced composite cannot be judged very accurately and similar uncertainties, perhaps more profound, surround the values of interfacial properties such as bond strength. For randomly distributed short fibre composite there is the further uncertainty regarding efficiency factors.

For continuous aligned fibre cement composites showing multiple cracking of the matrix the ACK model predicts properties

that are in good agreement with experimental results (see Fig 3).
The application of the computer model of Hughes(17) to predict the
enhanced matrix failure strain of cement in grc and in composites
containing networks of fibrillated polypropylene has also been
reasonably successful. The data obtained with the latter
composite are shown in Fig 2. However, Hannant et al(18) have
pointed out that the length of the relaxation zone on either side
of the crack calculated for these composites from Hughes's model
is often much larger than the observed crack spacing. Perhaps
further refinement of the model will overcome this difficulty.

Some of the major sources of uncertainty in evaluating
fracture processes in fibre cement composites are briefly
described below.

3.1 First Cracking Strain

The first cracking stress and strain of a fibre cement
composite are assumed to be those at the limit of proportionality
(LOP) on the tensile load-extension curves, the LOP being the
point where the curve deviates from linearity. LOP values are
subject to large variations, ~ 30% or more even with 'identical'
samples, and the possibility of operator bias certainly exists if
the deviation from linearity is determined by eye. As the
magnitude of the cracking strain of cement is rather small,
200-300 microstrain, and the predicted increase by fibre
incorporation in practical composites is also small, it is
essential that a reliable method is developed for locating LOP.
A statistical method based on linear regression has been
suggested recently(21) but in practice it also becomes arbitrary
where the material is inherently non-linear.

Because of the possible lack of reproducibility in the
determination of LOP, Proctor et al(22), initiated the use of the
'bend-over point' or BOP to mark the beginning of gross multiple
cracking of the matrix for grc stressed in tension. Hannant et
al(18) have used the measured strain values at BOP as the average
matrix cracking strain in cement composites. The uncertainties
that surround the location of the LOP also apply to BOP, although
to a lesser extent with some composites, and these uncertainties
are not entirely removed by the statistical treatment of the
data(21).

Some authors, Aveston et al(19) for example, obtained the
failure strain of the matrix by dividing the composite stress in
the middle of the multiple cracking region by the uncracked
composite modulus. They observed, for steel wire reinforced
cement, that the maximum acoustic emission coincided with the
minimum slope of the stress-strain curves. Since there could be
no wires breaking at this stage, the stress at that point was

taken as the mean cracking stress of the matrix. Lenain and Bunsell (23) have also used acoustic emission methods to detect matrix cracking in asbestos cement.

At present there is no consensus of opinion on how best to measure the matrix failure strain in fibre cement composites accurately. This makes verification of a fracture theory that treats fracture processes in quantitative terms all the more difficult.

In brittle-matrix composites visible multiple fracture of the matrix under load signals that the cracking strain of the matrix has been exceeded. Facilities have been developed recently that can stress cement composites in the specimen chamber of a scanning electron microscope (SEM) and examine fracture processes in situ. From such examinations Akers and Garrett (24)

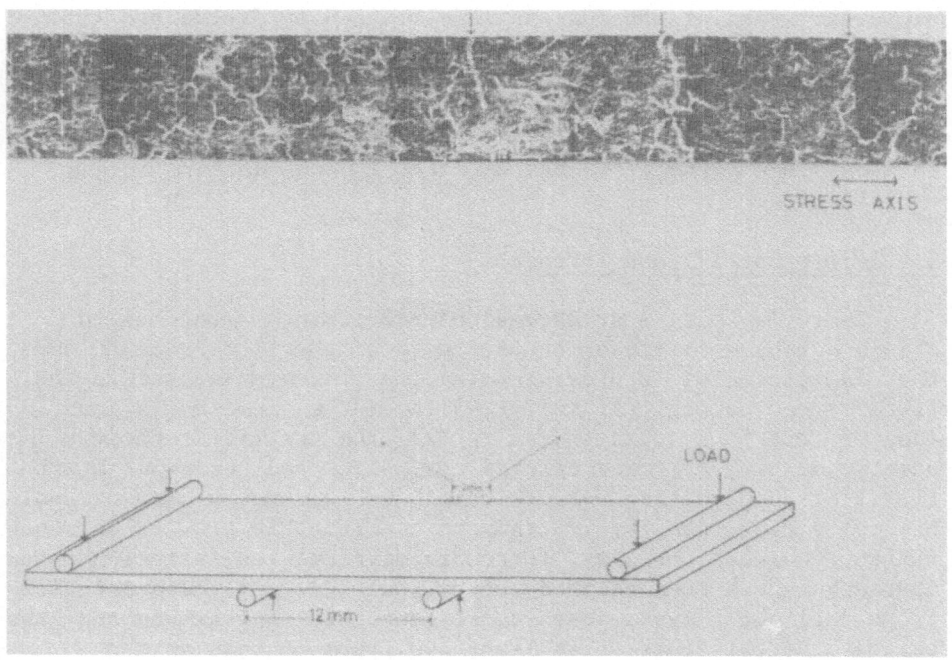

Figure 5 - Montage of multiple matrix cracking in asbestos cement, as observed in the SEM (after Akers and Garrett (24).

have concluded that multiple cracking of the matrix occurs
when asbestos-cement specimens are stressed (Fig 5).
Hannant et al(18), on the other hand, have suggested on
theoretical grounds that for aligned asbestos-cement composites
the ultimate failure strain of asbestos fibre is reached before
matrix cracks become unstable and therefore the breaking of the
specimen and first cracking of the matrix should occur
simultaneously. Although there is a degree of alignment of short
fibres in some commercial asbestos cement sheets, it is unlikely
that the materials studied by Akers and Garrett were similar to
those chosen by Hannant et al for their calculations.

Akers and Garrett(24) observed fibre pull-out in their
asbestos cement samples stressed in the SEM. For autoclaved
calcium silicates reinforced by short fibres of cellulose opinion
is divided(25,26) about whether fibre fracture or pull-out
constitutes the principal fracture process. Examples of in situ
fracture experiments in the SEM with cement composites that show
multiple fracture of the matrix unmistakably (eg carbon fibre
reinforced cement, Fig 3, or young grc) are not known to us.
However, Bentur and Diamond(27) have recently shown that the
presence of a few glass fibre strands in a cement sample produces
a very complicated pattern of cracks as the specimen is stressed.
Several different types of cracks can be recognised in the SEM
pictures. Similar conclusions were reached by Stucke and
Majumdar(28) in an earlier study in which they examined fracture
surfaces of grc in the SEM. It is clear that in cement composites
containing long fibres crack deflection by the fibre is a major
process. Much of the subsidiary cracking in the matrix may be due
to local inhomogeneity in the material and non-uniform stress
distribution.

3.2 Fibre/Matrix Bond Strength

There is still a great deal of uncertainty about how to
obtain a good estimate of the strength of the fibre/cement 'bond'
that is meaningful in understanding the fracture mechanisms in
fibre cement composites. Kelly(29) is of the view that a good
idea of this bond strength can be had from the crack-spacings
seen in multiply cracked aligned composites and recent work with
fibrillated polyolefin network reinforced cement by Hannant and
his colleagues (see Fig 2) seems to confirm this view. For short
fibre composites, however, there are difficulties in taking this
approach as has been pointed out by Laws(30). In these materials
fibre pull-out plays a most significant role in fracture and Laws
assumes the existence of at least two types of bond at the
fibre/cement interface – a static bond and a sliding or dynamic
frictional bond (see Fig 4), the average bond τ being an
underestimate of the former and overestimate of the latter.

For polypropylene fibre in cement the estimated values from pull-out experiments for the strength of static bond range from 1/2 to 1 MPa and that of the frictional bond from 0 to about 1/3 MPa(30). From crack spacings in aligned composites containing polypropylene network Hughes(17) estimated the average strength of the fibre/cement bond to be in the range 0.2 - 0.8 MPa.

There are several frictional forces that can operate on the fibres or fibre bundles as they slip through the matrix in short fibre composites after the failure of the static bond. Some of these forces arise from non-uniform geometry of the fibre as manufactured or from damage caused to them during movement through the matrix.

The effect of an increase in the polypropylene/cement bond strength from 0.2 to 0.8 MPa on the average matrix failure strain in composites containing fibrillated polypropylene as calculated from theoretical models is illustrated in Fig 2 by the vertical bars. It is seen that the effect is not negligible, particularly at large fibre volumes. Fibre/matrix bond strength in cement composites is likely to be affected somewhat as cement hydration continues and the environment of use may influence these results. Some experimental results obtained by Laws(31) with CemFIL 2 1) glass fibre strands are shown in Fig 6. It appears that in relatively dry air the bond strength remains fairly constant with time but in a wet environment there is an increasing trend. Such dependence of the glass/ cement bond on the conditions of use has obvious implications for fracture processes in grc.

Organic fibres such as polypropylene with Poisson's ratio of ∼0.45 which is much larger than that of hcp or mortar should show a large contraction across its diameter when aligned composites containing such fibres highly are stressed. Debonding at the interface resulting from this contraction should lead to catastrophic failure, but this is not observed and in many cement composites containing polymer fibres multiple cracking of the matrix is clearly visible. Irregularities on the surface of the fibre, variations in fibre diameter along its length, misalignment of the fibre, damage to the surface of the fibre during pull-out may all contribute in preventing the expected debonding but the shrinkage of cement is thought to be the most important factor in this context(29).

Microstructural studies of the interface in fibre cement composites indicate different features to be prominent with different fibres(32). In the case of steel wire there is good

---

1   Trade mark of Fibreglass Ltd

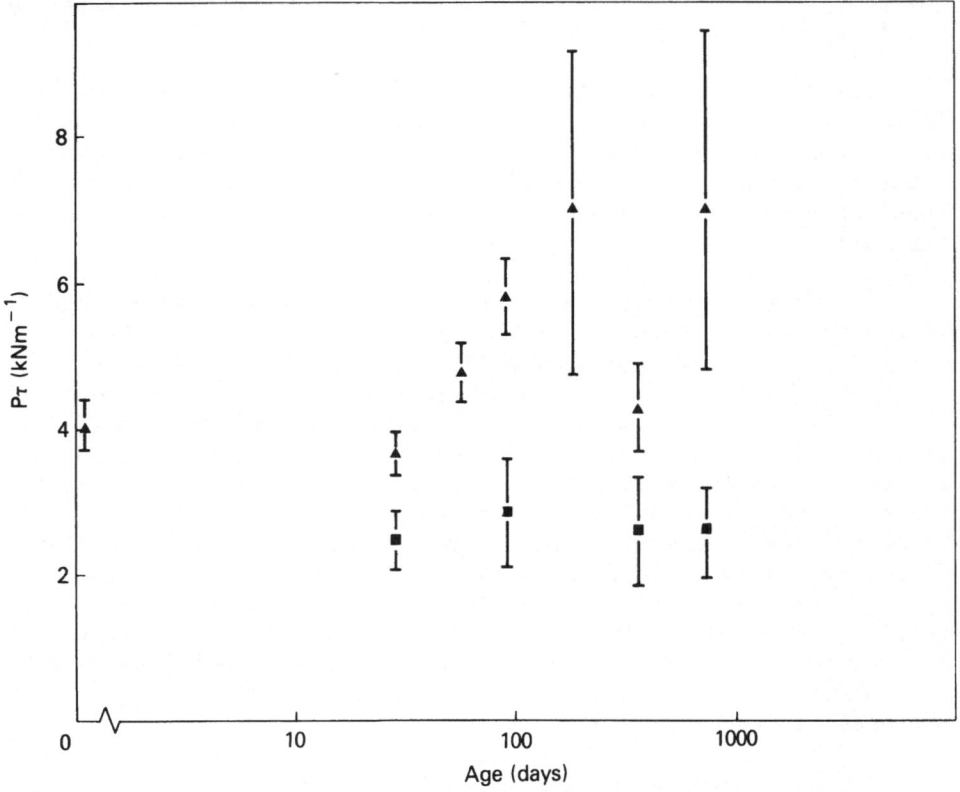

Figure 6 Variation of bond strength with time. CemFIL 2/cement
composite; ▲ cured in water at 20°C, ■ cured in air
at 40% r.h. and 20°C; P is the perimeter of the fibre
bundle.

evidence to suggest that interfacial failure is initiated a few
microns away from the interface. This has led Tabor(10) to
suggest that the inter-layer adhesion in $Ca(OH)_2$ is the weakest
link in fibre cement composites and unless this weakness can be
overcome, no enhancement of composite strength can be expected
from an increase in fibre/cement bond strength. Akers and
Garrett(24) on the other hand, found no evidence of the
$Ca(OH)_2$ - rich zone in asbestos cement.

It is more difficult to describe the fibre/matrix interface
microstructurally when the reinforcement is in the form of fibre

bundles. In some cases, for example grc and asbestos cement, the
open porosity in these bundles is soon filled up by the products
of cement hydration and there is some evidence to suggest that
Ca(OH)$_2$ crystals precipitate inside the glass fibre strands and
asbestos bundles. There may be fundamental crystallo-chemical
reasons for this affinity. Ca(OH)$_2$ crystals, however, are softer
than glass and cannot generate flaws on glass fibre surfaces
by pressing against them.

When the porosity of fibre cement composites is
substantially reduced, say by polymer impregnation, the matrix
phase shows a more linear elastic behaviour than hcp or mortar.

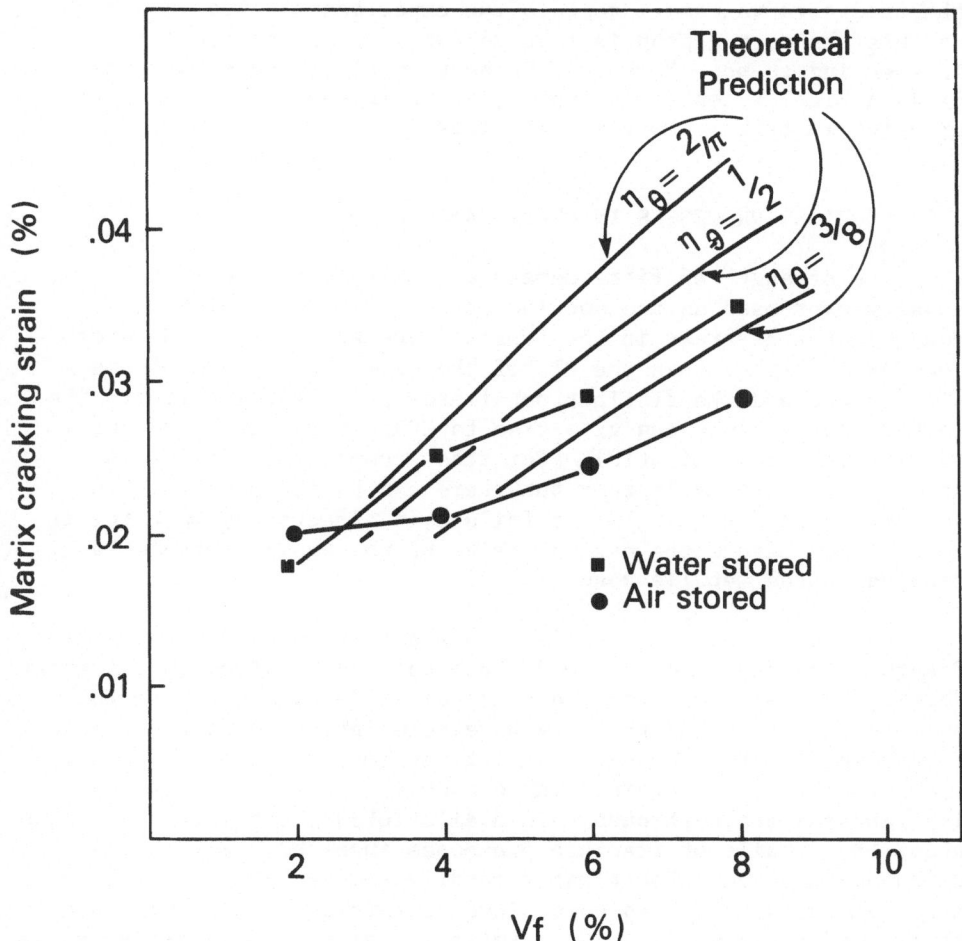

Figure 7   Relation between matrix cracking strain and fibre volume
fraction for grc composites at 28 days.

3.3 Efficiency Factors

For short fibre composites such as asbestos cement or grc
there is a further source of uncertainty - the so-called
efficiency factors - that the experimentalist has to deal with in
establishing agreement or otherwise with theories that attempt to
describe the fracture processes in these materials in
quantitative terms. Several theoretical values have been
proposed for these factors for different fibre orientations but
only a few examples exist that compare experimental results with
theoretical predictions. One set of results showing the increase
in the matrix failure strain in spray-dewatered grc as a function
of fibre volume fraction is shown in Fig 7 as an example. The
value of the fibre orientation efficiency factor (η_θ) that brings
the predicted values closest to the experimental ones is 3/8 but
is this the most appropriate value for such composites? It is
interesting to note here that Hannant et al[18] have used the
data in Fig 7 recently in support of their new theory on matrix
cracking in brittle matrix composites.

4.0 FRACTURE PROCESSES IN OTHER FAILURE MODES

The strength of fibre cement composites in bending is
usually expressed as the modulus of rupture (MOR) which is a
measure of the stress in the tensile face at failure. However,
the MOR is higher than the UTS of the material because of the
size effect and the distinct non-linearity of the material. The
latter factor alone can give rise to MOR/UTS ratios up to the
theoretical limit of 3[19]. For these composites failure is
usually in the tensile zone but there can be instances, eg in
composites with high V_f where failure is effected by buckling in
compression before the maximum value of the tensile stress is
reached in the tensile zone.

As fibre cement composite sheets are not used in compression,
fracture processes in this mode have not been studied in any great
detail, but certain fibres such as polyamides are known to buckle
under compression and this may have some influence in the fracture
mechanisms in the composite. A limited amount of information on
the strength of grc against various shear forces - interlaminar,
in-plane and punch-through - is available[33], but little is known
about the details of fracture processes involved. Fibres like
cellulose that can absorb water readily and swell may produce an
interlaminar weakness in composites containing them if moisture
is allowed to travel inward. Similarly, glass fibres are known to
be susceptible to stress corrosion effects in resin matrices and
although there is some evidence to suggest[33] that in grc such
phenomena do not affect composite strength when stressed below the
LOP, what happens above this stress level remains unknown.

5.0 WORK TO BREAK

Fracture processes, described in earlier sections, reflect the response to stress seen in the appropriate stress-strain diagrams. The response to tensile stresses is perhaps the most informative and it has been suggested recently that for fibre cement composites the area under the tensile stress strain curve when expressed as energy per unit volume of material is the most appropriate parameter to describe their energy absorbing capacity(18). It has also been suggested that when the fibres are short and randomly dispersed in the matrix in planar orientation, the fibrous mat model (see Fig 4) can be used to calculate the work to break(35). The 'toughness index' proposition described by Henager(34) is based on similar ideas but the area under a portion of the load deflection curve of the composite provides the measure of energy absorption in this case.

Estimates of the energy absorbed by the composite can be qualitatively treated as the shatter resistance of the material in the same way as results from the traditional Izod or Chapy impact or drop ball tests. Since the greatest benefit of adding fibres to cement is to improve its shatter resistance, an energy absorption parameter that can be easily defined and reproducibly measured may be useful in design.

For idealised continuous aligned composites the work done on a specimen in taking it through the complete stress-strain curve can be calculated from the expressions given by Hannant et al(18). Their experimental value for an aligned composite containing 6 vol % of fibrillated polypropylene was 790 kJ/m^3 compared with the theoretical estimate of 880 kJ/m^3. In a similar composite containing ~6 vol % of a high-modulus polyethylene fibre we have obtained a value of 500 kJ/m^3 based on the area under the curve up to maximum stress(21).

6.0 ELASTO-PLASTIC FRACTURE MECHANICS

The non-linear nature of cement-based composites has prompted many investigators in recent years to examine the applicability of several elasto-plastic methods to the study of concrete and fibre reinforced cement and concrete. Swamy(5) has described these methods in detail. The most common techniques used in these studies are (i) critical crack opening displacement method (COD), (ii) the J-integral approach, (iii) R-curve analysis and (iv) the 'fictitious crack' model. It is not entirely clear whether any one of these methods can provide a fundamental fracture parameter for fibre cement composites. The J-integral approach has been questioned on theoretical grounds(18). Both R-curve and fictitious crack models assume that the effective crack length in the composites is greater than the visible crack by a certain amount and an elastic analysis is performed using the extended crack length. All these methods, however, recognise that the energy dissipation processes in cement composite are many and most of them cannot be treated by LEFM.

R-curve analyses have ben carried out on asbestos cement by Lenain and Bunsell(23), on asbestos-cellulose cement composites by Mai et al(36) and on steel fibre reinforced concrete by Wecharatana and Shah(37). Using compact tension specimens of asbestos cement, Lenain and Bunsell were able to distinguish three stages of crack growth: (i) creation of a zone of microcracks, upto 28 mm long, in front of the visible crack, (ii) growth of this zone together with slow stable crack growth and (iii) extension of the principal crack while the size of the microcracked zone remained constant. These authors proposed that fibres which bridge the microcracks induce a closing pressure at the crack-tip and so increase the work of fracture.

At present it is not easy to assess the fundamental value of R-curve analyses to the study of fracture in cement composites. It has not been shown conclusively that the results obtained in these experiments are independent of specimen size and test methods. The treatment of the fracture process zone which can be very large, 75 mm in double cantilever specimens of concrete(38), as an elastic extension of the visible crack is an over-simplification. With fibre reinforced materials there is the added difficulty of stress transfer in the fibres which may spread the influence of the process zone over a considerable area surrounding the crack tip. Energy dissipation by crack deflection and subsidiary cracking is also pronounced in these materials. It is perhaps true to say that the limited success claimed so far for the application of R-curve analysis to cement based composites has been with materials for which final failure more or less coincides with cracking in the matrix. Earlier work by Brown(39) which is conceptually similar to R-curve analysis also employed grc samples that showed very little, if any, pseudoductility. For composites which display the phenomenon of multiple cracking and rising stress-strain relationships it would appear that fracture mechanics approaches currently under consideration will have little prospect of success in producing parameters that will be useful in design.

7.0   EFFECT OF ENVIRONMENT ON FRACTURE PROCESSES

The properties of the cement matrix change with time depending on the environment. In natural weather hydration of Portland cement continues for many years and its cracking strength and elastic modulus increase. The effect of these changes in

matrix properties on the properties of a composite containing
6 vol % of fibrillated polypropylene nets has been estimated by
Hannant et al(18). The critical fibre volume increases from
3% after 28 days to 4.4% after 3 years. The measured energy
absorption for this particular composite showed a small decrease
from a value of 790 kJ/m^3 at 28 days to 730 kJ/m^3 after 3 years
of natural weathering. Increases in critical fibre volumes on
weathering are likely to be greater in the case of grc as pointed
out by Hannant et al(18). This arises because the strength of
glass fibres including the alkali-resistant varieties such as
CemFIL is reduced substantially when grc is exposed to natural
weather. Typical mechanical properites of grc now in use
containing ∿ 4 vol % of a two-dimensional array of 34-38 mm
long CemFIL glass fibres up to 10 years of natural weathering
in the UK have been published by BRE(40) and from these data
Hannant et al have estimated that the critical fibre volume for
these composites would increase from 3.5% at 28 days to 7.4% after
10 years. Since the more attractive properties of grc arise from
the multiple fracture of the matrix and a rising stress-strain

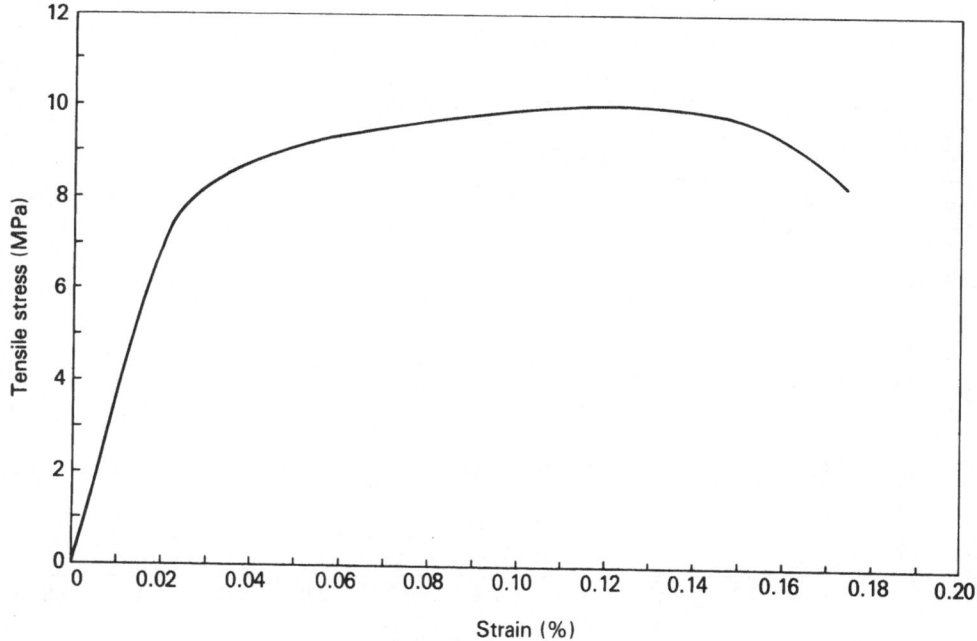

Figure 8   Stress-strain diagram of grc containing 6 vol % glass
            fibres after weathering for 10 years.

curve when the material is stressed, the loss of at least a large portion of its pseudoductility with time on weathering must be taken into consideration in design. The tensile stress-strain curve of a 10 year old grc sample containing 6 vol % of 30 mm long CemFIL fibre is shown in Fig 8. Although a major proportion of its initial pseudoductility (ultimate failure strain ~1%) has been lost, the nature of the stress-strain curve would suggest that failure did not take place by single fracture. In grc stored in relatively dry air, there is very little loss of pseudoductility over a period of 10 years. Only a little reduction in fibre strength with time due to limited hydration of cement in dry air is responsible for this behaviour. These differences in the fracture patterns in grc aged in different environments can be related to the developments in the microstructure of these composites(28).

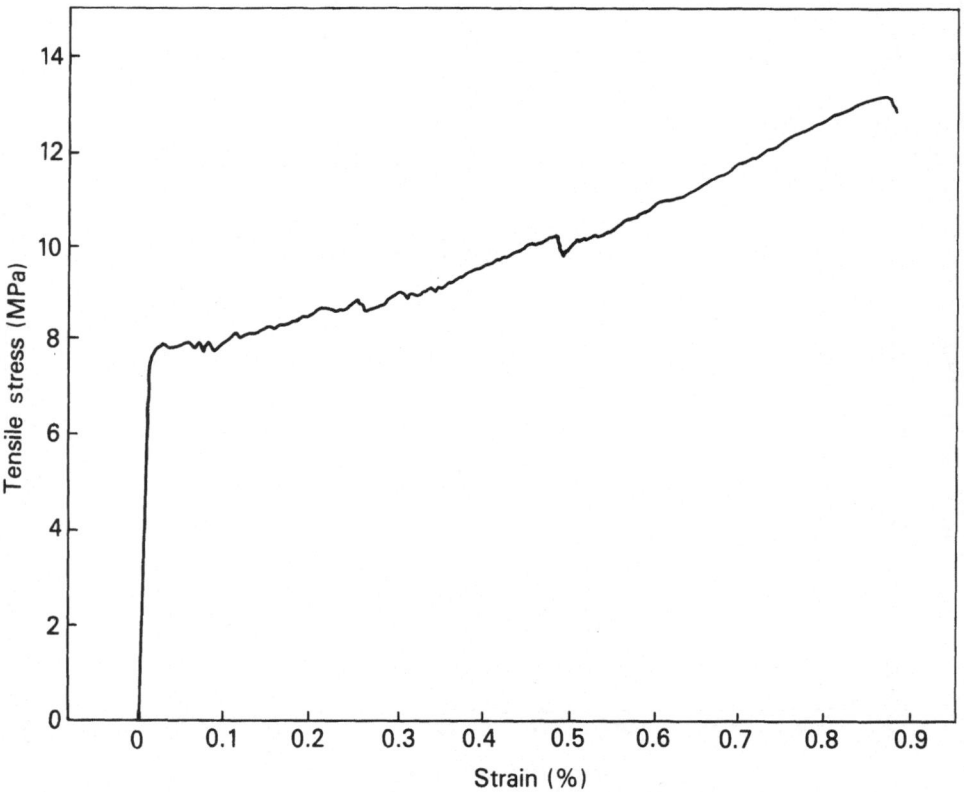

Figure 9   Stress-strain diagram of grc containing 4 vol % CemFIL 2
           glass fibres and 30% OPC, 70% granulated blastfurnace
           slag, after ageing under water at 50°C for 50 days.

The retention of the features of multiple cracking and a rising tensile stress-strain curve with ultimate failure strain of the order of 1% in long-term weathered grc has been the main objective of much recent research on grc. It has been shown that the addition of pozzolanas such as pfa or suitable polymer dispersions to grc formulations are useful in this respect(35,41), but a much greater benefit has resulted from the improvement in the alkali-resistance of the CemFIL glass fibre by special surface treatment. One matrix formulation that has given particularly encouraging results in conjunction with the new fibre is a mixture of 70 wt % ground granulated blast furnace slag and 30% Portland cement(35). The tensile stress-strain diagram of grc made from this matrix and containing ~4 vol % fibre and cured under water at 50°C for 50 days are shown in Fig 9. Although on much longer exposure to hot water the ultimate failure strain is reduced, the data in Fig 9 may correspond to weathering in the UK for 15 years or more(42). It appears that in this composite the alkalis released by the hydration of cement react with the glassy slag rather than the glass fibre and as long as a reservoir of fine slag particles is maintained the strength properties of the composite do not change much.

As has been mentioned already grc made from Portland cement and CemFIL glass fibre and kept in wet environments for a long time behaves in a substantially brittle manner. Specimens stressed in bending fail by the advance of a single crack, more or less plane and normal to the stress direction. When subjected to fatigue loading it has been found that the fracture surface is no longer plane but contains one or more steps parallel to the stress direction (Fig. 10). These steps appear to originate at the surface of or within the glass fibre strands. This phenomenon is not uncommon as a crack arresting mechanism in material which contains planes of weakness, but the transition from plane fracture surfaces with monotonic loading to stepped surfaces with fatigue loading requires explanation. It is possible that the coherence of the cement hydration products (which deposit on or inside the fibre strand) with the fibre is disrupted during fatigue cycling thus allowing cracks to be deflected when they meet the fibre.

The effect of environment on the long-term properties of fibre cement composites other than asbestos cement and grc is not well known. Some attention has been paid to the cellulose reinforced autoclaved calcium silicate(36b) which is a possible replacement for asbestos cement. The impact resistance of these materials increases somewhat in wet environments and it is possible that this may be due to a softening of the fibre by absorption of water and swelling. Obviously such a process is expected to affect the properties of the interface in a significant way. The exact nature of the wood pulp, for example whether bleached or unbleached, has also been shown to be important (36b).

Figure 10 - Fracture in 10 year old grc (a) tensile
            (b) flexural fatigue

8.0 FRACTURE OF FIBRE CEMENT COMPOSITE PRODUCTS

Apart from asbestos cement very few fibre cement composite
sheets have been used in the construction industry to-date in
external applications. Composites containing cellulose, polyvinyl
alcohol and polyacrylonitrile fibres have been introduced recently
as materials for roofing and cladding and grc has been used in
some applications for 10 years or more. There is, as yet, very
little contribution from field experience to support a definite
view about the long-term viability of these materials.

It was established many years ago (43) that the bending strength
of asbestos cement used in roofing increases with time on weathering but its resistance to forces of impact is reduced. The most
common form of failure observed in asbestos cement products is
caused by impact and the failure mode is predominantly brittle.
Fibre lengths used in these composites are very short, only a few

mm's, so that the scope for energy absorption by fibre pull-out is rather limited. The situation is further aggravated on exposure to natural weather due possibly to the consolidation of material inside fibre bundles and at the fibre matrix interface brought about by continuous hydration of the cement and carbonation. There is also the possibility that the strength of the fibres is reduced in the cement environment over a long period of time.

One particular type of component made in grc, a highly profiled and dark coloured architectural panel of sandwich construction having a core of polystyrene bead concrete faced on both sides by $\sim$10 mm thick skins of grc, used to clad office buildings has come under close scrutiny in the UK(44) as a few percent of these panels have developed cracks in the outer grc skin within a few years of installation. Analysis of the causes of fracture in the grc sheets of these panels is being carried out by BRE, by Pilkington Brothers PLC who are the manufacturers of the fibre used in these panels and by the Glass Fibre Reinforced Cement Association. It is becoming clear that quality control in the production and curing of the grc sheet and correct fixing of the panel to the substructure are essential if the panels are to function properly. Stress analysis has revealed that large stress concentrations exceeding the failure stress of the material are possible in deeply profiled panels.

From the point of view of grc properties the ultimate failure strain of the material vis a vis the strain induced by moisture and thermal movements is perhaps the most important parameter that requires critical examination. For a typical case of a temperature gradient of 30°C and a moisture movement of 1500 microstrain from wet to dry across a 100 mm thick panel, calculations show that a stress-induced tensile strain of the order of 300 microstrain can arise in the outer grc skin by the combined effects of thermal and moisture movement and wind load. When these strain values are compared with the ultimate tensile failure strain of a 10 year old grc similar to that used in these panels, which can be as low as 300 microstrain, it is easy to see that a potential for cracking exists. However, since single-skin grc claddings fixed to buildings and exposed to weather have not shown any cracking of this kind so far, it is fair to conclude that the problem of cracking may be restricted to large panels having a shape and type of construction that restricts free movement.

A comprehensive analysis of failure in grc sandwich panels requires a more precise knowledge of the dynamic thermal response, moisture gradients, imposed loads and possible external restraints caused by inappropriate fixings acting on the panels and how these forces change with time. This example strengthens

the view expressed by Wittmann[1] and others that fracture
processes in concrete and related materials such as fibre cement
composites need to be studied at different levels. To be of any
practical value interrelationships in the hierarchy need to be
established and in this respect fracture mechanics has a role to
play.

ACKNOWLEDGEMENT

The work described has been carried out as part of the
research programme of the Building Research Establishment of the
Department of the Environment and this paper is published by
permission of the Director.

REFERENCES

1.   Wittmann, F.H. Structure of Concrete with Respect to Crack
      Formation, in F.H. Wittmann, ed., Fracture Mechanics of
      Concrete (Amsterdam, Elsevier, 1983) pp 43-74.
2.   Parrott, L.J., R.G. Patel., D.C. Killoh and H.M. Jennings.
      Effect of Age upon Diffusion in Hydrated Alite Cement (To
      appear in Journal of the American Ceramic Society).
3.   Diamond, S. Cement Paste Microstructure - an Overview at
      Several Levels, in Hydraulic Cement Pastes : Their Structure
      and Properties (Slough, Cement and Concrete Association, 1976)
      pp 2-30.
4.   Mindness, S. The Application of Fracture Mechanics to Cement
      and Concrete : a Historical Review, in F.H. Wittmann, ed.,
      Fracture Mechanics of Concrete (Amsterdam, Elsevier, 1983)
      pp 1-30.
5.   Swamy, R.N. Linear Elastic Fracture Mechanics Parameters of
      Concrete. ibid, pp 411-462.
6.   Birchall, J.D., A.J. Howard and K. Kendall. Flexural
      Strength and Porosity of Cements. Nature 289 (1981) 388-390;
      292 (1981) 89-90.
7.   Kendall, K., A.J. Howard and J.D. Birchall. Improvements in
      the Early Strength Properties of Portland Cement.
      Philosophical Transactions of the Royal Society of London
      A 310 (1983) 139-153.
8a.   Higgins, D.D. and J.E. Bailey. Fracture Measurements on
      Cement Paste. Journal of Materials Science 11 (1976)
      1995-2003.
8b.   Eden, N.B. and J.E. Bailey. On the Factors Affecting
      Strength of Portland Cement. ibid 19 (1984) 150-158.
9.   Berger, R.L., F.V. Lawrence and J.F. Young. Studies on the
      Hydration of Tricalcium Silicate Pastes II, Strength
      Development and Fracture Characteristics. Cement and
      Concrete Research 3 (1973) 497-508.

10. Tabor, D. Principles of Adhesion – Bonding in Cement and Concrete, in P.C. Kreijger, ed., Adhesion Problems in the Recycling of Concrete (New York, Plenum Press, 1980) pp 63–87.

11. Mindess, S. and S. Diamond. A Preliminary SEM Study of Crack Propagation in Mortar. Cement and Concrete Research 10 (1980) 509–519.

12. Hannant, D.J. Fibre Reinforced Cement and Concrete, Part 1 Theoretical Principles, Concrete 18 (1984) 25–26.

13. Romualdi, J.P. and G.B. Batson. Mechanics of Crack Arrest in Cocnrete. Journal of Engineering Mechanics Division, American Society of Civil Engineers 89 (1963) 147–168.

14. Kelly, A. Some Scientific Points Concerning the Mechanics of Fibrous Composites. Conference Proceedings, National Physical Laboratory : Composites – Standards, Testing and Design. (Guildford, I.P.C. Science and Technology Press, 1974), pp 9–16.

15. Aveston, J., G.A. Cooper and A. Kelly. Single and Multiple Fracture. Conference Proceedings National Physical Laboratory : The Properties of Fibre Composites. (Guildford, I.P.C. Science and Technology Press, 1971) pp 15–26.

16. Korczynskyj, S.J., S.J. Harris and J.G. Morley. The Influence of Reinforcing Fibres on the Growth of Cracks in Brittle Matrix. Journal of Materials Science 16 (1981) 1533–1547.

17. Hughes, D.C. Fibrillated Polyalkene Films in Cement. Ph.D. Thesis, University of Surrey (1983).

18. Hannant, D.J., D.C. Hughes and A. Kelly. Toughening of Cement and of Other Brittle Solids with Fibres. Philosophical Transactions of the Royal Society of London A 310 (1983) 175–190.

19. Aveston, J., R.A. Mercer and J.M. Sillwood. Fibre Reinforced Cements – Scientific Foundations for Specifications. Conference Proceedings, National Physical Laboratory : Composites – Standards, Testing and Design. (Guildford, I.P.C. Science and Technology Press, 1974) pp 93–103.

20. Laws, V., M.A. Ali and R.W.B. Nurse. The Response to Stress of a Short-Fibre-Reinforced Brittle Matrix. Conference Proceedings, National Physical Laboratory : The Properties of Fibre Composites (Guildford, I.P.C. Science and Technology Press, 1971) pp 29–30.

21. Walton, P.L., A.J. Majumdar and T.J. Evans. Properties of Cement Reinforced with High-Modulus Polyethylene Fibrillated Tape. Journal of Materials Science Letters 3 (1984) 718–724.

22. Proctor, B.A., D.R. Oakley and W. Wiechers. Tensile Stress/ Strain Characteristics of Glass Fibre Reinforced Cement. Conference Proceedings, National Physical Laboratory : Composites – Standards, Testing and Design. (Guildford, I.P.C. Science and Technology Press, 1974) pp 106–107.

23. Lenain, J.C. and A.R. Bunsell. The Resistance to Crack Growth of Asbestos Cement. Journal of Materials Science 14 (1979) 321–332.
24. Akers, S.A.S. and G.G. Garrett. Observations and Predictions of Fracture in Asbestos-Cement Composites. Journal of Materials Science 18 (1983) 2209–2214.
25. Coutts, R.S.P. and M.D. Campbell. Coupling Agents in Wood Fibre - Reinforced Cement. Composites 10 (1979) 228–232.
26. Mai, Y.W., M.I. Hakeem and B. Cotterell. Effects of Water and Bleaching on the Mechanical Properties of Cellulose Fibre Cements. Journal of Materials Science 18 (1983) 2156–2162.
27. Bentur, A. and S. Diamond. Fracture of Glass Fibre Reinforced Cement. Cement and Concrete Research 14 (1984) 31–42.
28. Stucke, M.S. and A.J. Majumdar. Microstructure of Glass Fibre Reinforced Cement Composites. Journal of Materials Science 11 (1976) 1019–1030.
29. Kelly, A. Fibre Reinforced Cements in Context. Conference Proceedings, Materials Research Society: Advances in Cement-Matrix Composites (University Park, M.R.S, 1980) pp 3–16.
30. Laws, V. Micromechanical Aspects of the Fibre-Cement Bond. Composites 13 (1982) 145–151.
31. Laws, V Private Communication (1984).
32. Page, C.L. Microstructural Features of Interfaces in Fibre Cement Composites. Composites 13 (1982) 140–144.
33. Proctor, B.A. Properties and Performance of GRC. Symposium Proceedings, Concrete Society: Fibrous Concrete (Lancaster, Construction Press, 1980) pp 69–86.
34. Henager, C.H. A Toughness Index for Fibre Concrete, R.N. Swamy, ed., Testing and Test Methods of Fibre Cement Composites (Lancaster, Construction Press, 1978) pp 79–86.
35. Majumdar, A.J. and V. Laws. Composite Materials Based on Cement Matrices. Philosophical Transactions of the Royal Society London A 310 (1983) 191–202.
36a. Mai, Y.W., R.M.L. Foote and B. Cotterell. Size Effects and Scaling Laws of Fracture in Asbestos Cement. International Journal of Cement Composites 2 (1980) 23–34.
36b. Mai, Y.W. and M.I. Hakeem. Slow Crack Growth in Cellulose Fibre Cements. Journal of Materials Science 19 (1984) 501–508.
37. Wecharatana, M. and S.P. Shah. A Model for Predicting Fracture Resistance of Fibre Reinforced Concrete. Cement and Concrete Research 13 (1983) 819–829.
38. Wecharatana, M. and S P. Shah. Predictions of Non Linear Fracture Process Zone in Concrete. Journal of Engineering Mechanics, ASCE 109 (1983) 1231–1246.
39. Brown, J.H. The Failure of Glass-Fibre-Reinforced Notched Beams in Flexure. Magazine of Concrete Research 25 (1973) 31–38.

40. Building Research Establishment. Properties of Glass Reinforced Cement : Ten Year Results. BRE Information Paper IP 36/79 (1979).
41. Jacobs, M.J.N. Forton PGRC. A Many Sided Cosntruction Material. International Congress Proceedings : GRC in the 80's (Gerrards Cross, Glass Fibre Reinforced Cement Association, 1982) pp 31-49.
42. Proctor, B.A., D.R. Oakley and K.L. Litherland. Developments in the Assessment and Performance of GRC over 10 years. Composites 13 (1982) 173-179.
43. Jones, F.E. Weathering Tests on Asbestos Cement Roofing Materials. Building Research Technical Paper No 29 (London, H.M.S.O., 1947).
44. Moore, J.F.A. The Use of Glass Reinforced Cement in Cladding Panels. Building Research Establishment Report (1984).

**APPLICATION OF FRACTURE MECHANICS
TO CEMENTITIOUS COMPOSITES**
NATO-ARW - September 4-7, 1984
Northwestern University, U.S.A.
S. P. Shah, Editor

SOME ASPECTS OF EXPERIMENTAL TECHNIQUES

L. Cedolin

Dipartimento di Ingegneria Strutturale
Politecnico di Milano
Milano, Italy 20133

PRINCIPAL LECTURES

The first lecture, entitled "On the Cracking in Concrete and
Fiber Reinforced Concrete" was presented in two parts. The first
part, given by S. Diamond, focused on the observations of cracking
patterns in cement paste, mortar and concrete through a scanning
electron microscope (SEM). The limits of previous studies with
optical microscopy were apparent from the finding that the higher
resolution provided by SEM allowed to detect the presence of
cracks having a width of fractions of a μm. The tortuosity of the
crack path, the existence of branching, the regions of fine multiple cracking near the ends of each branch were documented by
micrographs of different magnification. The second part, given by
A. Bentur, illustrated the results obtained for fiber reinforced
cement composites. The different types of cracking patterns at
the intersection of a propagating crack and a fiber were illustrated. The complexity of the debonding phenomenon and its dependence on the type of fibers appeared clearly by the use of a
specially designed specimen.

The second lecture, entitled "Fracture and Damage Mechanics
of Concrete", authored by D. Francois, was given by P. Acker. An
experimental apparatus for testing very large double cantilever
specimen was presented, and the methodology adopted for measuring
the fracture toughness of concrete was illustrated. The role of
damage mechanics in explaining the fracture properties of smaller
specimen was also discussed.

The third lecture, entitled "Fracture Processes in Fiber
Reinforced Concrete Cement Sheets", by A. J. Majumdar and P. L.

Walton, presents a very complete review of the cracking behavior
of fiber reinforced cement composites. Various theoretical models
are discussed first in order to illustrate the additional mechanisms of resistance to crack growth produced by the presence of
fibers. The experimental observations are then described, point-
ing out the more important parameters which characterize the
behavior of the composite material.

DISCUSSION BY THE REPORTER

The reporter focused his attention on the need for a precise
experimental determination of the fracture process in concrete.

The scanning electron microscope, as shown by S. Diamond and
A. Bentur [1], provides an extremely detailed description of the
crack path, showing its tortuous character and the existence of
branching and multiple cracking. These irregularities and the
presence of discontinuities in the plane of observation show that
at this scale the penomenon is essentially three-dimensional and
so complex that it is not possible to discriminate between the
advancing crack and the fracture process zone ahead of it. The
information given by this technique appears then to be extremely
useful for understanding the microstructural aspects of crack
propagation, but cannot be used for interpreting it in terms of
stresses and strains.

Moiré interferometry with coherent (LASER) light [2], [3],
seems to overcome many of the limits of the microscopical observation, i.e. small dimensions of the specimen with respect to aggregate size and narrow observation field, and still seems to be able
to measure with sufficient sensitivity the strain localization
which characterizes the fracture process. This technique uses two
beams of LASER light to produce, by interference, a reference grid
of very high density (1000 lines/mm) which is also recorded on a
special emulsion on the specimen surface in the initial unloaded
configuration. When the load is applied, the grid recorded on the
specimen follows its deformation, while the reference grid remains
unaltered and gives rise, by interference between reflected and
diffracted light, to moiré fringes. Iori et al. [4] demonstrated
that this technique may be applied to concrete through the use of
an aluminum foil, glued on the specimen surface, acting as a
support for the sensitive emulsion.

The application of moiré interferometry with LASER light to
the study of the tensile behavior of notched and unnotched concrete specimen by L. Cedolin, S. Dei Poli and I. Iori [5], [6],
has shown that it is possible to obtain an accurate measurement of
the longitudinal component of strain in a region of 10 cm in
diameter. In a more recent investigation [7] the use of a stiffer

loading device has allowed to follow to a greater extent the
evolution of the fracture process. Fig. 1 illustrates two different stages of crack propagation in a notched concrete specimen of
1.2 cm aggregate size. The contour lines corresponding to values
of the longitudinal component of strain, ε_x, equal to 0.2×10^{-3},
0.3×10^{-3} and 0.6×10^{-3} indicate a high gradient of deformation
in a zone which moves ahead of the crack and corresponds to a
diffused microcracking of the matrix. Immediately behind this
zone two or more moiré fringes, merge, and this indicates a
discontinuity in the strain field and corresponds to the coalescence of the diffused microcracking into what can be macroscopically identified as a single crack. The evidence presented in the
above mentioned references suggests that the width of the fracture
zone is not affected by the aggregate size, and that the phenomenon is independent of the notch width or depth. The existence of
the notch causes a single fracture process to advance along the
cross-section, giving rise to different distributions of strains
which, as already pointed out in [5] and [6], allow the determination of the constitutive relation through an identification
process. This procedure appears more suitable for this task than
a direct tensile test of an unnotched specimen, which gives rise,
as shown in Fig. 2, to a stress and strain distribution not
uniform both longitudinally and transversely. As already remarked
in the same references, this finding makes the stress-strain
relations derived from direct tensile tests valid only in an
average sense, and explains the wide scatter found for strains at
peak stress and for the slope of the declining branch of the
stress-strain curve. Fig. 2 shows that in an unloaded specimen
the fracture zone develops from a weak point and advances through
the cross section, and that its size is similar to the one found
in notched specimens. The only difference is that in unnotched
specimens more than one fracture process may occur simultaneously.

The high density of the grid used in the above mentioned
investigation can be obtained only by isolating the testing apparatus from vibrations through an air cushion. This puts limitations on the weight and type of loading device. Variations of the
same technique, described in [8], [9], are less sensitive to these
effects and may be used for specimens of larger dimensions.

GENERAL DISCUSSION

The two principal lectures were followed by an exhaustive
discussion.

The problem of the determination of the actual geometry of
cracks in concrete was addressed by J. W. Dougill, who pointed out
that a crack could go around a fiber without being detected by

microscopic observation which is essentially two-dimensional. Z. P. Bazant observed that the complexity of the cracking pattern, as described by S. Diamond and A. Bentur, confirms the validity of the smeared crack concept in analytical descriptions. A. R. Ingraffea remarked that, on the scale of the experiment presented by D. Francois, only a discrete crack representation makes sense. S. E. Swartz pointed out that the epoxy coating used in the experiment presented by D. Francois could have a surface hardening effect. In reply to a question by A. Ingraffea, S. Diamond confirmed that almost always many branches could be observed along a propagating crack, and these branches could be interpreted as main cracks which came to an arrest. S. P. Shah expressed the opinion that these branches disappear when unloading, and R. N. Swamy pointed out that the opening of new cracks may cause stress relaxation and closing of old cracks.

The problem of the contribution of the presence of calcium hydroxide on crack propagation in fiber reinforced cement composites was raised by J. J. Beaudoin. S. Diamond warned about the possibility that microscopic observation may lead to wrong conclusions. Calcium hydroxide precipitates where it finds empty voids, which can also be incompletely filled by it. J. F. Young added that regardless of phases, there are zones of porosity at the interfaces. Slate pointed out that under pressure calcium hydroxide becomes more soluble, and it should diffuse and precipitate in empty zones, providing mechanical strength. J. J. Beaudoin confirmed that the mechanical properties of crystallized calcium hydroxide are good and equivalent to the ones of cement paste, so that it should give a good contribution to strength.

The session ended with two prepared short contributions.

R. Eligehausen illustrated the results of tensile tests on notched concrete specimen. The relative displacements between the two sides of the crack were measured with LVDT gages, and the results indicated a strong influence of the distance of the measuring points.

B. Cotterman reported the results of tests on double cantilever concrete beams with wedge loading, which allowed the consideration of fully developed cracks under plane strain conditions. In the following discussion S. P. Shah expressed the opinion that in concrete there should not be much difference between plane stress and plane strain. A. Ingraffea pointed out how the process zone may extend in the back of a visible crack tip. V. C. Li emphasized the dependence of the process zone on loading and geometry. In reply to a question from G. Sih, B. Cotterman explained that without a side grove, due to the inhomogeneity of concrete, the crack would deviate from a straight trajectory. A. S. Kobayashi stated that the effect of the side grove on the phenomenon under investigation should be minor.

The session ended with a visit to the laboratories of the Department of Civil Engineering involved in concrete research.

REFERENCES

1. Diamond, S. and A. Bentur. On the Cracking in Concrete and Fiber-Reinforced Cements. Application of Fracture Mechanics to Cementitious Composites, NATO Advanced Workshop, Sept. 4-7, 1984, Northwestern University.

2. Fu-Pen Chiang. Moiré Methods of Strain Analysis. Proc. S.E.S.A. Vol. 36 (1979) No. 2.

3. Boone, P. M. Surface Deformation Measurements Using Deformation-Following Holograms. Nouvelle Revue d' Optique Appliquée (1970) No. 1 p. 10.

4. Iori, I., H. Lu, C. A. Marozzi and E. Pizzinato. Metodo per la determinazione dei campi di spostamento nei materiali eterogenei. L' Industria Italiana del Cemento. No. 4 (1982).

5. Cedolin L., S. Dei Poli and I. Iori. Analisi sperimentale del processo di formazione della frattura nel calcestruzzo. Studi e Ricerche No. 3 (1981) Politecnico di Milano.

6. Cedolin, L., S. Dei Poli and I. Iori. Experimental Determination of the Fracture Process Zone in Concrete. Cement and Concrete Research Vol. 13 (1983) 557-567.

7. Cedolin, L., S. Dei Poli and I. Iori. Comportamento a trazione del Calcestruzzo. Studi e Ricerche No. 5 (1983) Politecnico di Milano.

8. Post, D. Moiré Interferometry at VPI & SW. Experimental Mechanics Vol. 23, 2 (June 1983).

9. Post, D. Developments in Moiré Interferometry. Optical Engineering (June 1982).

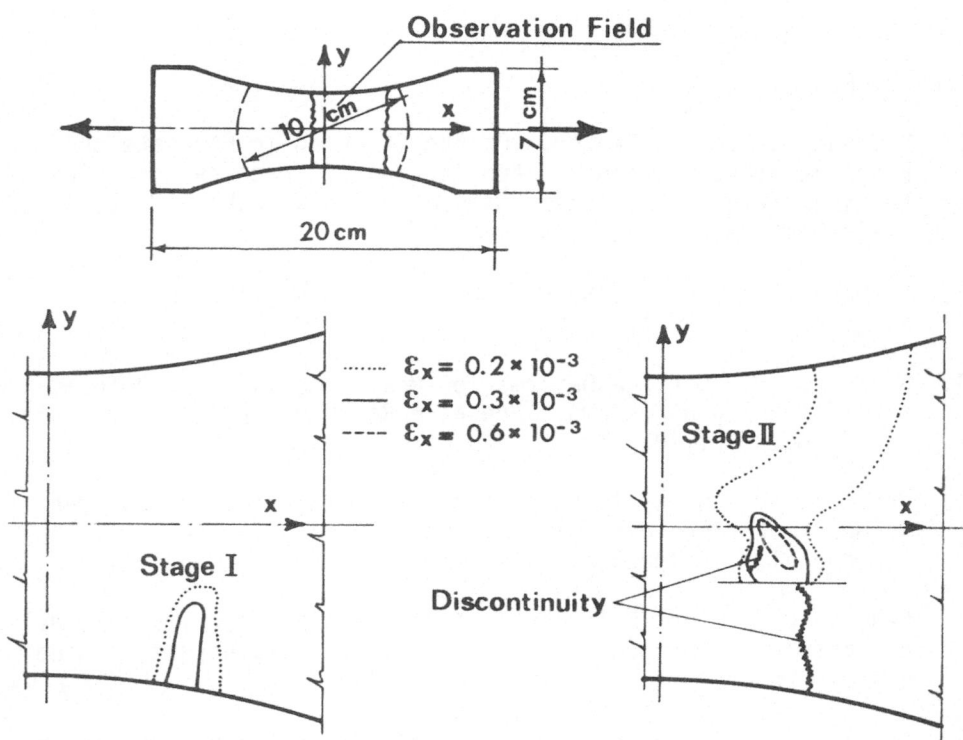

FIG. 2 : Unnotched Concrete Specimen under Tensile Load :
Propagation of the Fracture Process from a Weak Point

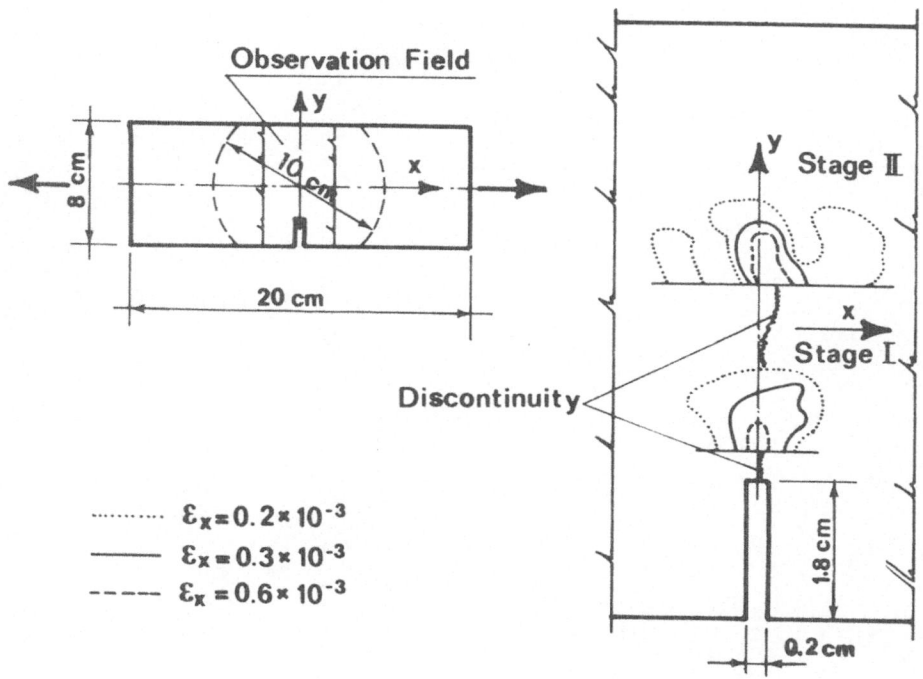

FIG. 1 : Notched Concrete Specimen under Tensile Load :
Contour Lines of Equal Longitudinal Strain for Two
Different Stages of Crack Propagation

SECTION III

NUMERICAL MODELING OF FRACTURE

APPLICATION OF FRACTURE MECHANICS
TO CEMENTITIOUS COMPOSITES
NATO-ARW - September 4-7, 1984
Northwestern University, U.S.A.
S. P. Shah, Editor

CONTINUUM MODEL FOR PROGRESSIVE CRACKING AND IDENTIFICATION OF
NONLINEAR FRACTURE PARAMETERS

Zdeněk P. Bažant[a], Jin-Keun Kim[b], and Phillip Pfeiffer[c]

Center for Concrete and Geomaterials
The Technological Institute, Northwestern Univeristy
Evanston, Illinois 60201, U.S.A.

ABSTRACT

Fracture of concrete as well as other materials such as rocks
or sea ice is preceded by progressive distributed cracking. On the
macroscale, this behavior calls for a continuum model, and the
crack tip blunting due to distributed cracking necessitates a
nonlinear fracture mechanics approach.

The first part of this lecture gives a review of a recently
formulated nonlocal continuum model which permits distributed
cracking to occur in a stable manner over finite-size zones of the
material, and summarizes the finite element crack band model, which
is a special case of the nonlocal continuum approach. The size
effect in blunt fracture is also briefly reviewed.

The second part of this lecture presents in detail a new
method of identifying the material parameters for propagation of
fractures blunted by a cracking zone. This method exploits the
recently derived size effect law for blunt fracture for determining
the parameters of the R-curve and the parameters of the finite

---

[a]Professor of Civil Engineering and Center Director
[b]Graduate Research Assistant
[c]Graduate Research Assistant; presently Resident Student
 Associate at Argonne National Laboratory

element crack band model (as well as Hillerborg's fictitious crack model). No measurements of the crack length or of the unloading compliance are needed, and it suffices to measure only the maximum load values for a set of geometrically similar notched specimens of different sizes. From these data, the parameters of the size effect law are identified by linear regression in certain transformed variables. The inverse slope of the regression line yields the fracture energy. The regression has further a two-fold benefit: it smoothes statistically scattered data according to a theoretically known law, and it extends the range of the data, so that fewer tests are needed than without the use of the size effect law. Using the experimentally calibrated size effect law, the R-curve may then be obtained as the envelope of family of curves representing fracture equilibrium for different specimen sizes. A simple algebraic formula for the R-curve, which closely agrees with the size effect law, is also presented. In the case of the crack band model, the size effect regression plot makes it again possible to determine all material parameters, particularly the fracture energy, the crack band width and the strain-softening modulus. Formulas for that purpose may be set up for each fracture specimen geometry, and some are presented here. The parameters of Hillerborg's fictitious crack model can be also easily identified from the size effect regression plot.

1. INTRODUCTION

Structures made of a heterogeneous brittle material such as concrete often exhibit brittle failures in which the material fractures progressively and the fracturing is distributed over a zone of finite size. In the macroscopic continumm approximation, the behavior of the fracturing zone is characterized by strain-softening, i.e., a stress-strain relation in which the maximum principal stress decreases at increasing strain. Strain-softening may be easily implemented in a finite element code, however, problems are encountered in convergence as the mesh is refined. Using a finite element discretization of the classical, local continuum, the failure zone always localizes into a zone of vanishing thickness, which means that in the limit of an infinitely small mesh size the structure is indicated to dissipate zero energy during failure. This aspect is obviously unrealistic, and it causes an incorrect spurious sensitivity of the results to the chosen element size [3,7,9.10,16].

An expedient remedy is possible with the crack band model, in which the cracking front is forced to have a fixed width which is a material property. This type of analysis has been shown to yield good agreement with all important fracture test data for concrete, as well as rock [3,16]. However, a disconcerting feature remains. The mesh cannot be refined to sizes smaller than a certain charactristic length, and so one does not have a limiting continuum which the finite element model is supposed to approximate. Furthermore, the size of the strain-softening zone and the strain distribution over the zone, which must be known if the energy dissipated by cracking should be correctly calculated, is unknown. To overcome these limitations, it is necessary to abandon the classical idea of a local continuum, as will now be shown in the first part of the lecture, in which a recently established [5,7,12,13] nonlocal continuum model is described.

The existence of a characteristic length in the nonlocal continuum model implies a size effect which differs from both the failure criteria of limit analysis and the linear elastic fracture mechanics. After reviewing the recently derived size effect law [6] that is associated with the nonlocal continuum approach, the second part of the lecture deals with the problem of determination of the material parameters for propagation of fractures blunted by a cracking zone. A new method of their determination, which uses experimental data on the size effect, is presented in detail.

2. CONTINUUM MODEL FOR DISTRIBUTED CRACKING

2.1 Imbricate Nonlocal Continuum

From the works of Kroner, Kunin, Krumhansl, Levin and others [21-2, 27-31], it is known that in a statistically heterogeneous medium which is not in a macroscopically homogeneous state of strain, the averaged (smoothed) stress at a certain point depends not only on the gradient of the averaged displacements at that same point (local properties), but also on the averaged displacements within a certain characteristic finite neighborhood of that point. The properties of such a medium cannot be said to be local, and the medium is, therefore, called nonlocal.

The nonlocal displacement gradient may be defined by the relation

$$D_i u_j(\underset{\sim}{x}) = \frac{1}{V} \int_{V(\underset{\sim}{x})} \frac{\partial u_j(\underset{\sim}{x}')}{\partial x_i} \, dV' = \frac{1}{V} \int_{S(\underset{\sim}{x})} u_j(\underset{\sim}{x}') \, n_i(\underset{\sim}{x}') \, dS' \quad (1)$$

in which u_j are the cartesian displacement components ($j = 1,2,3$), $\underset{\sim}{x}$ is the coordinate vector of the given point characterized by cartesian coordinates x_i, $V(\underset{\sim}{x})$ is the characteristic volume (Fig. 1) of the material centered at point $\underset{\sim}{x}$, $S(\underset{\sim}{x})$ is the surface of this volume, $n_i(\underset{\sim}{x}')$ is the unit normal of this surface at point $\underset{\sim}{x}'$, and D_i is the gradient averaging operator. The surface integral in Eq. 1 follows from the volume integral by application of the Gauss integral theorem. More generally, a weighting function can be introduced in Eq. 1. Using the gradient averaging operator,

200

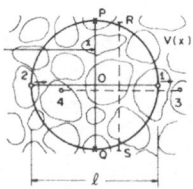

Fig. 1 - Representative Volume of an Aggregate Material

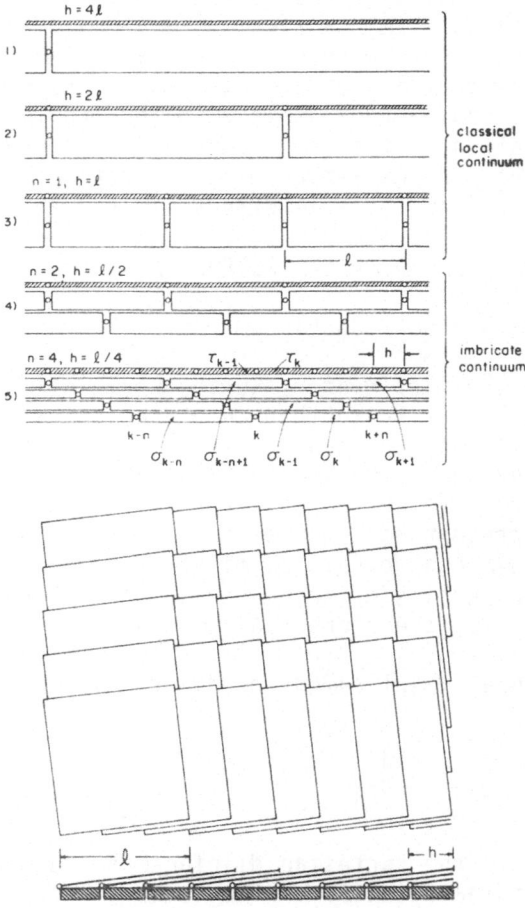

Fig. 2 - Finite Element Discretization of Imbricate Nonlocal
Continuum for One Dimension and Two Dimensions (the
continuum is the limiting case as the element size
tending to zero).

the mean strains may be defined as

$$\bar{\varepsilon}_{ij} = \frac{1}{2} (D_i u_j + D_j u_i) \tag{2}$$

In previous works dealing with nonlocal continua, it has been generally assumed that the continuum equation of motion has the form

$$\frac{\partial}{\partial x_j} \bar{C}_{ijkm}(\tilde{\varepsilon}) D_m u_k = \rho \ddot{u}_i \tag{3}$$

in which $\bar{C}_{ijkm}$ are secant elastic moduli which, in general, depend on the mean strain, ρ is the mass density, and superior dots refer to time derivatives. It is found, however, that Eq. 3 is incapable of describing a strain-softening continuum. It always leads to unstable response as soon as strain-softening begins. The difficulty has been traced to the asymmetry of these equations due to the combinatin of partial derivatives $\partial/\partial x_j$ with the gradient averaging operator D_m. This feature gives rise to nonsymmetric finite element matrices even if $\bar{C}_{ijkm}$ are constant, i.e., if the medium is elastic. Such a nonsymmetry is certainly an unacceptable characteristic.

For this reason, a systematic derivation of the continuum equation of motion on the basis of Eq. 1 has been attempted, using the calculus of variations. It has been found [5,7] that the proper form of the continuum equation of motion is

$$(1-c)D_j \bar{C}_{ijkm}(\bar{\varepsilon})D_m u_k + c \frac{\partial}{\partial x_j} C_{ijkm}(\tilde{\varepsilon}) \frac{\partial}{\partial x_m} u_k = \rho \ddot{u}_i \tag{4}$$

in which c is an empirical coefficient between 0 and 1, and C_{ijkm} are the local secant moduli. In contrast to Eq. 3, each term of the last equation has a symmetric structure, and consequently, discretization by finite elements leads to symmetric stiffness matrices if the elastic moduli $\bar{C}_{ijkm}$ and C_{ijkm} are symmetric.

Eq. 4 can be also written in the form

$$(1-c)D_j \sigma_{ij} + c \tau_{ij,j} = \rho \ddot{u}_i \tag{5}$$

in which

$$\sigma_{ij} = \bar{C}_{ijkm}(\bar{\varepsilon})\varepsilon_{km} = \bar{C}_{ijkm}(\bar{\varepsilon})D_m u_k \tag{6}$$

$$\tau_{ij} = C_{ijkm} \varepsilon_{km} = C_{ijkm} \frac{\partial u_k}{\partial x_m} \tag{7}$$

in which τ_{ij} are the usual, local stresses, and σ_{ij} are the stresses characterizing the stress state in the entire representative volume of the material and are called the broad-range stresses [5,7].

When the continuum defined by Eq. 4 is discretized by finite elements the size of which is smaller than the size ℓ of the representative volume, one obtains a system of imbricated (regularly overlapping) finite elements visualized in Fig. 2. Therefore, the present type of nonlocal continuum has been called the imbricate continuum [4]. The finite elements keep a constant size ℓ as the mesh is refined, and the number of imbricated finite elements that cross a given point is inversely proportional to the mesh size, while the cross section of these elements diminishes so that all the imbricated finite elements have the same total cross section for any mesh size. It can be also shown that the limiting case of the finite difference equations describing such an imbricated system of finite elements is the differential equation in Eq. 4 [5,7]. If the finite element size h is larger than the characteristic length ℓ, then the finite element model of the imbricate nonlocal continuum becomes identical to that for the classical local continuum.

To assure convergence and stability, the local stress-strain relations (Eq. 7) may not exhibit strain-softening, or else unstable response and spurious sensitivity to mesh size, along with incorrect convergence, may be obtained. The strain-softening properties must be described solely by the broad-range stress-strain relation in Eq. 6.

Fig. 3 reproduces some of the results of explicit dynamic finite element calculations from Ref. 7, in which wave propagation in a strain-softening bar of length ℓ was analyzed [7]. Both ends of the bar are subjected to a constant outward velocity d beginning at time t = 0. This loading produces step waves of strain propagating inward. When these waves meet at midlength, the strain suddenly increases and strain-softening ensues. If this problem is analyzed with the usual finite element method for local continuum, it is found that strain-softening is always limited to a single-element width. Thus, the width of the strain-softening zone reduces to zero as the element mesh is refined (Fig. 4). As a consequence, the energy W consumed by failure decreases with decreasing mesh size and approaches zero as the mesh size tends to zero (Fig. 5). Moreover, the finite element model of local continuum exhibits a discontinuous dependence of response on the prescribed end velocities as well as on the slope E_t of the strain-softening branch. The solution, however, converges to a unique exact solution, although this solution is unrealistic from the physical point of view.

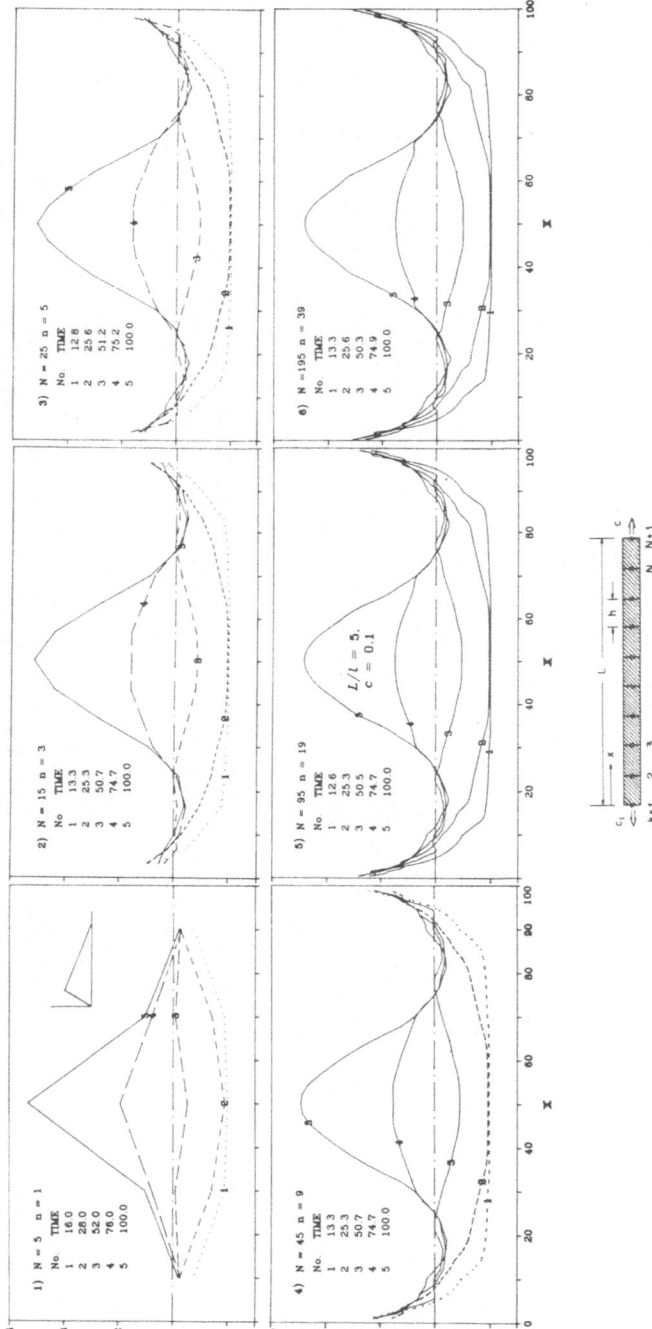

Fig. 3 - Numerical Results for Strain Distributions at Various Times for Wave Propagation in a Strain-Softening Bar, obtained for Subdivisions with 5 to 195 Elements (characteristic length of material = L/5; element arrangement shown in Fig. 3) (after Bazant, Chang and Belytschko, 1983).

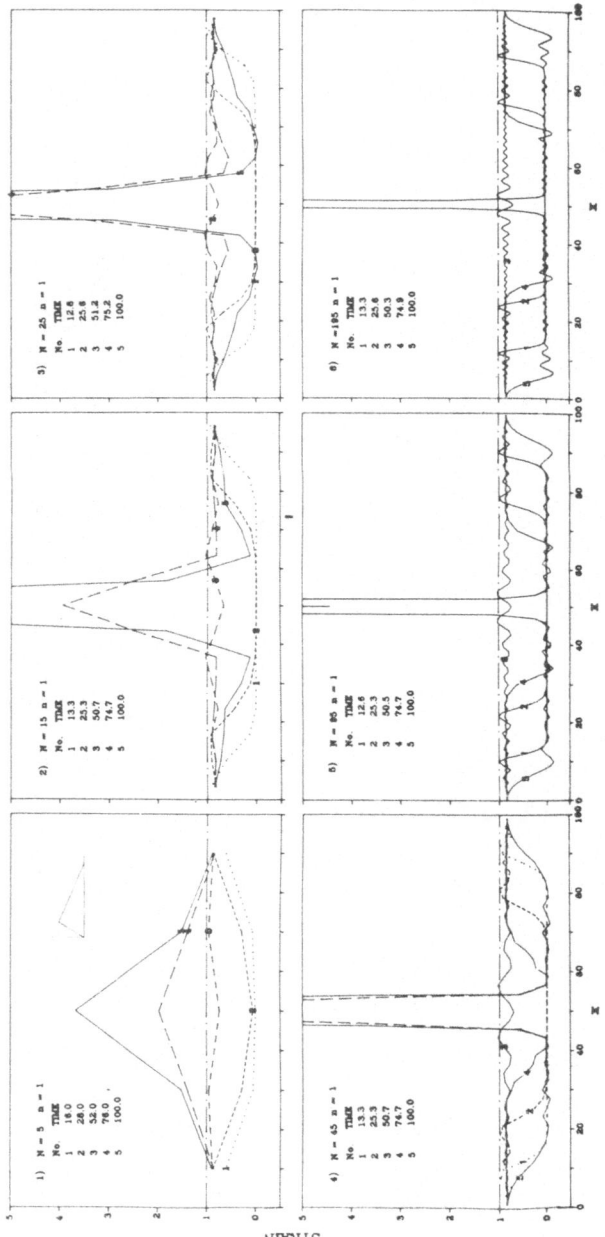

Fig. 4 – Numerical Results for the Same Problem as in Fig. 3
Obtained with Finite Elements of Classical Local
Continuum.

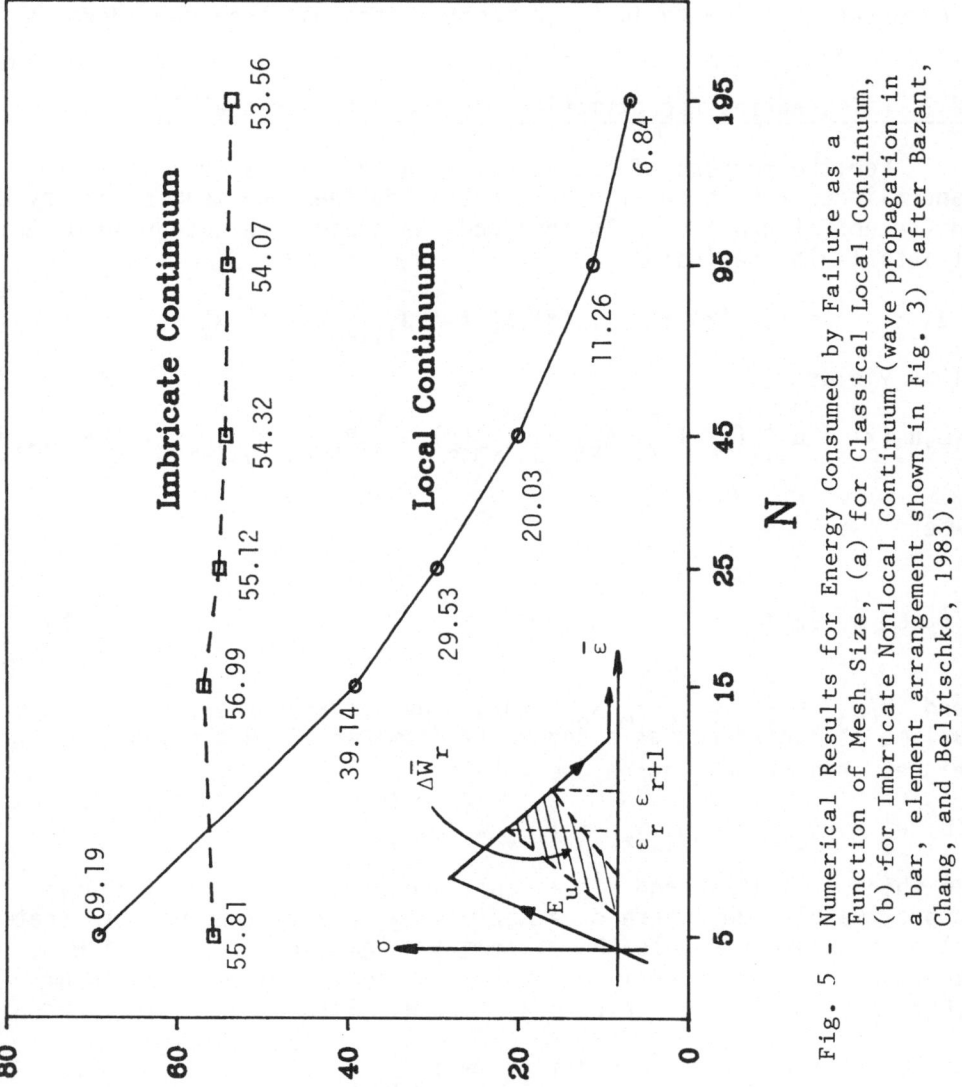

Fig. 5 - Numerical Results for Energy Consumed by Failure as a Function of Mesh Size, (a) for Classical Local Continuum, (b) for Imbricate Nonlocal Continuum (wave propagation in a bar, element arrangement shown in Fig. 3) (after Bazant, Chang, and Belytschko, 1983).

For the present imbricate continuum, by contrast, the solution of wave propagation in the strain-softening bar (Fig. 3) exhibits correct convergence, with a strain-softening zone of a finite size in the limit [7]. Also, the energy consumed by failure in the bar converges to a finite value, as shown in Fig. 5. The characteristic length in these computations has been considered as $\ell = L/5$.

2.2 Differential Approximation of Imbricate Nonlocal Continuum

For the purpose of analytical solutions it may be useful to approximate the integral operator that defines the mean strain by a differential operator. To this end, we expand the integrand of Eq. 1 into Taylor series:

$$u_{j,i}(\underset{\sim}{x}') = u_{j,i}(\underset{\sim}{x}) + u_{j,ik}(\underset{\sim}{x}) \, x_k' + \frac{1}{2} u_{j,ikm}(\underset{\sim}{x}) \, x_k' \, x_m' \, x_m'$$

This yields

$$D_i u_j(\underset{\sim}{x}) = u_{j,i}(\underset{\sim}{x}) + \frac{1}{2!} A_{km} u_{j,ikm}(\underset{\sim}{x}) + \frac{1}{4!} B_{kmpq} u_{j,ikmpq}(\underset{\sim}{x}) + \ldots \quad (8)$$

where

$$A_{km} = \frac{1}{V} \int_V x_k \, x_m \, dV = \frac{\ell^2}{20} \delta_{km} \quad (9)$$

and $B_{kmpq} = \frac{1}{V} \int_V x_k \, x_m \, x_p \, x_q \, dV$, provided that the representative volume is considered as a sphere of diameter ℓ. Neglecting terms of higher than second degree, we obtain

$$D_i u_j = u_{j,i} + \lambda^2 u_{j,ikk} = (1 + \lambda^2 \nabla^2) \frac{\partial u_j}{\partial x_i} \quad (10)$$

in which $\lambda = \ell^2/40$ and ∇^2 = Laplace operator. Since ℓ equals approximately $3d_a$ where d_a = maximum aggregate size, we note that λ approximately equals the maximum aggregate radius. In view of Eq. 10, the field equations for the imbricate nonlocal continuum (Eqs. 5-6) may now be written as follows [5]:

$$(1-c) \, (1 + \lambda^2 \nabla^2) \, \sigma_{ij,j} + c \, \tau_{ij,j} = \rho \, \ddot{u}_i \quad (11)$$

$$\sigma_{ij} = \bar{C}_{ijkm}(\underset{\sim}{\varepsilon}) \, \bar{\varepsilon}_{km} \, , \qquad \tau_{ij} = C_{ijkm}(\underset{\sim}{\varepsilon}) \, \varepsilon_{km} \quad (12)$$

$$\bar{\varepsilon}_{km} = (1 + \lambda^2 \nabla^2) \varepsilon_{km}, \quad \varepsilon_{km} = \frac{1}{2} \, (u_{k,m} + u_{m,k}) \quad (13)$$

The principal of virtual work for a body (whose domain is B

and surface is S) made of the imbricate continuum may be stated as follows:

$$\delta W = \int_B \sigma_{ij} \, \delta\bar{\varepsilon}_{ij} \, dV - \int_S p_i \, \delta u_i \, dS + \int_B \rho\ddot{u}_i \, \delta u_i \, dV = 0 \qquad (14)$$

in which $\delta u_i(\underline{x})$ is any kinematically admissible displacement variation, and p_i are the given distributed surface loads. Substituting Eq. 13 for $\bar{\varepsilon}_{ij}$ and applying repeatedly Gauss integral theorem, one can derive the field equations (Eqs. 11-13) from the virtual work relation (Eq. 14). Moreover, the variational procedure yields the boundary conditions at surface S;

either

$$u_i = 0 \text{ or } \bar{\sigma}_{ij} \, n_j = p_i \qquad \text{(on S)} \qquad (15)$$

where

$$\bar{\sigma}_{ij} = (1 + \lambda^2 \, \nabla^2) \, \sigma_{ij} \qquad (16)$$

In the classical nonlocal continuum theory, the mean strain is more generally defined with the help of a certain given weighting function $w(\underline{r})$ where $\underline{r} = \underline{x}' - \underline{x}$, i.e.,

$$D_i u_j(\underline{x}) = \frac{1}{V} \int_{V(\underline{x})} w\,(\underline{r}) \, \frac{\partial u_i(\underline{x}')}{\partial x_j'} \, dV' \qquad (17)$$

in which $\int_V w(\underline{r}) \, dV' = 1$ (normalized weights). Introducing again the Taylor series expansion of $\partial u_i/\partial x_j'$ and truncating it after the quadratic term, one finds that

$$D_i u_j = (1 + \alpha \, \lambda^2 \, \nabla^2) \, \frac{\partial u_i}{\partial x_j} \qquad (18)$$

in which $\alpha = \frac{1}{2} \int_V w(\underline{r}) \, x_k' \, x_k' \, dV'$. However, as long as λ is to be calibrated empirically, one can determine only the product $\alpha \lambda^2$, and not α and λ^2 separately. Thus, it does not matter which weighting function is used, and the simplest case $w(\underline{r}) = 1/V = \text{const.}$ may be chosen.

It is interesting to compare the equation

$$\bar{\varepsilon}_{ij} = (1+\lambda^2\nabla^2)\varepsilon_{ij} = \varepsilon_{ij} + \frac{1}{2} \lambda^2 \, (u_{i,jkk} + u_{j,ikk}) \qquad (19)$$

with the well-known couple stress theories or micropolar theories. In them, only first and second derivatives appear and the second derivatives are skipped. Moreover, there is no need to associate with the higher displacement derivative any special type of stress tensor of a higher rank, such as the couple stress tensor. Only one, second-rank stress tensor is used here.

Let us now check stability of the continuum. Consider linearly elastic properties, characterized by Young's modulus E, and the one-dimensional case, with $x_i = x$, $u_i = u$. From Eqs. 11-13 we obtain the differential equation of motion

$$(1 - c) \left(1 + \lambda^2 \frac{\partial^2}{\partial x^2}\right)^2 \frac{\partial^2 u}{\partial x^2} = \frac{\rho}{E} \frac{\partial^2 u}{\partial t^2} \tag{20}$$

Now seek a solution of the form $u = A \exp [i \omega(x - vt)]$ where v = wave velocity, ω = frequency. Substitution in Eq. 20 provides the condition

$$v^2 = \frac{E}{\rho} [(1 - c) (1 - \lambda^2 \omega^2)^2 + c] \tag{21}$$

For stability, v must always be real and positive, and so we must have $c > 0$.

The chief advantage of the approximation by derivatives is that it facilitates analytical solutions, for which the boundary layer method known from fluid mechanics may be utilized. For computer programming, the use of imbricated finite elements (Fig. 1) seems, however, the simplest approach, since ordinary finite elements may be used and the nonlocal properties are entirely taken care of by the element imbrication (regular overlapping). Existing finite element codes and the usual element types can be used and the only change to be made in the existing finite element codes is to properly define the integer matrix giving the nodal numbers corresponding to each element number.

2.3 Crack Band Theory for Progressive Fracturing

For very fine meshes for which the element size h is less than the characteristic length ℓ of the medium, the fracture front may be many elements in width. However, for many practical applications it is sufficient to use finite elements whose size is equal to the characteristic length or is larger. In such a case, the cracking zone is of a single-element width at its front, and the finite element model of the imbricate continuum then coincides with that of the classical local continuum. The fracture analysis then becomes identical to what has been previously developed as the crack band theory [3,16].

Distributed cracking has been modeled in finite element analysis by adjustments in the material stiffness matrix since 1967 when Rashid [35] introduced this approach. Recently it has been demonstrated this approach yields consistent results, independent of the mesh size, only if the stress-strain relation with strain-softening is associated with a certain fixed finite element size, ℓ. For concrete, this size appears to be roughly $\ell = 3d_a$ where d_a = the maximum size of the aggregate. This size of finite elements is too small for many practical purposes. In the crack band theory it has been proposed and verified that consistent results can be obtained with larger finite elements provided that the tensile strain-softening relation is adjusted so that it yields the same fracture energy regardless of the mesh size [10,16]. The fracture energy is expressed as

$$G_f = w_c \int \sigma_{33} \, d\varepsilon_{33} = \frac{w_c}{2} f_t'^2 \left(\frac{1}{E_0} - \frac{1}{E_t}\right) \tag{22}$$

in which w_c now represents the width of the cracking front, $w_c \approx \ell$, σ_{33} and ε_{33} are the stress and strain in the finite element normal to the direction of cracking, f_t' is the direct tensile strength of the material, E_0 is the initial elastic Young's modulus, and E_t is the mean downward slope of the strain-softening segment of the stress-strain diagram, which is negative (Fig. 6). If the finite element size is $h = \ell$, then Eq. 8 with ℓ replaced by h must yield the same value of G_f. This may be achieved by adjusting, first, the downward strain-softening slope E_t, and second, if the slope becomes vertical, by reducing the actual tensile strength f_t' to a certain equivalent strength f_{eq}' [3,8-10,16].

The crack band theory has been shown to agree with essentially all fracture test data for concrete, including the maximum load data and the R-curve data [3,16].

It may be noted that approximately the same results may also be obtained if the cracking strain accumulated across the width of the crack band is expressed as a single cracking displacement, and a certain stress-displacement relation in the connections between the finite elements is introduced into the analysis. This has been the approach followed by Hillerborg, et al. [23].

2.4 Constitutive Relations for Strain-Softening

In the analysis of many practical situations, including all fracture tests, the principal stress direction in the fracture process zone remains constant during fracturing. Triaxial strain-softening can then be introduced in the form

$$\underset{\sim}{\varepsilon} = \underset{\sim}{D} \, \underset{\sim}{\sigma} + \underset{\sim}{\xi} \tag{23}$$

in which $\underset{\sim}{\varepsilon}$ and $\underset{\sim}{\sigma}$ are the column matrices of the components of strain and stress, $\underset{\sim}{D}$ is the 6 x 6 matrix of elastic constants, and $\underset{\sim}{\xi}$ is a column matrix representing additional smeared-out strains due to cracking, $\underset{\sim}{\xi} = (\xi_{11}, \xi_{22}, \xi_{33}, 0, 0, 0)^T$. The normal

210

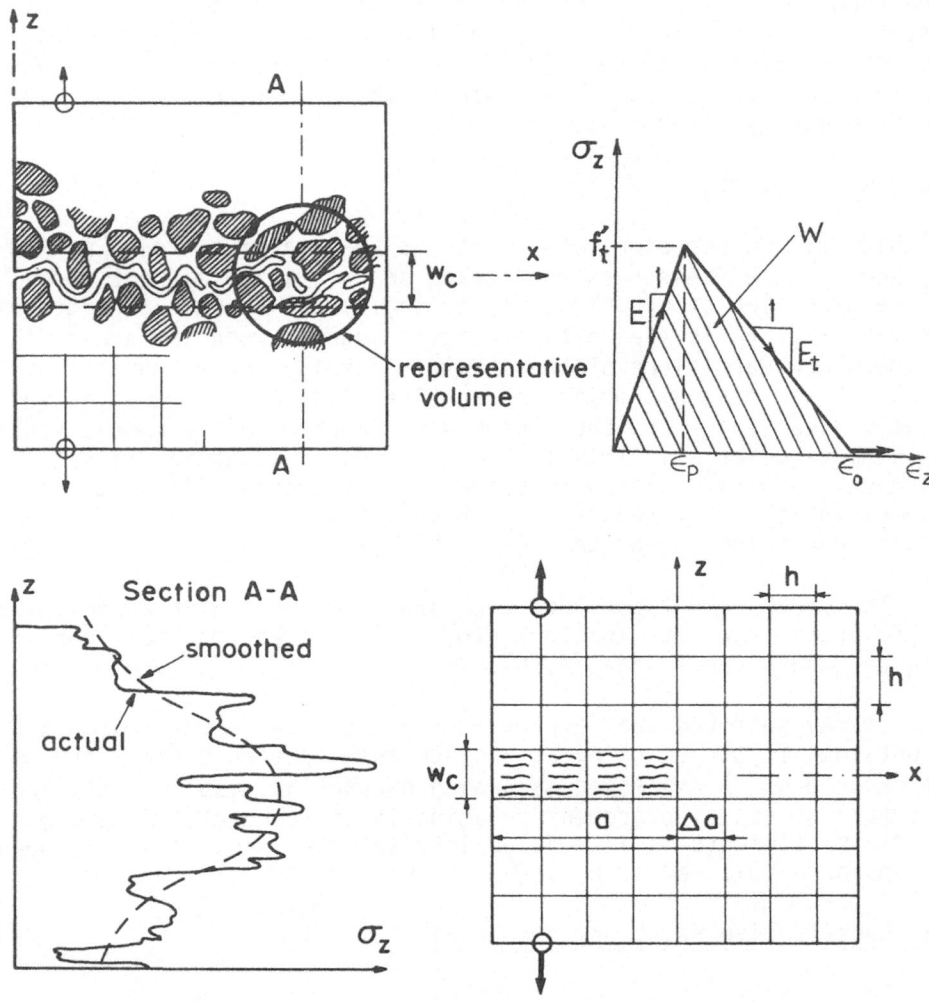

Fig. 6 – Crack Band Model and Corresponding Tensile Stress-
Strain Relation with Strain-Softening.

stresses may be assumed to be uniquely related to their associated cracking strains,

$$\sigma_{11} = C(\xi_{11}) \; \xi_{11}, \quad \sigma_{22} = C(\xi_{22}) \; \xi_{22}, \quad \sigma_{33} = C(\xi_{33}) \; \xi_{33} \quad (24)$$

in which C is the secant modulus which reduces to zero at very large cracking strains and may be calibrated from direct tensile test data which cover strain-softening [1,4,19,33,34,36,39,41]. Different algebraic relations must, of course, be used for unloading.

For some situations, especially in dynamics, it is necessary to describe progressive formation of fracture during which the principal stress directions rotate. In such a case, the foregoing model is inadequate. A satisfactory formulation can be obtained with an analog of the slip theory of plasticity, which is called the microplane model [4,18]. In this model it is assumed that the strain on a plane of any inclination within the macroscopic smoothing continuum consists of the resolved components of one and the same macroscopic strain tensor ε_{ij}. Using the condition that the energy dissipation calculated in terms of the stresses and strains on all such planes and in terms of the macroscopic stress and strain tensors must be equal, one may obtain the stress-strain relation

$$d\sigma_{ij} = D^C_{ijkm} \; d \; \varepsilon_{ij} \quad (25)$$

in which

$$D^C_{ijkm} = \int_0^{2\pi} \int_0^{\pi/2} n_i n_j n_k n_m \; F'(e_n) \; \sin\phi \; d\theta d\phi \quad (26)$$

This equation superimposes contributions to inelastic stress relaxations from planes of all directions within the material, defined by spherical coordinates θ and ϕ; n_i are the direction cosines for all such directions, and $F(e_n)$ is a function characterizing the constitutive properties and representing the stress-strain relation for one particular microplane within the material; $e_n = n_i \; n_j \; \varepsilon_{ij}$ = normal strain on a plane with direction cosines n_i.

It has been demonstrated that the microplane model allows describing tensile strain-softening under general stress or strain histories and always leads to a reduction of stress to zero at sufficiently large tensile strain.

3. SIZE EFFECT AND ITS USE IN DETERMINATION OF FRACTURE PARAMETERS

3.1 Problem of Experimental Determination of Material Parameters

Analysis of distributed fracturing by nonlocal continuum, crack band theory or other methods is feasible only if the material parameters involved can be identified from test data. This question will now be addressed and a novel method which exploits the size effect will be presented in detail. Since this method can be closely linked with the concept of R-curves, the use of R-curves for characteizing nonlinear fracture associated with progressive cracking will now be analyzed.

Fracture analysis of brittle heterogeneous materials, as well as ductile metals, must take into account the blunting of the crack front caused by microcracking or yielding. In consequence, the fracture properties of these materials are not completely described by a single parameter, the fracture energy, and at least two further parameters are required. Several mathematical models with additional parameters have been recenty formulated [3,4,11,16,23,34,41] and shown capable of closely representing the available experimental evidence. These models, however, are practically useful only if their fracture parameters can be easily determined from tests of a given material.

The simplest, although crudest, method consists of an approximate linearly elastic fracture analysis using an equivalent crack length (which is unrelated to the actual crack length) and a function describing how the energy, R, required for crack growth (per unit crack length and unit thickness) depends on the length c of the crack extension from the notch. Irwin and Krafft, et al., [24,26], proposed that this function, called the resistance curve or the R-curve, may be considered to be unique even though this is not exactly true. Shah and co-workers [39,40], introduced the R-curve concept to fracture analysis of concrete.

The existing method for determining the R-curve utilizes the relation $R = k_1 P^2 a/(E_c b^2 d^2)$ in which k_1 is a known coefficient for a given specimen geometry, E_c is the Young's elastic modulus, b = specimen thickness, d = characteristic dimension of the specimen, a = the length of crack plus notch, and P = load at which the crack extends. A series of R-values is determined either on a single specimen from the crack lengths, a, corresponding to various loads P, or on a series of specimens from their critical values of P and the corresponding critical values of crack length a. In both cases, however, the crack length needs to be measured. This is a considerable obstacle in the case of a material like concrete, for two reasons: First, the crack length is hard to define since the crack tip is blurred by a microcracking zone, and second, even if one succeeds to measure the location of the crack tip, the mesurement is of dubious significance since the R-curve is actually a function of a certain equivalent crack length which yields the correct remote elastic stress field rather than the actual crack length.

In view of these difficulties, it has been attempted to determine the crack length indirectly, by measuring specimen compliance, either at unloading or at reloading. However, this approach is also questionable because at unloading or reloading the microcracks within the fracture process zone do not completely close (due to rubble and fragments within the crack space, as well as irreversibility of material deformation at microcrack tips). Thus, the compliance for unloading and reloading is smaller than the compliance for continued loading, which, however, cannot be measured since the crack growth cannot be arrested. Therefore, the compliance measurements tend to yield crack lengths which are much too small.

The R-curve is in essence a device to make possible an approximate linearly elastic solution even though the material behaves nonlinearly [11]. A nonlinear fracture analysis, which is more realistic, may be carried out with the finite element blunt crack band model [3,16,17], in which a certain fixed triaxial tensile strain-softening constitutive relation is used, and the crack front is assumed to have a certain characteristic width w_c which is a material property (and equals about three maximum aggregate sizes of concrete). A similar finite element model for nonlinear fracture analysis of concrete, due to Hillerborg, et al.[23,24], utilizes, instead of a softening stress-strain relation, a softening stress-displacement relation for the relative displacement between two finite elements. The material parameters for these finite element models can be determined [16,17] by optimization or trial-and-error procedures using a finite element program; but this is not exactly simple for every-day applications. The fracture energy can, in theory, be also determined by measuring the area under the complete load-deflection diagram of a single specimen [34,38]. This approach, however, has certain disadvantages (see Section 3.9), and it does not yield the nonlinear fracture parameters other than the fracture energy. Thus, there is a need for another model, for which we propose here to exploit the size effect, following an idea briefly outlined in Ref. 6.

3.2 Review of Structural Size Effect

For blunt fracture one can generally introduce the hypothesis [4,7] that the total potential energy release W caused by fracture in a given structure depends on both:

1) the length a of the fracture, and

2) the area traversed by the fracture process zone, such that the size of the fracture process zone at failure is constant, independent of the size of the structure.

Dimensional analysis and similitude arguments then show that the structural size effect for geometrically similar specimens or structures made of the same material (and having the same thickness) is governed by the simple law [6]:

$$\sigma_N = B f_t' \left(1 + \frac{d}{d_0}\right)^{-1/2} \tag{27}$$

in which σ_N = P/bd = nominal stress at failure, P = maximum load (i.e., failure load, b = thickness, d = characteristic dimension of the specimen or structure (e.g., the beam's depth), f_t' = direct tensile strength; and B, d_0 = empirical constants, d_0 being a certain multiple of the maximum aggregate size, d_a. The values of B and of ratio $\lambda_0 = d_0/d_a$ depend on the geometrical shape of the structure, but not on its size. In the graph of log σ_N versus log d, Eq. 27 is plotted in Fig. 7a.

If the structure is very small, then the second term in the parenthesis in Eq. 27 is negligible comprared to 1 and then $\sigma_N = B f_t'$ in the failure condition, which represents the strength (or yield) criterion and corresponds to a horizontal line in Fig. 7a. If the structure is very large, then 1 is negligible compared to the second term in the parenthesis of Eq. 27, and then $\sigma_N = const./\sqrt{d}$. This is the type of a size effect typical of linear elastic fracture mechanics; it corresponds to the inclined straight line in Fig. 7a, having the slope - 1/2.

The size effect law according to Eq. 27 represents a gradual transition from the strength (or yield) criterion to the energy criterion of linear elastic fracture mechanics. This law is approximate because the hypothesis of a constant size of the fracture process zone at failure cannot be considered to be exact. However, the errors due to this approximation appear to be insignificant compared to inevitable random scatter of material properties. Statistical errors due to this scatter are, of course, superimposed on Eq. 27, which describes only the mean behavior.

Let us now summarize the dimensional analysis from Ref. 6 that leads to Eq. 27. To take the dispersed and progressive nature of cracking at the fracture front into account, the following

215

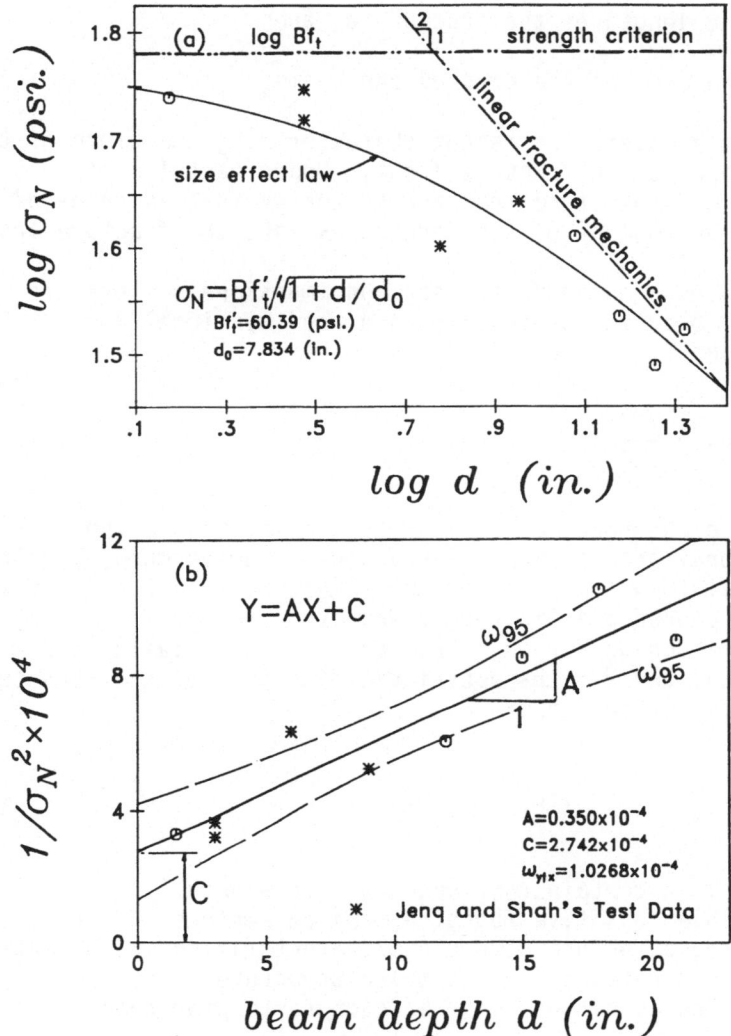

Fig. 7 - Size Effect Law and Regression Analysis of Maximum
Load Data.

hypothesis may be introduced:

The total potential energy release W caused by fracture in a given structure is a function of both:

(1) the length of the fracture a, and

(2) the area of the cracked zone amd_a.

Here m = material constant characterizing the width of the cracking zone at the fracture front. Under part (1) of this hypothesis we understand the part of energy that is released from the uncracked regions of the structures into the fracture front.

Variables a and amd_a are not nondimensional. They are, however, allowed to appear only in a nondimensional form. This form is given by the following variables

$$\alpha_1 = \frac{a}{d}, \quad \alpha_2 = \frac{amd_a}{d^2} \tag{28}$$

representing the nondimensional fracture length and the nondimensional area of the cracked zone. Furthermore, W must be proportional to volume $d^2 b$ of the structure (where b = thickness) and to the characteristic energy density $\sigma_N^2/2E_c$ in which $\sigma_N = P/bd$ = nominal stress at failure, P = given applied load, and d = characteristic dimension of the structures. Consequently, we must have

$$W = \frac{1}{2E_c} \left(\frac{P}{bd}\right)^2 bd^2 f(\alpha_1, \alpha_2, \xi_i) \tag{29}$$

in which f is a certain continuous and continuously differentiable positive function, and parameters ξ_i represent ratios of the structure dimensions characterizing the geometrical shape of the structure. For similar structures, ξ_i are constant. The condition for the fracture to propagate is

$$\frac{\partial W}{\partial a} = G_f b \tag{30}$$

in which G_f is the fracture energy - - a material property characterizing the energy consumed per unit extension of the fracture, per unit thickness.

Consider now geometrically similar structures, for which

parameters ξ_i are constant and only the characteristic dimension d varies. According to the chain rule of differentiation, $\partial f/\partial a = f_1 (\partial \alpha_1/\partial a) + f_2 (\partial \alpha_2/\partial \alpha)$, in which we introduce the notations $f_1 = \partial f/\partial \alpha_1$, $f_2 = \partial f/\partial \alpha_2$. Thus, substitution of Eq. 29 into Eq. 30 yields

$$\left(\frac{f_1}{d} + \frac{f_2 \, md_a}{d^2}\right) \frac{P^2}{2bE_c} = G_f \, b \tag{31}$$

Furthermore, the fracture energy may be expressed as the area under the complete tensile stress-strain curve, including the strain-softening down to zero stress, times the width of the cracking front md_a:

$$G_f = md_a \left(1 - \frac{E_c}{E_t}\right) \frac{f_t'^2}{2E_c} \tag{32}$$

in which E_c is the initial elastic modulus of concrete, E_t is the mean strain-softening modulus, which is negative, and f_t' is the direct tensile strength of concrete. Substituting Eq. 32 and $P = \sigma_N \, bd$ into Eq. 31, we may obtain:

$$\sigma_N = B f_t' \left(1 + \frac{d}{\lambda_0 d_a}\right)^{-1/2} \tag{33}$$

in which $B = [(1 - E_c/E_t)/f_2]^{1/2}$ and $\lambda_0 = mf_2/f_1$. B and λ_0 are constants when geometrically similar structures of different sizes are considered. Thus, Eq. 33 proves our starting equation, Eq. 27.

3.3 Equivalent Linear Fracture Analysis and R-Curves

Frequently the nonlinear fracture process zone is small compared to the structure dimensions and the stress and strain fields remote from the fracture process zone are then almost the same as the elastic ones. The fracture may then be approximately treated as an equivalent line crack which produces the same remote elastic stress and strain fields. For this equivalent crack, however, the energy, R, required for crack growth (also called the fracture resistance) may not be assumed constant (as is done in linear elastic fracture mechanics) but must be considered as a function of the crack extension c from the notch or smoothed surface, i.e., R = R(c) in which $c = a - a_0$, a_0 = length of the notch, and a = total length of the crack plus notch; see Fig. 8. The energy that must be supplied to the structure to produce the crack is $U = b \int R(c) \, da - W(a)$ where W(a) = the total release of

218

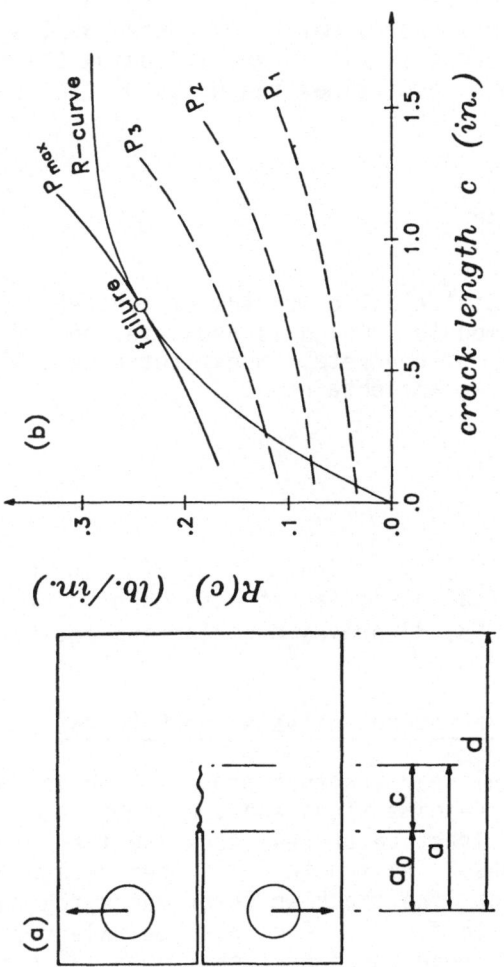

Fig. 8 – R-Curve.

strain energy from the structure. An equilibrium state of fracture occurs if no energy needs to be supplied to change a by δa and none is released, i.e., if $\delta U = 0$. Since $\delta U = (\partial U/\partial a)\,\delta a = 0$, in which $\partial U/\partial a = b(R - G) = 0$ and $bG = W' = \partial W/\partial a$, it follows that fracture equilibrium takes place if

$$G(a) = R(c) \qquad\qquad \text{(equilibrium)} \qquad\qquad (34)$$

in which $a = a_0 + c$, G is the energy release rate of the structure and R is its critical value characterizing the material. The equilibrium fracture state is stable if the second variation $\delta^2 U$ is positive. Since
$\delta^2 U = (\partial^2 U/\partial a^2)\delta a^2$ where $\partial^2 U/\partial a^2 \equiv b[dR/dc - \partial G(a)/\partial a] = b[R'(c) - G'(a)]$, $bG'(a) = W''(a) = \partial^2 W/\partial a^2$, the following conditions ensue

$$\begin{aligned} R'(c) - G'(a) &> 0 & \text{(stable)} \\ &= 0 & \text{(critical)} \end{aligned} \qquad (35)$$

Considering structures that are geometrically similar but could have dissimilar notches, and using dimensional analysis for the equivalent linear problem, one can show that

$$G(a) = \frac{W'(a)}{b} = \frac{P^2 g(\alpha)}{E_c b^2 d} \qquad\qquad (36)$$

$$G'(a) = \frac{W''(a)}{b} = \frac{P^2 g'(\alpha)}{E_c b^2 d^2} \qquad\qquad (37)$$

in which d is the characteristic dimension of the structure, $\alpha = a/d$, $g'(\alpha) = dg(\alpha)/d\alpha$, E_c = Young's elastic modulus; and $g(\alpha)$ is a nondimensional function that characterizes the geometry of the structure. It can be determined by linear finite element analysis and it can be also found for typical shapes in handbooks [37], which usually give the stress intensity factor K_I, from which $G = K_I^2/E'_c$, $E'_c = E_c$ for plane stress and $E'_c = E_c(1 - \nu^2)$ for plane strain ν = Poisson ratio; since $\nu^2 \approx 0.03$ the distinction between plane stress and plane strain is not very important for concrete. Plane stress usually describes test conditions better than plane strain.

Substitution of Eqs. 34, 36 and 37 into Eq. 35 leads to the conditions

$$R(c) \ g'(\alpha) - R'(c) \ g(\alpha) \ d > 0 \qquad \text{(stable)}$$

$$= 0 \qquad \text{(critical)}$$

Alternatively, since $dg/dc = dg/da = g'(\alpha)/d$, the last condition is equivalent to

$$\frac{R(c)}{g(\alpha)} = \text{Max} \qquad (39)$$

in which $\alpha = (a_0 + c)/d$. So the critical state of failure may be found by maximizing the ratio $R(c)/g(\alpha)$. This can be easily done by evaluating Eq. 39 for many values of c.

The foregoing relations are generally true for any equivalent linear analysis of nonlinear fracture. However, the function R(c) is generally not the same for different specimen shapes and different notch length. The R-curve concept of fracture analysis rests on the hypothesis [24,26] that function R(c) may be approximately considered as a unique material property, the same for different notch length and different specimen shapes.

3.4   Identification of Fracture Energy from Size Effect

As an example, consider now three-point bent fracture tests on specimens of various depth d, all with b = 1 in., span to depth ratio L/d = 4, and initial notch depth $a_0 = d/3$; also, $E_c = 4.3 \times 10^6$ psi and $f'_c = 3650$ psi [25]. According to Tada's handbook [37], for $L/d \cong 4$:

$$g(\alpha) = 16\pi \ \alpha(1.635 - 2.603 \ \alpha + 12.30 \ \alpha^2 - 21.27 \ \alpha^3 + 21.86 \ \alpha^4)^2 \qquad (40)$$

The following experimental data for the maximum loads are considered for specimens of various depths d:

d= 1.5,   3[*],   6[*],   9[*],   12,   15,   18,   21 in.

P=82.4, 156.9[*]and 167.4[*], 239.6[*], 395.6, 490.0, 514.5, 555.5, 700 lb.

$$(41)$$

(asterisks label the values taken from Jenq and Shah's measurements, Ref. 25). Inevitably, the data are statistically scattered. They may be smoothed out using the size effect law in Eq. 27. This equation may be algebraically transformed to a linear form,

$$Y = AX + C \qquad (42)$$

in which

$$X = d, \quad Y = \frac{1}{\sigma_N^2} = \frac{b^2 d^2}{P^2}, \quad A = \frac{C}{d_0}, \quad C = \frac{1}{B^2 f_c'^2} \tag{43}$$

Thus, coefficients A and C can be determined by linear regression, either by computer or by hand; see Fig. 7b. This yields $C = 2.742 \times 10^{-4}$, and $A = 0.350 \times 10^{-4}$. The coefficient of variation ω of the deviations from the regression line and the corresponding 95% confidence limits are also shown in Fig. 7b.

As mentioned before, for very large specimen sizes $(d \to \infty)$ linear elastic fracture mechanics applies, and the corresponding value of R, called the fracture energy G_f, may be obtained from the inclined asymptote in Fig. 1a, i.e., from Eq. 27 when 1 is neglected, in which case $Y = AX = \sigma_N^{-2}$. Noting that $\alpha \to \alpha_0$ for failure of very large specimens, and substituting here $X = d$ and $\sigma_N^2 = G_f E_c / g(\alpha_0) d$, which follows from Eq. 36 by seting G $= G_f$ and $P = \sigma_N bd$, we obtain:

$$G_f = \frac{g(\alpha_0)}{A E_c} \tag{44}$$

in which $\alpha_0 = a_0/d$. So the fracture energy is inversely proportional to the slope, A, of the regression line of size effect. Substituting A and $g(\alpha_0) = 43.76$ from Eq. 40, Eq. 44 yields the result $G_f = 0.291$ lb./in.

The test data used (Eq. 41, asterisk labeled) are those of Jenq and Shah [25]. They measured the R-curve by a certain novel method not exploiting the size effect, and the asymptotic value of their measured R-curve was $G_f = 0.42$ lb./in. This is not very different from the value found here, in view of the radically different methods of evaluation which must be influenced by different sources of errors.

Evaluating Eq. 42 for various values of d, we may further obtain smoothed values of maximum load data

d= 1.5,   3,   6,   9,   12,   15,   18,   21   in.
$$\tag{45}$$
P=83.0, 154.1, 272.7, 370.8, 455.4, 530.6, 598.6, 661.0 lbs.

(This same smoothed data would be obtained even if the measurements included only the three asterisk-marked values in Eq. 41.) Thus, the size effect law in Eq. 42 allows substantially increasing the range of d-values compared to the range of measurements, which will be useful for determining the R-curve.

Geometric similarity is required only in two dimensions, and thus the specimens could have different thicknesses b. However, the fracture energy is not constant along the front edge of the crack, and thus the thickness has some effect on the mean G_f for the whole thickness. There are two effects causing the thickness effect. One is the disturbance of the free surface boundary conditions due to Poisson effect and surface point singularity of elastic solution [14], and another one is the different influence of the aggregate size near the surface and the interior on the microcracking zone size. The former effect is eliminated if b/d is constant, and the latter one if b is constant. Thus, no perfect answer exists for the choice of thickness. The condition b = constant seems, nevertheless, preferable.

Eq. 44 is valid for all equivalent linear analysis of fracture. In particular, it does not depend on the hypothesis that the R-curve is unique.

The fact that the size effect plot of log σ_N versus log d possesses an inclined straight line asymptote implies that the R-curve must have a horizontal asymptote, a property previously sometimes regarded as uncertain. The fracture energy, G_f - a term reserved here strictly for the final asymptotic value of the energy required for crack growth (per unit crack length and unit thickness) - is uniquely defined by the straight line asymptote of the size effect plot of log σ_N versus log d (Fig. 7a). Thus, the value of the fracture energy must be considered size independent. Worrying about its size dependence would be meaningless, just like saying that a value of some function F(x) at x_1 depends on $x_2(\neq x_1)$ or the values of F(x) at x_2. It is important to realize this with regard to the current debates on the size dependence of fracture energy. In these debates, other determinations of fracture energy that are not based on the size effect law have been considered. For such other definitions, the fracture energy values are in general different from the present one and can indeed depend on the size of the structure; but this cannot be so for the present definition. In this light the present definition of G_f, based on the asymptote of the size effect regression plot, seems to be preferable and circumvents the question of size dependence of fracture energy for its other definitions. The present definition relates most directly to the failure loads at different sizes of a structure - usually the main concern of a structural analyst.

If three-point bent fracture specimens are sufficiently slender, i.e., L/d is sufficiently large, the failure is governed primarily by the bending moment M in the notched cross section, and the effect of the shear force may be neglected. Then it is not necessary for the specimens of different sizes to be geometrically similar. It suffices if the notched cross sections are similar, i.e, hae the same ratios a_0/d.

Let the reference specimen be characterized by $d = d_1$ and $L = L_1$. Then the maximum load P_2 measured for a dissimilar specimen of dimensions L_2, d_2 such that $L_2/d_2 \neq L_1/d_1$ must be transformed to the maximum load P_2 corresponding to span $L_2^* = L_1 d_2/d_1$. Equating the bending moments, $P_2 L_2/4 = P_2^*(L_1 d_2/d_1)/4^2$ we obtain

$$P_2^* = P_2 \frac{L_2 d_1}{L_1 d_2} \qquad (46)$$

specimen of span L_2^* and the actual, dissimilar specimen of span L_2 have the same cross section, and therefore, they may be assumed to fail at the same crack length c, i.e., at the same a or the same α ($\alpha = a/d$). Thus, the energy release rates G at maximum load must be the same for both specimens, and so (according to Eq. 36)

$$G = P_2^2 g(\alpha_2)/E_c b^2 d_2^2 = P_2^{*2} g^*(\alpha_2)/E_c b^2 d_2^2 \text{ where } \alpha_2 = a_2/d_2 = a_2^*/d_2:$$

this yields

$$P_2^* = P_2 \sqrt{\frac{g(\alpha_2)}{g^*(\alpha_2)}} \qquad (47)$$

It may be checked numerically that if both L_2/d_2 are large Eq. 47 reduces to Eq. 46.

Eq. 47 has an inconvenient feature in that the crack length c at failure must be estimated before P_2^* can be calculated ($\alpha_2 = (a_0 + c)/d_2$). The estimate, however, can be improved iteratively. For this purpose one has to solve also the R-curve as described in the next section, and then solve α_2 from Eq. 37, upon which one may obtain an improved value of P_2^* from Eq. 47.

3.5 Determination of R-Curve as an Envelope

Consider again that a series of geometrically similar specimens (of the same thickness b) has been tested and the maximum load $P = P_{max}$ has been measured for each of them. However, the crack length c corresponding to each failure load in each of these specimens is unknown (its measurement, no matter how careful, would not really help since c should be considered as the equivalent crack length giving the same remote elastic stress field, rather than the actual crack length). Using function $g(\alpha)$, which is common to all specimens, we can plot according to Eq. 36 the equilibrium curves of G versus c for each of these specimens, as is shown in Fig. 9a. Only one point on each of these equilibrium

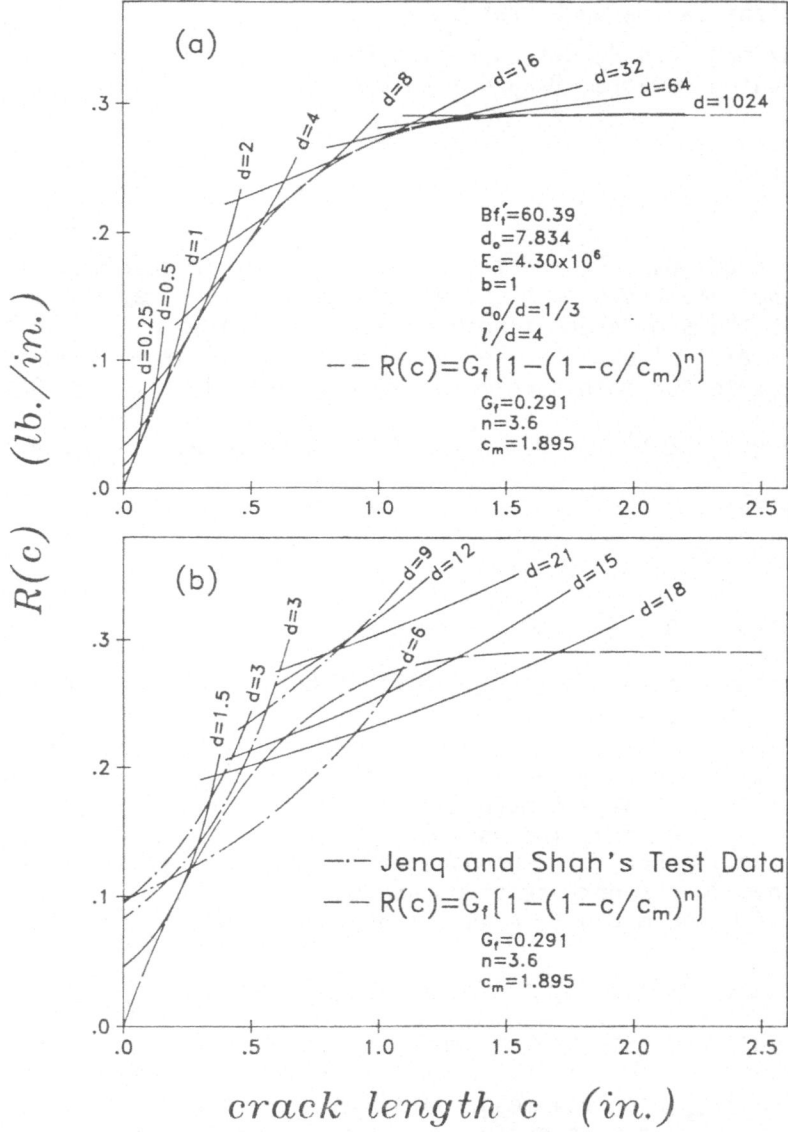

Fig. 9 - R-Curve as an Envelope for (a) Smoothed Data
(b) Unsmoothed Data.

curves is the failure state. Now, each equilibrium curve of G versus c must contain a failure point, and so this curve should be tangent to the R-curve, as seen in Fig. 9a. This leads to the following conclusion:

The R-curve is the envelope of all equlibrium curves of G versus c for the failure loads of specimens of different sizes.

To prove it rigorously, the curves in Fig. 9a are considered as a one-parameter family of curves (with parameter λ) described by the equation

$$f(c, G, \lambda) = 0 \tag{48}$$

in which

$$f(c, G, \lambda) = R(c) - G (a_0 + c, \lambda) \tag{49}$$

Differentiation of Eq. 48 yields

$$\frac{\partial f}{\partial c} + \frac{\partial f}{\partial G} G' + \frac{\partial f}{\partial \lambda} \frac{\partial \lambda}{\partial c} = R' - G' + \frac{\partial f}{\partial \lambda} \frac{\partial \lambda}{\partial c} = 0 \tag{50}$$

The envelope of the family of all curves is given by Eq. 50 with the condition $\partial f/\partial \lambda = 0$. This leads then to the relation $R'(c) - G' = 0$, which is the same as Eq. 35 for the critical state. So the envelope of the curve represents the critical states.

It may now seem that the R-curve could be constructed as an envelope directly from measured data (Eq. 41) by plotting G versus c according to Eq. 36. In practice, however, such an approach does not work. This is because of inevitable statistical scatter of test results (see subsequent comments on Fig. 9b). Therefore, one needs some simple law for the mean behavior which could be used to smooth out the experimental data before construction of the envelope is attempted. Eq. 27 for the size effect provides such a law. So we use the smoothed data from Eq. 45, and for each pair d and P we plot the equilibrium curve of G versus c according to Eq. 36. This yields the family of curves shown in Fig. 9a. The envelope, representing the R-curve, may be graphically plotted with ease. To obtain the envelope numerically, it is simplest to locate from the graph the approximate c-values of the points on the envelope, then evaluate for each of them the G-value from Eq. 36, and then fit these values of G with a suitable formula.

Although, from the viewpoint of representation of the original measured data, many different formulas are about equally good [5] (due to large scatter of data), the choice of formulas becomes more limited if one wants to closely match the size effect law (Eqs. 27 or 42). From this viewpoint, the following formula appears to work best

$$R(c) = G_f[1 - (1 - \frac{c}{c_m})^n] \text{ for } c < c_m, \quad R(c) = G_f \text{ for } c > c_m \quad (51)$$

Its parameters are found, according to the values in Eq. 45, as G_f = 0.291 lb./in., c_m = 1.895 in., and n = 3.6. This formula, plotted graphically in Fig. 9a, appears to be a perfect envelope.

Alternatively, one can also use the formula

$$R(c) = G_f \ (1 + \frac{k_0}{c + k_1 \ c^n})^{-m} \quad (52)$$

with coefficients G_f, k_0, k_1, m and n. It so happens that the approximations of the envelope remain quite good if m and n in Eq. 52 are fixed as m = 0.19 and n = 4.24. Then Eq. 52 has the advantage that parameters k_0 and k_1 may be determined by linear regression. Indeed, Eq. 52 can be transformed to the form y = A'x + C', in which

$$x = c^{n-1}, \ y = [(G_f \ R^{-1})^{1/m} - 1]^{-1} /c, \ A' = k_1 \ C', \ C' = 1/k_1.$$

For the values in Eq. 45, one obtains k_0 = 1.53 x 10^{10}, and k_1 = 3.58 x 10^{10}.

For comparison, let us see what we would get if the size effect law (Eq. 1) were not available to us. Then, plotting the curves of G vesus c (Eq. 36) on the basis of the measured maximum loads (Eq. 41), we would get the family of curves shown in Fig. 9b. This family of curves has no common envelope, and since the failure points (i.e., the values of c at failure) for each of the curves are unknown, it is not even possible to deduce any R-curve by statistical regression. Thus, knowledge of the size effect law (Eq. 27) is crucial for being able to determine the R-curve without having to measure the crack lengths at failure states, a task notorious for its difficulty and ambiguity. For comparison, Fig. 9b also shows the R-curve which is obtained if the measured maximum loads are first smoothed with the size effect law, Eq. 27. (The fact that the R-curve is only an approximate concept and is not strictly unique for different specimen shapes is not the cause of the lack of a common envelope in Fig. 9b; indeed, if the curves in Fig. 9b are calculated from fixed material properties, a smooth common envelope always exists.)

Instead of constructing the envelope graphically (Fig. 9a), one can define it analytically. For this purpose, we insert $P = \sigma_N$ bd in Eq. 36, and we set equal 0 the partial derivative of this equation with respect to λ at constant G, which is the condition for an envelope. This yields

$$G = \frac{d_0}{E_c} \lambda \, \sigma_N^2(\lambda) \, g(\alpha), \qquad \alpha = \frac{a_0 + c}{\lambda d_0} \tag{53}$$

$$\frac{a_0 + c}{d_0} \, \frac{1}{g(\alpha)} \, \frac{dg(\alpha)}{d\alpha} = \lambda + \frac{2\lambda^2}{\sigma_N(\lambda)} \, \frac{d\sigma_N(\lambda)}{d\lambda} \tag{54}$$

in which functions $\sigma_N(\lambda)$ and $g(\alpha)$ are defined by Eqs. 27 and 40. Eqs. 53-54 represent a parametric equation of the R-curve, with λ as a parameter. To calculate points on the R-curve, a series of λ-values is chosen, and for each λ the value of c is solved from Eq. 54 by Newton iterative method. G then results by substitution in Eq. 51 (in which $c = \alpha\lambda d_0 - a_0$).

It is interesting to calculate the size effect curve $\sigma_N(\lambda)$ from the R-curve R(c) that has previously been calculated from the size effect curve according to Eq. 27. Since Eqs. 41 or 42 for the R-curve are only approximate, the resulting curve $\sigma_N(\lambda)$ cannot be exactly the same as Eq. 27, but it should be almost the same. The calculations are carried out for the present example (Eq. 50), and the R-curve in Eq. 51 is obtained by maximizing the ratio in Eq. 38 in which $d = \lambda \, d_0$, with $g(\alpha)$ given by Eq. 40. After solving c from Eq. 39, G is obtained from Eq. 34 and P (or σ_N) is obtained from Eq. 36. The results are plotted in Fig. 10 and tabulated in Table 1. We see that the size effect curves thus obtained are indeed very close to the size effect law (Eq. 27).

As is well known, the R-curve concept is only approximate. For different specimen geometries different R-curves must be obtained, in theory, although usually the differences are not large (especially when compared to the scatter of test data). To check it, we follow the procedure from Eq. 34 to Eq. 51 and calculate the R-curves for three-point bent specimens of various span-to-depth ratios, various notch length-to-depth ratios, and also for a different type of specimen - the compact tension specimen. The corresponding functions for these geometries were taken from Tada's handbook [37]. The shapes of the resulting R-curves R(c) are plotted in Fig. 4b (the R-curves were scaled both vertically and horizontally so that the final value be 1.0 and the point where one-half of the final value is reached be common to all the R-curves). It is interesting to note how small are the differences among the shapes of the R-curves for different specimen shapes.

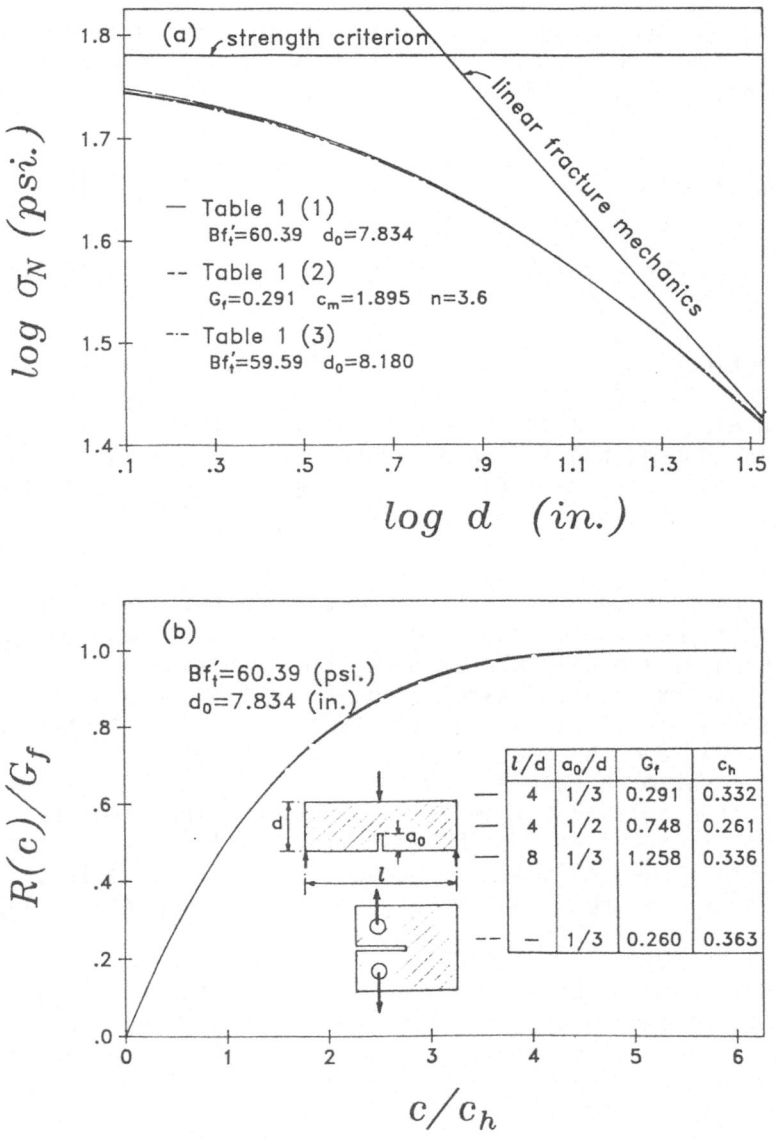

Fig. 10 - Shape of R-Curve and Corresponding Size Effect Curve

They are so small that they are hardly distinguishable graphically. Therefore, the R-curves from Fig. 10 are also tabulated in Table 2. (The parameters for the three columns in this Table are G_f = 0.291, 0.748, 1.258, 0.260 lb./in., c_m = 1.895, 1.491, 1.919, 2.280 in., c_h = 0.332, 0.261, 0.336, 0.363 in., respectively.)

In consequence of these considerations, it may be concluded that the relationship between the size effect law and the R-curve for a given specimen geometry may be considered as approximately unique.

In the size effect law (Eq. 27), there are two independent parameters, Bf_t' and d_0. For the relative values of R-curve $R(c)/G_f$ in Eq. 41, there are also only two parameters, n and c_m (G_f is obtained by linear regression from Eq. 44). Hence, for all specimens for which the values of Bf_t' and d_0 are the same, the values of nondimensional parameters n and c_m should also be the same. This means that all geometrically similar specimens of different sizes should be characterized by the same values of n and c_m/d_0. This property is verified by numerical examples. We may, therefore, construct a table of n and c_m/d_0 for various typical specimen geometries, see Fig. 11. For the specimen geometries included in Fig. 11, it is not necessary to construct the R-curve as an envelope. It suffices to carry out the linear regression shown in Fig. 7b, and then take n and c_m/d_0 from Fig. 11.

3.6 Determination of Crack Band Model Parameters

The size effect law (Eq. 27) may be also exploited to identify the material parameters for the finite element crack band model [3,16]. This may be done in two ways - either with or without an equivalent linear elastic fracture analysis. Consider the latter approach first.

In the crack band model, the cracking, assumed to be uniformly distributed throughout the finite element, is described by a triaxial constitutive relation with tensile strain-softening. Under the assumption that the principal stress directions do not rotate significantly during the passage of the fracture process zone through a given point, the strain-softening may be defined by a total stress-strain relation $\varepsilon = C \sigma + \xi$, in which σ = column matrix of stress components = $(\sigma_{11}, \sigma_{22}, \sigma_{33}, \sigma_{12}, \sigma_{23}, \sigma_{31})^T$, where T denotes a transpose of a matrix and the subscripts refer to cartesian coordinates x_i (i = 1, 2, 3), of which x_1 is the direction of crack propagation and x_3 is normal to the crack plane; ε = similar column matrix of strain components, C = 6 x 6 matrix of elastic compliances (constants), and ξ = column matrix of

Specimen Type	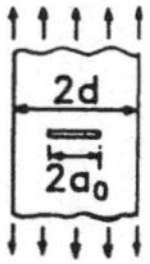	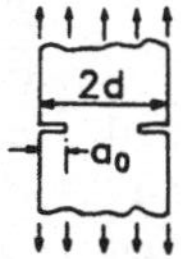	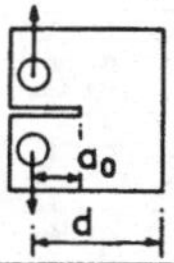	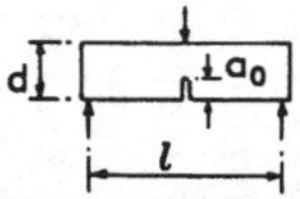 l/d=4	l/d=8	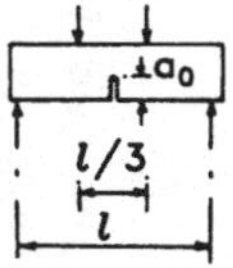
a_0/d=1/3 n	2.8	2.7	4.0	3.6	3.6	3.5
c_m/d_0	0.2930	0.3569	0.2911	0.2419	0.2450	0.2427
a_0/d=1/2 n	3.7	4.3	4.1	3.6	3.6	3.8
c_m/d_0	0.3542	0.5037	0.2190	0.1903	0.1944	0.2062

Fig. 11 – Parameters of R-Curve Corresponding to Size Effect Law.

additional averaged strains due to cracking. Only the normal components of ξ are nonzero and they are defined as $\xi_{11} = \phi(\sigma_{11})$, $\xi_{22} = \phi(\sigma_{22})$, $\xi_{33} = \phi(\sigma_{33})$, in which ϕ is a certain function. The precise shape of this function is not important, and in the preceding work [3,16] this function is taken as bilinear. It is calibrated according to the bilinear uniaxial tensile stress-strain diagram shown in Fig. 12, characterized by negative tangent modulus E_t for the strain-softening segment, and by tensile strength f_t'. For unloading (decreasing strain), a different stress-strain diagram is considered.

It must be emphasized that the foregoing stress-strain relation is valid only for a certain element size which corresponds to the representative volume of the heterogeneous material and to a certain width w_c of the cracking front. If a different element size has to be used, the stress-strain relation must be adjusted so as to yield the same fracture energy, G_f.

The fracture energy is equal to the area under the tensile uniaxial stress-strain diagram in Fig. 12a, times the width of cracking front, w_c;

$$G_f = \frac{w_c}{2E} f_t'^2 \left(1 - \frac{E}{E_t}\right) \tag{55}$$

Approximately, $w_c = 3d_a$ where d_a = maximum aggregate size.

For geometrically similar specimens, the finite element solution of the nominal stress at failure σ_N should depend only on the material parameters. There are four of them, f_t', E, G_f, and w_c. Parameter E_t is not independent since it must satisfy Eq. 55. Consequently, for geometrically similar specimens

$$\frac{P}{bd} = \sigma_N = \phi_1 (f_t', E, G_f, w_c) \tag{56}$$

in which ϕ_1 is a certain function. Now, according to Buckingham's Π theorem of dimensional analysis [2], the number n_s of independent nondimensional governing parameters should be $n_p - n_d$ where $n_p = 4$ = number of governing parameters in Eq. 56 and $n_d = 2$ = number of independent dimensions in these parameters. Since there are among them only two independent dimensions, namely those of length and of force, $n_s = 4 - 2 = 2$. So there can be only two independent nondimensional governing parameters. Along with the nondimensional function, they may be introduced as

$$s = \frac{\sigma_N}{f_t'}, \qquad \theta = \frac{d}{w_c}, \qquad \kappa = \frac{E \, G_f}{w_c f_t'^2} \tag{57}$$

and the governing Eq. 56 reduces to

$$s = \Phi(\theta, \kappa) \tag{58}$$

Eqs. 57-58, describing the similitude of blunt fracture, greatly reduce the number of cases that have to be solved by finite elements in order to cover all possible situations.

For identifying the material parameters of the crack band model, the following approach may be adopted. We choose a certain specimen geometry, and by testing specimens of different sizes we determine, by linear regression, parameters B and d_0 of the size effect law, Eq. 27. Then, by carrying out finite element solutions for specimens of different sizes and similar geometry, we determine the size effect law as a function of the governing parameters θ and κ (Eq. 57). Finally, we determine those material parameter values for which both size effect laws are matched. The interjection of the size effect law not only facilitates analysis, but also has the effect of smoothing and extending randomly scattered measured data. The detailed procedure is as follows.

1. Set $w_c = E = f'_t = b = 1$.

2. Fix the value of κ ($\kappa = G_f$).

3. Fix the value of θ ($\theta = d$) and solve by a finite element program with incremental loading the maximum load P (the load-point displacements are prescribed and P is calculated as the reaction). From this, calculate $s = P/\theta$.

4. Repeat step 3 for various values of θ and construct the plot of s^{-2} versus θ.

5. Now, according to the size effect law in Eq. 27, this plot should ideally agree with $s = B[1 + (\theta/r)]^{-1/2}$, which is equivalent to $y = Ax + C'$ where $x = \theta$, $y = s^{-2}$, $A = (B^2 r)^{-1}$, $C' = B^{-2}$ and $r = d_0/w_c$. Determine the regression line of this plot; its y-intercept is C', from which $B = 1/\sqrt{C'}$, and its slope is A, from which $r = C'/A$. These values of B and r correspond to the previously fixed value of κ.

6. Repeat steps 2 - 5 for various values of κ and construct the graphs $B(\kappa)$ and $r(\kappa)$ to be used for interpretation of test data.

The calculation results show that the graphs of

$B(\kappa)$ and $r(\kappa)$ are linear. Thus, calculation of only two points on each graph is sufficient, and the values may generally be calculated as

$$B(\kappa) = k_1 + k_2 \kappa, \qquad r(\kappa) = k_3 + k_4 \kappa \tag{59}$$

Fig. 13 shows the calculation results for several typical fracture specimens.

We are now ready to give an example using the data from Eq. 45 for similar specimens of different sizes. The test results are plotted as $1/\sigma_N^2$ versus d (Fig. 7b), and from the slope and the intercept of the regression line we get the values of B and d_0 for the size effect law as measured (Eq. 27). For the data in Fig. 7b we have $B = 0.1817$ and $d_0 = 7.834$. Then, for this value of B, we calculate from Eq. 59, using the values from Fig. 13, that $\kappa_m = 8.9$ and $r = 11.9$. From this we finally obtain

$$w_c = \frac{d_0}{r(\kappa_m)}, \qquad G_f = \frac{w_c}{E} f_t'^2 \kappa_m, \qquad E_t = \left[\frac{1}{E} - \frac{2G_f}{w_c f_t'^2}\right]^{-1} \tag{60}$$

For specimen geometries other than those in Fig. 13 the analyst needs to calculate first (with the help of finite element solutions) the values of k_1 and k_2.

Exploiting the size effect law (Eq. 27) makes it possible to do with a lesser amount of measurements. If the values of maximum loads for only a few specimens are fitted directly with the finite element program for the crack band theory, the values of material parameters which give good fits of data are quite ambiguous; even very different material parameter values yield equally good fits. This ambiguity and uncertainty is removed by the size effect law, which has the effect of smoothing and extrapolating the measured data.

If a smoothly curved tensile strain-softening constitutive equation is used, one may calibrate it according to the value G_f obtained as above. Instead of calculating E_t one needs to adjust the tensile strain-softening uniaxial curve so that the area under the strain-softening segment and under the unloading diagram emanating from the peak stress point would be equal to G_f.

Construction of the graphs in Fig. 13 requires the use of a nonlinear finite element program with step-by-step loading for the crack band theory. Such a program is not needed, however, if equivalent linear fracture analysis is used. The value of G_f may then be calculated from Eq. 12 on the basis of the slope of the size effect regression plot and the value of $g(\alpha_0)$ obtained by an analytical or finite element solution according to linear elastic fracture mechanics.

In the preceding procedure based on Eq. 60, the value of w_c has been considered as unknown. As a rough approximation, however, $w_c \sim 3d_a$ where d_a = maximum aggregate size [16]. If this approximation is adopted, and if f_t' is known, then the identification of material parameters of crack band model becomes much simpler; G_f is obtained from Eq. 44 on the basis of the slope of the regression plot for the size effect (Fig. 7b), and E_t is then simply solved from the relation $G_f = w_c$ $(E^{-1} - E_t^{-1}) f_t'^2/2$, i.e.

$$E_t = (\frac{1}{E} - \frac{2G_f}{w_c f_t'})^{-1} \qquad (61)$$

3.7 Fracture Models with Stress-Displacement Relation

Through a simple extension of the foregoing analysis, it is possible to determine the material parameters for nonlinear fracture models, such as Hillerborg's model for concrete [23,34], in which a sharp line fracture is assumed and a relation between the relative displacement δ and normal stress σ across the line is introduced as a material property. The displacement δ lumps into a line the accumulated normal strain due to cracking across the crack band width w_c, and so

$$\delta(\sigma) = w_c [\varepsilon(\sigma) - \frac{\sigma}{E}] \qquad (62)$$

where $\varepsilon(\sigma)$ describes the tensile stress-strain diagram for the equivalent crack band model. Most simply, the $\sigma - \delta$ relation may be considered as a straight line of negative slope C_f (Fig. 12c), described as $\delta(\sigma) = (f_t' - \sigma)/C_f$ if $\sigma \geqslant 0$ and $\delta \geqslant 0$, and $\sigma = 0$ if $\delta \geqslant f_t'/C_f$.

Equating the total relative displacement across the crack band according to both models and assuming the stress at the crack front (in a thin plate) to be approximately uniaxial, we have the relation

$$w_c \frac{\sigma}{E} + \frac{f_t' - \sigma}{C} = w_c \frac{f_t' - \sigma}{E} + \frac{f_t'}{E} \qquad (63)$$

This relation is satisfied for any σ if

$$\frac{1}{C_f} = w_c \ (\frac{1}{E} - \frac{1}{E_t}) \tag{64}$$

Thus, if E_t has been determined from measurement, C_f can be found also.

Alternatively, we may use the condition of equal fracture energy for both theories: $G_f = f_t'^2/2C_f = w_c \ (E^{-1} - E_t^{-1}) \ f_t'^2/2$. From this, Eq. 64 is again obtained.

3.8 On Material Parameter Identification Without the Size Effect Law

To further illustrate the advantage derived from the use of the size effect law, consider a series of tests of specimens of the same dimension d but different notch lengths a_0. If one wishes to use the envelope property, one would plot for each pair of a_0 and P the curve of G versus c where $G = P^2 g(\alpha)/(E_b b^2 d)$ and $\alpha = (a_0 + c)/d$. If the measurements were perfect, with no error, then these plots would yield a family of curves such as illustrated in Fig. 14a, for which the envelope representing the R-curve can be, in theory, constructed. In practice, this does not work, for two reasons: 1) If d is constant, only a small portion of the R-curve is covered by the failure states for various a_0-values. 2) There is always statistical scatter, which causes that these plots yield a family of curves such as illustrated in Fig. 14b. Obviously, no envelope can be constructed for this family, and thus smoothing of the data is imperative before the envelope could be traced. However, for the effect of the notch length a_0 at a constant cross section dimension d, there is no simple law which could be used for smoothing the data. Such a law is known only for the size effect.

Nevertheless, a more sophisticated procedure similar to that employed in Ref. 11, can be used. It consists of the following steps:

1. Choose a certain formula for the R-curve, such as Eq. 51 or 52.

2. Choose certain values of the material parameters in the formula.

3. Choose a certain value of initial notch length a_0.

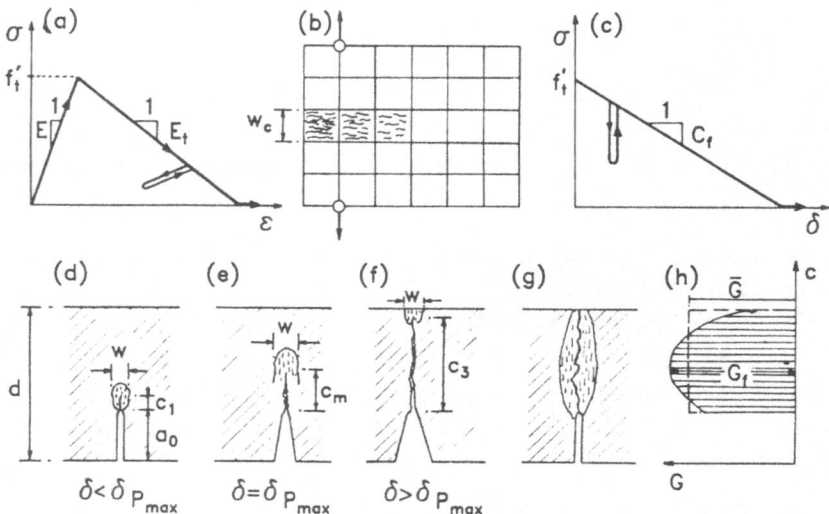

Fig. 12 - (a) Uniaxial Stress-Strain Relation for Crack Band
Model (b) Finite Element for Crack Band Model,
(c) σ - δ Relation for Fictitious Crack Model of
Hillerborg, (d-h) Explanation of Energy Dissipation
During Crack Growth.

Specimen Type	$\pm d/3$		.275d .6d $\odot$ / .5d $\odot$ .6d / d/4 ⊢ d ⊣	d/10 3d ⊢ d ⊣
	$l/d=2.67$	$l/d=4$	$d/4$ ⊢ d ⊣	
k_1	-26.81	-17.20	56.00	49.10
k_2	154.04	160.00	-450.00	-51.28
k_3	-19.69	-18.88	20.00	10.92
k_4	103.44	150.00	-155.00	-11.54

Fig. 13 - Coefficients for Determining Crack Band Model
Parameter.

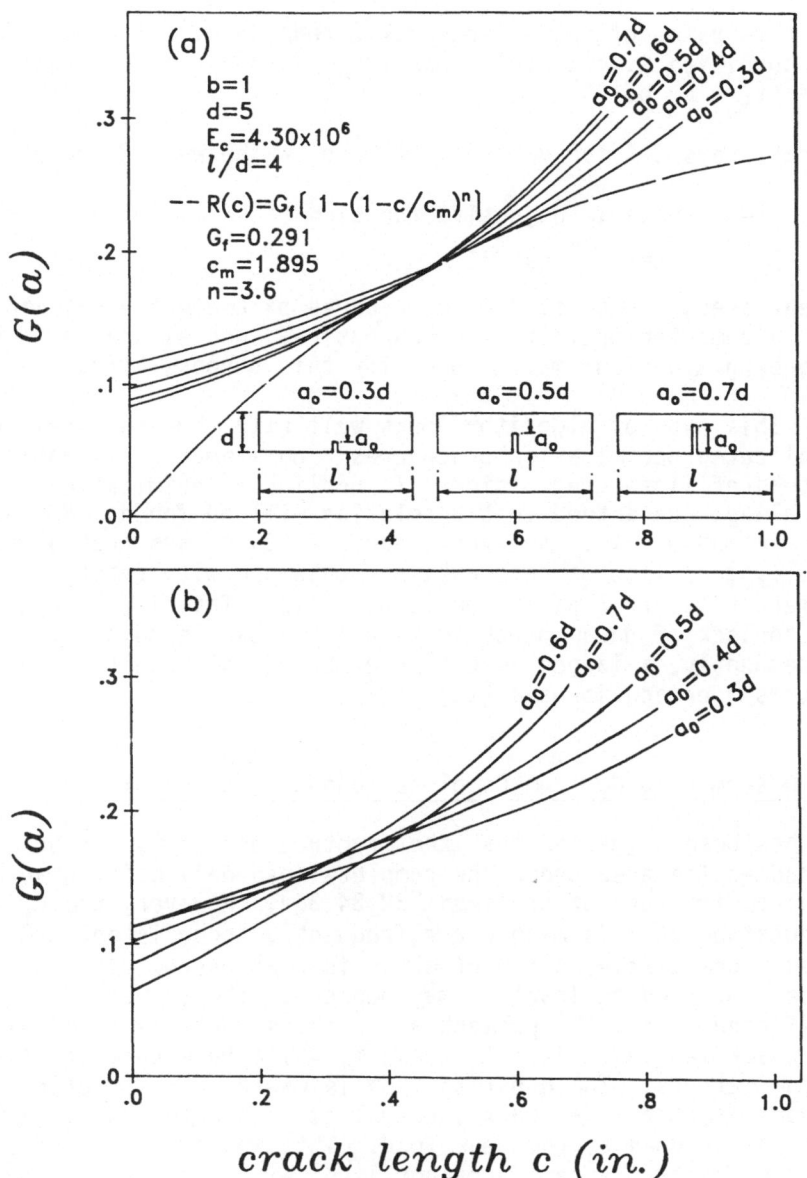

Fig. 14 - R-Curve as an Envelope for Various Notch Lengths in
Specimen of One Size; (a) Smoothed, (b) Unsmoothed
Data.

4. For this value of a_0, solve crack extension at failure, c, from the condition $G_c(c)g(\alpha)$ = Max (=f_{max}), and then evaluate $P = (f_{max} E_c bd)^{1/2}$.

5. Repeat steps 2-4 for various values of a_0 between 0 and d.

6. Evaluate a suitable objective function to be minimized, such as $\Phi = \Sigma[(P - P_{test})/P_{test}]^2$.

7. Repeat steps 1-6 to find which R-curve parameters yield minimum Φ. A computer optimization subroutine, such as the Marquardt-Levenberg algorithm may be used for this purpose [11].

        This type of algorithm works well [11], but requires more complicated optimization procedures - nonlinear optimization instead of linear regression. In nonlinear optimization it is not always guaranteed that a solution will be found. Moreover the minimum of Φ is not very sharp, which causes that almost equally good fits of test data are obtained with rather different material parameter values [11]. This is, of course, due to lack of data smoothing by a law known in advance. Consequently, a larger amount of experimental results is needed if this approach is used [11].

3.9 On Determining G_f from Complete Load-Deflection Diagram

        It has been suggested that the fracture energy G_f can be determined as the area under the complete load-deflection diagram for complete fracture of specimen [32,34,38]. However, the G_f - values obtained in this manner are frequently inconsistent and scattered. One likely source of error is that energy dissipation which does not produce fracture may happen in the system. The basic difference from the present approach is that the load values for all crack lengths affect the result, while here only the peak load value matters. The question then is whether the fracture pocess zone at other than peak load values dissipates energy at the same rate as it does at the peak load. This would certainly be true if the fracture were a straight line, with a sharp tip and a fracture process zone of negligible size. This is not so, however. It is likely that the width and the length of the microcracking zone at the fracture front are different than they are at the peak load, and the fact that G_f is far larger than double the theoretical specific surface Gibbs' free energy of the solid serves as the proof that much energy dissipation must occur due to microcracking on the sides of the final continuous crack. The width of the microcracking zone ahead of the fracture front is assumed to be constant in the crack band model as well as

Hillerborg's fictitious crack model, but this is no doubt a simplification (the nonlocal continuum approach points to that, too). In fact it is likely that the width w of this zone varies (Fig. 12g), and especially that the widths at crack initiation (Fig. 12d) and at crack termination (Fig. 12f) may be quite different from the width at the peak load (Fig. 12), at which the fracture front is remote from both the notch and the opposite face. Consequently, G along the crack path is variable (Fig. 12h) and the mean value $\overline{G}$ (obtained by dividing the area under the load-deflection curve by the ligament length $d-a_0$) need not be the same as the value G_f at maximum load (Fig. 12h).

Now, of the values $\overline{G}$ and G_f (Fig. 12h) which one is more useful? That depends. If the goal is to predict the peak loads or the response near the peak loads, then it is more reasonable to use only peak load values for determining G_f. Besides, they are easier to measure.

In the light of Fig. 14, there might also be another difficulty. From this figure it is apparent that the use of a single size specimen with different crack lengths cannot, due to inevitable random scatter, give information on the complete R-curve, or the complete size-effect curve. It pertains only to a portion of these curves, and does not indicate unambiguously the limiting value of the R-curve, which represents G_f (Fig. 14). If tests on single size specimens (without crack length measurements) do not give sufficient data on the R-curve, how can they unambiguously yield the R-curve asymptote?

R-Curves for Different Specimen Shapes. - Although this is not the main objective of this study, the use of the present results for the R-curves requires knowing to what extent the R-curve may be considered unique. This is a strictly theoretical question, which is hard to answer experimentally because the random scatter of material properties and other measurement difficulties obfuscate the comparisons of various experimentally determined R-curves. Therefore, it is preferable to make comparisons of R-curves which are calculated for specimens of different shapes using the same material properties. Such calculations have been carried out as described in Ref. 16 for specimens of various typical geometries. The calculated R-curves are plotted and compared in Fig. 15. The differences between some of these curves may seem large, however, they are not large compared to the inevitable statistical scatter of measured R-curve values. Therefore, the hypothesis of a unique R-curve appears to be an acceptable approximation for the purposes of crude structural analysis.

4. CONCLUSIONS

1. The macroscopic continuum description of dispersed cracking should properly be based on the nonlocal continuum concept. A suitable type of nonlocal continuum is the imbricate continuum, representing the limit of an imbricated element system. This model is capable of describing stable strain-softening zones of a finite size. For large finite elements, this model reduces to the previously formulated crack band theory. The existance of a characteristic length implies a simple size effect law for failures due to progressive cracking, which results from dimensional analysis and represents a transition from limit analysis (strength or yield criterion) to linear elastic fracture mechanics.

2. The size effect law of blunt fracture (Eq. 27) is useful for identifying the material parameters for nonlinear fracture, regardless of whether the R-curve approach, or the strain-softening crack band model, or the stress-displacement relation (Hillerborg's model) is used. The basic idea is to transform the size effect law to a linear plot and determine in this plot the regression line for the measured data obtained by tests of geometrically similar specimens of different sizes. The slope of this regression line then yields the fracture energy (the value of which is, by definition, size-independent). The method can be also extended to certain dissimilar specimens of similar cross sections. The remaining nonlinear fracture parameters for the R-curve or the crack band model (or Hillerborg's model) may then be identified by finding a matching size effect regression line for this model.

3. The R-curve may be obtained as the envelope of a family of fracture equilibrium curves determined on the basis of maximum load data smoothed with the size effect law. Without this smoothing, no envelope exists.

4. The size effect law and the parameters of the crack band model are uniquely related. If one of them is specified, the other one may then be calculated. The same is true of the R-curve for specimens of given shape.

5. Exploiting the size effect law has important advantages: Statistically scattered measurements are smoothed with a known law permitting linear regression (Fig. 1b). The range of the test data is extended, thus reducing ambiguity of data fitting and uncertainty in the material parameter values. Consequently, the experimentalist can get by with fewer tests covering a narrower range of conditions. Data smoothing enables constructing an envelope. A simpler measurement

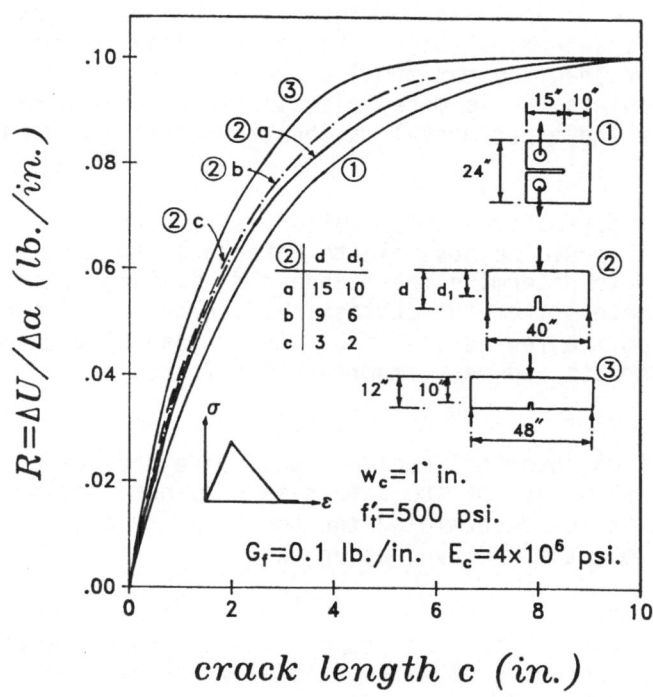

Fig. 15 - R-Curves Calculated by Crack Band Model for Same
Material Parameters and Different Specimens.

procedure than with the existing methods is made possible.
Only maximum load values are needed. They can be obtained even
in a laboratory with the most rudimentary equipment. There is
no need to measure the crack length, which avoids the ambiguity
in defining the location of the crack tip and the difficulty of
its observation. No measurement of unloading or reloading
compliance is needed. Since the size effect law for blunt
fracture is applicable to diverse materials such as concrete,
rock, or ductile metals [6], the present method of
identification of material parameters should be applicable to
all these materials as well.

6. When the purpose of applying nonlinear fracture mechanics in
   practice is the determination of the maximum load for monotonic
   loading, rather than the maximum load after a series of
   previous unloadings, it is more relistic to use only maximum
   loads also for the experimental calibration of the mathematical
   model.

7. Since material parameter identification is reduced to linear
   regressions, it would be possible to introduce statistics, and
   in particular, to determine the standard deviations of the
   material parameters from the statistical characteristics of the
   deviations from the regression line. As is generally agreed,
   statistical aspects are very important for fracture
   predictions.

8. Measuring maximum loads of specimens that have notches of
   various lengths but are of the same size does not provide
   sufficient basis for determining the nonlinear fracture
   properties, e.g., the R-curve, accurately.

Acknowledgment. - Partial financial support under AFOSR Grant No.
830009 to Northwestern University is gratefully acknowledged.
Mary Hill deserves thanks for her dedicated secretarial assistance.

REFERENCES

1.   ASCE State-of-the-Art Report on "Finite Element Analysis of
     Reinforced Concrete," Prepared by a Task Committee chaired by
     A. Nilson, Am. Soc. of Civil Engrs., New York, 1982.

2.   Barenblatt, G. I., "Similarity, Self-Similarity and
     Intermediate Asymptotics," Consultants Bureau, New York 1979
     (transl. from Russian).

3.   Bazant, Z. P., "Crack Band Model for Fracture of
     Geomaterials," Proc., 4th Intern. Conf. on Numerical Methods
     in Geomechanics, held in Edmonton, Alberta, Canada, June 1982,
     ed. by Z. Eisenstein, Vol. 3.

4.   Bazant, Z. P., "Fracture in Concrete and Reinforced Concrete,"
     Chapter 4 in "Mechanics of Geomaterials: Rocks, Concretes,
     Soils," John Wiley & Sons, London, in press (Proc. of IUTAM W.
     Prager Symposium held at Northwestern University, Sept. 1983).

5.   Bazant, Z. P., "Imbricate Continuum and Its Variational
     Formulation," Journal of Engineering Mechanics ASCE, Vol. 110,
     No. 12, Dec. 1984, pp. 1693-1712.

6.   Bazant, Z. P., "Size Effect in Blunt Fracture: Concrete,
     Rock, Metal," J of Engineering Mechanics, ASCE, Vol. 110,
     1984, pp. 518-535.

7.   Bazant, Z. P., Belytschko, T. B., and Chang, T. P., "Continuum
     Theory for Strain Softening," Journal of Engineering Mechanics
     ASCE, Vol. 110, No. 12, Dec. 1984, pp. 1666-1692.

8.   Bazant, Z. P., and Cedolin, L., "Blunt Crack Band Propagation
     in Finite Element Analysis," Journal of the Engineering
     Mechanics Division, ASCE, Vol. 105, No. EM2, April 1979, pp.
     297-315.

9.   Bazant, Z. P., and Cedolin, L., "Fracture Mechanics of
     Reinforced Concrete," Journal of the Engineering Mechanics
     Division, ASCE, Vol. 106, No. EM6, Proc. Paper 15917, December
     1980, pp. 1287-1306; with Discussion and Closure in Vol. 108,
     1982, EM., pp. 364-471.

10. Bazant, Z. P., and Cedolin, L., "Finite Element Modeling of
     Crack Band Propagation," Journal of Structural Engineering,
     ASCE, Vol. 109, No. ST2, Feb. 1983, pp. 69-92.

11. Bazant, Z. P., and Cedolin, L., "Approximate Linear Analysis of Concrete Fracture by R-Curves,' Journal of Structural Engineering, ASCE, Vol. 110, 1984, pp. 1336-1355.

12. Bazant, Z. P., "Imbricate Continuum and Progressive Fracturing of Concrete and Geomaterials," Meccanica (Italy), Vol. 19, 1984, pp. 86-93.

13. Bazant, Z. P., "Numerical Simulation of Progressive Fracture in Concrete Structures: Recent Developments," Proceedings of the Sept. 1984 International Conference on "Computer-Aided Analysis and Design of Concrete Structures, held in Split, Yugoslavia, ed. by E. Hinton, R. Owen and F. Damjanic, Univ. of Wales, Swansea, U.K., Riveridge Press, Swansea, pp. 1-18.

14. Bazant, Z. P., and Estenssoro, L. F., "Surface Singularity and Crack Propagation," International Journal of Solids and Structures, Vol. 15, 1979, pp. 405-426 and Vol. 16, 1980, pp. 479-481.

15. Bazant, Z. P., and Kim, J. K., "Size Effect in Shear Failure of Longitudinally Reinforced Beams," Amer. Concrete Inst. Journal, Vol. 81, No. 5, 1984, pp. 456-468.

16. Bazant, Z. P., and Oh, B. H., "Crack Band Theory for Fracture of Concrete," Materiaux et Constructions (Materials and Structures) RILEM, Paris, Vol. 16, 1983, pp. 155-177.

17. Bazant, Z. P., and Oh, B. H., "Rock Fracture via Strain-Softening Finite Elements," Journal of Engineering Mechanics ASCE, Vol. 110, No. 7, July 1984, pp. 1015-1035.

18. Bazant, Z. P., and Oh, B. H., "Microplane Model for Fracture Analysis of Concrete Structures," Proc. Symp. on the "Interaction of Nonnuclear Munitions with Structures," U. S. Air Force Academy, Colorado Springs, May 1983, pp. 49-55.

19. Broek, D., "Elementary Engineering Fracture Mechanics," Noordhoff International Publishing, Leyden, Netherlands, 1974.

20. Eringen, A. C., and Edelen, D. C. B., "On Nonlocal Elasticity," International Journal of Engineering Science, Vol. 10, 1972, pp. 233-248.

21. Eringen, A. C., and Ari, N., "Nonlocal Stress Field at Griffith Crack," Cryst. Latt. Def. and Amorph. Mat., Vol. 10, 1983, pp. 33-38.

22. Evans, R. H., and Marathe, M. S., "Microcracking and Stress-Strain Curves for Concrete in Tension," Materials and Stuctures (RILEM, Paris), No. 1, Jan.-Feb. 1968, pp. 61-64.

23. Hillerborg, A., Modeer, M., and Petersson, P. E., "Analysis of Crack Formation and Crack Growth in Concrete by Means of Fracture Mechanics and Finite Elements," Cement and Concrete Research, Vol. 6, 1967, pp. 773-782.

24. Irwin, G. R., Report of a Special Committee, "Fracture Testing of High Strength Sheet Material," ASTM Bulletin, Jan. 1960, p. 29 (also G. R. Irwin "Fracture Testing of High Strength Sheet Materials under Conditions Appropriate for Stress Analysis," Report No. 5486, Naval Research Laboratory, July 1960).

25. Jenq, Y. S., and Shah, S. P., "Nonlinear Fracture Parameters for Cement Based Composites: Theory and Experiments", Preprints, NATO Advanced Research Workshop on "Applications of Fracture Mechanics to Cementitious Composites," ed. by S. P. Shah, Northwestern Univeristy, Evanston, IL. 60201, U.S.A., Sept. 4-7, 1984, pp. 213-253.

26. Krafft, J. M., Sullivan, A.M., Boyle, R. W., "Effect of Dimensions on Fast Fracture Instability of Notched Sheets," Cranfield Symposium 1961, Vol. 1, pp. 8-28.

27. Kroner, E., "Elasticity Theory of Materials with Long-Range Cohesive Forces," International Journal of Solids Structures, Vol. 3, 1967, pp. 731-742.

28. Kroner, E., "Interrelations Between Various Branches of Continuum Mechanics," Mechanics of Generalized Continua, ed. by E. Kroner, Springer-Verlag, 1968, pp. 330-340.

29. Krumhansl, J. A., "Some Considerations for the Relation Between Solid State Physics and Generalized Continuum Mechanics," Mechanics of Generalized Continua, ed. by E. Kroner, Springer-Verlag, 1968, pp. 298-311.

30. Kunin, I. A., "The Theory of Elastic Media With Microstructure and the Theory of Dislocations," Mechanics of Generalized Continua, ed. by E. Kroner, Springer-Verlag, 1968, pp. 321-328.

31. Levin, V. M., "The Relation Between Mathematical Expectation of Stress and Strain Tensors in Elastic Microheterogeneous Media," Prikladnaya Matematika i Mekhanika, Vol. 35, 1971, pp. 694-701 (in Russian).

32. Mai, Y. W., "Fracture Measurements of Cementitious Composites," Preprints, NATO Advanced Research Workshop on "Applications of Fracture Mechanics to Cementitious Composites," ed. by S. P. Shah, Northwestern University, Evanston, Il., Sept. 1984, pp. 289-319.

33. Petersson, P. E., "Fracture Energy of Concrete," Cement and Concrete Research, Vol. 10, 1980, pp. 78-89 and 91-101.

34. Petersson, P. C., "Crack Growth and Development of Fracture Zones in Plain Concrete and Similar Materials," Doctoral Dissertation, Lund Institute of Technology, Lund, Sweden, Dec. 1981.

35. Rashid, Y. R., "Analysis of Prestressed Concrete Pressure Vessels," Nuclear Engng. and Design, Vol. 7, No. 4, April 1968, pp. 334-344.

36. Reinhardt, H. W., and Cornelissen, H. A. W., "Post-Peak Cyclic Behavior of Concrete in Uniaxial Tensile and Alternating Tensile and Compressive Loading," Cement and Concrete Research, Vol. 14, 1984, pp. 263-270.

37. Tada, H., Paris, P. C., and Irwin, G. R., "The Stress Analysis of Cracks Handbook," Del Research Corp., Hellertown, Pa., 1973.

38. Tattersall, H. G., and Tappin, G., "The Work of Fracture and Its Measurement in Metals, Ceramics and Other Materials," Journal of Materials Science, Vol. 1, 1966, pp. 296-301.

39. Velazco, G., Visalvanich, K., and Shah, S. P., "Fracture Behavior and Analysis of Fiber Reinforced Concrete Beams," Cement and Concrete Research, Vol. 110, pp. 41-51, 1980.

40. Wecharatana, M., and Shah, S. P., "Slow Crack Growth in Cement Composites," Journal of the Structural Division, ASCE, Vol. 108, June 1982, pp. 1400-1413.

41. Wittmann, F. H. (Editor), "Fracture Mechanics of Concrete," Elsevier, Netherlands, 1983.

APPLICATION OF FRACTURE MECHANICS
TO CEMENTITIOUS COMPOSITES
NATO-ARW - September 4-7, 1984
Northwestern University, U.S.A.
S. P. Shah, Editor

NON-LINEAR FRACTURE MODELS
FOR DISCRETE CRACK PROPAGATION

Anthony R. Ingraffea, Associate Professor
Walter H. Gerstle, Graduate Research Assistant

Department of Structural Engineering
Cornell University, Ithaca, New York, U.S.A.

1. INTRODUCTION

The proper fracture mechanics to be applied to crack propagation in concrete is determined by scale effects. To ascertain whether linear elastic fracture mechanics (LEFM) or a non-linear approach is more applicable to a particular problem, one must answer the following question:

- How large is the process zone compared to the smallest critical dimension of the structure under consideration?

The purpose of this paper is to address this question from the particular point-of-view of discrete representation of a crack in a finite element model. Before proceeding to outline the paper's approach and methods, it is essential that key terms used in the above question be defined and that important assumptions be stated:

- The process zone is that area accompanying crack propagation in which inelastic material response is occurring.

- The term crack is not used here in its classical sense, as a complete discontinuity in both traction and displacement fields. Rather, it is used to describe an effective crack which consists of a length of true crack (in the classical sense) preceded by its process zone.

- The critical dimension might be the length of the crack itself, including its process zone, or, if it is smaller, the distance from the true crack tip to the nearest free surface or reinforcing bar.

247

These definitions are in the spirit of the approach to the fracture mechanics of metals first proposed by Irwin (1). In fact, throughout this paper comparisons and analogies will be drawn between the well-accepted formulations for process zone size in ductile metals and the estimates derived here for brittle non-metallics.

The basis for any process zone size estimate, whether it be for a zone of plastic deformation in a metal or for some form of inelastic response in concrete, is a constitutive model for the material in this zone. Whereas for metals a model might be composed of the von Mises yield criterion, to relate the effective stress to all the principal stresses, and a normality flow rule, to relate the effective stresses to elastic and plastic strains, a much simpler, yet analogous, constitutive model will be used in this paper. To describe the inelastic behavior in the process zone of a crack propagating in concrete the following assumptions are made:

1. The only constitutive modeling required for process zone description in pure Mode I is the stress-versus-crack-opening-displacement (COD) relation which can be obtained from a displacement-controlled direct tension test (2). This relation is, in fact, the post-peak stress-COD curve measured in such a test. A range of such process zone softening models used in the present analyses is shown in Figure 1.

2. The previous assumption implies that normal stress continues to be transferred across a displacement discontinuity which may or may not be visible to the naked eye. It is assumed that this stress transfer is due to aggregate bridging and the undulating, three-dimensional nature of the opposing crack surfaces (3).

3. It is assumed that the process zone localizes, due to the rapid softening behavior shown in the models of Figure 1, into a very narrow band ahead of the true crack tip. In fact, for the purposes of the present finite element analysis, all softening is confined to one-dimensional interface elements lying in the crack plane ahead of the true crack tip (4).

4. Although for metals the process zone size is especially influenced by the principal stress parallel to the true crack front, it is assumed that this stress has no influence on the process zone in geomaterials such as concrete and rock (5,6).

Schematically, these assumptions combine to paint the picture shown in Figure 2. The constitutive model is defined by the direct

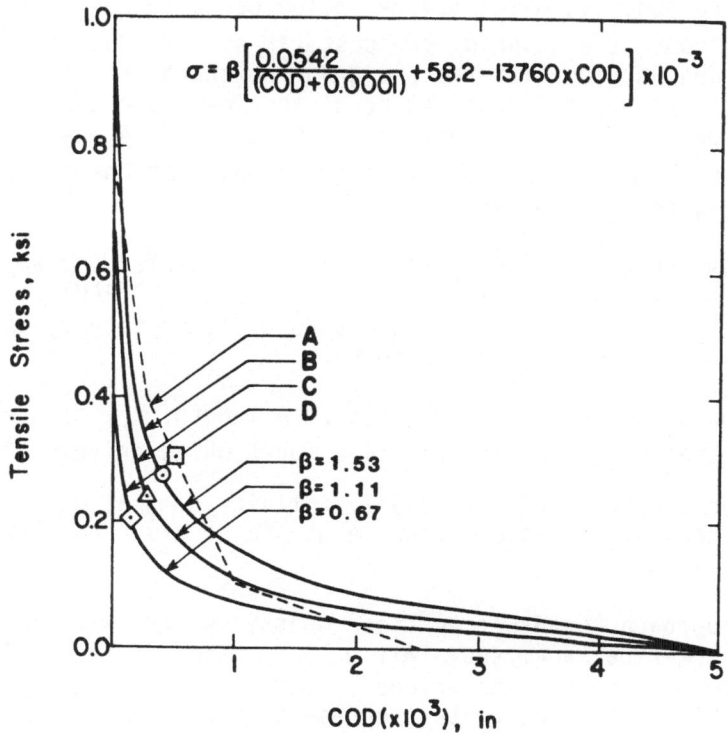

Figure 1. Various constitutive models for a discrete representation of the process zone in concrete.

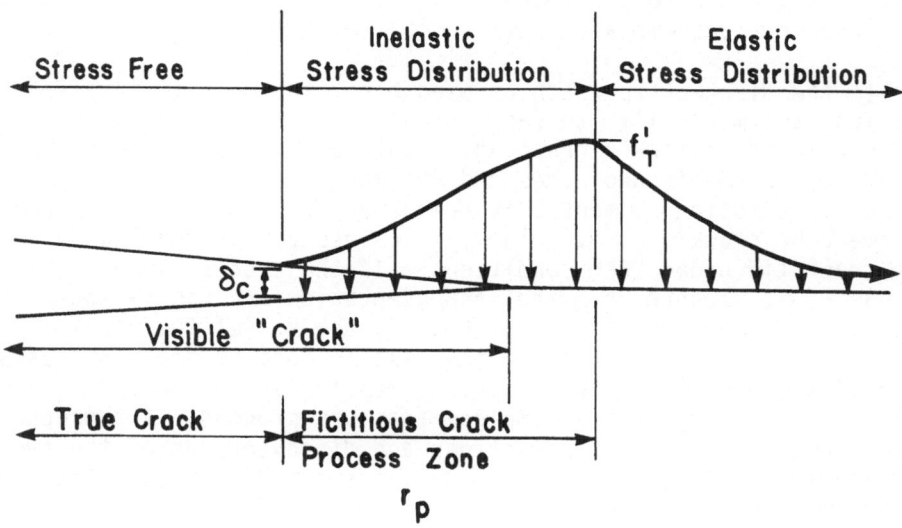

Figure 2. A schematic of the hypothesized process zone in concrete.

tensile strength, f_T', which may be influenced by stress normal to the crack front, the shape of the postpeak stress-COD relationship, which, in conjunction with the COD-gradient characteristic of the problem at hand, produces the shape of the inelastic stress distribution, and finally, the characteristic COD, δ_c, which will occur at the true crack tip. Given these observations and assumptions, it is natural to ask:

- How large must the critical dimension be for the application of linear elastic fracture mechanics (LEFM) to be valid? Or, from the perspective of assumption 3, above, how long is the process zone?

- How sensitive is the process zone length, r_p, to structural geometry and the constitutive model which drives it?

- For a problem of effectively infinite domain, what is the steady-state process zone length for a given constitutive model?

The approach to answering these questions here will be a series of numerical analyses. All will be based on the well-known finite element method. The unique feature of all the analyses, however, will be the "discrete" representation of the crack in the mesh. The current state-of-the-practice for representing a crack in a finite element mesh is the "smeared" approach, originally proposed by Rashid (7). In that approach, the constitutive model is used to simulate the cracking process: the mesh is not changed as arbitrary cracking progresses, nor is a crack with a trajectory known a priori accommodated with special meshing.

In the discrete approach as used here, the mesh is itself modified, automatically, to represent the cracking process (8,9). In the case of arbitrary cracking, local remeshing, the introduction of new elements and nodes and the modification of some previously existing elements, is performed for each crack at each increment of cracking. Singularity elements are introduced at a true crack tip under LEFM conditions. If the process zone, because of its length, requires representation, it too is modeled discretely as mentioned in assumption 3, above. If the crack trajectory is known, as will be the case in most of the test cases analyses to be reported here, special meshing is introduced to facilitate modeling of the complete fracture process. A thorough discussion of the general advantages and disadvantages of the two approaches is given, in an historical perspective, by Ingraffea and Saouma (9).

The differences between the "discrete" and "smeared" approaches are stark when true crack modeling is performed. However, the distinctions begin to blur when a process zone is being represented. It will be shown here that, under this condition, the only difference between the approaches lies in assumption 3, above.

Before beginning presentation of the finite element analyses which address the questions presented above, it is necessary to begin more simply, with hand calculations. These will enable one to relate process zone effects in strain-softening geomaterials to parallel processes in metals. Further they will offer a simple check for the acceptability of the finite element calculations to follow.

2. PROCESS ZONE SIZE: SOME QUALITATIVE ASSESSMENTS

Figure 3, a complete stress-strain curve from a strain-controlled, direct-tension test on a strain-softening material, is the basis for the introductory calculations of this section. The constitutitve models shown in Figure 1 can be derived from such a test. Evans and Marathe were first to observe (10), that, once the fracturing process of the test specimen has begun, nearly all the strain is due to the formation of a crack. That is, if the gage length were the specimen length, then the measured strain would be approximately equal to the COD divided by the gage length.

Alternatively, one can groove or so shape the test specimen so that the location of the fracture plane is known. The COD can then be measured directly (2) as a function of applied stress.

For comparison, let's assume that two tests of the type shown schematically in Figure 3 are performed, one on a concrete specimen leading to the response shown, the other on a metal specimen, with uniaxial yield stress f_Y, exhibiting elastic-perfectly plastic behavior. We shall first seek to compare measures of process zone size for these materials based on very simple assumptions.

2.1 Comparisons of Strain-Softening Versus Perfectly-Plastic Behavior in the Process Zone

Focus first on the region ahead of a crack tip in the metal. It is straightforward to show that a second-order measure of process zone size for the metal r_{p_m}, is (11)

$$r_{p_m} = \frac{1}{\pi} \left(\frac{K_{Ic}}{f_Y}\right)^2 \tag{1}$$

where,

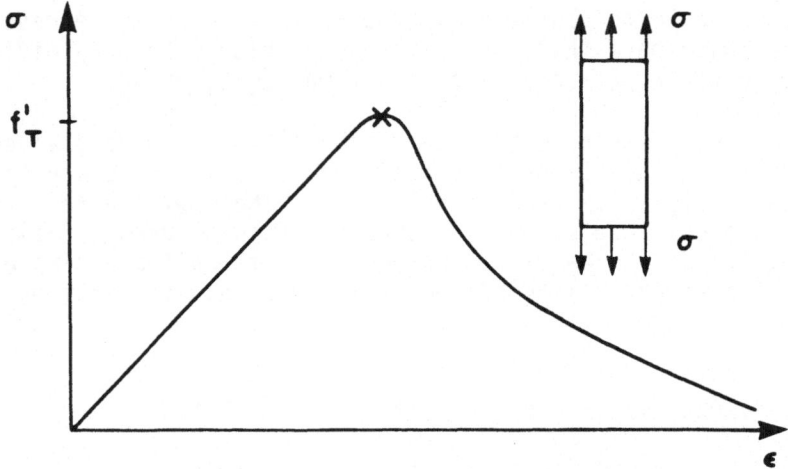

Figure 3. Idealized stress-strain curve for concrete loaded in uniaxial tension under strain control.

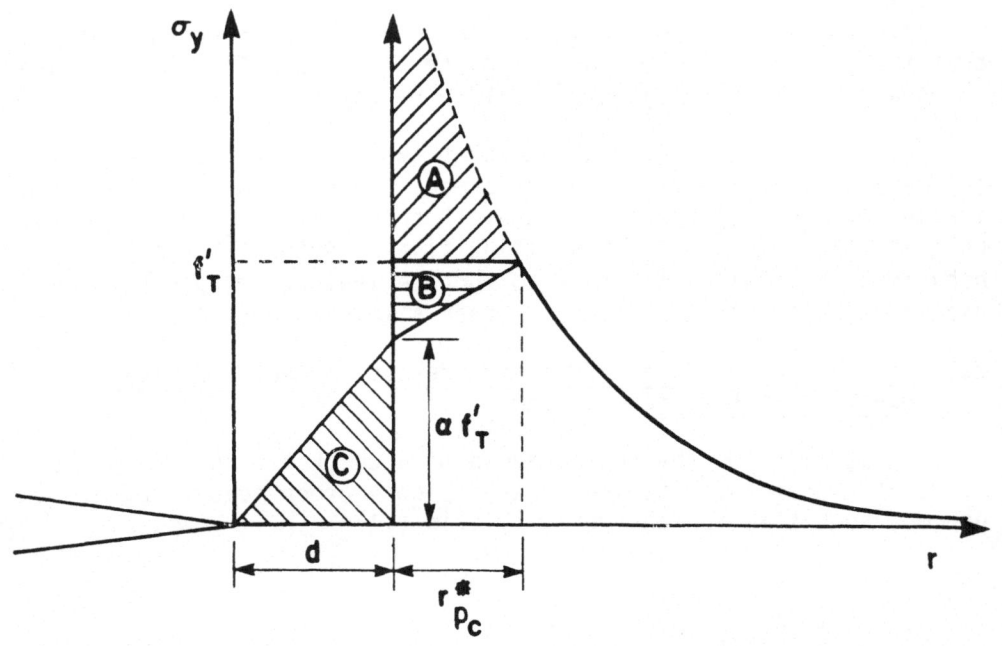

Figure 4. Hypothesized stress redistribution ahead of a true crack tip in concrete. For equilibrium, area C must equal area A plus area B.

K_{Ic} = plane strain fracture toughness

This measure is the process zone length along the crack direction. It assumes that the effective yield stress is the uniaxial yield stress, that is, stress multiaxiality is neglected, and that approximate stress redistribution has occurred to satisfy equilibrium in the y-direction (see Figure 4 for coordinate system).

Focus now on Figure 4, a simplified schematic of the region ahead of a true crack tip in concrete. The first difference which arises between this and the analogous situation in the elastic, perfectly-plastic metal is that one does not know the shape of the inelastic stress distribution in the process zone because the COD profile is problem dependent. However, to continue the analysis, we will assume a bilinear distribution as shown, with α as a shape parameter, and seek to find the concrete's process zone length, r_{p_c}, from

$$r_{p_c} = d + r^*_{p_c} \tag{2}$$

where,

$$r^*_{p_c} = \frac{1}{2\pi} \left(\frac{K_{Ic}}{f'_T}\right)^2 \tag{3}$$

It can be easily shown (11) that $r^*_{p_c}$ is a first-order measure of process zone length which does not account for any stress redistribution. To satisfy equilibrium in the y-direction it is necessary that stress redistribution occur such that,

$$\frac{1}{2} d\alpha f'_T = \int_0^{r^*_{p_c}} \frac{K_I}{\sqrt{2\pi r}} \, dr - f'_T r^*_{p_c} + \frac{1}{2}(1-\alpha) f'_T r^*_{p_c} \tag{4}$$

Solving Equation (2) for d yields,

$$d = r^*_{p_c} \left(\frac{3}{\alpha} - 1\right) \tag{5}$$

It is interesting to compare now the second order process zone size estimates for the two materials using Equations (1) and (2) and (5). The comparison is done in Table 1 and it reveals the second

Table 1

Comparison of Process Zone Size Measures for Concrete, r_{p_c}, and an Elastic, Perfectly-Plastic Metal, r_{p_m}

α	$d/r^*_{p_c}$	$r_p/r^*_{p_c}$	r_{p_c}/r_{p_m}
0.05	59	60	30
0.20	14	15	7.5
0.50	5	6	3
0.75	3	4	2
1.00	2	3	1.5

difference in process zone characteristics of the two materials: when the y-stress is allowed to decay as the true crack tip is approached, r_{p_c} becomes significantly larger than r_{p_m}. Some representative inelastic stress distributions are plotted in Figure 5 in terms of α. Again, it is important to emphasize that the rate of decay, as indicated by α, will depend on the shape of the constitutive curve and on the COD profile characteristic of the particular structural geometry and boundary conditions under consideration.

As an example of the interaction of COD profile and constitutive model, consider the following example. Assume the process zone material model shown in Figure 6. Let us now solve for the COD profile in the inelastic zone. First, write the COD profile in terms of the stress distribution using Figure 6,

$$COD(\sigma) = \delta_c - \frac{\delta_c}{f'_T} \sigma(r) \tag{6}$$

Next, using the bilinear stress representation shown in Figure 4, solve for the COD in terms of the r-coordinate,

For $0 \leqslant r \leqslant d$,

$$\sigma(r) = \frac{\alpha f'_T}{\left(\frac{3}{\alpha} - 1\right) r^*_{p_c}} r \tag{7}$$

and,

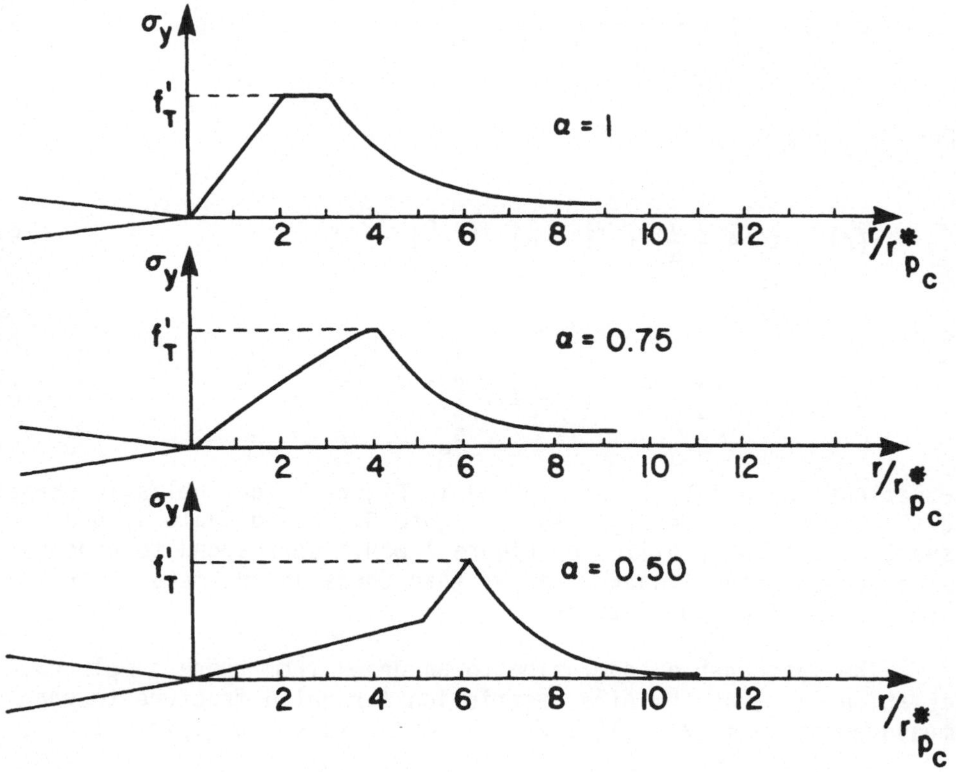

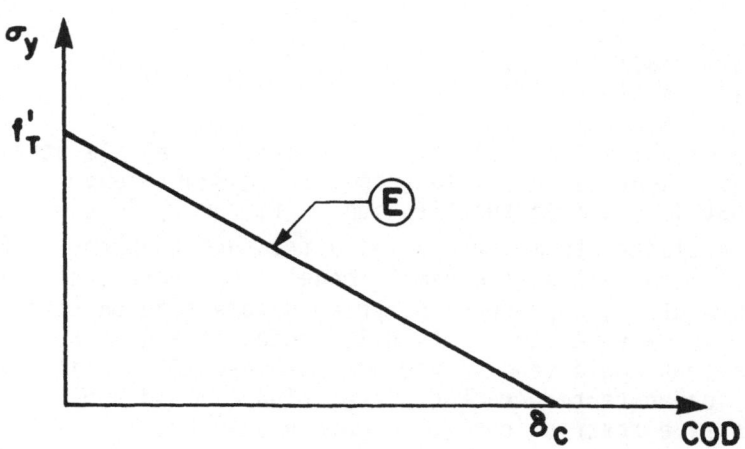

Figure 5. Representative stress distributions ahead of the true crack tip for various values of α.

Figure 6. A simple constitutive model for the discrete process zone.

$$COD(r) = \delta_c\left[1 - \frac{\alpha}{(\frac{3}{\alpha} - 1)} \left(\frac{r}{r^*_{p_c}}\right)\right] \tag{8}$$

For $d \leq r \leq r_{p_c}$,

$$\sigma(r) = f'_T\left(4 - \frac{3}{\alpha}\right) + (1 - \alpha)\, f'_T\left(\frac{r}{r^*_{p_c}}\right) \tag{9}$$

and,

$$COD(r) = \delta_c\left[\frac{3}{\alpha} - 3 + (\alpha - 1)\left(\frac{r}{r^*_{p_c}}\right)\right] \tag{10}$$

Equations (8) and (10) are plotted in Figure 7 for inelastic stress distributions previously seen in Figure 5. Using Table 1, one can see that the COD-profiles of Figure 7 would correspond to process zone lengths significantly larger than those in the metal with the same order of approximations.

The inescapable conclusion to be drawn from these simple calculations is that the size restriction for valid fracture toughness measurement in metals (12), written in terms of multiples of r_{p_m},

$$a, W-a > 2.5\left(\frac{K_{Ic}}{f_Y}\right)^2 \tag{11}$$

where,

    a = crack length,
    W-a = remaining ligament

is unconservative for materials which exhibit tensile strain softening. That is, crack lengths and ligaments longer than would be indicated by simple substitution of f'_T for f_Y in Equation (11) will be necessary to measure a valid fracture toughness. Further, and a bit more subtly, the simple models used here indicate that there is a strong dependence of process zone size on specimen type. For example, assume the constitutive model of Figure 6. For a specimen that would tend to produce a linear COD-profile, such as a wedge loaded center-cracked plate, Figure 7 and Table 1 show that specimen size restrictions would more nearly be,

$$a, W-a > 5\left(\frac{K_{Ic}}{f'_T}\right)^2 \tag{12}$$

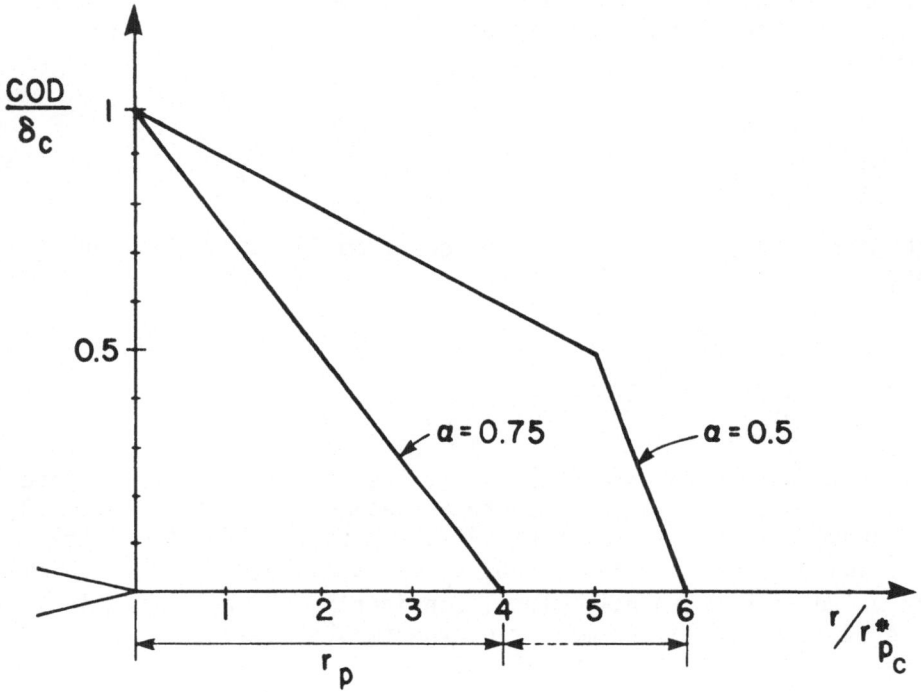

Figure 7. COD profiles for the constitutive model of Figure 6 and the stress distributions shown in Figure 5.

If the COD profile were closer to the elliptical shape exhibited by, say, a center cracked plate loaded in remote tension, the restriction would become,

$$a, \; W-a > 7.5\left(\frac{K_{Ic}}{f'_T}\right)^2 \tag{13}$$

At this point five material characteristics have been introduced: E, the Young's modulus, K_{Ic}, the fracture toughness, and the f'_T and δ_c pair which, together with its shape, define the σ-COD curve. In fact, not all of these parameters are independent. Assume again the constitutive model of Figure 6. Since every material element in the process zone must ride down this curve before it cracks, in the LEFM sense, the critical energy absorption rate is the area under this curve,

$$G_{Ic} = \frac{1}{2} \delta_c f'_T \tag{14}$$

If a test on a large enough specimen were performed, it would yield a critical stress intensity factor related to G_{Ic} by the familiar expression,

$$G_{Ic} = \frac{K_{Ic}^2}{E} \tag{15}$$

with plane stress assumed. It follows from Equations (14) and (15) that,

$$\delta_c = \frac{2K_{Ic}^2}{Ef_T'} \tag{16}$$

It is interesting to note that if a value of critical stress intensity typical of the largest concrete specimens tested to date (13, 14), about 2.5 ksi $\sqrt{in}$ (2.75 MNm$^{-3/2}$), is used in Equation (16), the predicted characteristic COD, δ_c, is about 0.01 inch (0.25 mm). This value is surprisingly close, considering the approximations employed here, to values measured experimentally (2,10), about 0.005 to 0.010 inch (.13 to .25 mm).

In the next section we investigate the implications of one of these approximations on the assumption that the process zone width can be viewed, from the numerical modeling standpoint, as no more than δ_c.

2.2 Stress Biaxiality Effects

Recall that the calculations of the previous section neglected stress multiaxiality in the process zone. Equation (1), for example, is based on a uniaxial yield criterion despite the fact that tensile stress triaxiality theoretically occurs ahead of a crack tip. A third-order measure of process zone length directly ahead of the crack, and an estimate of the shape of the process zone can be obtained by admitting stress multiaxiality into the yield criterion (11).

The same process will be used here to make qualitative assessments of the influence of stress multiaxiality on process zone shape in geomaterials. In what follows no stress redistribution is performed. Rather, we seek only the implications of stress biaxiality on the tensile strength parameter, f_T', in the process zone constitutive model.

Consider the family of curves shown in Figure 8. They are analogous to yield surfaces in plasticity and indicate, from

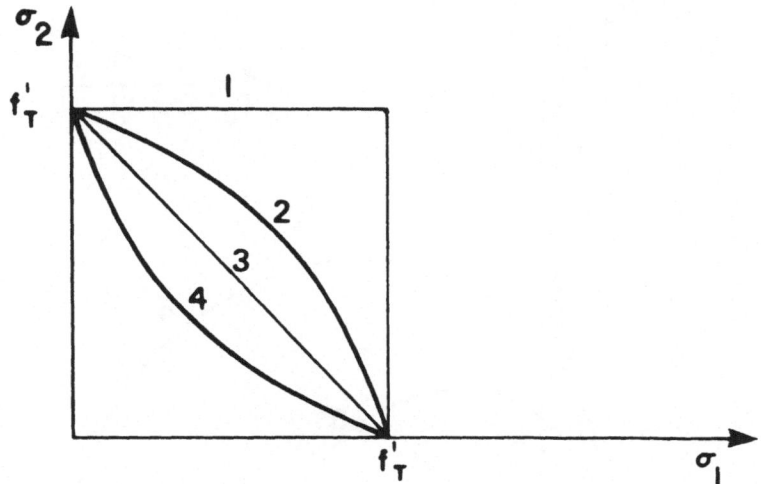

Figure 8. A family of tensile strength interaction curves.

numbers 1 to 4, an increasing degree of principal stress inter-action effect on tensile strength according to,

$$1 \qquad \sigma_1 \quad \text{or} \quad \sigma_2 = f'_T$$

$$2 \qquad \sigma_1^{1.5} + \sigma_2^{1.5} = f'^{1.5}_T$$

$$3 \qquad \sigma_1 + \sigma_2 = f'_T \qquad\qquad (17)$$

$$4 \qquad \sigma_1^{0.7} + \sigma_2^{0.7} = f'_T$$

where,

σ_1, σ_2 = principal stresses in the plane normal to the crack plane

Note that the third principal stress has been neglected for simplicity, although its influence is probably minimal because of the relatively low value of Poisson's ratio for concrete (see assumption 4 in the INTRODUCTION). It can easily be shown that substitution of Equations (14) into the well known expressions in polar coordinates for the principal stresses near a true crack tip (11) leads directly to the envelop curves of Figure 9.

The conclusion we wish to draw from Figure 9 is that, as the degree of biaxial stress influence on tensile strength increases,

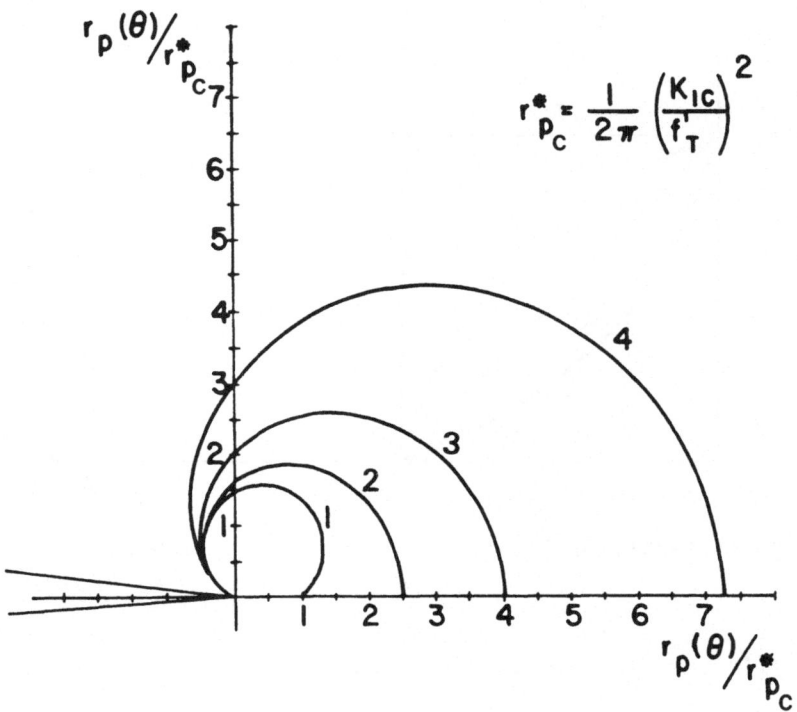

Figure 9. Process zone shapes predicted by LEFM assumptions, no stress redistribution, and the tensile strength interaction curves of Figure 8.

the length of the process zone along the crack direction grows faster than on any other radius. The qualitative implication here is quite different from what one sees in a plastic zone at a crack tip in a metal. It is generally assumed in elasto-plastic fracture mechanics that post-yield, distortional strain energy absorption accounts for the inelastic behavior in the process zone. This energy is absorbed in zones of high shear, well off the crack axis, and produces the familiar kidney-shaped plastic zones ahead of a crack tip in Mode I.

When, however, we posit that the inelastic energy absorption mechanism is a function of the normal stresses, as implied by the constitutive models of Figures 2 and 8, the process zone localizes along the crack axis. This observation is supported experimentally (13,14,15). Attempts to observe process zone shape in mortar and concrete using interferometry (14) and scanning electron microscopy (15), and in rock (13) using epoxy impregnation, have all indicated that the width of the process zone, if at all measurable, is less than its length. In the computer simulations to follow we will take the approach that the process zone width is zero, that the process zone is the "fictitious crack" (18) shown in Figure 2. The inelastic energy sink is the normal traction on this "crack" moving through its COD: every point in this zone rides down a curve like one of those in Figure 1 on its way to becoming part of a true crack surface. For each unit of crack advance, the area under such a curve is the energy absorbed in the cracking process per unit of crack front length.

3. EXAMPLE PROBLEMS

In this section a series of three fracture tests on concrete specimens will be simulated using the discrete cracking, finite element approach and the constitutive models previously described. There are some characteristics common to these tests. These are:

1. Unless otherwise noted, it is assumed that,

$$E = 3 \times 10^3 \text{ ksi} (20.7 \text{ MPa})$$
$$\nu, \text{ Poisson's ratio} = 0.2$$

2. Plane stress is assumed.

3. Simulations are performed with the Finite Element Fracture Analysis Program (FEFAP) (8,9,19). All elements are iso-parametric and of quadratic displacement order.

4. The symbol r_{p_c} is replaced by r_p as it is understood that we refer only to concrete hereafter.

3.1 Example 1: A Very Large Center-Cracked Plate

The first example problem is shown in Figure 10. The structure simulates an infinite plate with a central crack normal to a remote tensile stress, the problem whose stability was first investigated by Griffith (20) in formulating the basis for LEFM. Here, however, we shall assume at first that LEFM is not applicable and perform a non-linear fracture analysis using constitutive model D, shown in Figure 1. The objectives of this simulation are:

1. Obtain the relationship between applied load and the COD at the plate's center to investigate structural stability.

2. Compute COD profiles at various load levels to investigate the relationship among total crack length, $a + r_p$, visible crack length (based on an assumed minimum, visible crack width of 0.001 in. (0.025 mm)), and process zone length, r_p.

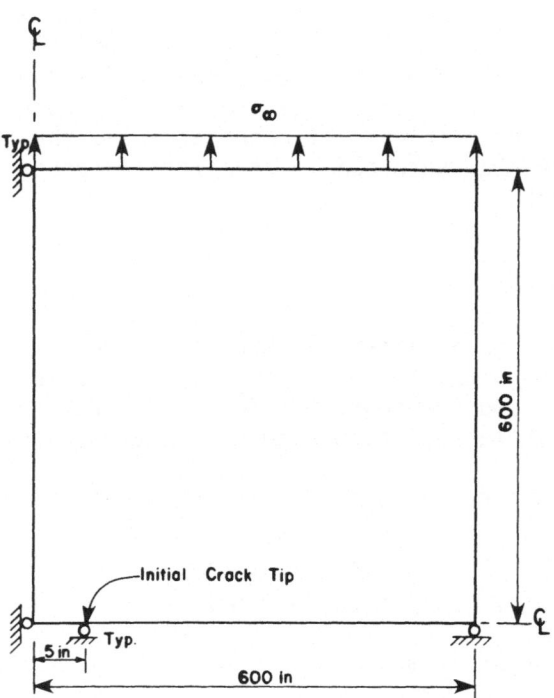

Figure 10. Example Problem #1.

3.   Investigate the evolution of r_p, and ascertain whether it reaches a steady state length.

The meshes used in this simulation are shown in Figures 11 and 12. Figure 11b, a detail of the initial crack region, shows quarter-point singular elements arrayed around the initial, true crack tip, the only place they were employed. A typical global displacement pattern is shown in Figure 13, with the darkened area indicating the process zone.

Results of this simulation are shown in Figures 14 through 16, and they should be studied together. For example, Figures 14 and 15 show that at the peak load, 142 psi (0.98 MPa), softening had occurred for about 5 inches (127 mm) ahead of the initial, true crack tip. Additional visible cracking began to occur between the peak load and the next analysis step at 127 psi (0.88 MPa). The true crack tip did not begin to extend, however, until the load had dropped to about 84 psi (0.58 MPa). True crack length had increased to about 120 inches (3.05 m) by the last analysis step. Simulation was halted at this point because it was felt that, with additional fracturing, the structure would cease to represent an infinite plate to the crack.

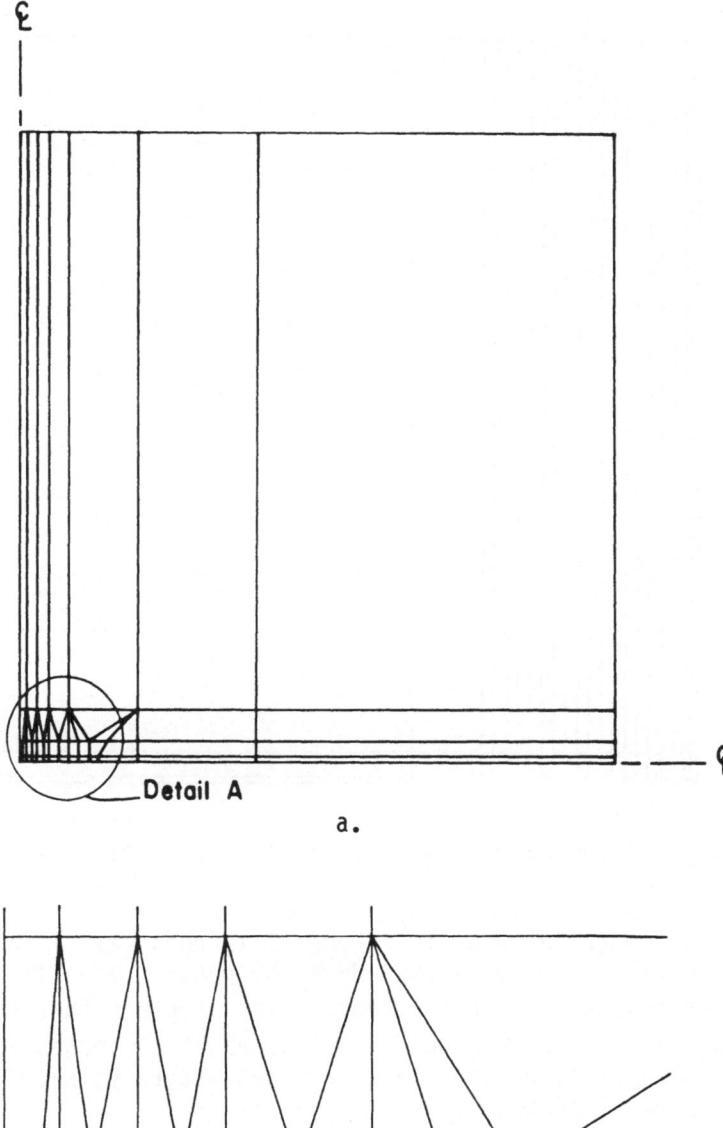

a.

b.

Figure 11. a. First mesh used in Example Problem #1. b. Detail A of Figure 11a.

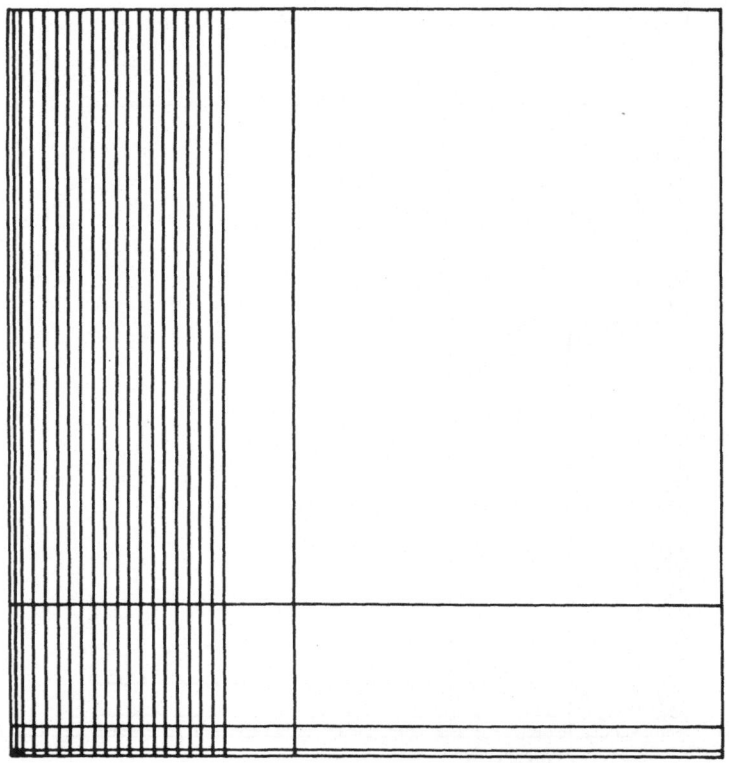

Figure 12. Second mesh used in Example Problem #1.

The information shown in Figure 15 is replotted in a most interesting form in Figure 16. Here the process zone length, r_p, is plotted against true crack length. It is clear that r_p first grows rapidly, reaching a peak value of about 60 inches (1.52 m), and then decays asymptotically to a steady-state value of no more than 36 inches (914 mm).

We are now in a position to make some direct comparisons of the behavior of this simulation with the qualitative predictions of the previous section and with the implications of LEFM. First, is the steady-state r_p of 36 inches reasonable? To answer this question, we must first estimate K_{Ic} for this material. The area under the Model D curve, Figure 1, yields a G_{Ic} of about 0.23 lb/in. (0.04 N/mm). This translates, using Equation (15), into a K_{Ic} of about 830 psi $\sqrt{in}$ (914 MNm$^{-3/2}$). Now, it was previously shown that

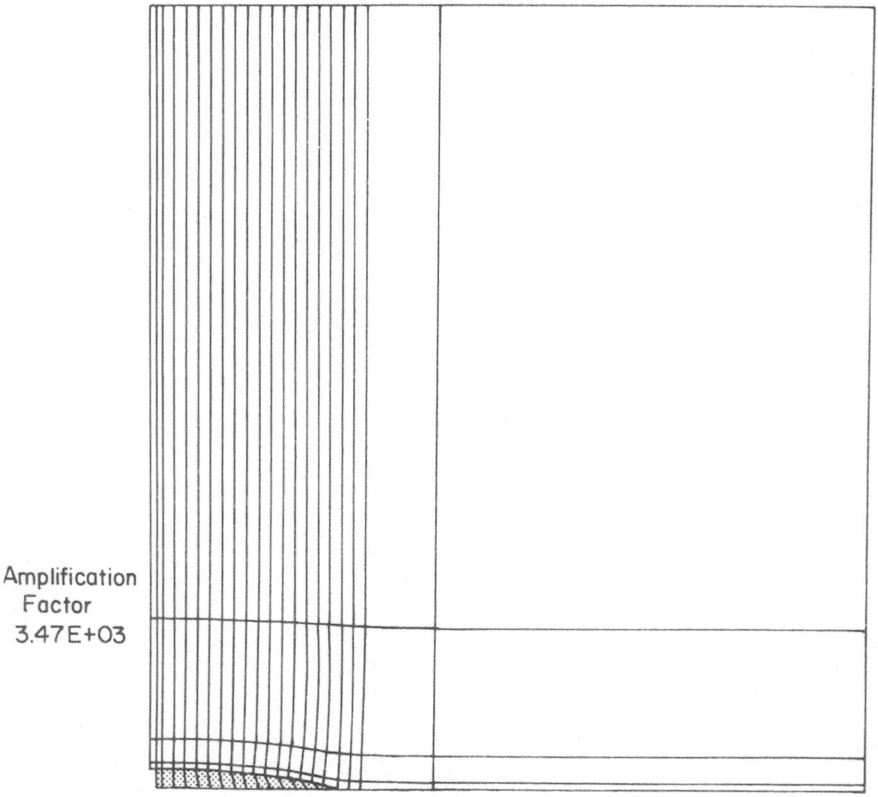

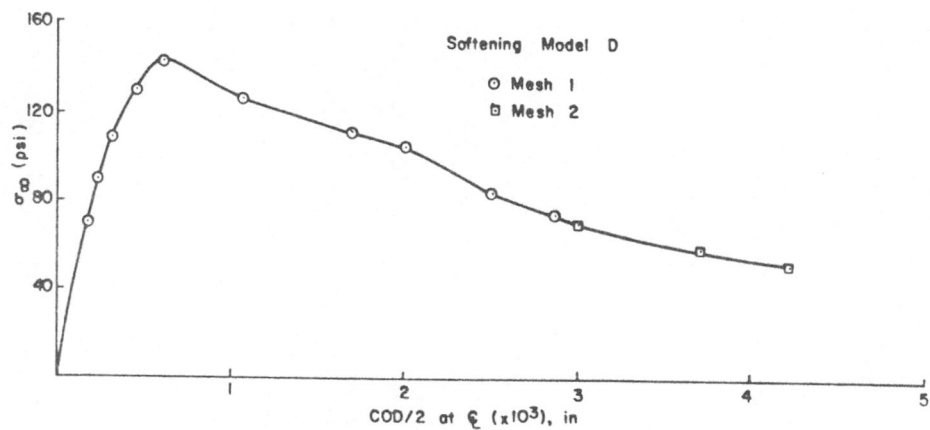

Figure 13. Amplified displaced shape of second mesh at a load
level of σ_∞ = 54 psi (0.38 MPa). Amplication factor 3.47E+03.

Figure 14. Predicted load versus displacement curve for Example
Problem #1.

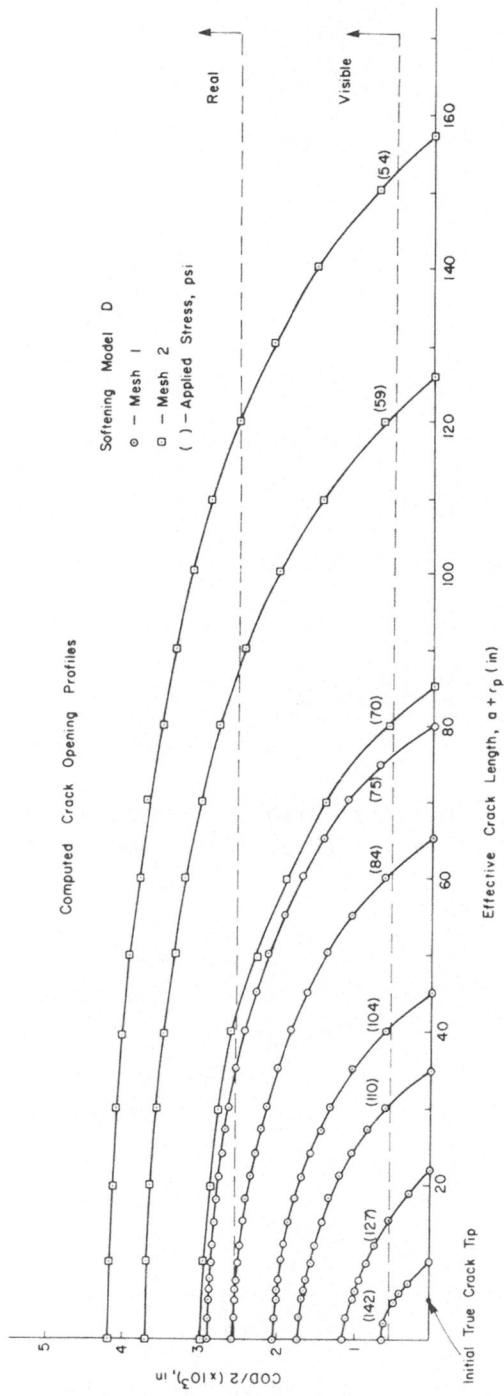

Figure 15. Evolution of computed COD profiles for Example Problem #1.

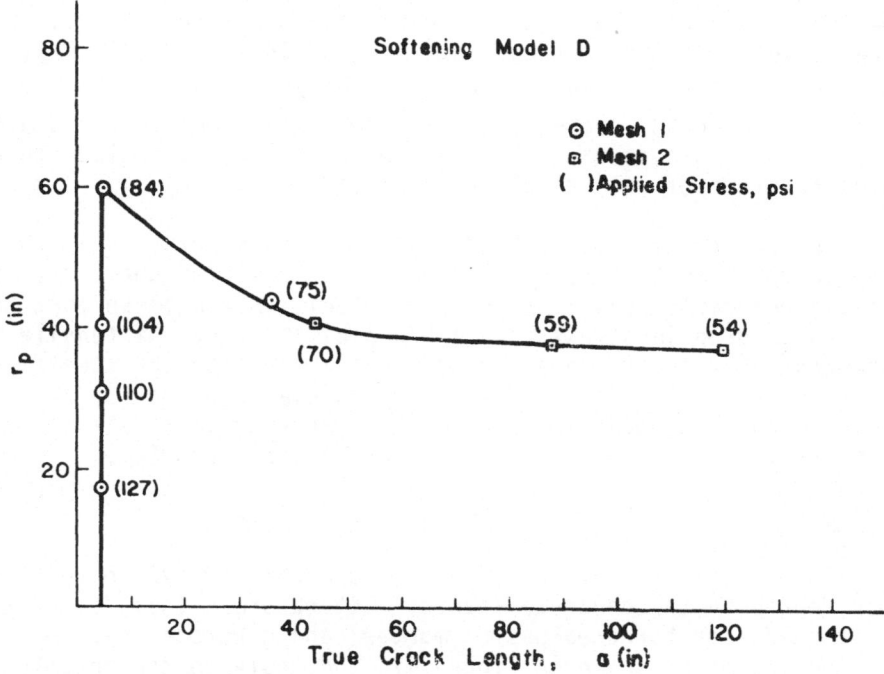

Figure 16. Development of process zone length for Example
Problem #1.

$$r_p \alpha \; (\frac{K_{Ic}}{f'_T})^2 \qquad\qquad (18)$$

With the above estimate of K_{Ic} and the f'_T of Model D, and using the
second-order, plasticity-based estimate, Equation (1), the process
zone length would be computed to be,

$$r_p \simeq 1.37 \text{ inches (34.8 mm).}$$

However, Table 1 shows that, with a softening constitutive model,
this process zone must be longer than this. In fact, it is easy
to show, using the Model D stress-COD curve and the COD profiles
of Figure 16, that the inelastic stress distribution in the process
zone is concave upwards. It falls very steeply from f'_T to less
than one-half f'_T over the first two inches of the process zone.
Table 1 and Figure 5 show that for such distributions the concrete
process zone can be many times the estimate of Equation (19), the
factor of about 26 computed here certainly being believable.

A second question to ask concerning this simulation is: to how long would the true crack have to have grown for LEFM to be acceptable? Assuming a steady-state r_p of 36 inches (914 mm), and a ratio of r_p to true crack length of 25 for acceptability, a growth to about 900 inches (22.9 m) would have been required with Model D. At that point, LEFM would predict a critical load on an infinite plate of only about 16 psi (0.11 MPa).

The key conclusion to be drawn from this simulation is this: despite the fact that concrete seems more "brittle" than, say, an elastic-perfectly plastic structural steel, in the sense that there is a rapid drop in stress carrying capability past the tensile strength, its process zone can be much larger than the steel's. It is neither accurate nor sufficient to use Equation (3) to assess the applicability of LEFM to concrete. Further implications of this conclusion are pursued in the next example problem.

3.2 Example 2: A Finite, Center-Cracked Plate

The experiments of Kesler, Naus, and Lott (21,22) have stirred much debate in the concrete fracture literature. They tested a large number of hardened paste, mortar, and concrete specimens in the configuration shown in Figure 17, and analyzed the results using an approximate stress-intensity factor calibration. Since all their results indicated a strong dependence of apparent toughness, K_Q, on specimen size, they concluded that LEFM is not applicable to these materials.

Recently, however, the first author re-analyzed their test results (23) with a more accurate stress-intensity factor calibration and reached the opposite conclusion, with K_Q being fairly independent of specimen size.

These tests, however, will just not go away. Bazant and Oh (24) have more recently shown that strain readings made near the initial true crack tip during the tests could not be explained using LEFM. What then is the explanation for this contradictory behavior?

One of the concrete test specimens (Test 12, Series LC-2-AD-C of (22)), the same one analyzed by Bazant and Oh (24), is analyzed here with the hope of explaining this contradiction. Further, this example problem seeks these additional objectives:

1. To investigate the difference in response to two constitutive models, D of Figure 1, and E of Figure 6, f_T' and δ_c remaining the same.

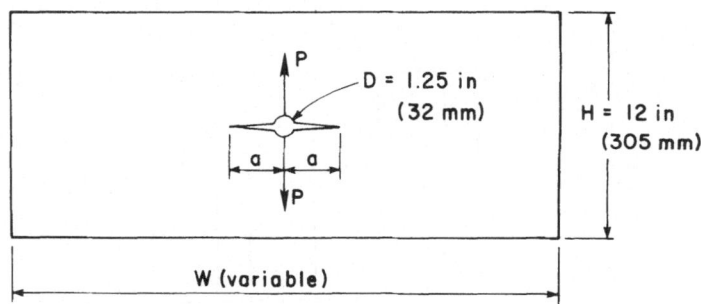

Figure 17. Specimen configuration used in the tests of References 21 and 22. For Example Problem #2, W = 18 inches (457 mm), a = 2.5 inches (63.5 mm).

2. To investigate the shape of the inelastic stress distributions in the process zone on a problem for which the COD profile is theoretically linear, and compare to the previous example in which it was theoretically elliptical.

The mesh used for this simulation is shown, in one of its deflected states, in Figure 18. The linear strain interface (LSI) elements used to model the process zone inelastic stress distribution are clearly seen.

The first result to be shown is the load-displacement response, shown in Figure 19. To interpret this result properly, it is necessary to understand the constitutive models used in the simulations. The σ-COD relationship for the material used in the test is not known. The only material property reported in (22) is the splitting tensile strength, about 620 psi (4.28 MPa) at the age of testing. It is well known that the direct tensile strength is less than the splitting strength, so an f'_T of 400 psi (2.76 MPa) is low but not unreasonable. Tests by Petersson (2) show a strong dependence of both δ_c and the shape of the σ-COD curve on maximum aggregate size, Figure 20. The shape and δ_c of Model D are close to the curve in Figure 20 corresponding to a maximum aggregate size of 2 mm. The maximum aggregate size used in the specimen under consideration, however, was 3/4 inch (19 mm). It is probable that δ_c for such a size is at least 0.008 inch (200 μm), assuming the same bond strength and aggregate angularity as in Petersson's tests. Also, the shape of the post-peak σ-COD curve would be expected to be straighter than the curves shown in Figure 20. Therefore, the δ_c for both models is probably too small, and the shape for Model D much too steep.

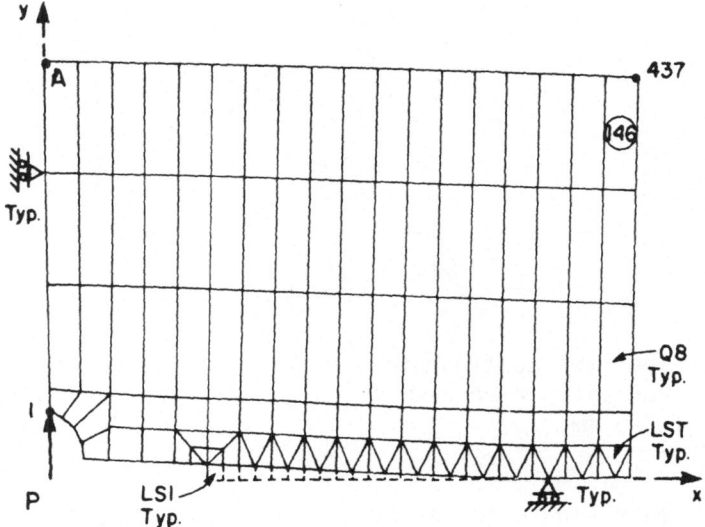

Figure 18. Amplified deflected shape of mesh used in Example Problem #2.

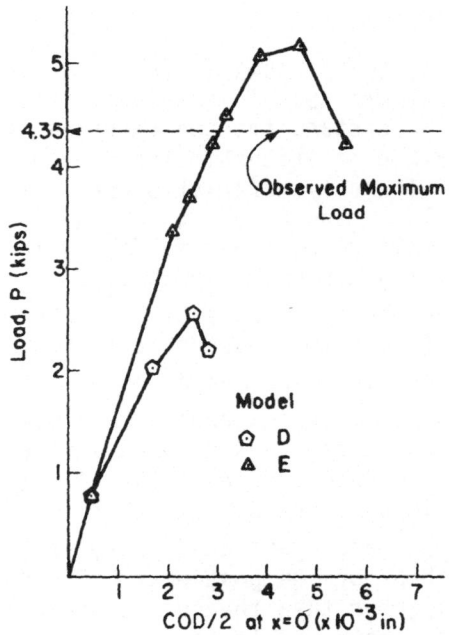

Figure 19. Predicted load-displacement responses for Example Problem #2.

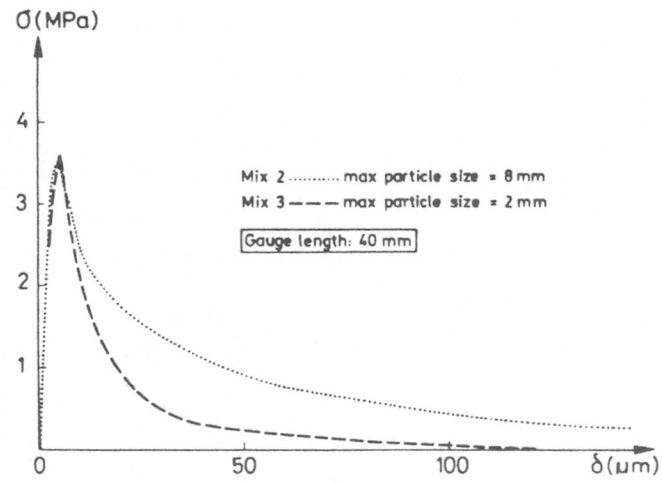

Figure 20. Measured σ-COD relationships for two maximum aggregate sizes. From Ref. 2.

Given these observations, the results shown in Figure 19 are not surprising. With Model D, the process zone unloads much too rapidly, and the peak load is underestimated by over 40 percent. Model E, underestimating toughness because of too small a δ_c, but overestimating because of its linear shape, yields a peak load prediction about 18 percent too high. Note, however, that the ratio of the two peak load predictions is about 2, while the G_{Ic} ratio is about 4. This is exactly what would be expected from an LEFM point of view. But is the specimen actually behaving according to LEFM? Let us next investigate process zone length.

The process zone is described by Figures 21 and 22 for Model D, and 23 and 24 for Model E. Comparisons of Figures 21 with 23 and 22 with 24 show the effect of the shape of the constitutive model for the process zone on its COD profile and inelastic stress distribution, respectively. Under LEFM assumptions, the COD profile for this structure is linear except for the very near crack tip region. For Model D, it is slightly concave downwards, Figure 22, while for Model E it progresses from slightly concave upwards to nearly linear with progressive fracturing, Figure 24. Comparison of Figures 22 and 24 with Figure 15 clearly shows how different specimen loading arrangements can produce markedly different COD profiles even for the same process zone constitutive models. This rigorous numerical observation was predicted by the approximate techniques used earlier in this paper, as in Figure 7.

272

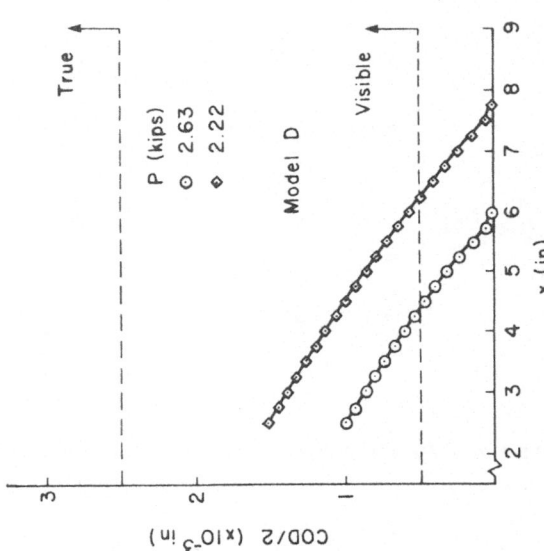

Figure 22. Evolution of COD-profiles
for Example Problem #2, constitutive
Model D.

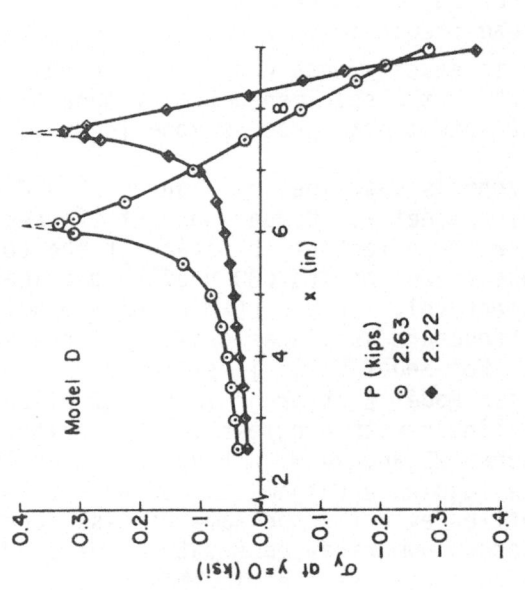

Figure 21. Evolution of normal stress
distribution on remaining ligament for
Example Problem #2, constitutive Model D.

273

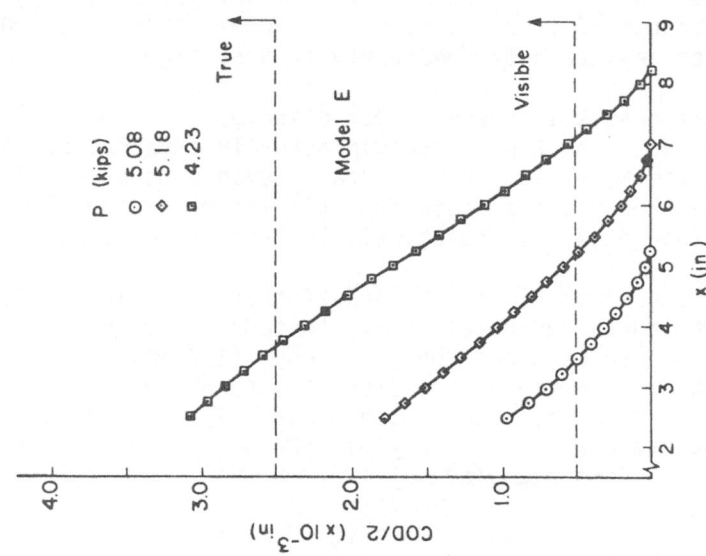

Figure 24. Evolution of COD-profiles for Example Problem #2, constitutive Model E.

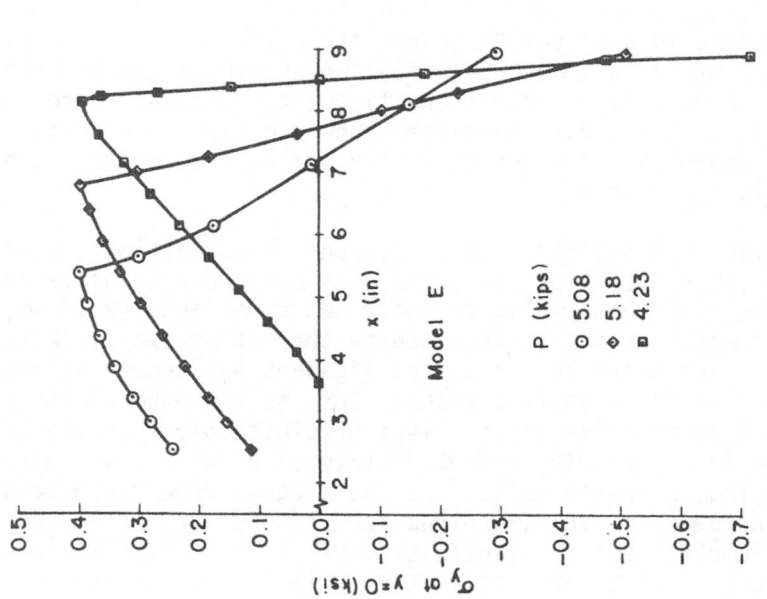

Figure 23. Evolution of normal stress distribution on remaining ligament for Example Problem #2, constitutive Model E.

The direct result of these COD profile differences is the shape of the inelastic stress distribution in the process zone. The distributions for Model D, Figure 21, and Model E, Figure 23 mirror the shapes of these constitutive mdoels since the COD profiles both deviate only moderately from straight lines.

The differences in these stress distributions, however, are significant. The true crack tip actually advanced during the simulation with Model E. No true crack advance occurred with Model D even though the process zone tip was pushed nearly as far. Instead, a long "tail" of relatively low stress evolved.

It is clear from Figures 22 and 24 that no steady-state process zone length had developed in these simulations. In both cases the zone had extended about 5 inches (127 mm) ahead of the true crack tip. The tip of the process zone had arrived about 1 inch (25 mm) from the edge of the plate. This observation suggests an important question: Why did the process zone become so long in the previous example?

The answer is that the effectively infinite extent of the plate in that problem allowed virtually uninhibited stress redistribution ahead of the advancing true crack tip. In the present example, boundary effects are felt almost immediately. Consider the stress distributions shown in Figure 23. The first distribution, for $P = 5.08$ kips (22.6 kPa), shows 3 regimes of stress behavior. Beginning at the initial, true crack tip and extending to about $x = 5.5$ inches (140 mm) is the inelastic stress distribution of the process zone. Starting from the edge of the plate and proceeding in a direction toward the crack to about $x = 6.5$ inches (165 mm) is a linear distribution arising from the bending moment which must exist across the ligament. Between these two distributions is a stress-concentration-like rise, not unlike what one would expect in approaching a true crack tip, as the process zone tip is approached.

However, the distribution corresponding to the last, post-peak load level, $P = 4.23$ kips (18.8 kPa), is considerably different. At this stage in the simulation, only about one inch (25.4 mm) of ligament remains to try to accommodate the latter two regimes just described. The distribution in the ligament has become essentially linear, except for a small distance close to the edge of the plate wherein the compressive stress rises precipitously. In the previous example, no bending moment distribution, with its compressive stress region, needs to exist. As the process zone tip extends there is no need to rapidly unload the process zone as in the present example. Further manifestations of the non-LEFM behavior of this specimen are revealed in the following discussion.

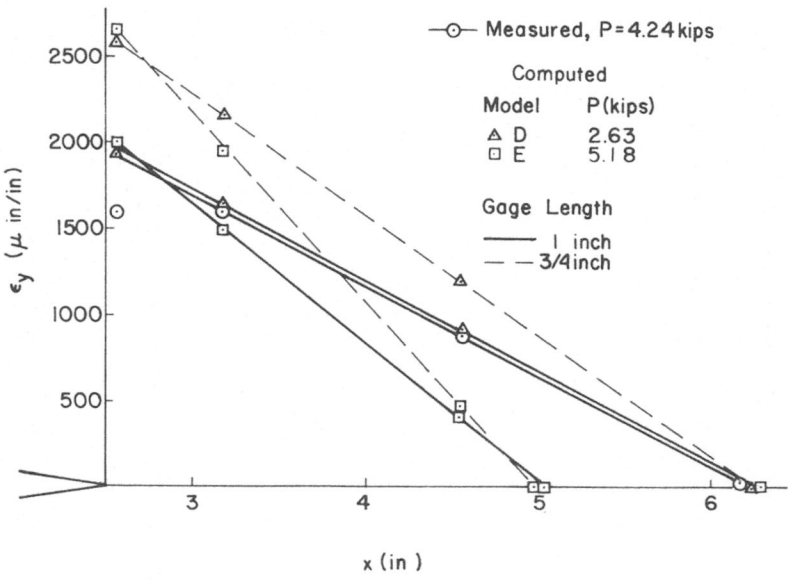

Figure 25.   Comparison of measured and computed strain for Example Problem #2.

As mentioned above, Naus (22) made strain measurements during his tests. Wire strain gages oriented to measure ϵ_y were arrayed ahead of the initial, true crack tip. Strains measured just before specimen failure are shown in Figure 25. Strain is seen to be very high compared to the expected tensile failure value of about 200 μin/in. Further, it is seen to increase <u>linearly</u> except for the last reading close to the initial true crack tip.

Comparison of these observations with predictions from the present simulations, Figures 22 and 24, is encumbered by the fact that gage length was not reported in (21,22). Figure 25 shows the COD-profiles for peak load from Figures 22 and 24 converted to strain profiles via assumed gage lengths. The predicted profiles bracket the measured profile, except for the reading close to the initial crack tip, and are nearly linear. It is likely that the gage closest to the crack tip began to slip at this very high strain; measurements from this gage location on other specimens also showed this anomalous behavior.

Figure 25 shows that Bazant and Oh (24) are correct. The measured strains can only be explained by considerable extension of a process zone before peak load. From the perspective of the methods of the present simulation, the gages were reading highly localized strain: a crack-opening-displacement.

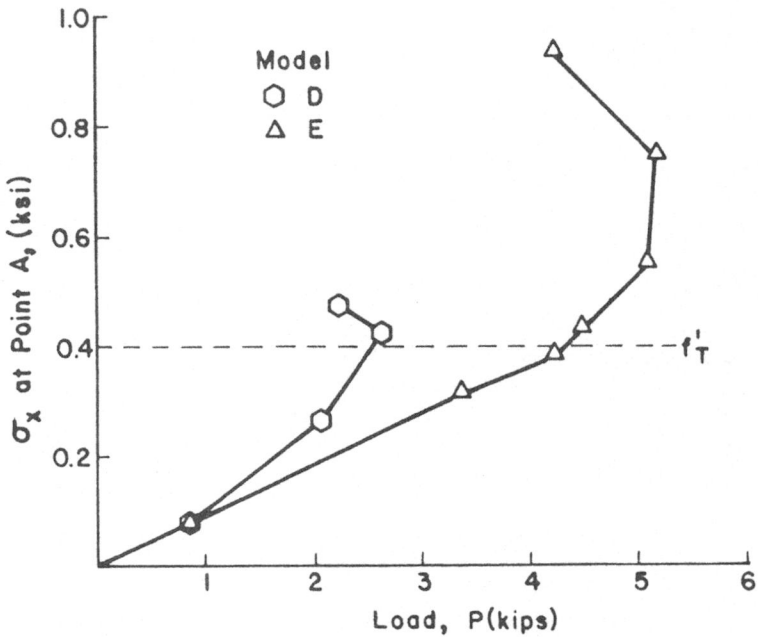

Figure 26. Predicted variation of tensile stress at point A, Figure 18, with applied load, Example Problem #2.

All of the previous discussion concerning the process zone in this example points decidedly towards this being a non-LEFM test, and yet an enigma remains: why the apparently LEFM response of this and companion tests when viewed on the basis of K_Q independence from specimen size and crack length (23)? A very plausible explanation, one that in fact requires our attention to be shifted away from the initial crack and its propagation, proceeds as follows.

Consider the location of point A, Figure 18, and the displaced shape shown in the same figure. This shape shows that the top and bottom halves of this specimen behave like deep beams with partial moment restraint at their ends produced by remaining ligament. Point A lies at the location of highest fiber stress in tension of such a beam. The writers find it more than just coincidental that the x-direction tensile stress at point A for the specimen analyzed in this example varies as shown in Figure 26: as the computed peak load is approached, the concrete tensile strength is exceeded at point A! Consistency in application of the constitutive model would dictate that a second set of cracks would initiate at points A at approximately the peak load. This phenomenon was not included in the present simulations. Note that it is not obvious that these cracks would have been visible after specimen failure, but their existence would certainly have caused stress redistribution in the specimen. Is it possible that the apparently LEFM response of the tests was the result of two obscuring non-linear effects? This question could be definitively answered and these tests finally put to rest by a study which included both nonlinearities and a range of specimen sizes and crack lengths.

In the next example problem we relax another assumption by extending our non-linear fracture modeling technique to a problem in which crack growth is not self-similar.

3.3 Example 3: Mixed-Mode Fracture of a Plain Concrete Beam

The last example problem is the structure shown in Figure 27. Mortar and concrete beams in this configuration were tested by Arrea and Ingraffea (25). The antisymmetric loading produces mixed-mode, K_I-K_{II}, stress-intensity at the tip of the initial, true crack tip.

The purpose of this example is to show that the techniques used in the previous examples can be extended to curvilinear crack propagation in geomaterials. Details of the algorithms necessary for this extension can be found in References 4 and 19.

Because this example involves both crack sliding and crack opening displacements, CSD and COD, respectively, there exists not only normal stress transfer across the process zone but shear stress as well. Consequently, the so-called "aggregate interlock" model of Fenwick and Pauley (26) in which shear transfer across a crack is related to the COD was employed in this simulation.

Two analysis phases were employed. The first was a parameter study in which the crack trajectory observed in testing was modeled in the mesh, as shown in Figure 28. The σ-COD constitutive model was then varied in an attempt to reproduce the observed load versus crack-mouth-sliding-displacement, CMSD. This approach is the same as the "fictitious crack" method (18), except that the crack is discretely modeled.

The results of this parameter study are shown in Figure 29, with the lettered models shown in Figure 1. The experimental results shown in this figure and from two different tests on mortar beams (25) with a maximum aggregate size of about 0.375 inch (9.5 mm) and a compressive strength f'_c, of about 6.6 ksi (45.5 MPa). These results indicate a trend towards a constitutive model with a δ_c much less than that used in the previous example. This is to be expected since the maximum aggregate size is much less in this case. A typical displaced shape of the structure is shown in Figure 30.

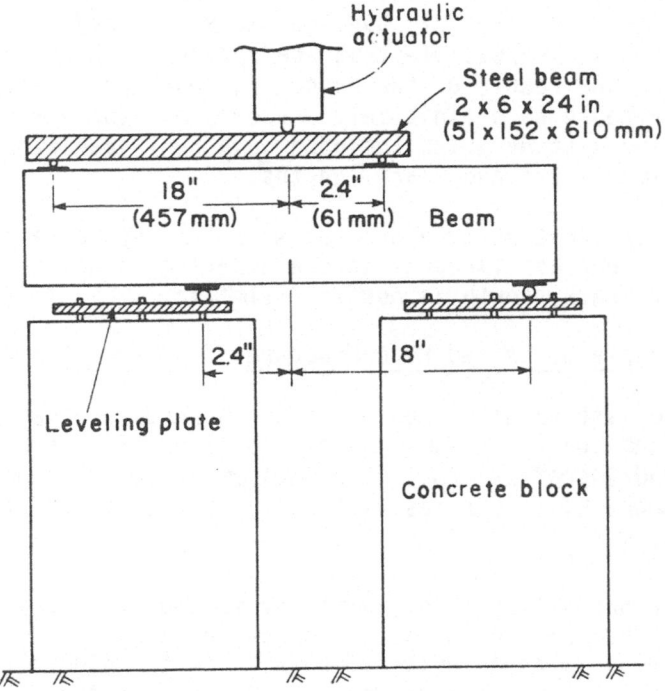

Figure 27. Test specimen and loading arrangement for Example
Problem #3. From Ref. 25.

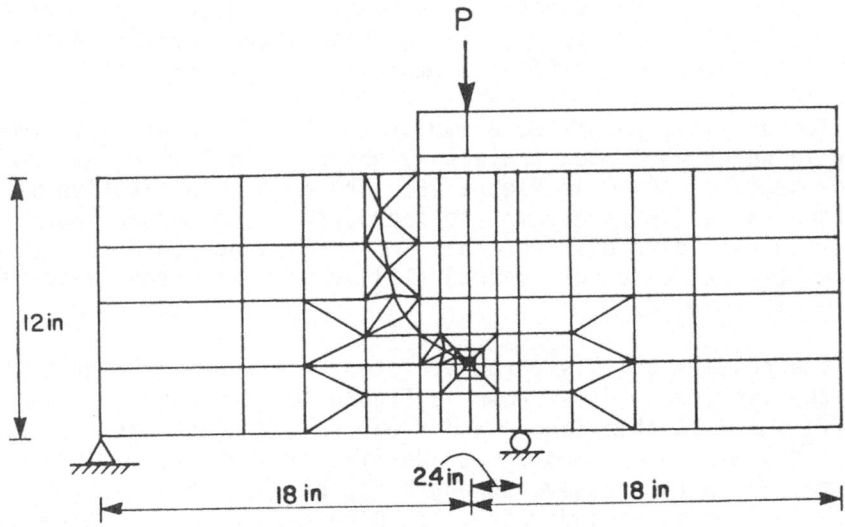

Figure 28. Mesh used for first phase, parameter study for Example
Problem #3.

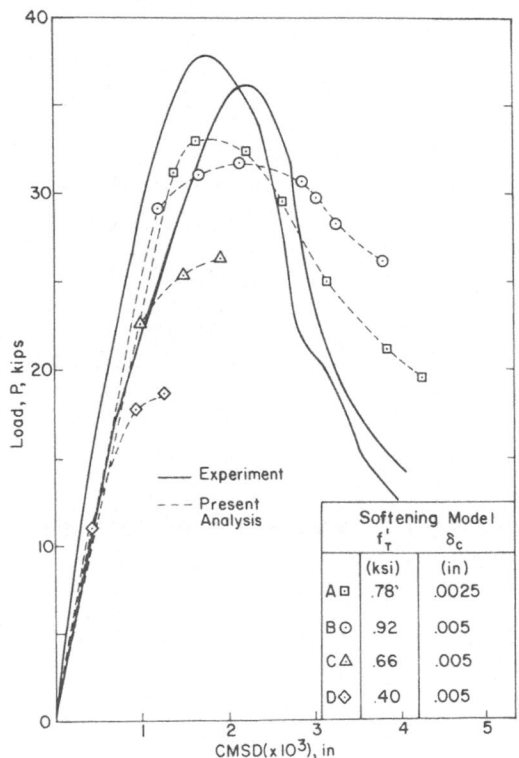

Figure 29. Results of parameter study for first phase of Example Problem #3.

Although none of the models used proved completely satisfactory, Model A was chosen for the second phase of this simulation. In this phase only the cast-in starter crack was modeled in the initial mesh, Figure 31. A discrete propagation analysis was performed with automatic remeshing occurring at each crack increment. The final mesh configuration and displaced shape are shown in Figure 32.

The predicted trajectory of the crack was very close to the observed, and the computed load versus CMOD response, Figure 33, was very similar to that obtained during the parameter study phase.

Experience with this simulation strongly suggests that the response is sensitive to the shear transfer model across the process zone and across the true crack itself. Although the shear and normal stress transfer models used in these simulations were uncoupled, some degree of coupling through dilatency is certain. These are fertile areas for further research.

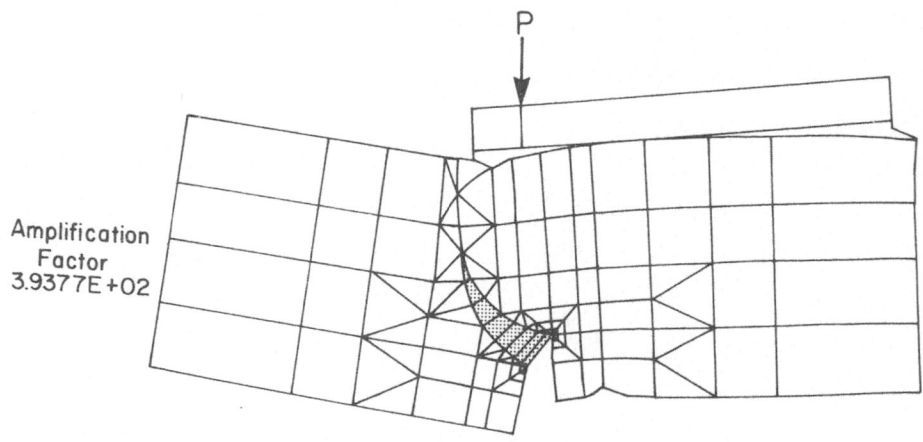

Figure 30. Amplified displaced shape of mesh shown in Figure 28. Amplification factor = 394.

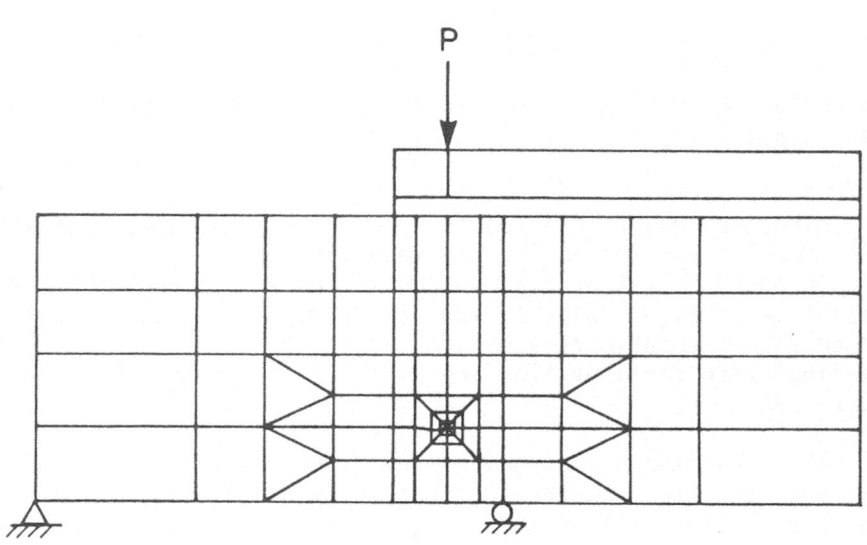

Figure 31. Initial mesh for second-phase, crack propagation study of Example Problem #3.

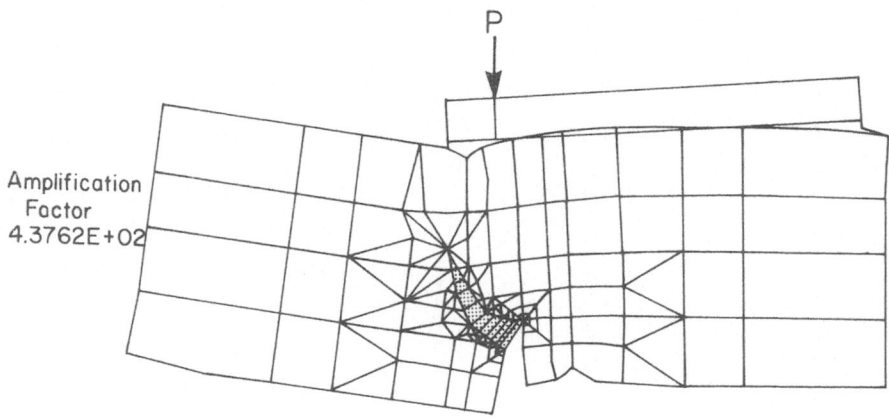

Figure 32. Amplified displaced shape after 5 crack increments through the mesh shown in Figure 31. Amplification factor = 438.

4. CONCLUSIONS

This paper has shown how non-linear fracture processes can be modeled in finite element simulations involving discrete crack representations. The most important assumption involved is that the process zone is no more than an extension of the true crack itself: the zone is not an area, it is a length.

The constitutive modelling for such a zone is then considerably simplified. For Mode I propagation, one needs only a relationship between the normal tensile stress transmitted across the process zone and the opening-displacement of that part of the crack in this length.

We first showed how, with these simplifications, one could arrive at quantitative assessments of process zone length in geomaterials using only hand calculations. It was quickly seen that the strain softening character of such materials produces process zone lengths much larger than those in more ductile materials.

These simple calculations were then supported by finite element analyses of three example problems. In the first, hypothetical problem, the growth of the process zone ahead of a true crack was studied in a very large structure. It was concluded that, as predicted by hand calculation and LEFM considerations, the process zone grew to a steady state length and that this length was far in excess of that predicted by models based on elastic-perfectly plastic constitutive models.

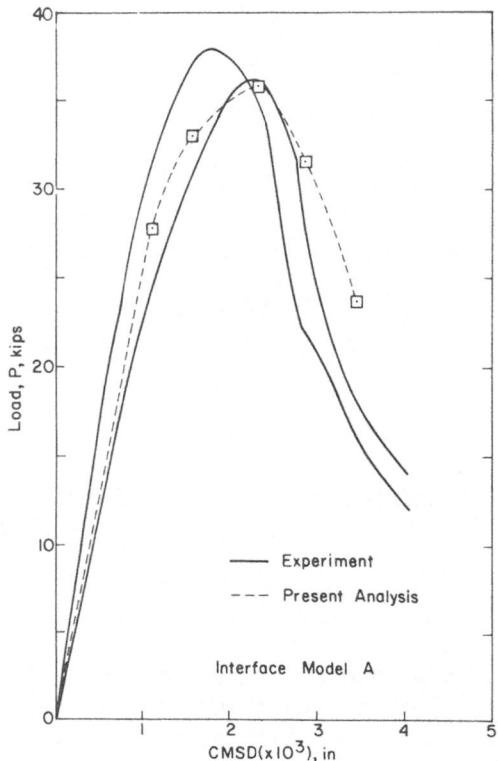

Figure 33. Comparison of crack propagation model prediction with experimental results for Example Problem #3.

The next two problems involved simulations of actual experiments. The tests of Kesler, Naus, and Lott (21) were again analyzed, but for the first time using a non-linear fracture model in a discrete crack representation. It was concluded that these tests were not valid measurements of K_{Ic}. Moreover, it was further shown that they still have not been properly analyzed.

The final analysis involved curvilinear crack propagation. The process zone constitutive modeling was extended to include shear transfer across the zone as a function of its opening-displacement. The automatic rezoning feature of FEFAP (19) was employed here to permit cracking to evolve as predicted by the non-linear stress transfer models without being constrained by meshing considerations.

Parameter studies performed during these simulations clearly show the dependence of process zone length on the crack opening profiles characteristic of the structure, and on the σ-COD relationship characteristic of the geomaterial.

Using the modeling tools described here, it is now possible to make a rigorous, a priori assessment of the applicability of LEFM to the fracture process occurring over a wide range of structure scale.

ACKNOWLEDGEMENTS

The work reported here was sponsored by the National Science Foundation under Grants PFR-7900711 and CME80-20925.

The authors would like to thank Mr. Walid Najjar for performing some of the analyses, and Mrs. Oneita Weeks for typing the manuscript.

REFERENCES

1. Irwin, G.R. Plastic Zone Near a Crack and Fracture Toughness. Proceedings 7th Sagamore Conference (1960) IV-63.

2. Petersson, P-E. Crack Growth and Development of Fracture Zones in Plain Concrete and Similar Materials. Report TVBM-1006, Division of Building Materials, Lund Institute of Technology, Lund, Sweden (1981).

3. Catalano, D. and A.R. Ingraffea. Concrete Fracture: A Linear Elastic Fracture Mechanics Approach. Department of Structural Engineering Report 82-1, School of Civil and Environmental Engineering, Cornell University, Ithaca, NY (1982).

4. Ingraffea, A.R., W. Gerstle, P. Gergely, and V. Saouma. Fracture Mechanics of Bond in Reinforced Concrete. Journal of the Structural Division, American Society of Civil Engineers, Vol. 110, No. 4 (April 1984) 871-890.

5. Mindess, S. and J.S. Nadeau. Effect of Notch Width on K_{Ic} for Mortar and Concrete. Cement and Concrete Research, Vol. 6 (1976) 529-534.

6. Schmidt, R. and T. Lutz. K_{Ic} and J_{Ic} of Westerly Granite-- Effects of Thickness and In-Plane Dimensions. American Society for Testing and Materials, STP 678 (1979) 166-182.

7. Rashid, Y.R. Analysis of Prestressed Concrete Pressure Vessels. Nuclear Engineering and Design, Vol. 7, No. 4 (April 1968) 334-344.

8. Saouma, V., A.R. Ingraffea, P. Gergely, and R.N. White. Interactive Finite Element Analysis of Reinforced Concrete: A Fracture Mechanics Approach. Department of Structural Engineering Report 81-5, School of Civil and Environmental Engineering, Cornell University, Ithaca, NY (1981).

9. Saouma, V.E. and A.R. Ingraffea. Fracture Mechanics Analysis of Discrete Cracking. Proceedings, IABSE Colloquium on Advanced Mechanics of Reinforced Concrete, Delft (June 1981) 393-416.

10. Evans, R.H. and M.S. Marathe. Microcracking and Stress-Strain Curves for Concrete in Tension. Matériaux et Constructions, Vol. 1, No. 1 (1968) 61-64.

11. Broek, D. Elementary Engineering Fracture Mechanics (The Hague, Martinus Nijhoff Publishers, 1982).

12. Annual Book of Standards, American Society for Testing and Materials, Section 3, E399-83 (1983) 518-553.

13. Sok, C., J. Baron, and D. Francois. Fracture Mechanics Applied to Concrete. Cement and Concrete Research, Vol. 9, No. 5 (1979) 641-648 [In French].

14. Entow, V. and V. Yagust. Experimental Investigation of Laws Governing Quasi-Static Development of Macrocracks in Concrete. Mekhanika Tverdogo Tela, Vol. 10 (1975) 93-103.

15. Cedolin, L., S. Dei Poli, and I. Iori. Experimental Analysis of the Process of Fracture Formation in Concrete. Studi e Ricerche, No. 3, Politecnico di Milano, Italia (1981) [In Italian].

16. Mindess, S. and S. Diamond. The Cracking and Fracture of Mortar. In Fracture in Concrete, W.F. Chen and E.C. Ting, eds. American Society of Civil Engineers (1980) 15-27.

17. Hoaglund, R., G. Hahn, and A. Rosenfield. Influence of Microstructure on Fracture Propagation in Rock. In Rock Mechanics, Vol. 5 (1973) 77-106.

18. Hillerborg, A., M. Modeér, and P-E. Petersson. Analysis of Crack Formation and Crack Growth in Concrete by Means of Fracture Mechanics and Finite Elements. Cement and Concrete Research, Vol. 6, No. 6 (1976) 773-782.

19. Ingraffea, A.R. and V. Saouma. Numerical Modeling of Discrete Crack Propagation in Reinforced and Plain Concrete. To appear in Application of Fracture Mechanics to Concrete Structures, G.C. Sih and A. DiTommaso, eds. (The Hague, Martinus Nijhoff Publishers, 1984).

20. Griffith, A.A. The Phenomena of Rupture and Flow in Solids. Phil. Trans. Royal Soc. of London, A221 (1921) 163-197.

21. Kesler, C., D. Naus, and J. Lott. Fracture Mechanics — Its Applicability to Concrete. Proceedings of the 1971 International Conference on Mechanical Behavior of Materials, Vol. IV, Japan (1972) 113-124.

22. Naus, D. Applicability of Linear-Elastic Fracture Mechanics to Portland Cement Concrete. Ph.D. thesis, University of Illinois at Urbana-Champaign (1971).

23. Saouma, V., A.R. Ingraffea, and D. Catalano. Fracture Toughness of Concrete: K_{Ic} Revisited. Journal of the Engineering Mechanics Division, ASCE, Vol. 108, No. EM6 (1982) 1152-1166.

24. Bazant, Z. and B.H. Oh. Crack Band Theory for Fracture of Concrete. Materiaux et Constructions, Vol. 16, No. 93 (1983) 155-177.

25. Arrea, M. and A.R. Ingraffea. Mixed-Mode Crack Propagation in Mortar and Concrete. Department of Structural Engineering Report 81-13, School of Civil and Environmental Engineering, Cornell University, Ithaca, NY (1981).

26. Fenwick, R.C. and T. Paulay. Mechanics of Shear Resistance of Concrete Beams. Journal of the Structural Division, American Society of Civil Engineers, Vol. 94, No. ST10 (1968) 2325-2350.

APPLICATION OF FRACTURE MECHANICS
TO CEMENTITIOUS COMPOSITES.
NATO-ARW - September 4-7, 1984
Northwestern Univeristy, U.S.A.
S. P. Shah, Editor

INTERPRETATION OF THE GRIFFITH INSTABILITY
AS A BIFURCATION OF THE GLOBAL EQUILIBRIUM (*)

Alberto Carpinteri

Istituto di Scienza delle Costruzioni, University of Bologna,
40136 Bologna, Italy.

Summary.

The process zone at the crack tip of a concrete-like material can be simulated in two different alternative ways: (a) with a *damage zone* in front of the stress-free crack tip, or (b) with a *cohesive force distribution* behind a fictitious crack tip. Both these numerical models are able to simulate the slow crack growth and to reproduce the scale effects of fracture toughness testing. With large structural sizes the softening structural behaviour disappears and the global ductility drastically decreases. In the *damage model*, this is due to the priority of the crack instability over the traditional structural instability. On the other hand, in the *cohesive model*, this is revealed by a bifurcation of the global equilibrium, the stress-singularity being not included in such a model.

1. INTRODUCTION

The non-linear and dissipative phenomena occurring at the crack tip of concrete-
-like materials are different from those occurring in metals and theory of plasticity
is not able to describe them in a consistent manner. Microcracking of mortar and
debonding between mortar and aggregates are the principal damage mechanisms in
the vicinity of a crack tip. When a sufficiently high number of microcracks and
debondings coalesces, a macroscopical growth of the macrocrack occurs and this is
often only a stable stage of the crack propagation process. The firstly partial and then
total stress relaxation at the crack tip is the result of the strain-softening constitutive
law of concrete-like materials.

(*)   The results of section 2 were obtained in a joint research program between Lehigh University
and University of Bologna, while the results of section 3 in a research convention between University of Bologna and ENEL-CRIS-Milano.

The crack represented in Fig. 1-a cannot be analyzed by Linear Elastic Fracture Mechanics, unless the damage zone is much smaller than the crack length, $a_d \ll a_o$. It is possible to use two alternative models to describe the non-linear behaviour of this crack:

(a) to consider the *real crack* of length a_o with a *damage zone* in front of the crack tip where the effective Young's modulus E^* is lower than the initial one, E (Fig. 1-b); (b) to consider a *fictitious crack* of length $(a_o + a_d)$ with *cohesive forces* in the damage zone representing the aggregate interlocking and bridging (Fig. 1-c). While the former numerical model has recently been developed in a joint research program between Lehigh University (Pennsylvania) and University of Bologna (Italy) [1], the latter was firstly realized at the Lund Institute of Technology (Sweden) [2] and then, with some modifications, at Northwestern University (Illinois) [3, 4], Cornell University (New York State) [5] and ENEL-CRIS (Italy) [6].

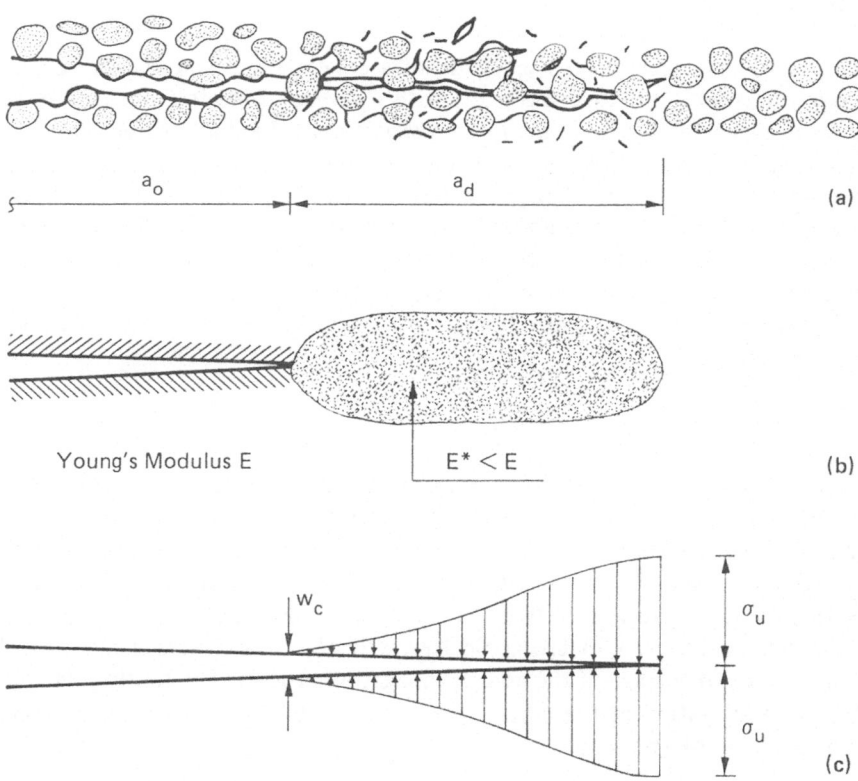

Figure 1 - Process zone in front of the crack tip (a): damage model (b) and cohesive model (c).

Both numerical models will be described in the present paper, on the basis also of numerical results obtained by the writer. They both are able to simulate the slow crack growth and to reproduce the scale effects of fracture toughness testing. With large structural sizes the softening structural behaviour disappears and the global ductility drastically decreases. In the *damage model,* this is due to the priority of the crack instability over the traditional structural instability [1]. On the other hand, in the *cohesive model* this is revealed by a bifurcation of the global equilibrium, the stress-singularity being not included in such a model.

2. DAMAGE MODEL

2.1 Mechanical Damage and Stress-Relaxation

Decrease in elastic modulus and stress relaxation at the crack tip are closely connected. As proposed by Janson and Hult [7-9], they can be jointly described through a simple analytical model. A sheet containing a crack and loaded in mode I can be studied with a one-dimensional method. The material is assumed to exhibit time-independent continuous damage when loaded and the damage to affect the stress in the   narrow region ahead of the crack,which corresponds to the Dugdale zone [10]. The stress field in the crack tip region is approximated as uniaxial, whereby a simplified analysis is possible. The following constitutive variables can be defined:

Damage   : $\omega = \ln (A/A_{eff})$,   (1-a)

Stress   : $\sigma = P/A$,   (1-b)

Net stress : $\sigma_{eff} = P/A_{eff}$.   (1-c)

Here P is the transmitted load and A is the macroscopic area, while A_{eff} is the microscopically load carrying area. The damage definition (1-a) is due to Broberg [11]. For small damage values the Kachanov definition [12]:

$$\omega = \frac{A - A_{eff}}{A} \quad ,$$ (2)

is approached.

Eqs (1) lead to the relation:

$$\sigma = \sigma_{eff} \exp (- \omega) .$$ (3)

Damage is then assumed to depend on strain according to a power relation:

$$\omega = K \, \epsilon^m .$$ (4)

Damage can also be defined in relation to the locally decreased elastic modulus E_{eff}:

$$\omega = \ln (E/E_{eff}) ,$$ (5)

so that:

$$E_{eff} = E \exp (- K \epsilon^m) .$$ (6)

On the other hand:

$$\sigma = E_{eff} \, \epsilon = E \, \epsilon \exp (- K \epsilon^m) ,$$ (7)

and, since in the crack tip vicinity we have very high strains, the stresses vanish there:

$$\lim_{\epsilon \to \infty} \sigma = \lim_{\epsilon \to \infty} E\epsilon \exp(-K\epsilon^m) = 0 . \tag{8}$$

This means that the above described model is able to reproduce the strain-softening material behaviour (Fig. 2).

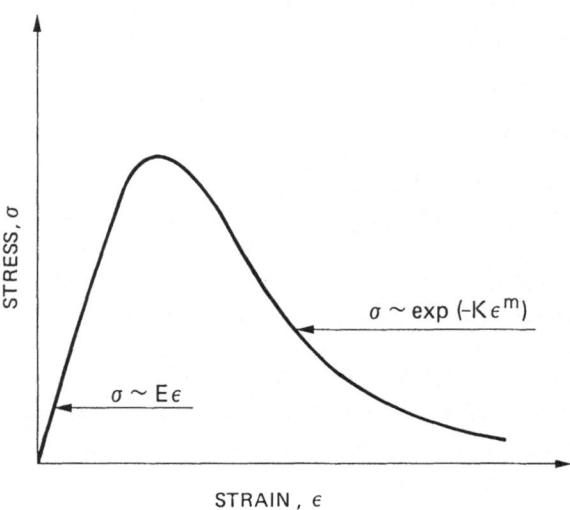

Figure 2 - Softening stress-strain law.

We now assume that the damage does not alter the strain distribution around the crack. This means that both strain ϵ and net stress:

$$\sigma_{eff} = E\epsilon , \tag{9}$$

vary as $r^{-1/2}$ and the damage ω as $r^{-m/2}$, where r is the distance from the crack tip. Close to the crack tip, the stresses are represented by a uniaxial distribution. From eq. (7), in fact, it follows:

$$\sigma \propto r^{-1/2} \exp(-Kr^{-m/2}), \tag{10}$$

and stress σ shows a course similar to that of Fig. 3. A zone can be identified close to the crack tip, within which the stress σ decreases, while σ_{eff}, ϵ and ω increase.

A numerical model very similar to the elementary model by Janson and Hult will be described in the next section [1]. Damage will depend on the absorbed strain energy density and not on the uniaxial strain ϵ, as assumed in eq. (4). On the other hand, damage will be measured by the Kachanov ratio in eq. (2) and will achieve the unit value when the critical value $(dW/dV)_c$ of the strain energy density is absorbed in the elementary volume. At that time, the elastic modulus is zero and the load carrying capacity of the material vanishes. In the same way, even the slow crack growth process will be simulated.

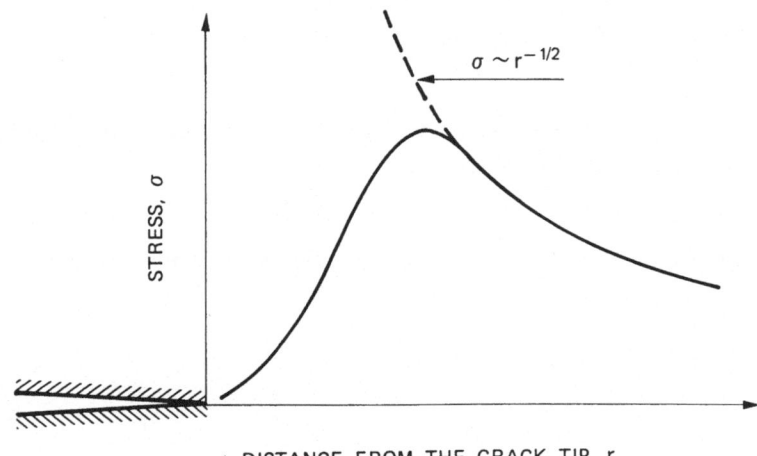

DISTANCE FROM THE CRACK TIP, r

Figure 3 - Stress-relaxation at the crack tip.

2.2 Mechanical Damage and Strain-Softening Constitutive Behaviour

Damage of the material at the crack tip and crack growth increments will be computed on the basis of a uniaxial bilinear elastic-softening stress-strain relation (Fig. 4-a). If the loading is relaxed when the representative point is in A, the unloading is assumed to occur along the line AO, so that the new bilinear constitutive relation is the line OAF. No permanent deformation is allowed by such a model, but only the degradation of the elastic modulus. The present model simulates the mechanical damage by decreasing elastic modulus E and strain energy density which can be absorbed by a material element. In fact, while for a non-damaged material element the critical value of the strain energy density, $(dW/dV)_c$, is equal to the area OUF (Fig. 4-a), for a damaged material element with representative point in A, the decreased critical value, $(dW/dW)_c^*$, is equal to the area OAF. In addition, as is shown in Fig. 4-a, the area OUA represents the dissipated strain energy density, $(dW/dV)_d$, OAB the recoverable strain energy density, $(dW/dV)_r$, and BAF the additional strain energy density, $(dW/dV)_a$.

The described model will be extended to the three-dimensional stress conditions using the current value of the absorbed strain energy density, (dW/dV), as a measure of damage. In other words, the effective elastic modulus E^* and the decreased critical value of strain energy density, $(dW/dV)_c^*$, will be considered as functions of the absorbed strain energy density (dW/dV), being:

$$\left(\frac{dW}{dV}\right) = \left(\frac{dW}{dV}\right)_d + \left(\frac{dW}{dV}\right)_r \tag{11}$$

Such functions in the uniaxial case are:

$$OUAB \text{ area} \longrightarrow AO \text{ slope, i.e., } \left(\frac{dW}{dV}\right) \longrightarrow E^* , \tag{12-a}$$

$$OUAB \quad \text{area} \longrightarrow OAF \quad \text{area,} \quad \text{i.e.,} \quad \left(\frac{dW}{dV}\right) \longrightarrow \left(\frac{dW}{dV}\right)_c^*. \tag{12-b}$$

The relations (12) will be discretized using 25 different values of the elastic modulus:

$$E^*(n) = \frac{(26-n)}{25} E \quad , \quad \text{for} \quad n = 1, 2, \ldots, 25 . \tag{13}$$

In Fig. 4-b, the results of different tensile test numerical simulations are reported.

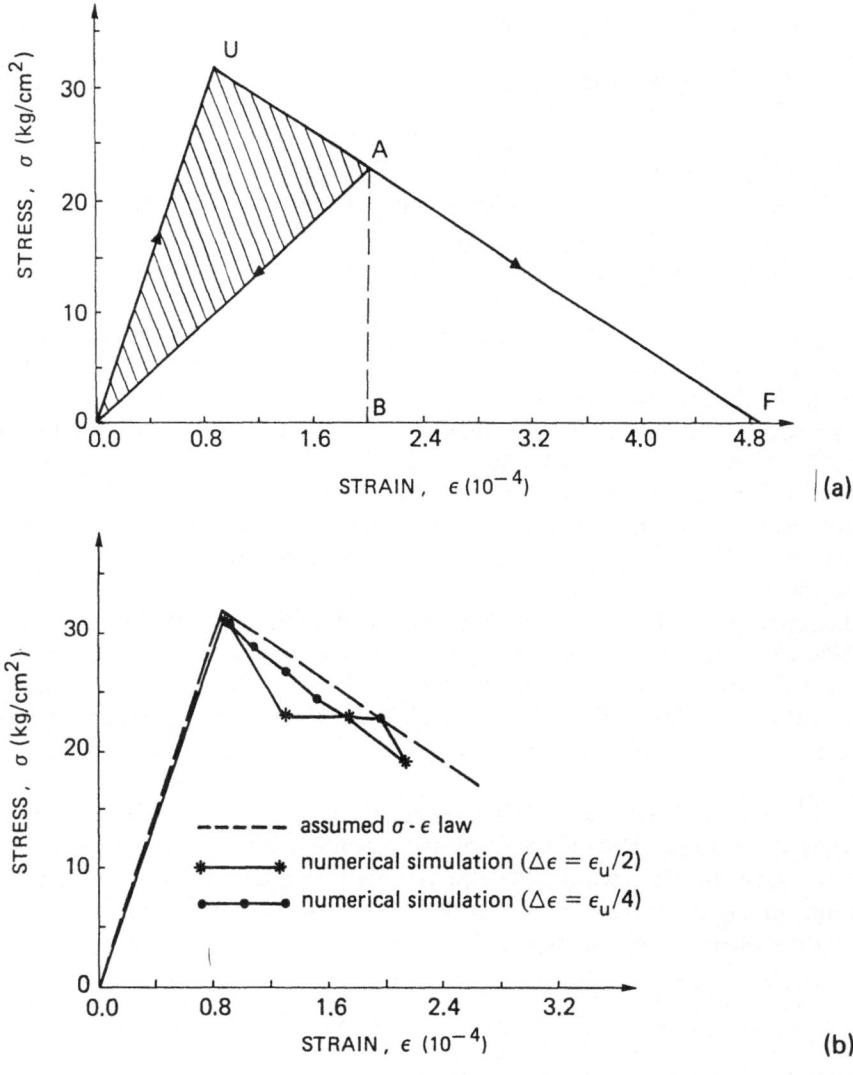

Figure 4 - Strain-softening constitutive law for concrete subject to tensile loading: assumed bilinear relation (a) and numerical damage simulation (b).

Two strain-controlled loading processes are carried out and the strain increment in one case is twice as much as in the other case. The discretized stress-strain relations are shown and the comparison with the assumed $\sigma - \epsilon$ constitutive law is also shown. It can be proved that when the effective elastic modulus E^* varies continuously and the strain increment $\Delta\epsilon$ tends to zero, the assumed bilinear $\sigma - \epsilon$ variation (dashed line in Fig. 4-b) is exactly reproduced by the numerical damage simulation.

2.3 Strain Energy Density Fracture Criterion

In order to evaluate the crack growth increment at each loading step, the Strain Energy Density Theory will be applied, as proposed by Sih [13, 14]. It is based on the following fundamental assumptions:

(1) the stress field in the vicinity of the crack tip cannot be described in analytical terms because of the relative heterogeneity of the material. A minimum distance r_o does exist below which it is a non-sense to study the mechanical behaviour of the material from a "continuum-mechanics" point of view and to consider macroscopic crack growth increments;

(2) outside such a core region of radius r_o, the strain energy density field can always be described by means of the following general relationship:

$$\left(\frac{dW}{dV}\right) = \frac{S}{r} \ , \tag{14}$$

where the strain energy density factor S is generally a function of the three space coordinates;

(3) according to Beltrami's criterion, all the material elements in front of the crack tip, where the strain energy density is higher than the critical value $(dW/dV)_c^*$, fail;

(4) when the following condition holds:

$$\Delta a = r_o^* = S_o/(dW/dV)_c^* \ , \tag{15}$$

the crack may be considered as arrested, at least from a macroscopical point of view. On the other hand, when the crack growth increment is:

$$\Delta a = r_c^* = S_c/(dW/dV)_c^* \ , \tag{16}$$

the unstable crack propagation takes place. S_c is a material constant and represents the strength of the material against rapid and uncontrollable crack propagation. S_c is connected with the critical value of the stress-intensity factor K_{IC} through the following equation (plane strain condition) [13]:

$$S_c = \frac{(1 + \nu)(1 - 2\nu)}{2\pi E} K_{IC}^2 \ . \tag{17}$$

2.4 Numerical Results According to the Damage Model

A *center cracked slab in tension* (Fig. 5-a) is analyzed by using the Axisymmetric/ Planar Elastic Structures (APES) finite element program [15]. It is a computer program which incorporates 12-noded quadrilateral isoparametric elements allowing for cubic displacement fields and quadratic stress and strain fields within each element.

The r^{-1} strain energy density singularity in the vicinity of the crack tip is embedded in the solution through the use of 1/9 to 4/9 nodal spacing on the element sides adjacent to the crack tip. The idealization of Fig. 5-b utilizes 309 nodes and 52 elements and is considered in a condition of plane strain.

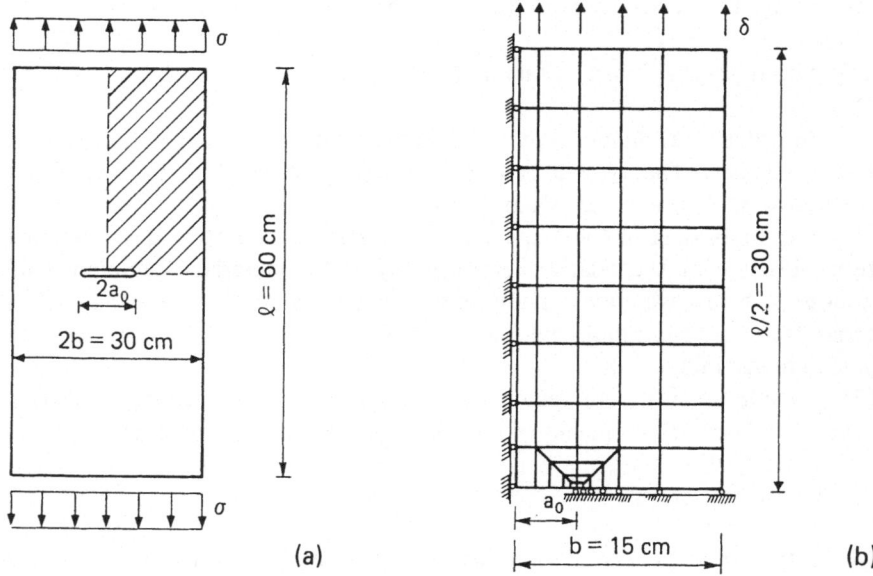

Figure 5 - Center cracked slab in tension.

The strain energy density criterion is applied to the tension test specimen considering a strain-controlled loading process. The stress-strain responses for three different initial crack lengths are shown in Fig. 6. The load carrying capacity decreases by increasing the initial crack length.

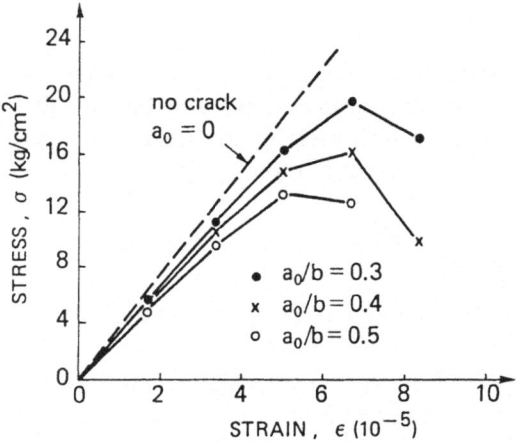

Figura 6 - Stress-strain responses for three different initial crack lengths.

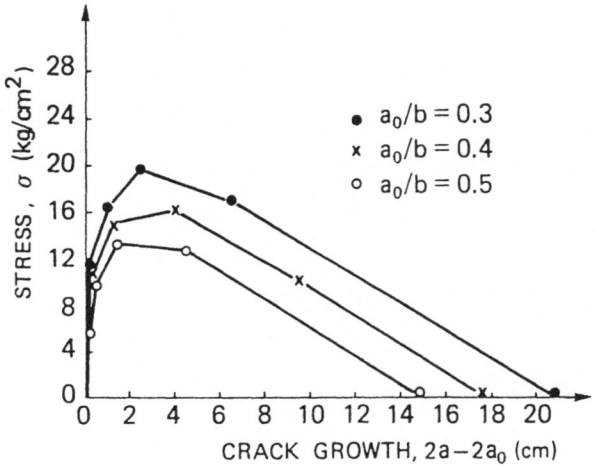

Figure 7 - Stress vs. crack growth plots for three different initial crack lengths.

In Fig. 7, the stress σ is represented against the crack growth $2(a - a_o)$. At the first steps the stress increases while the crack grows. Then, after reaching a maximum, the stress decreases and attains the value zero when the whole ligament is separated. The maximum represents the transition between stable and unstable structural behaviour. On the other hand, the transition between stable and unstable crack propagation depends on the achievement of the critical value of the strain energy density factor, S_c. Such a value has physical dimensions different from those of $(dW/dV)_c$ and produces the scale effects recurrent in Fracture Mechanics. As a matter of fact, the crack instability may precede or follow the structural instability. This mainly depends on the structural size scale [1].

Because material damage and crack growth occur in a non self-similar fashion for each step of loading, specimens of different size appear to behave differently. The application of Buckingham's theorem for physical similitude and scale modelling gives [16 - 18]:

$$\sigma = \pi \left(\epsilon, \frac{E}{\left(\dfrac{dW}{dV}\right)_c}, \frac{\sigma_u}{\left(\dfrac{dW}{dV}\right)_c}, \nu, \frac{\ell}{b}, \frac{t}{b}, \frac{a_0}{b} \right), \qquad (18)$$

where material toughness $(dW/dV)_c$ and specimen width b have been used as the fundamental quantities. The stress σ may be regarded as a function of the strain ϵ only, if all other ratios are kept constant.

In the same way, it is possible to define di dimensionless strain energy density factor:

$$\frac{S}{\left(\dfrac{dW}{dV}\right)_c b} = \sum\left(\frac{a}{b}, \frac{E}{\left(\dfrac{dW}{dV}\right)_c}, \frac{\sigma_u}{\left(\dfrac{dW}{dV}\right)_c}, \nu, \frac{\ell}{b}, \frac{t}{b}, \frac{a_0}{b}\right). \tag{19}$$

Function Σ can be regarded as linear in a/b [1]:

$$\frac{S}{\left(\dfrac{dW}{dV}\right)_c b} = \frac{dS/da}{\left(\dfrac{dW}{dV}\right)_c}\frac{(a - a_0)}{b} + \frac{S_0}{\left(\dfrac{dW}{dV}\right)_c b}, \tag{20}$$

which may obviously be rearranged in the form:

$$\frac{S}{\left(\dfrac{dW}{dV}\right)_c b} = A\left(\frac{a}{b}\right) + B. \tag{21}$$

The constants A and B are dimensionless and scale independent. It follows that the slope of the $S - a$ diagram is constant varying the scale and the intercept S_0 is proportional to the scale b.

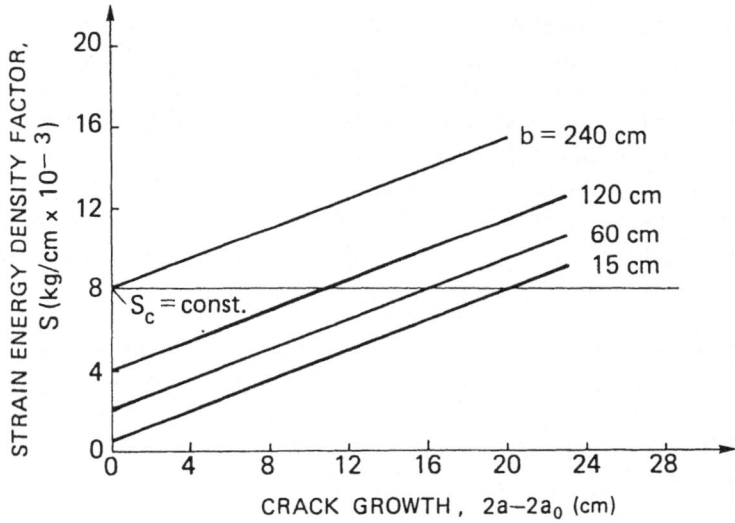

Figure 8 - Strain energy density factor vs. crack growth plots by varying size, $b\,(a_0/b = 0.3)$.

Fig. 8 shows the straight line plots of S versus crack growth for increasing size $b(a_0/b = 0.3)$. The critical crack growth decreases with increasing specimen size.

For example, with the critical value $S_c = 8 \times 10^{-3}$ kg/cm, the limiting size is $b = 240$ cm. Beyond this size, stable crack growth ceases to occur and failure corresponds to unstable crack propagation or catastrophic fracture.

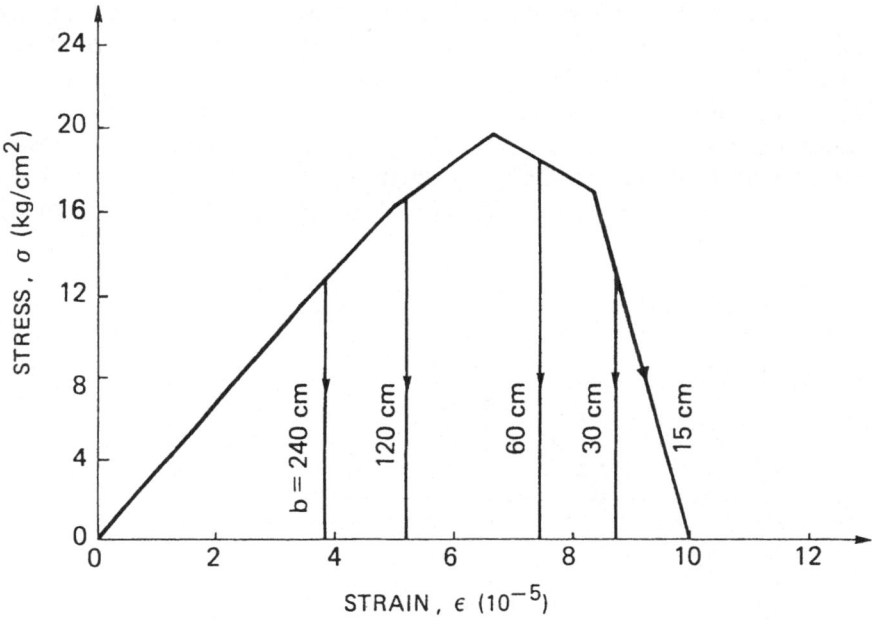

Figure 9 - Stress-strain relations by varying size, $b(a_o/b = 0.3)$.

Fig. 9 presents the relations between stress and strain ($a_o/b = 0.3$). The vertical lines with arrows indicate the limiting values of ϵ as the critical strain energy density factor, $S_c = 8 \times 10^{-3}$ kg/cm, is reached. Crack instability occurs for smaller strains as the size b increases. This is obvious, since the initial crack length a_o also increases, the ratio $a_o/b = 0.3$ being constant. The structural instability occurs before crack instability only for $b \leqslant 80$ cm (Fig. 9). When, for example, $b = 120$ cm, softening behaviour is not present and the crack starts to spread in an unstable manner, stress σ still being in the ascending stage.

It is also interesting to consider the maximum stress $\sigma_{max}^{(2)}$ resulting from the Linear Elastic Fracture Mechanics (LEFM) solution:

$$K_I = \sigma \sqrt{\pi a_o} \left(\sec \frac{\pi a_o}{2b} \right)^{1/2} . \tag{22}$$

On the other hand, the predicted values of $\sigma_{max}^{(3)}$ coming from the limit analysis at the ligament are given by:

$$\sigma_{max}^{(3)} = \sigma_u \left(1 - \frac{a_o}{b} \right) . \tag{23}$$

Normalizing the strengths $\sigma_{max}^{(1)}$ and $\sigma_{max}^{(3)}$ obtained by the present approach and the limit analysis respectively, with $\sigma_{max}^{(2)}$ obtained through LEFM we can evaluate the interaction between the different failure modes (Fig. 10). The horizontal line $\sigma_{max}^{(1)}/\sigma_{max}^{(2)}$ and $\sigma_{max}^{(3)}/\sigma_{max}^{(2)}$ equal to 100% represents the case when failure coincides totally with *brittle fracture,* while the dashed line represents the case when failure coincides totally with *plastic collapse.* When the specimen size is small, the simple formula in eq. (23) gives good prediction based on the *ultimate strength* alone. On the other hand, when the specimen size is large, eq. (22) gives good prediction based on the *stress-intensity factor* alone. The two extreme situations are then connected by a transition. For intermediate sizes, the ratio of Fig. 10 appears higher than one. This means that a fictitious critical stress-intensity factor $K_{IC}^{(f)}$ larger than the true K_{IC} may be assumed. In this way, the energy-absorbing damage process at the crack tip is taken into account.

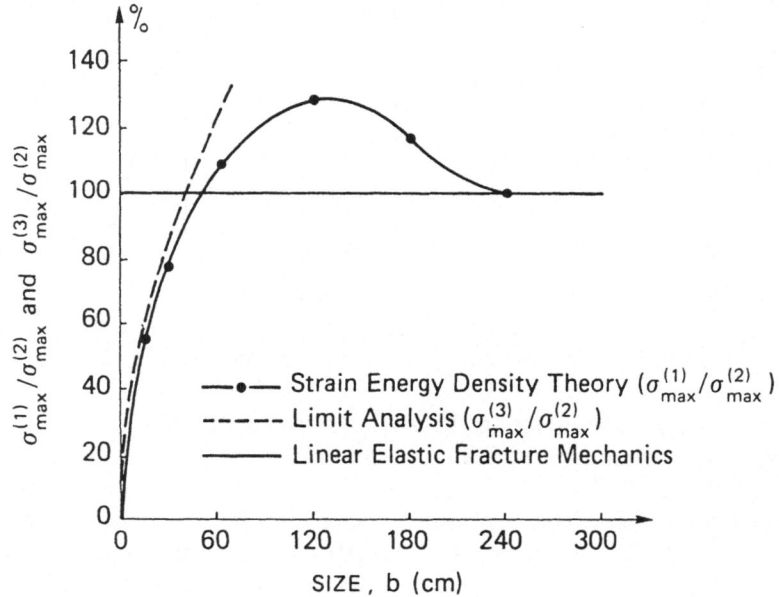

Figure 10 - Interaction between plastic collapse and brittle fracture by varying size, $b (a_0/b = 0.3)$.

A *plain or reinforced concrete beam in bending* is considered (Fig. 11) and now the input-parameter for the loading process is the deflection δ. In the case of a plain concrete beam, the numerical results mostly present the same trends as those previously illustrated for the tension test. They are widely reported in [1].

On the other hand, for a reinforced concrete beam the curves of Fig. 11 do not describe the usual transition occurring without reinforcement. For small sizes, the strain energy density theory prediction is very close to that of limit analysis, whereas, for large sizes, it tends asymptotically to a limit, which is lower than the LEFM load [19 , 20]. It then follows that the maximum load, for $b \to \infty$, is a constant fraction of the LEFM solution. Such a fraction increases for decreasing steel percentages, A_S/A, and tends to unity when A_S/A tends to zero (Fig. 11).

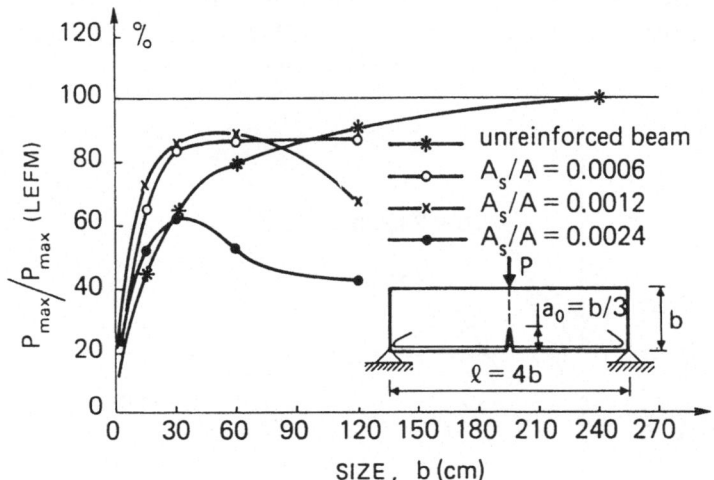

Figure 11 - Comparison between linear and non-linear fracture mechanics prediction for a reinforced concrete beam in bending.

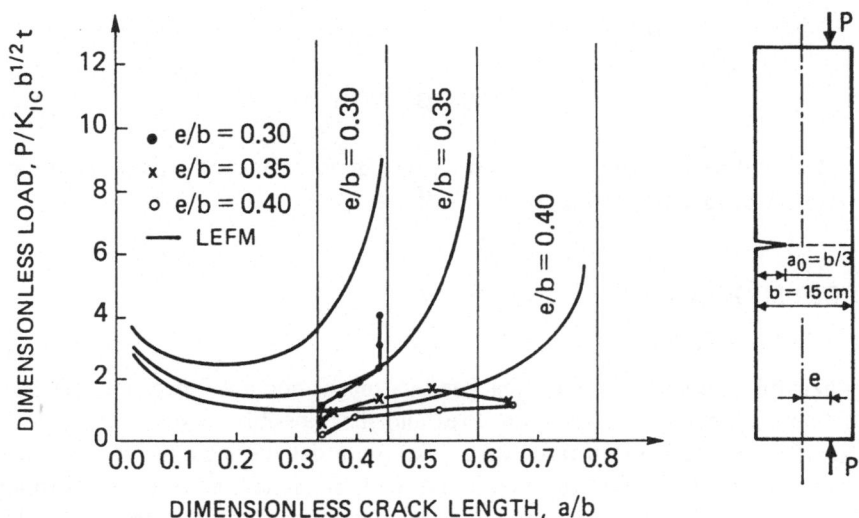

Figure 12 - Load vs. crack length plots for a wall subject to eccentric compression.

Fig. 12 shows the dimensionless load vs. the relative crack length for a *wall loaded by eccentric compression* and with a lateral growing crack. For large eccentricities ($e/b = 0.35$ and 0.40), these diagrams present a softening behaviour instead of the stable one predicted by *LEFM* [21, 22]. On the other hand, it is interesting to observe that the crack arrest predicted by *LEFM* (see vertical asymptotes of Fig. 12) is confirmed by the strain energy density theory for small eccentricities ($e/b = = 0.30$).

Fig. 13 shows the bilinear $S - a$ variations for different sizes b and eccentricity $e/b = 0.30$. If, for example, the critical value of the strain energy density factor is $S_c = 8 \times 10^{-3}$ kg/cm, the crack instability occurs only for $b \geqslant 53$ cm.

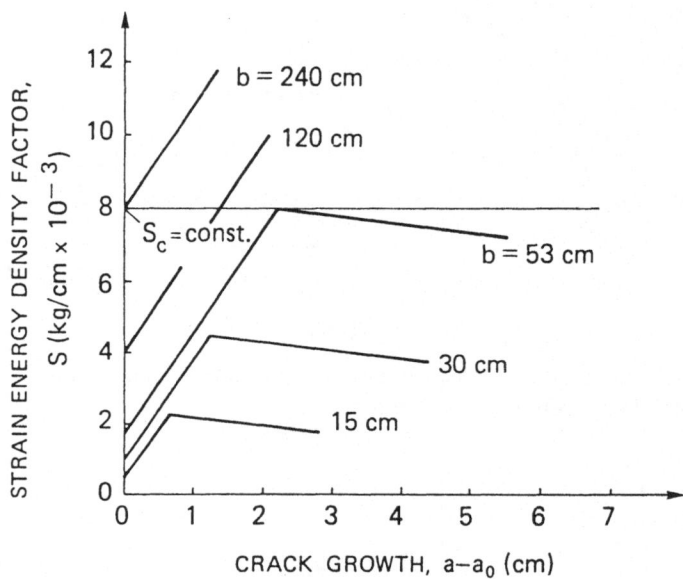

Figure 13 - Bilinear strain energy density factor vs. crack growth plots for a wall subject to eccentric compression ($e/b = 0.30$).

3.   COHESIVE MODEL

3.1   Strain Localization

When the maximum stress is reached in a tensile concrete specimen, the weakest cross-section is unable to carry more load and it is possible to verify that the development of damage is concentrated on a small material volume close to this cross-section. This means that, after the maximum load is reached, additional deformations will take place in the microcracked material volume, or fracture zone, while the material outside the fracture zone will be elastically unloaded. The load decreases when the fracture zone develops and consequently only a single zone develops [23]. This process occurs even in metallic materials, where the yielding is usually localized in a necking zone. An experimental confirmation regarding strain localization in direct tensile tests on concrete was given by Heilmann, Hilsdorf and Finsterwalder [24].

As the width of the fracture zone in the stress direction seems to be small, it should be possible to describe the tensile test by a simple model according to Fig. 14. In the model the fracture zone is replaced by a slit that is able to transfer stress and the stress transferring capability depends on the width w of the slit. Then, the total deformation of the specimen $\Delta\ell$ becomes:

$$\Delta\ell = \epsilon_o \ell + w \, , \tag{24}$$

where ϵ_o is the strain in the material outside the fracture zone and ℓ is the specimen length. From eq. (24) it follows that the mean strain ϵ_m is (Fig. 14):

$$\epsilon_m = \frac{\Delta\ell}{\ell} = \epsilon_o + \frac{w}{\ell} \, . \tag{25}$$

After the maximum stress is reached, the deformation of the fracture zone affects the mean strain and consequently the stress-strain curve of concrete may appear dependent on the specimen length. For this reason, Hillerborg, Modéer and Petersson [2] considered unsuitable to use the stress-strain curve as a material property. They proposed to use two relations: one between *stress* and *strain* for the undamaged material outside the fracture zone (Fig. 15-a) and the other between *stress* and *width of the fracture zone* (Fig. 15-b).

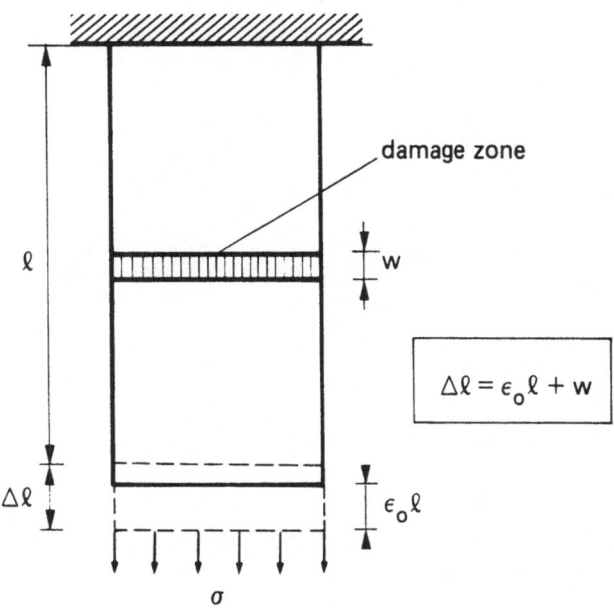

Figure 14 - Strain localization in a tensile test specimen.

On the other hand, it is important to observe that the strain localization does not imply the change from a $\sigma - \epsilon$ to a $\sigma - w$ constitutive law, as suggested by Hillerborg, Modéer and Petersson [2]. It was proved by Rice [25, 26] that the strain localization implies only a low hardening modulus or a softening behaviour in the constitutive law, which can therefore be of the $\sigma - \epsilon$ type even in the descending part. As pointed out by Bažant and 0h [3], what is important is to consider the fracture zone of a finite width w_o also at the beginning of the loading process. Such a width may be considered a material property and connected with the characteristic size D_{max} of the aggregates: $w_o \simeq 3 D_{max}$. In this case, eq. (24) becomes:

$$\Delta \ell = \epsilon_o \, (\ell - w_o) + \epsilon_d w_o \, , \tag{26}$$

where ϵ_d is the strain in the damage zone, which increases monotonically during the loading process, while ϵ_o decreases after reaching the ultimate stress σ_u. At the fracture condition, eq. (24) gives:

$$\Delta \ell = w_c \, , \tag{27}$$

while eq. (26) gives:

$$\Delta \ell = \epsilon_f w_o \quad . \tag{28}$$

The fundamental relation follows from eqs (27) and (28):

$$w_c = \epsilon_f w_o \, , \tag{29}$$

which connects critical value of the crack opening displacement, w_c, fracture strain, ϵ_f, and characteristic crack band width, w_o.

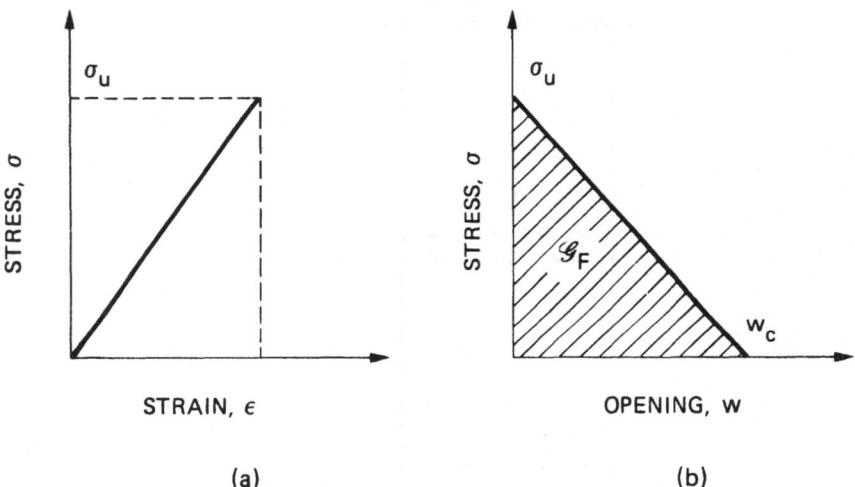

Figure 15 - Stress-strain law of the material outside the damage zone (a) and stress-displacement law of the damage zone (b).

3.2 Fictitious Crack Model

The fictitious crack model was firstly proposed by Hillerborg, Modéer and Petersson [2] in 1976 in order to study crack propagation in concrete. On the other hand, a similar cohesive model had already been presented by Barenblatt [27], Dugdale [10] and Rice [28] and applied by Palmer and Rice [29] to shear band propagation in overconsolidated clay soils. Hillerborg, Modéer and Petersson made the following assumptions:

(1) the fracture zone starts developing at one point when the highest principal stress reaches the tensile strength σ_u ;

(2) the fracture zone develops perpendicular to the highest principal stress;

(3) the material in the damage zone is partly destroyed but is still able to transfer stress. In the calculations the damage zone is replaced by a stress-transferring crack and the stress transferring capability depends on the width of the crack in the stress direction according to a $\sigma-w$ curve (Fig. 15-b);

(4) the width of the damage zone is zero when it starts developing;

(5) the properties of the material outside the damage zone are given by a $\sigma-\epsilon$ curve (Fig. 15-a).

The fictitious crack model is not a pure fracture mechanics model but initially unnotched structures can also be analyzed.

Several authors have assumed a crack propagation model similar to that by Hillerborg, Modéer and Petersson. Gerstle, Ingraffea and Gergely [5] generalized the Fictitious Crack Model by Hillerborg to mixed mode crack problems and considered both normal and shear stresses on the crack surface as a function of both discontinuities in normal and tangential displacements. This model was programmed into a crack propagation algorithm. As a result of this analytical investigation, the cracking behaviour of bond was clearly understood. In particular, the interaction between primary and secondary cracks in reinforced concrete structures became apparent.

Chappell and Ingraffea [30] applied a discrete crack modelling based on Linear Elastic Fracture Mechanics to an actual dam cracking problem. The crack mechanism was a combination of thermal expansion and thermally induced concrete growth. Theories of mixed-mode fracture propagation were then incorporated into a two-dimensional finite element analysis to perform a fracture analysis on a cross-section of the cracked portion of the dam. In this way, an exact prediction of the crack trajectory was obtained. It was not necessary to use non-linear strain-softening models owing to the large size of the dam.

Saouma [31] developed a computer program capable of automatically simulating discrete crack nucleation and extension through a finite element mesh as governed by an appropriate analysis. Continuous modification of the mesh and nodal renumbering were included in the procedure and a parametric study of reinforced concrete beams was performed. The effects of fracture toughness, aggregate interlock and concrete non-linearity were assessed. It was found that the aggregate interlock did play a crucial role.

Bažant and Oh [3] developed a fracture theory for a heterogeneous material for which the damage zone is not small if compared to structural dimensions and exhibits a gradual strain-softening due to microcracking. Simple triaxial stress-strain relations which model the strain-softening and describe the effect of crack formation and extension on the deformation were derived. In substance, they applied a method similar to that by Hillerborg and used eq. (29) in the implementation of a finite element code. A statistical analysis of the errors exhibited by test data compared to the theory revealed a drastic improvement compared to Linear Elastic Fracture Mechanics as well as to the strength theory.

Wecharatana and Shah [4] recently developed a numerical iterative procedure based on the fictitious crack model to include the effect of large scale crack tip non-linearity and to predict the extent of this non-linear process zone. They proved that the value w_c of the critical crack opening displacement strongly affects the length of the process zone, whereas the latter is non-sensitive to the shape of the $\sigma - w$ diagram. Visalvanich and Naaman [32] proposed a generalized $\sigma - w$ law for fibre reinforced mortar and plain concrete, which reproduces the experimental results very well. Gjørv and Løland [33] combined the damage model by Janson and Hult [7] with the fictitious crack model by Hillerborg, Modéer and Petersson [2].

Bressi and Ferrara [34] carried out two series of three point bending tests and ring tests to study thickness, size and shape effects on the evaluation of toughness parameters. The experimental results were in agreement with those obtained by a linear elastic finite element method, once taken into account the interaction between ultimate strength collapse at the ligament and brittle fracture. Then, at the "Centro di Ricerca Idraulica e Strutturale" of the "Ente Nazionale Energia Elettrica" (ENEL-CRIS) the computer code FRANA was developed by Colombo and Limido [6] according to the fictitious crack model [23]. Such a code was recently used with success in a joint research program pursued by the "Istituto di Scienza delle Costruzioni" of the University of Bologna and by ENEL-CRIS.

3.3 Description of the Code FRANA [6]

The closing stresses acting on the crack surfaces (Fig. 16-a) are replaced by nodal forces (Fig. 16-b). The intensity of these forces depends on the opening of the fictitious crack, w, according to the $\sigma - w$ curve of the material. When the tensile strength is achieved at the fictitious crack tip (Fig. 16-b), the top node is "opened" and a cohesive force starts acting across the crack at this point, while the fictitious crack tip moves to the next node. In this way it is possible to follow the crack growth through the material [23].

With reference to the three point bending test of Fig. 17, the nodes are distributed along the fracture line. It is impossible to extend the node pairs to the whole cross-section depth, as would be required to follow the fracture process up to the final collapse, since a sufficiently large ligament is needed to guarantee a correct structural analysis. A ligament equal to $b/10$ is assumed.

The coefficients of influence in terms of node pair opening and central beam deflection are computed by a finite element analysis where the fictitious structure of Fig. 17 is subjected to $(n + 1)$ different loading conditions. Let us consider the three point bending test of Fig. 18-a with the initial crack tip in the node k. The crack opening at each node pair may be expressed as follows:

$$\{w\} = [K]\{F\} + \{C\}P \quad , \tag{30}$$

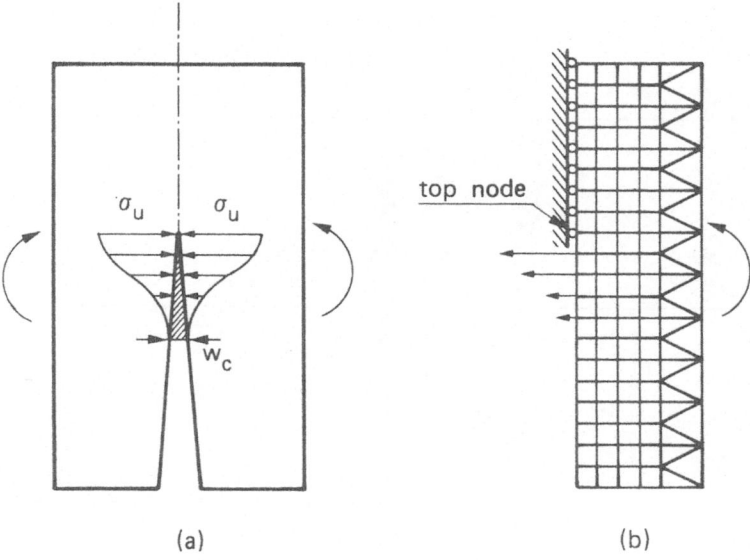

Figure 16 - Stresses acting across the fictitious crack (a) and equivalent nodal forces in the finite element mesh (b).

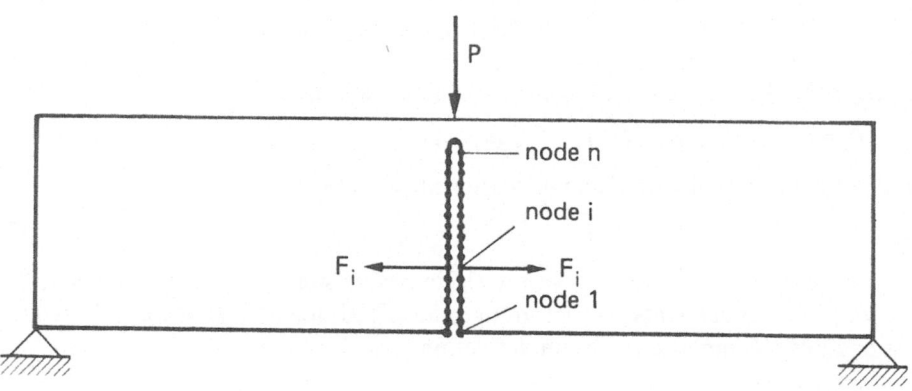

Figure 17 - Finite element nodes along the fictitious crack line.

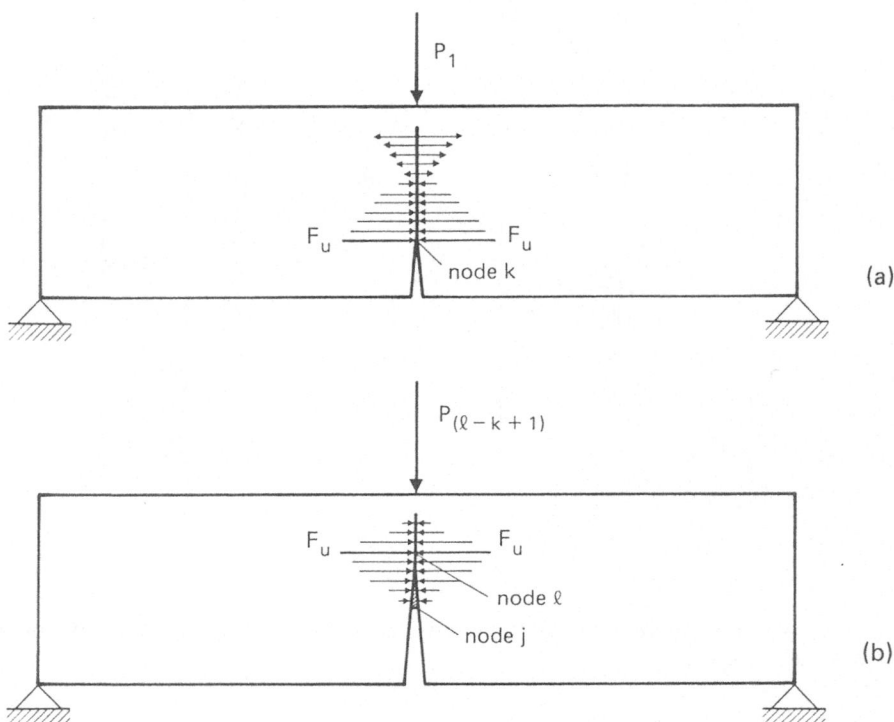

Figure 18 - Fictitious crack line at the first loading step (a) and at $(\ell - k + 1)$ th loading step (b).

where: $\{w\}$ = vector of the crack openings at the n nodes;
        $[K]$ = matrix of the coefficients of influence $(F_i = 1)$;
        $\{F\}$ = vector of the n closing forces;
        $\{C\}$ = vector of the coefficients of influence $(P = 1)$;
        P = external load.

On the other hand, the initial crack is stress-free, and therefore:

$$F_i = 0 \quad , \quad \text{for} \quad i = 1, \ldots, (k-1) , \tag{31-a}$$

while at the ligament there is no displacement discontinuity:

$$w_i = 0 \quad , \quad \text{for} \quad i = k, \ldots, n . \tag{31-b}$$

Eqs (30) and (31) constitute a linear algebraical system of $2n$ equations and $2n$ unknowns, i.e. the elements of vectors $\{w\}$ and $\{F\}$. If load P and vector $\{F\}$ are known, it is possible to compute the beam deflection, δ:

$$\delta = \{C\}^T \{F\} + D_p P , \tag{32}$$

where D_p is the deflection when $P = 1$.

The code FRANA is an iterative procedure consisting in a repeated solution of eqs (30), conditions (31) being modified step by step to follow the crack growth. In fact, after the first step, a process zone forms in front of the crack tip (Fig. 18-b), say between nodes j and ℓ; then, eqs (31) are replaced by:

$$F_i = 0 \qquad\qquad , \text{ for } \quad i = 1, \ldots, (j-1), \tag{33-a}$$

$$F_i = F_u \left(1 - \frac{w_i}{w_c}\right) \qquad , \text{ for } \quad i = j, \ldots, \ell, \tag{33-b}$$

$$w_i = 0 \qquad\qquad , \text{ for } \quad i = \ell, \ldots, n, \tag{33-c}$$

where F_u is the force corresponding to the ultimate strength:

$$F_u = 0.9\, b\, \sigma_u / n . \tag{34}$$

Eqs (30) and (33) constitute a linear algebraical system of $(2n + 1)$ equations and $(2n + 1)$ unknowns, i.e. the elements of vectors $\{w\}$ and $\{F\}$ and the load P .

At the first step the process zone is missing $(\ell = j = k)$ and the load P_1 producing the ultimate force F_u at the crack tip (node k) is computed. Such a value, P_1, together with the related deflection δ_1 computed through eq. (32), gives the first point of the curve $P - \delta$. At the second step, the process zone is between the nodes k and $(k + 1)$, and the load P_2 producing the ultimate force at the new crack tip (node $k + 1$) is computed. Eq. (32) then gives the deflection δ_2. At the third step, the fictitious crack tip is in the node $(k + 2)$, and so on. The program stops with the un-tieing of the nth node pair and, consequently, with the determination of the last pair of values F_n and δ_n. In this way, the complete load-deflection curve is automatically plotted by the computer.

3.4 Numerical Results According to the Cohesive Model

The unreinforced three point bend specimen analyzed in section 2 is considered again (Fig. 11). The material presents ultimate tensile strength $\sigma_u = 31.9$ kg/cm^2, Young's modulus $E = 365,000$ kg/cm^2, Poisson ratio $\nu = 0.1$ and fracture energy $\mathcal{G}_F = 0.05$ kg/cm. These values correspond to the material of Fig. 4. In that case, the critical parameter for rapid crack propagation was the strain energy density factor, $S_c = 8 \times 10^{-3}$ kg/cm, which is related to the fracture energy $\mathcal{G}_F$ through the following relation [13]:

$$\mathcal{G}_F = \mathcal{G}_{IC} \simeq 2\pi S_c . \tag{35}$$

On the other hand, the damage model presented in section 2 requires an additional material constant: the fracture strain ϵ_f, which is connected with the critical crack opening displacement w_c, and the crack band width w_o, through equation (29). Recalling that (Fig. 15-b):

$$w_c = 2\, \mathcal{G}_F / \sigma_u , \tag{36}$$

we obtain the value of the crack band width:

$$w_o = \frac{2 \, \mathcal{G}_F}{\sigma_u \, \epsilon_f} \; . \tag{37}$$

Recalling eq. (35), eq. (37) becomes:

$$w_o = \frac{2\pi S_c}{\left(\dfrac{dW}{dV}\right)_c} = 2\pi r_c \; . \tag{38}$$

Eq. (38) directly connects strain energy density theory [35, 36] and crack band theory [3] and, in the case of the material considered, it gives: $w_o = 6.47$ cm, which corresponds to about three times the aggregate size ($D_{max} \simeq 2$ cm).

The load-deflection curves are shown in Fig. 19-a for different initial crack depths. Of course, stiffness and loading capacity of the specimen decrease by increasing the initial crack depth. Even the slope of the softening branch decreases, so that the uncracked specimen reveals considerable instability and an abrupt drop in its loading capacity, whereas the cracked specimens appear much more ductile and controllable in the $P - \delta$ descending stage.

The load-deflection curves in Fig. 19-b describe the case when all the data are the same as in the preceding case except the fracture energy, $\mathcal{G}_F = 0.01$ kg/cm. For small crack depths ($a_0/b \leqslant 0.2$) a very interesting bifurcation of the global equilibrium occurs. That is, a softening branch is revealed for which: $dP/d\delta > 0$. However, it is distinct from the ascending branch and could be defined "stable" according to the Drucker's postulate. If the loading process is controlled by the deflection, the $P-\delta$ curve will show a discontinuity in its loading capacity and the representative point will drop on the fourth branch of the curve, where: $dP/d\delta < 0$ (Fig. 20).

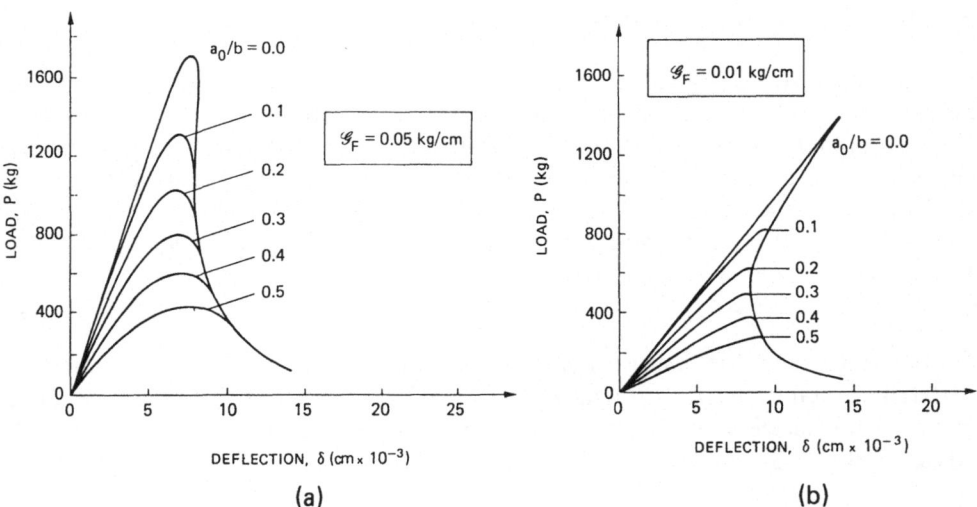

Figure 19 - Load-deflection plots by varying initial crack depth:
(a) $\mathcal{G}_F = 0.05$ kg/cm ; (b) $\mathcal{G}_F = 0.01$ kg/cm.

Probably, a way for evidencing the stable softening branch (third branch of the curve in Fig. 20) is to control the crack mouth opening displacement, w_1, instead of the beam deflection, δ. In fact, as is shown in Fig. 21, the crack mouth opening displacement increases while the load P and the beam deflection δ decrease.

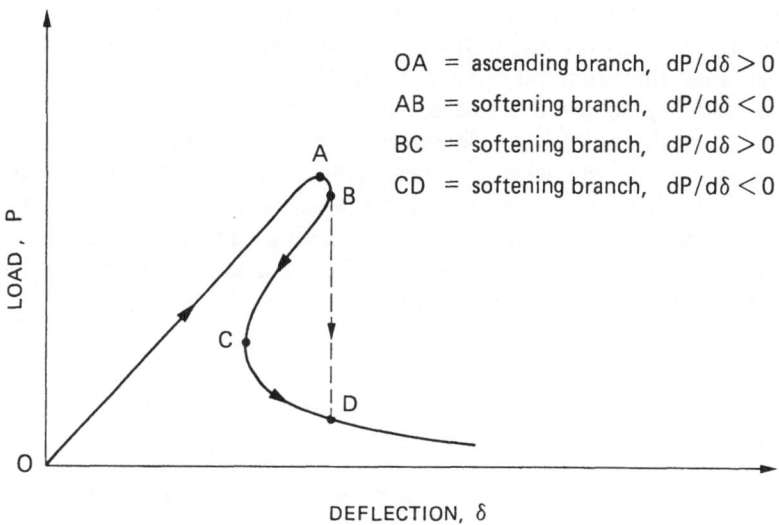

OA = ascending branch, $dP/d\delta > 0$
AB = softening branch, $dP/d\delta < 0$
BC = softening branch, $dP/d\delta > 0$
CD = softening branch, $dP/d\delta < 0$

Figure 20 - Bifurcation of the global equilibrium.

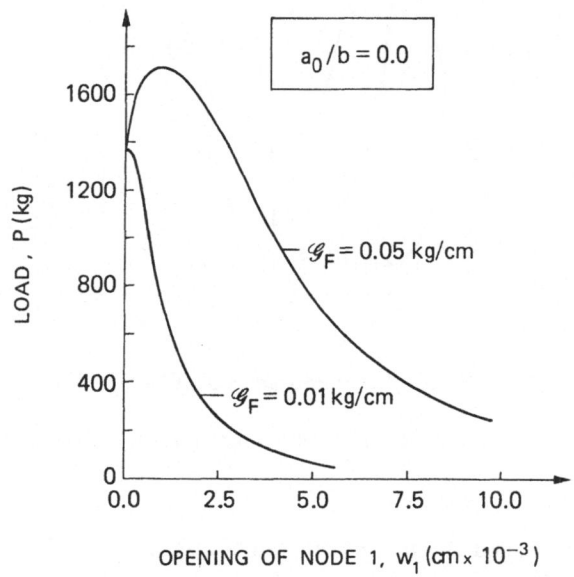

Figure 21 - Load vs. crack mouth opening displacement.

The non-dimensional load-deflection curves are represented in Fig. 22 for $a_o/b = 1/3$ and varying the non-dimensional number:

$$s_E = \frac{\mathscr{G}_F}{\sigma_u \, b} = \frac{w_c}{2b} \quad . \tag{39}$$

Such a *brittleness number* [16] governs the scale effects of fracture mechanics when a cohesive model is assumed, e.g. the fictitious crack model. When the fracture energy $\mathscr{G}_F$ is very high as, for example, in the fibre reinforced concretes, or when the beam size b is very small, the structural behaviour is ductile. It is very brittle in the opposite cases and a tendency to reveal stable softening ($dP/d\delta > 0$) appears. On the other hand, above certain sizes and/or below certain fracture energies, i.e. below certain brittleness numbers s_E, the mesh becomes unable to give reliable numerical results. This is due to the fact that, in these cases, the cohesive forces at the crack tip are confined to a very small region and become inessential. Therefore, the crack growth is governed by the stress-singularity instead of by the process zone. On the other hand, the mesh is too coarse to reproduce the stress-singularity.

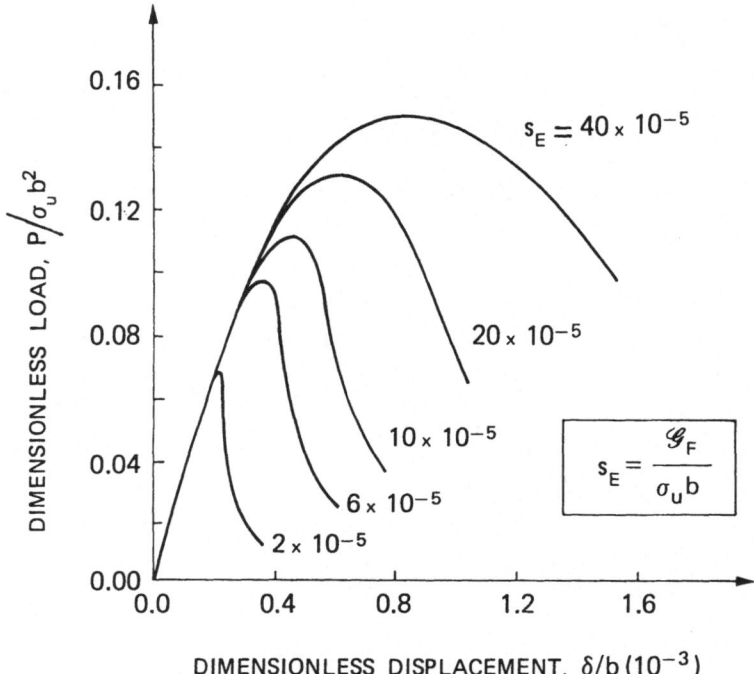

Figure 22 - Non-dimensional load-deflection plots by varying the brittleness number, $s_E = w_c/2b$.

If the mesh is refined the numerical results improve and the threshold value of s_E, below which the results are unacceptable, decreases. For example, the mesh of Fig. 23-a imposes the condition: $s_E \gtrsim 4 \times 10^{-5}$, whereas for the mesh of Fig. 23-b,

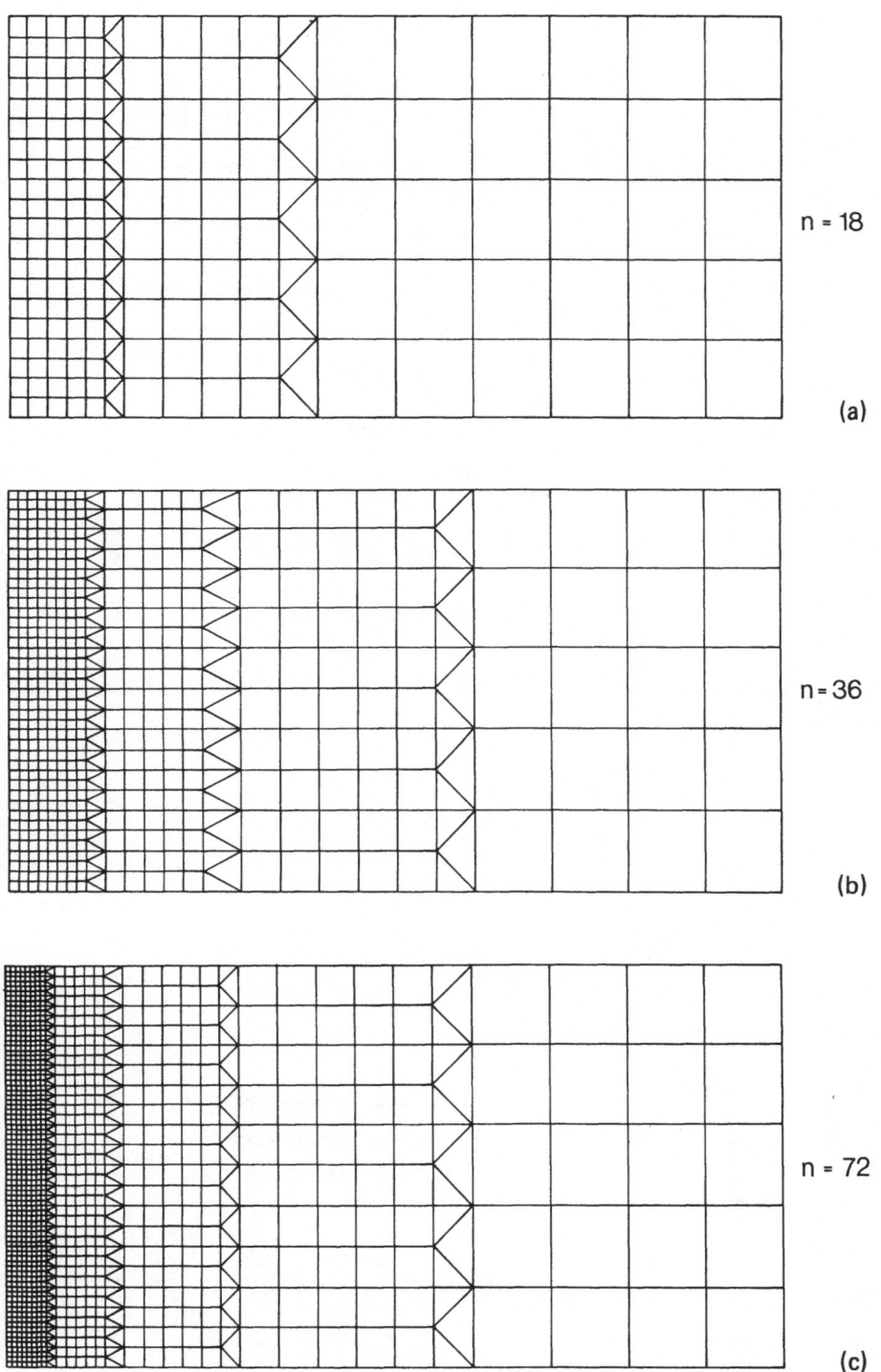

n = 18

(a)

n = 36

(b)

n = 72

(c)

Figure 23 - Refinement of the finite element mesh.

312

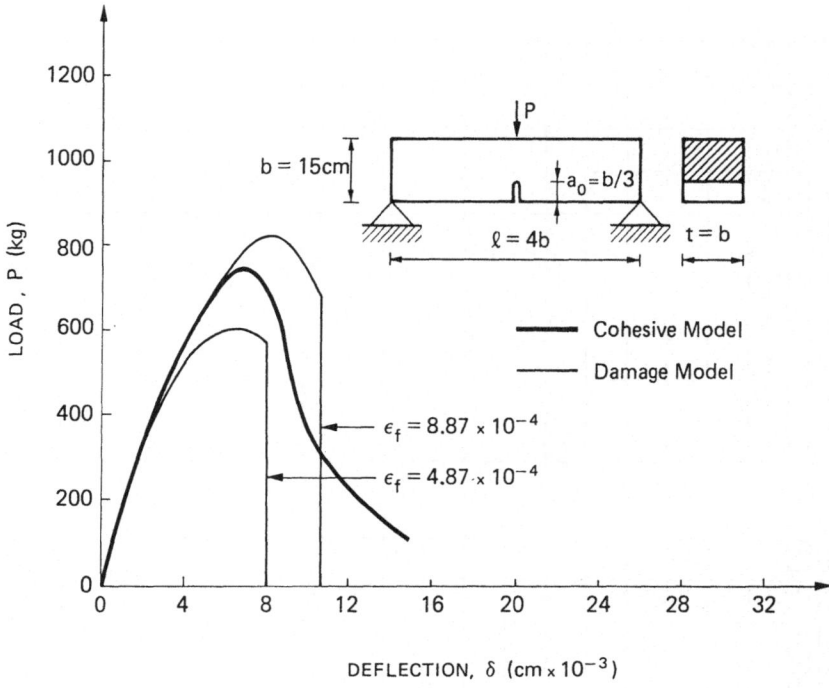

Figure 24 - Load-deflection plots obtained through the application of damage and cohesive models.

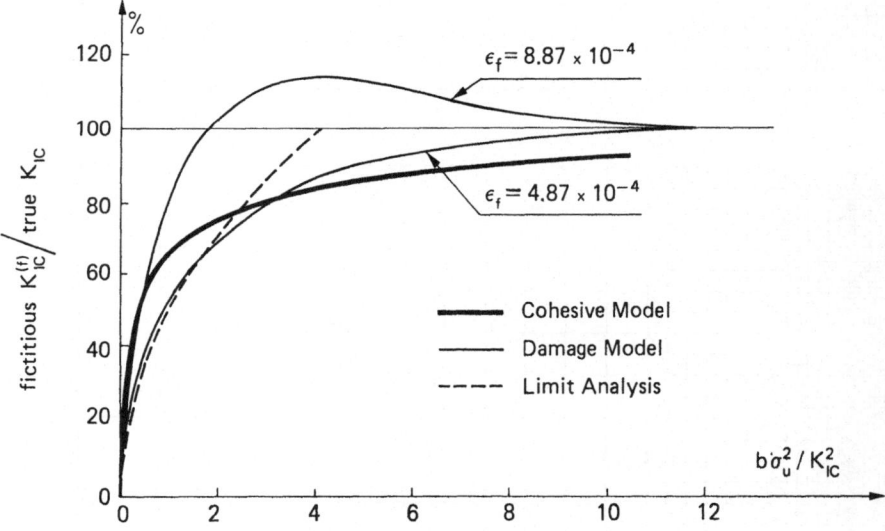

Figure 25 - Transition from structural instability to crack instability by varying the size scale.

with twice the number of crack nodes, it has to be: $s_E \gtrsim 2 \times 10^{-5}$. Finally, the mesh of Fig. 23-c imposes: $s_E \gtrsim 1 \times 10^{-5}$.

On the other hand, when $s_E \rightarrow 0$, Linear Elastic Fracture Mechanics does work better than any cohesive fracture model. In this case, the process zone is much smaller than the area where the stress-singularity is dominant, and this is true even for concrete!

If there is not the possibility of embedding the stress-singularity into the crack tip region, the mesh must be refined. Recalling eq. (39) and that the distance d between two consecutive fracture nodes is equal to $b/20$ in the mesh of Fig. 23-a, to $b/40$ in that of Fig. 23-b and to $b/80$ in that of Fig. 23-c, the lower bound to the brittleness number, s_E:

$$\frac{w_c}{160\, d} \gtrsim 10^{-5}, \tag{40}$$

becomes an upper bound to the finite element size:

$$d \leqslant 625\, w_c. \tag{41}$$

For a normal concrete we have approximately $w_c \simeq 0.1$ mm and then eq. (41) provides: $d \leqslant 60$ mm.

4.   CONCLUDING REMARKS

The process zone at the crack tip of a concrete-like material can be simulated in two different ways (Fig. 1):
(1)   with a *damage zone* in front of the stress-free crack tip, or
(2)   with a *cohesive force distribution* behind the fictitious crack tip.

The former model includes and additional constitutive parameter if compared to the latter one: the fracture strain ϵ_f. In Fig. 24 the load-deflection curves are shown according to both non-linear models and considering the concrete parameters previously assumed. It is clear that the damage model is sensitive to the variation in the fracture strain ϵ_f, while the cohesive model does not contain this parameter. For common values of ϵ_f, however, the cohesive solution is substantially in accordance with the damage ones (Fig. 24).

The ratio between maximum load computed according to non-linear models and maximum load obtained through the application of Linear Elastic Fracture Mechanics, is reported in the diagram of Fig. 25 as a function of specimen size. Such a ratio is equal to that between fictitious $K_{IC}^{(f)}$ and true K_{IC}, where the fictitious $K_{IC}^{(f)}$ is the value of the stress-intensity factor obtained at the maximum load bearable by the specimen, whereas the true K_{IC} is the critical value of the stress-intensity factor

314

considered as a constant material property.

From Fig. 25 it appears clear that the two extreme situations of ultimate strength collapse at the ligament (small size) and brittle fracture (large size) are connected by a transition, which can be described by both cohesive and damage models.

ACKNOWLEDGEMENTS

The author wishes to thank Prof. George C. Sih for the very stimulating and fruitful discussions he had with him at Lehigh University, and Prof. Angelo Di Tommaso for providing valuable comments and support in writing this manuscript. Special gratitude is due to Prof. Michele Fanelli for permission to make use of numerical results obtained during the research convention between University of Bologna and ENEL–CRIS–Milano.

REFERENCES

[1] Carpinteri, A. and Sih, G.C., Damage accumulation and crack growth in bilinear materials with softening: application of strain energy density theory, *Report IFSM-83-115, Institute of Fracture and Solid Mechanics, Lehigh University* (1983).

[2] Hillerborg, A., Modéer, M. and Petersson, P.E., Analysis of crack formation and crack growth in concrete by means of fracture mechanics and finite elements, *Cement and Concrete Research,* 6, pp. 773-782 (1976).

[3] Bažant, Z.P. and Oh, B.H., Concrete fracture via stress-strain relations, *Report 81-10/665c, Center for Concrete and Geomaterials, Northwestern University* (1981).

[4] Wecharatana, M. and Shah, S.P., Predictions of nonlinear fracture process zone in concrete, *Journal of Engineering Mechanics,* 109, pp. 1231-1246 (1983).

[5] Gerstle, W.H., Ingraffea, A.R. and Gergely, P., The fracture mechanics of bond in reinforced concrete, *Report 82-7, Department of Structural Engineering, Cornell University* (1982).

[6] Colombo, G. and Limido, E., Un metodo numerico per l'analisi di prove TPBT stabili; confronto con alcuni dati sperimentali, *XI Convegno Nazionale dell'Associazione Italiana per l'Analisi delle Sollecitazioni,* Torino, pp. 233-243 (1983).

[7] Janson, J. and Hult, J., Fracture mechanics and damage mechanics - a combined approach, *Journal de Mécanique Appliquée,* 1, pp. 69-84 (1977).

[8] Janson, J., Dugdale-crack in a material with continuous damage formation, *Engineering Fracture Mechanics,* 9, pp. 891-899 (1977).

[9] Janson, J., Damage model of crack growth and instability, *Engineering Fracture Mechanics,* 10, pp. 795-806 (1978).

[10] Dugdale, D.S., Yielding of steel sheets containing slits, *Journal of Mechanics and Physics of Solids,* 8, pp. 100-104 (1960).

[11] Broberg, H., Damage measures in creep deformation and rupture, *Swedish Solid Mechanics Reports* (1974).

[12] Kachanov, L.M., Time to failure under creep conditions, *AN SSSR, OTN,* 8 (1958).

[13] Sih, G.C., Some basic problems in fracture mechanics and new concepts, *Engineering Fracture Mechanics*, 5, pp. 365-377 (1973).

[14] Sih, G.C., and Macdonald, B., Fracture mechanics applied to engineering problems — Strain energy density fracture criterion, *Engineering Fracture Mechanics*, 6, 361-386 (1974).

[15] Hilton, P.D., Gifford, L.N. and Lomacky, O., Finite element fracture mechanics of two dimensional and axisymmetric elastic and elastic-plastic cracked structures, *Report 4493, Naval Ship Research and Development Center* (1975).

[16] Carpinteri, A., Notch sensitivity in fracture testing of aggregative materials, *Engineering Fracture Mechanics*, 16, pp. 467-481 (1982).

[17] Carpinteri, A., Application of fracture mechanics to concrete structures, *Journal of the Structural Division* (ASCE), 108, pp. 833-848 (1982).

[18] Carpinteri, A., Plastic flow collapse vs. separation collapse in elastic-plastic strain-hardening structures, *Materials & Structures* (RILEM), 16, pp. 85-96 (1983).

[19] Carpinteri, A., A fracture mechanics model for reinforced concrete collapse, *IABSE Colloquium on Advanced Mechanics of Reinforced Concrete*, Delft, pp. 17-30 (1981).

[20] Carpinteri, A., Stability of fracturing process in R.C. beams, *Journal of Structural Engineering* (ASCE), 110, pp. 544-558 (1984).

[21] Carpinteri, A., Di Tommaso, A. and Viola, E., Sulla capacità portante limite di pareti lapidee lesionate, *5 th Italian Congress of Theoretical and Applied Mechanics*, Palermo, pp. 93-104 (1980).

[22] Carpinteri, Al. and Carpinteri, An., Softening and fracturing process in masonry arches, *6th International Brick Masonry Conference*, Roma, pp. 502-510 (1982).

[23] Petersson, P.E., Crack growth and development of fracture zones in plain concrete and similar materials, *Report TVBM-1006, Division of Building Materials, Lund Institute of Technology* (1981).

[24] Heilmann, H.G., Hilsdorf, H.H. and Finsterwalder, K., Festigkeit und Verformung von Beton unter Zugspannungen, *Deutscher Ausschuss für Stahl-beton*, 203 (1969).

[25] Rice, J.R., The localization of plastic deformation, *Theoretical and Applied Mechanics*, Proceedings of the 14th IUTAM Congress, Delft, pp. 207-220 (1976).

[26] Needleman, A. and Rice, J.R., Limits to ductility set by plastic flow localization, *Mechanics of Sheet Metal Forming*, Edited by Donald P. Koistinen and Neng-Ming Wang, Plenum Publishing Corporation, pp. 237-265 (1978).

[27] Barenblatt, G.J., The mathematical theory of equilibrium crack in the brittle fracture, *Advance in Applied Mechanics*, 7, pp. 55-125 (1962).

[28] Rice, J.R., Path independent integral and approximate analysis of strain concentration by notches and cracks, *Journal of Applied Mechanics*, 35, pp. 379-386 (1968).

[29] Palmer, A.C. and Rice, J.R., The growth of slip surfaces in the progressive failure of overconsolidated clay, *Proceedings of the Royal Society*, London, A332, pp. 527-548 (1973).

[30] Chappell, J.F. and Ingraffea, A.R., A fracture mechanics investigation of the cracking of Fontana dam, *Report 81-7, Department of Structural Engineering, Cornell University* (1981).

[31] Saouma, V.E., Interactive finite element analysis of reinforced concrete: a fracture mechanics approach, *Report 81-5, Department of Structural Engineering, Cornell University* (1981).

[32] Visalvanich, K. and Naaman, A.E., Fracture modeling of fiber reinforced cementitious composites, *Report 82-1, Department of Materials Engineering, University of Illinois at Chicago Circle* (1982).

[33] Gjørv, O.E. and Løland, K.E., Ductility of concrete and tensile behaviour, *Rapport BML 80.613, Institutt for Bygnings-materiallaere, Universitetet i Trondheim* (1980).

[34] Bressi, D.B. and Ferrara, G., Valutazione sperimentale e numerica dei parametri di tenacità alla frattura per un microcalcestruzzo, *IX Congresso AIAS,* Trieste (1981).

[35] Sih, G.C., Mechanics of crack growth: geometrical size effect in fracture, *Fracture Mechanics in Engineering Application,* Edited by Sih, G.C. and Valluri, S.R., Sijthoff and Noordhoff, pp. 3-29 (1979).

[36] Sih, G.C., Analytical aspects of macro-fracture mechanics, *Analytical and Experimental Fracture Mechanics,* Edited by Sih, G.C. and Mirabile, M., Sijthoff and Noordhoff, pp. 3-15 (1981).

SECTION IV

EXPERIMENTAL METHODS OF DETERMINING
FRACTURE PARAMETERS

APPLICATION OF FRACTURE MECHANICS
TO CEMENTITIOUS COMPOSITES
NATO-ARW - September 4-7, 1984
Northwestern University, U.S.A.
S. P. Shah, Editor

NONLINEAR FRACTURE PARAMETERS FOR CEMENT BASED COMPOSITES:
THEORY AND EXPERIMENTS

Y. S. Jenq and S. P. Shah

Department of Civil Engineering
Northwestern University
Evanston, Illinois 60201

Abstract

Many have attempted to apply linear elastic fracture
mechanics to quantitatively express the fracture toughness of
concrete. For example, by testing notched beams one can calculate,
using the formulas developed from LEFM, fracture toughness in
terms of K_{Ic}, the critical stress intensity factor for Mode I
opening. Unfortunately, it has been observed that when K_{Ic} is
calculated from the measured values of the maximum load and the
initial notch length, its value is dependent on the dimensions of
the beams. This size dependency of the single   parameter LEFM
based fracture criterion can be attributed primarily to the non-
linear effects associated with crack propagation in concrete.
In this paper, a simple method is suggested to calculate a size-
independent fracture toughness parameter.

Introduction

It was Griffith who first proposed the basic concept of
fracture mechanics in 1921. However, it was not until 1961 that
Kaplan [1] applied the theory of fracture mechanics to concrete.
He performed three and four point bending tests on two different
sizes of specimens and determined the strain energy release rate,
G_c. Kaplan found that G_c varied greatly with the specimen size.
Slow crack growth was pointed out to be the possible factor that
influenced the results. However, this effect was not included in
the analysis. In 1964, Lott and Kesler [2] tested notched-beams
by 4-PB. The specimens were 4" x 4" x 12". They mentioned that

the resistance of concrete to the propagation of an existing crack
is a function of the cement paste matrix and also of the heterogeneity of the concrete. They proposed the concept of pseudo-
stress-intensity factor and concluded the puseudo-stress-intensity
factor is independent of the water-cement ratio and only varies
directly with the coarse aggregate.

Moavenzadeh and Kuguel [3] measured the critical strain
energy release rate, G_c; the critical stress intensity factor, K_c;
and the fracture surface energy, γ; of notch-beam specimens. The
specimen sizes were 1" x 1" x 12" and were subjected to 3-PB. The
actual fracture area was determined by microscopic technique.
They concluded that the true surface energies of cement paste,
mortar, and concrete were approximately the same and the fracture
of concrete specimens was associated with substantial amounts of
side cracking which did not happen in cement paste and mortar
specimens.

Naus and Lott [4] reported the influence of water-cement
ratio, air content, curing time, fine aggregate content, and gravel
content to the effective fracture toughness $\bar{K}_c$. The specimens'
sizes were 2" x 2" x 14" for cement paste and mortar, and 4" x 4"
x 12" for concrete. Four point-bend tests was used. The authors
concluded that the effective fracture toughness increased with
the increasing age, maximum size of aggregate, gravel-cement
ratio (concrete), and water-cement ratio (paste and mortar), and
decreased with the increasing air content. However, the water-
cement ratio had no apparent effect on the effective fracture
toughness of concrete.

In 1971, Shah and McGarry [5] tested 2" x 2" x 22" notched
beams using 3-PB. Different notch-depth ratios and mixed proportions were used. They concluded that cement paste is notch-
sensitive, while the concrete and mortar are notch-insensitive.
The fracture toughness was reported to increase with increasing
volume of aggregate and size of aggregate.

Walsh [6] studied the size effect on fracture toughness.
Three different specimen sizes were tested with the same notch-
depth ratio. The beam depths were ranging from 3" to 15". He
concluded a minimum depth of 9" should be used for fracture test
of concrete.

4-PB and DCB (double cantilever beam) tests were performed
by Brown [7] to measure the fracture toughness of cement paste
and mortar. Tests of both materials showed that the fracture
toughness of cement paste was independent of crack growth but
the toughness of mortar increased as the crack propagated.

Moreover, the stress intensity factor required to initiate crack growth was less than that to maintain crack growth at the loading rates used. The crack lengths of these tests were determined by compliance method.

Higgins and Bailey [8] reported the 3-PB test results of cement pastes. They found that the fracture toughness calculated from LEFM varied with specimen sizes, and suggested that the concepts of LEFM are not readily applicable to hardened cement paste.

Mindess and Naudeau [9] investigated the effect of specimen width on the fracture toughness of concrete and mortar. The specimen sizes were 8 in. long and 2 in. deep, but with the width varying from 2 in. to 10 in. It was concluded that within the size range studied, there was no dependence of fracture toughness upon the length of the crack front.

Gjørv, Sørensen, and Arnesen [10] studied the applicability of LEFM in mortar and concrete. 2" x 2" x 22" notched beams were tested with 3-PB. They found that the fracture toughness of concrete and mortar were decreasing with the increasing of notch-depth ratio. Only toughness of cement paste and lightweight concrete were more or less constant.

Swartz, Hu, and Jones [11] used compliance technique to calibrate the crack length of beams precracked by fatigue. Swartz, Hu, and Huang [12] tested a series of experiments to compare the fracture toughness of natural cracks and artificial cracks (i.e., casting cracks). The fracture toughness of naturally cracked beams were found to be larger than those of beams with artificial cracks.

Strange and Bryant [13] examined the fracture toughness of cement paste, and concrete using beam test and tension test. They concluded that fracture toughness was dependent on the specimen sizes.

Recently, Wecharatana and Shah [14, 15, 16, 17], proposed a model by considering the effect of slow crack growth and nonlinear fracture process zone. The results reported showed that only by taking the slow crack growth and nonlinear process zone into the analysis work, can one get a true fracture toughness that is independent of geometries of specimens.

It is the purpose of this paper to explain the inconsistency of results reported by former investigators.

INELASTIC DISPLACEMENTS

A plot of load vs. crack mouth opening displacement (CMOD) of a concrete beam obtained from the experiments to be described later is shown in Fig. 1. A significant amount of inelastic displacement can be observed when the specimens are unloaded immediately after the peak load. At any given load, the total displacement u^T (load-point deflection or CMOD) is composed of a sum of the elastic displacement (u_0^e), inelastic displacement (u^*), and the elastic displacement due to slow crack growth (u_s^e) (as shown in Fig. 1). The nonlinear displacement can be due to creep, microcracking and slow crack growth. For the short-term loading considered here, it is assumed that creep is negligible. It is clear that in order to apply LEFM, the inelastic displacement should be extracted from the total displacement. It was observed that the inelastic displacement at the peak load can be related to the total displacement at the peak load as:

$$u^T = \alpha \, u^* \tag{1}$$

where the constant α is considered a material property and was established from the cyclic tests to be reported later.

Inelastic displacement during crack growth in cement composites appear primarily due to friction associated with roughness of cracks and geometrical interlock (aggregate interlock, fiber bridging, etc.). To incorporate this nonlinearity, an actual crack is represented by an effective Griffith crack and a crack-closing pressure as shown in Fig. 2. Further away from the crack tip at a large crack opening displacement the interlocking effect should diminish. To obtain the relationship between the crack opening displacement and crack closing pressure, double-edged notched specimens were tested by Gopalaratnam and Shah [22]. They reported that the relationship between average uniaxial tensile stress (σ) and the average crack width (w) can be given by:

$$\sigma = \sigma_p \, (e^{-Kw^\lambda}) \tag{2}$$

where K and λ are constants which depend on the material composition, σ_p is the uniaxial tensile strength of the material, and w is the crack width. The above expression was used to relate the crack opening displacement and crack closing pressure to derive fracture toughness parameter from the results of the notched beam tests to be reported later. It should be noted that the elastic crack opening displacement must be transformed into total crack opening displacement prior to using Eq. (2).

GLOBAL ENERGY CONSIDERATION

From global energy balance concept, the energy required for crack formation can be expressed as:

$$W = F - U \tag{3}$$

where W = total energy consumed for crack formation, F = work done by external load, and U = elastic strain energy.

The crack resistance or the strain energy release rate (G_R) can be expressed as:

$$G_R = \frac{dW}{da} = \frac{d}{da} (F - U) \tag{4}$$

where a = the effective crack length. For a crack with closing pressure and stress singularity at the crack tip, the total energy release rate can be separated into two parts: The contribution due to stress singularity and the energy absorbed in the nonlinear zone (Fig. 2).

The strain energy release rate for the critical section can be derived as:

$$G_R = \frac{d}{da} (W = W_e + W_p) = \frac{K_I^2}{E'} + \int_0^{CTOD} \sigma(w)dw \tag{5}$$

Where W_e and W_p are the elastic and inelastic energies consumed during formation of new surface; K_I = stress intensity factor due to applied loads and closing pressure, CTOD = crack opening displacement at the original crack tip (see Fig. 2), and $\sigma(w)$ is the closing pressure (see Eq. 2) and $E' = E/1-v^2$ for plain strain, E = Young's modulus of elasticity and v = Poisson's ratio. The partition of G into elastic and plastic components has been suggested for metals by Liebowitz and Jones [23].

Note that Eq. (5) is a general expression of strain energy release rate. If K_I is specified to be zero, Barenblatt or Dugdale model will result. However, if we assume zero closing pressure, the conventional expression for strain energy release rate in LEFM is derived.

DETERMINATION OF K_{Ic}

The phenomenon of slow crack growth prior to the peak load has long been noticed. In addition to the bridging phenomena resulting from the heterogeneous nature of concrete, the differences in the stress state along the crack front can also account for this observed slow crack growth. It has been noted that the length of the crack front at the surface of the beams is longer than that at the interior. Thus, a channel shaped slow crack growth profile was observed by Swartz and Go [19]. Therefore, it is difficult to precisely measure the length of the crack at the peak load.

To overcome these difficulties in measuring the exact length of the crack an effective crack length is defined in this paper. The effective crack length is the sum of the initial notch plus an effective crack extension at the peak load. The effective crack length is determined so that the crack mouth opening displacement calculated using LEFM at the peak load is equal to the measured elastic CMOD. Once the effective crack length is determined, then one can calculate using LEFM the value of K_{Ic}. This value of K_{Ic} was found to be specimen size independent. In addition, using this constant value of K_{Ic} it was possible to accurately predict the entire load vs. CMOD (as well as load vs. load-point deflection) relationship for beams tested in this investigation as well as by other researchers.

To theoretically determine CMOD due to the applied load and due to the closing pressure, principle of LEFM and superposition scheme were used (Fig. 3a). The calculation of the stress intensity factor (K_I^E) and the corresponding CMOD were based on already available equations [26]. For determining the stress intensity factor and the corresponding CMOD due to the closing pressure, a solution based on an infinite strip subjected   an unit load is used as Green's function (Fig. 3b). Gauss-Chebyshev integration   scheme was employed to integrate the contribution of closing pressure. More details about calculating K_{Ic} were given in Ref 31.

TEST PROGRAM AND EXPERIMENTAL DETAILS

Three point bend test was used to verify the validity of the proposed fracture criteria. Three different sizes of beams were used (Fig. 4). These beams were designated as large (L), medium (M),

and small (S). The span (s), depth (b), and thickness (t) for
these beams were respectively; 36 in. (914mm), 9 in. (229mm), and
3.375 in. (85.7mm); 24 in. (609mm), 6 in. (152mm), 2.25 in. (57.2mm);
and 12 in. (305.mm), 3 in. (76mm), and 1.125 in. (28.6mm). Four
series of different mixes were prepared for studying the effect of
the maximum aggregate size and water-cement ratio on the fracture
toughness K_{Ic}. The mix proportions are listed in Table 1. All
beams had an initial notch-depth which was equal to one-third of
their depth. These notches were precast with a thin steel plate
whose thickness was 1/32 in. (0.8mm). The beams were tested after
curing for approximately 90 days in an environment with 96% relative humidity and 80°F (26.7°C).

The large and medium size specimens were tested in a closed-
loop servo controlled testing machine with a capacity of 120 Kips
(534 KN) while the small beams were tested in a servo controlled
closed-loop testing machine with a capacity of 20 Kips (89KN).
All beams were tested so as to maintain a constant rate of increase of CMOD, which was measured by a clip gage. The
peak load was reached in about 10 minutes. During testing the
deflection of the beam was measured by an LVDT located at about
1.5 to 2 in. (38 to 51mm) away from the notch and right under the
center line of the beam. The surface crack length was monitored
by a microscope with an accuracy of 0.001 in. (0.0025mm). The
overall set-up is shown in Fig. 5. There were at least two beams
for each size and mix proportion; a total of 40 beams were tested
and analyzed. For each size and each mix proportion at least one
beam was loaded cyclically (Fig. 6). For the sake of brevity only
the results of concrete beams are given in somewhat more details
in the following section.

DISCUSSION OF TEST RESULTS

The test results of beams made with concrete (C1 series) are
summarized in Table 2 and Figs. 7, 8, and 9. Table 2
shows values of fracture toughness K_{Ic} calculated using the procedure described earlier. Also shown are the values of conventionally calculated fracture toughness ($\bar{K}_{Ic}$). The former is
calculated based on the effective crack length (a) while the
latter is calculated based on the initial notch length (a_0). It
can be seen that unlike the values of $\bar{K}_{Ic}$, the values of K_{Ic} as
suggested here are independent of the size of the beam. The same
observation was made for beams with the two types of mortar and
paste.

To further validate the idea that the crack propagation in concrete can be predicted by using a modified fracture toughness index K_{Ic}, the entire load-CMOD curves were theoretically calculated using this constant value of K_{Ic}.

To calculate the total crack mouth opening displacement from the predicted elastic CMOD^e, the following relationship based on cyclic tests (Fig. 6) was derived for loads beyond the peak loads:

$$u^T = u^e \left(\frac{\beta}{\beta - 1}\right) + u^T_{max} \left(\frac{\beta - \alpha}{\alpha \beta - \alpha}\right), \text{ for } u^T \geq u^T_{max} \tag{6}$$

where u = CMOD or load-point deflection as defined earlier, and u^T_{max} = CMOD or load-point deflection at peak load.

The values of α and β are considered material properties and for the four compositions tested here their values are given in Table 3. Note that it was observed that for load-point deflection values, the value of β was equal to 1. Eq. 6 is not valid in determining the total load-point deflection. Global energy balance concept was used to derive this value.

The predicted values of load-CMOD curves are compared with those obtained experimentally. For all the beams tested here, a satisfactory comparison was obtained.

A comparison of microscopically observed crack extension and the corresponding values of theoretically predicted values is shown in Fig. 10. Theoretically values are the effective crack extension. The optically measured surface crack extensions are slightly larger than the theoretical crack extension. This seems reasonable since the crack growth is likely to be channel-shaped [19].

The theoretically calculated plot of crack resistance (G_R) vs. effective crack extension for three sizes of concrete beams are shown in Fig. 11. The strain energy release rate was calculated using Eq. (5). Note that the first term in Eq. (5) remains constant since K_{Ic} remains constant during the descending part of the P-CMOD curve. The variation in strain energy is due to the variation in the inelastic energy absorbed during the crack growth. This term appears to depend on the geometry of the specimen as can be seen in Fig. 11.

From the theoretically calculated R-curves it is possible to calculate the relationship between the load and load-line deflection [31]. This was done for all three sizes of beams for the descending part of the curves (Figs. 7, 8 and 9).

FRACTURE ENERGY - G_F

The fracture energy (G_F) is defined as

$$G_F = \frac{W_0 + mg\delta_0}{(b - a_0)t} \tag{7}$$

where W_0 = area under P - δ curve, mg = self weight of the beam, and δ_0 = deflection when the load is equal to zero in the descending part. This value is reported in Table 2 for the beams in Series C1. It can be observed from Table 2 that there is size effect on G_F values. Other investigators [27, 28] have also observed that. The factors that can cause size effect include: (1) The settlement of the support during loading. This settlement should be proportioned to the amount of load which in turn will depend on the size of the beam and the size of the notch. A more accurate measurement of deflection can reduce this kind of size effect; (2) Energy absorbed by the non-critical section. This non-linear energy dissipation will be proportional to the amplitude of stresses of noncritical section and the volume of the specimen that is subjected to this stress. It can be shown that the non-linear energy absorbed in the non-critical section will be lower and as a result the measured value of G_F will be lower, the larger the value of a/b (for a constant b). For beams with a constant a/b, the more energy will be absorbed the larger the specimens and the G_F values will be higher (see Table 2).

AN APPROXIMATE AND SIMPLE METHOD
OF CALCULATING FRACTURE TOUGHNESS

The precise calculation of K_{IC} is rather involved as a result of the closing pressure. Since the closing pressure is dependent on the crack opening displacement, an iterative solution using nonlinear integral is necessary [29]. A further simplification can be made if it is assumed that closing pressure is zero and adjusting the effective crack length to yield the correct value of $CMOD^e$.

To determine fracture toughness from the notched beam tests with this approximation, an initial value of $a = a_0 + \Delta\ell_e$ was assumed. For this value of a, the measured value of maximum load and using LEFM, $CMOD^e$ was calculated. This procedure was repeated until the calculated and measured values agree. The simplified, approximate values of fracture toughness (K_{IC}^S) can then be calculated. These values are shown in Table 2. It can be seen that K_{IC}^S values are independent of size and are approximately close to more accurate values of K_{Ic}.

The values of K_{IC}^S for all four matrix compositions are shown in Table 4. Size independency of K_{IC}^S for all four materials can be seen in Table 4.

The load-CMOD curves (for the descending part) were calculated using a constant value of K_{IC}^S and ignoring the closing pressure. For a given beam, a given value of K_{IC}^S and given value of a_i. P and $CMOD^e$ can be simply calculated using LEFM. The value of $CMOD^T$ was computed using Eq. (6). These predicted values compare quite well with the experimental values (Figs. 7-9).

COMPARISON WITH OTHER EXPERIMENTAL DATA

To predict fracture toughness (K_{Ic} or K_{IC}^S) using the method suggested here, one needs to know the value of the peak load and the elastic portion of the crack mouth opening displacement at the peak load. This later value can be determined from the cyclic P-CMOD curves. Results from such investigations were analyzed.

Catalano and Ingraffea [24] tested two series of concrete beams which were 6" x 12" x 18" (t x b x s) (152 x 304.8 x 121.9mm). The inelastic coefficients of α and β were determined from their P - CMOD curves to be 6.0 and 1.73 respectively. The value of K_{IC}^S was determined from the value of their peak load, CMOD at peak load and Young's modulus (calculated from the initial slope of the P-CMOD curve). The assumed notch depth was 1.3 in. (33mm) to replace the Chevron notch depth used by them. These calculated values of K_{IC}^S were essentially the same for every beam in each series. From this K_{IC}^S value, the post-peak P-CMOD curves were calculated using the simple procedure. The accuracy of the theoretical prediction for all their beams was quite acceptable as can be seen in Fig. 12.

Fig. 13 showed a comparison of the theoretically calculated P-CMOD curve with the experimental data of Gylltoft [30]. The dimensions of the beams and other pertinent data are shown in Fig. 13. The comparison is judged quite satisfactory.

Two load - CMOD curves of different specimen sizes were reported by Imperato [32]. The dimensions of these two beams are 19.69" (50cm) x 19.69" (50cm) x 78.74" (200cm) (t x b x s) for the large beam and 6.3" (16cm) x 6.3" (16cm) x 25.2" (64cm) (t x b x s) for the small beam (Fig. 4). From the photographs provided by the author, it was estimated the CMOD was measured at a distance 1" (25.4mm) below the bottom of the beam (distance H0 in Fig. 4). The results of theoretical prediction were plotted in Figs. 14 and 15 along with the experimental results. The theoretical prediction for the large beam matches well with the experimental results. For smaller beam (6.3" deep) the predicted values of CMOD were lower than the measured ones. This may be because a relatively large gauge length (10 in.) was used to record CMOD. The recorded values is likely to be greater than the actual CMOD because of the rotational effects. This may also explain lower calculated values of K_{Ic}^{S} for the smaller beam.

The results reported by Eligehausen [33] were also analyzed. The measurement point of CMOD was assumed to be 1.8" (4.57cm) above the bottom of the concrete beams (H0 = - 1.8"). Load-CMOD curve was calculated and is compared with the data in Fig. 16. The predicted values of CMOD are in good agreement with the experimental results.

Parametric Study

Using the proposed model it is possible to explain some of the experimentally observed trends. To demonstrate this, a theoretical parametric study was conducted. Values of conventionally calculated fracture toughness ($\bar{K}_{Ic}$) were obtained using the proposed model. For this parametric study it was assumed that concrete and cement paste had the same Young's modulus (3×10^{6} psi) and the same K_{Ic}^{S} value (1200 psi $\sqrt{in}$.). Partly based on the test results reported here it was assumed that the elastic crack tip opening displacements for concrete and cement paste were respectively 0.0006 in. and 0.0001 in. From proposed model, the peak load, and hence the value of $\bar{K}_{Ic}$ can be calculated. Fig. 17 shows the ratio of $\bar{K}_{Ic}/K_{Ic}^{S}$ for different specimen sizes with notch-depth ratios ranging from 0.1 to 0.8 and depth-span ratio of four. It can be seen that $\bar{K}_{Ic}$ reaches its maximum values at $a_{0}/b = 0.2 \sim 0.3$. Such a trend has been reported from the test data by several investigators [1, 2, 8, 10]. Fig. 18 expresses the size-effect on $\bar{K}_{Ic}$ for a fixed notch-depth ratio. It can be seen that for large beams, the values of $\bar{K}_{Ic}$

become essentially independent of the geometry. This trend has been reported from test results by Walsh [6]. Fig. 19 shows the size-effect on $\bar{K}_{Ic}$ for concrete as well as cement paste. This figure shows that although the size-effect is negligible for the cement paste specimens, concrete specimens exhibit a noticeable size dependency. Such different behavior between cement paste and concrete has also been noted from 3 - PB notched beam tests by many other investigators [1, 2, 5, 7, 8, 10]. Fig. 20 gives the predicted results of notch-sensitivity, (i.e., the ratio of net failure flexure strength of notched-specimen and failure flexure strength of unnotched specimens), for concrete and cement paste. It can be noticed that cement paste is more notch-sensitive than concrete. Similar notch-sensitivity results were also reported by Shah and McGarry [5] (Fig. 21) and Gjørv, Sørenson and Arnesen [10].

Size effect on modulus of rupture of concrete is shown in Fig. 22. Note that M.O.R. decreased with the increasing specimen depth as has been commonly observed from test results.

CONCLUSIONS

1. Nonlinear effect, i.e., slow crack growth, closing pressure in the process zone, and inelastic behavior of concrete, need to be considered in order to apply LEFM to concrete.

2. The stress intensity factor, K_{Ic}, calculated as proposed here, is found to be size-independent and ramains constant during further crack propagation. It is concluded that K_{Ic} is a size-independent fracture toughness in the sense of stress.

3. Work done to break the beam specimens into two halves (or G_F) are found to be size-dependent for concrete, but more or less constant for mortar and cement paste. Crack surface formations mechanisms and energy absorbed by non-critical section are considered to be the factors causing this size-effect on G_F.

ACKNOWLEDGMENT

The notched beam tests reported here were made possible by a grant from the Air Force Office of Scientific Research, [Lt. Col. Lawrence D. Hokanson, Program Manager]. The test results for the closing pressure vs. COD were obtained from an investigation supported by National Science Foundation [Dr. Michael P. Gaus, Program Manager].

REFERENCES

1. Kaplan, M. F., "Crack Propagation and the Fracture of Concrete," Journal of the American Concrete Insatitute, Vol. 58, No. 5, pp. 591-610, November 1961.

2. Lott, J. L., and C. E. Kesler, "Crack Propagation in Plain Concrete," T. & A. M. Report. No. 648, University of Illinois at Urbana-Champaign, August 1964.

3. Moavenzadeh, F., and R. Kuquel, "Fracture of Concrete," Journal of Material, JMLSA, Vol. 4, No. 3, September 1969, pp. 497-519.

4. Naus, D. J., and J. L. Lott, "Fracture Toughness of Portland Cement Concretes," ACI Journal, June 1969, pp. 481-489.

5. Shah, S. P., F. J. McGarry, "Griffith Fracture Criterion and Concrete," Journal of Engineering Division (ASCE), Dec. 1971, pp. 1663-1675.

6. Walsh, P. F., "Fracture of Plain Concrete," Indian Concrete Journal, Vol. 46, No. 11, 469, 470, 476, Nov. 1972.

7. Brown, J. H., "Measuring the Fracture Toughness of Cement Paste and Mortar," Magazine of Concrete Research, Vol. 24, No. 81, 1972, pp. 185-196.

8. Higgins, D. D., and J. E. Bailey, "Fracture Measurements on Cement Paste," Journal of Material Science, Vol. 11, 1976, pp. 1995-2003.

9. Mindess, S., and J. S. Nadeau, "Effect of Notch Width on K_{Ic} for Mortar and Concrete," Cement and Concrete Research, Vol. 6, 1976, pp. 529-534.

10. Gjørv, O. E., S. I. Sørensen, and A. Arnesen, "Notch Sensitivity and Fracture Toughness of Concrete," Cement and Concrete Research, Vol. 7, 1977, pp. 333-344.

11. Swartz, S. E., K. K. Hu, and G. L. Jones, "Compliance Monitoring of Crack Growth in Concrete," Journal of the Eng. Mech. ASCE, Vol. 104, No. EM4, August 1978, pp. 789-800.

12. Swartz, S. E., K. K. Hu, and C. M. J. Huang, "Stress Intensity Factor for Plain Concrete in Bending - - Prenotched versus Precracked Beams," Experimental Mechanics, Nov. 1982, pp. 412-417.

13. Strange, P. C., and A. H. Bryant, "Experimental Tests on Concrete Fracture," Technical Notes, Journal of Eng. Mech. Div., ASCE, Vol. 105, No. EM2, April 1979, pp. 337-342.

14. Wecharatana, M., and S. P. Shah, "Double Torsion Tests for Studying Slow Crack Growth of Portland Cement Mortar," Cement and Concrete Research, Vol. 10, 1980, pp. 833-844.

15. Wecharatana, M., and S. P. Shah, "Slow Crack Growth in Cement Composites," Journal of Structural Div., ASCE, June 1982, pp. 1100-1113.

16. Wecharatana, M., and S. P. Shah, "Prediction of Nonlinear Fracture Process Zone in Concrete,' journal of EMD, ASCE, Vol. 109, No. 5, October 1983, pp. 1231-1246.

17. Wecharatana, M., and S. P. Shah, "A Model for Predicting Fracture Resistance of Fiber Reinforced Concrete," Cement and Concrete Research, Vol. 13, 1983, pp. 819-829.

18. Wecharatana, M., "Fracture Resistance in Cementitious Composites," Ph.D. Dissertation, Department of Materials Engineering, University of Illinois, Chicago, Chicago, Illinois, March 1982.

19. Go., C. G., and S. E. Swartz, "Fracture Toughness Techniques to Predict Crack Growth and Tensile Failure in Concrete," Report 150, College of Engineering, Kansas State University, July 1983.

20. Hillerborg, A., M. Modeer, and P. E. Petersson, "Analysis of Crack Formation and Crack Growth in Concrete by Means of Fracture Mechanics and Finite Elements," Cement and Concrete Research, Vol. 6, 1976, pp. 773-782.

21. Hillemier, B., and H. M. Hilsdorf, "Fracture Mechanics on Concrete Composites," Cement and Concrete Research, Vol. 7, 1971, pp. 523-536.

22. Gopalaratnam, V. S., and S. P. Shah, "Softening Response of Plain Concrete in Direct Tension," Accepted for publication, ACI Journal, (MS 5652).

23. Liebowitz, H., and D. L. Jones, "On Test Methods for Nonnlinear Fracture Mechanics," Proc. 10th Annual Meeting, Society of Engineering Science, Raleigh, N. C., 1973.

24. Catalano, P. M., and A. R. Ingraffea, "Concrete Fracture: A Linear Elastic Fracture Mechanics Approach," Report No. 82-1, Dept. of Structural Engineering, Cornell University, Nov. 1982.

25. Velazco, G., K. Visalvanich, and S. P. Shah, "Fracture Behavior and Analysis of Fiber Reinforced Concrete Beams," Cement and Concrete Research, Vol. 10, pp. 41-51, 1980.

26. Tada, H., P. C. Paris, and G. R. Irwin, The Stress Analysis of Cracks Handbook, Del Research Corporation, Hellertown, Pennsylvania 1973.

27. Hillerborg, A., "Additional Concrete Fracture Energy Tests Performed by 6 Laboratories According to a Draft RILEM Recommendation," Report to RILEM TC50-FMD.

28. Nallathambi, P., B. L. Karihaloo, and B. S. Heaton, "Various Size Effect in Fracture of Concrete".

29. Ballarini, R., S. P. Shah, and L. M. Keer, "Crack Growth in Cement Based Composites," Engineering Fracture Mechanics, (To appear).

30. Gylltoft, K., "Fracture Mechanics Models for Fatigue in Concrete Structures," Doctoral thesis, LULEA University of Technology, 1983.

31. Jenq, Y. S., and S. P. Shah, "A Fracture Toughness Criterion for Concrete," Submitted for publication, August 1984.

32. Imperato, L., private communication, ISMES, Bergamo, Italy.

33. Eligehausen, R., private communication, University of Stuttgart, Germany.

34. Jenq, Y. S., and S. P. Shah, " A Two Parameter Fracture Model For Concrete," (in preparation).

Series	Cement	Fine Aggregate	Coarse Aggregate	Water	Max. Aggregate size (in.)
Concrete C1	1.0	2.6	2.6	0.65	0.75
Mortar M1	1.0	2.6	0	0.65	0.1875
Mortar M2	1.0	2.6	0.	0.45	0.1875
Paste* P1	1.0	0.5	0.	0.45	———

*A small amount of sand was added to reduce the possibility of shrinkage cracking.

Table 1 – Mix-proportion and Maximum Aggregate Size of the Series Tested in this Program

Size	Max. Load (lbs.)	CTOD$_c$[***] at Max. Load $\times 10^{-3}$in.	K_{Ic} psi√in.	K_{Ic}^S psi√in.	$\bar{K}_{Ic}$ psi√in.	G_{Ic} lb./in.	G_F lb./in.	a_0 in.	a[*] in.
Large C1L1	1494	0.30	998.1	1006.9	978.5	0.268	0.591	3.0	3.131
Large C1L2	1176	0.48	831.6	855.3	770.2	0.247	0.456	3.0	3.471
Medium C1M1	580	0.35	688.3	716.4	668.4	0.173	0.507	1.9	2.274
Medium C1M2	498	1.06	859.1	910.6	583.8	0.362	0.506	1.94	3.055
Small C1S1	193	0.77	789.7	880.9	592.5	0.303	0.302	0.88	1.49
Small C1S2	160	0.71	704.5	811.2	491.2	0.272	0.432	0.88	1.64
Small C1S3	178	0.78	799.2	899.9	546.5	0.313	0.378	0.88	1.60
Small C1S4	198.6	0.91	931.6	1025.9	609.7	0.381	0.362	0.88	1.59

[*] $a = a_0 + \ell_{ec}$ = initial notch length + effective crack extensions at peak load.

[**] Two other small beams were damaged before test.

[***] Values of CTOD$_c$ were calculated based on simple model (see Ref. 34).

Table 2 - Some Results for Concrete Beams

Series	P – CMOD		P – δ	
	α	β	α	β
Concrete C1	4.5	1.35	4.89	1.0
Mortar M1	5.0	1.35	7.82	1.0
Mortar M2	8.5	1.6	9.79	1.0
Paste P1	10.0	1.6	12.8	1.0

Table 3 – Inelastic Coefficient of α and β Determined from P – CMOD and P – δ Curves.

Series	Water-Cement Ratio	Maximum Aggregate Size (in.)	Fracture toughness, K_{Ic}^S (psi $\sqrt{in.}$)			
			Large	Medium	Small	Average
Concrete C1	0.65	0.75	930.8	813.5	904.5	882.9
Mortar M1	0.65	0.1875	631.1	724.2	644.1	666.5
Mortar M2	0.45	0.1875	918.9	879.3	816.3	871.5
Paste P1	0.45	—	595.5	544.8	546.1	562.1

Table 4 - Influence of water-cement ratio and maximum aggregate size on the fracture toughness.

338

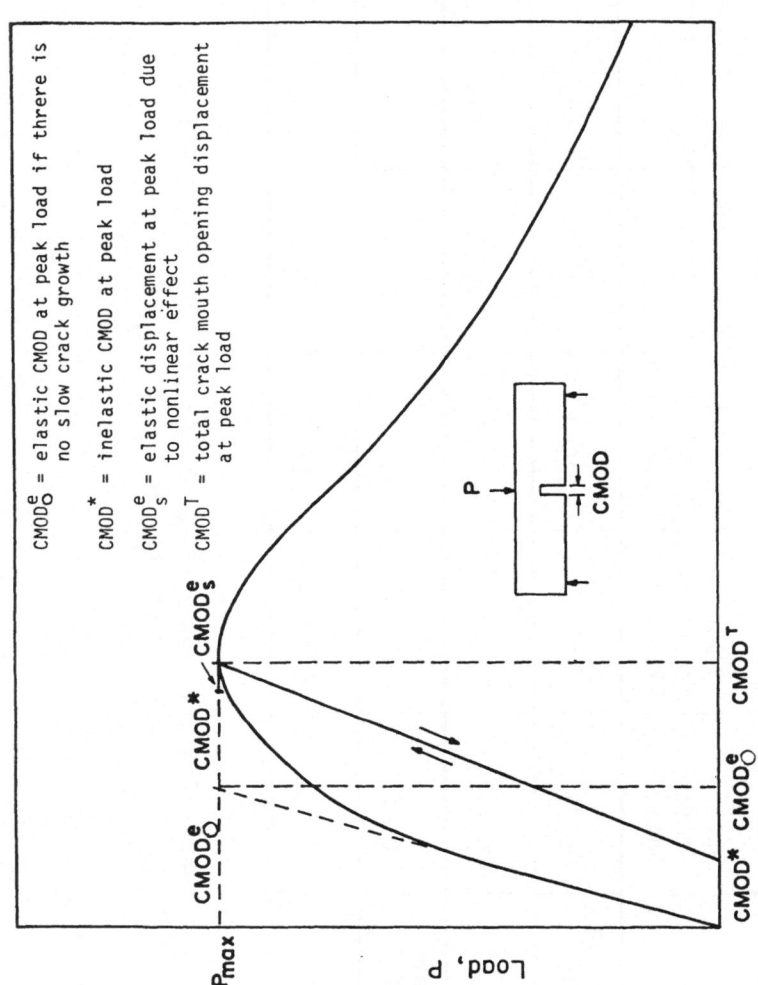

CMOD$_O^e$ = elastic CMOD at peak load if threre is no slow crack growth

CMOD* = inelastic CMOD at peak load

CMOD$_S^e$ = elastic displacement at peak load due to nonlinear effect

CMOD^T = total crack mouth opening displacement at peak load

Fig. 1 - Composition of CMOD Due to Nonlinear Effect

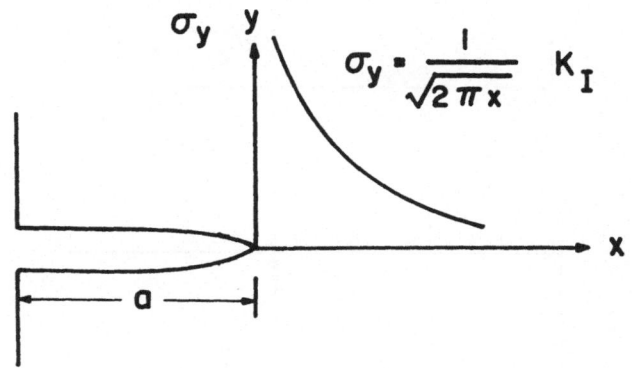

(a) Effective Griffith crack

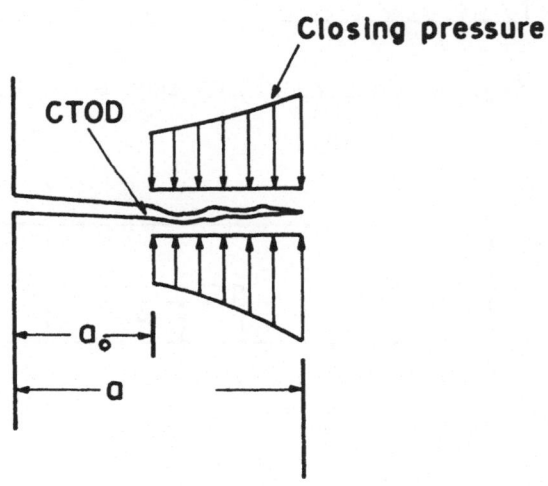

(b) Crack with closing pressure

Fig. 2 - Energy Consideration of Stress Singularity
and Closing Pressure.

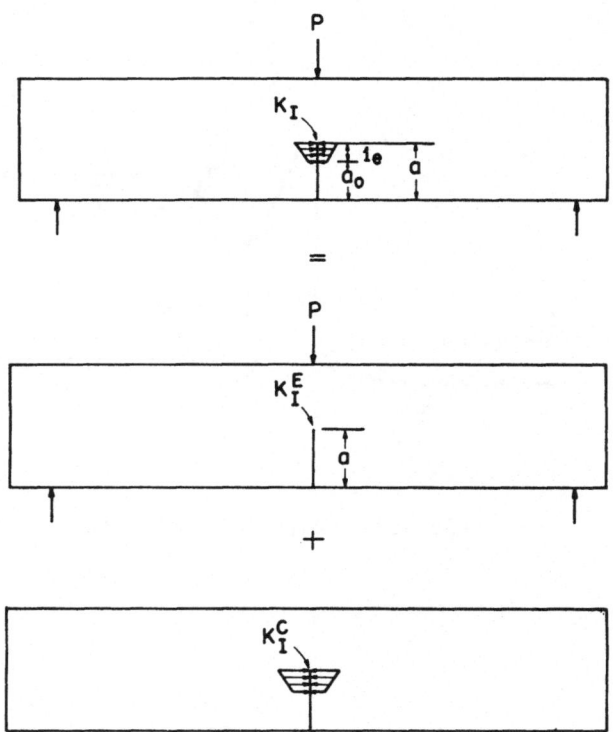

Fig. 3a - Superposition scheme for K_I

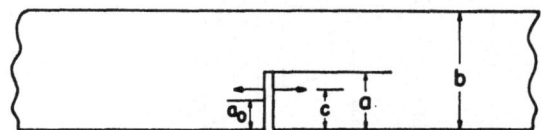

Fig. 3b - Infinite Strip Subjected Unit Point Load

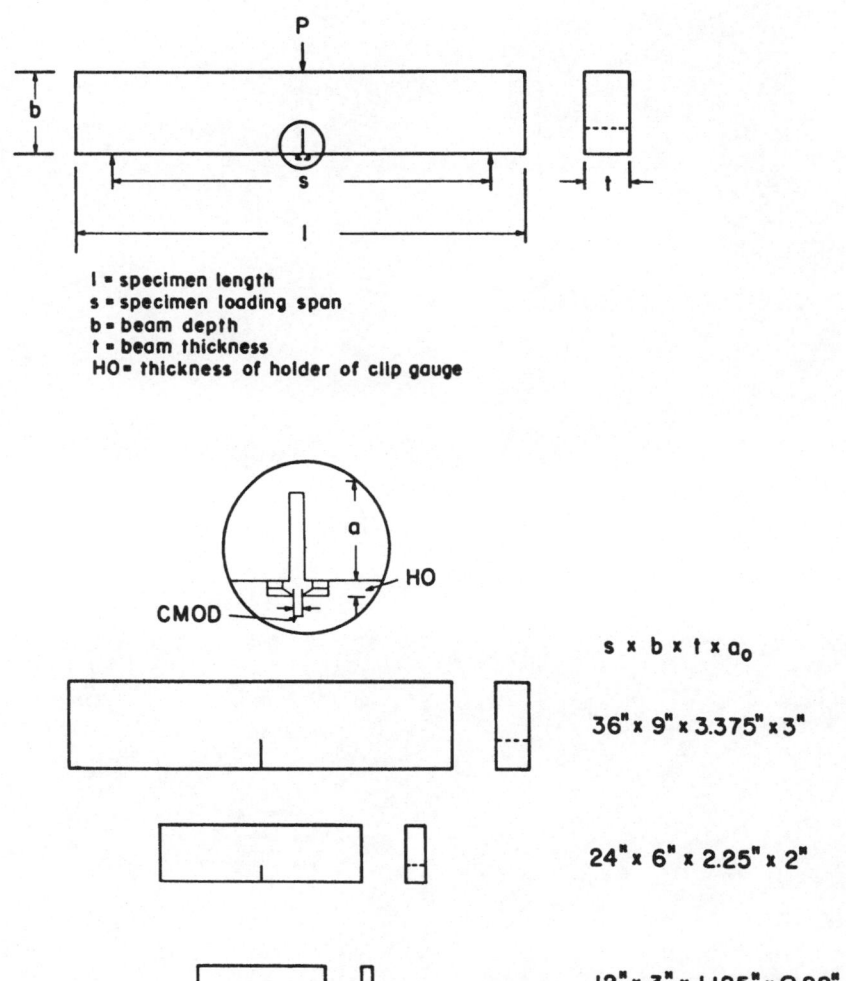

I = specimen length
s = specimen loading span
b = beam depth
t = beam thickness
HO = thickness of holder of clip gauge

$s \times b \times t \times a_0$

36" x 9" x 3.375" x 3"

24" x 6" x 2.25" x 2"

12" x 3" x 1.125" x 0.88"

Fig. 4 - Dimensions of Specimens

(a)

(b)

Fig. 5 - Experimental Set-Up

(a) Large and Medium Specimens
(b) Small Specimens

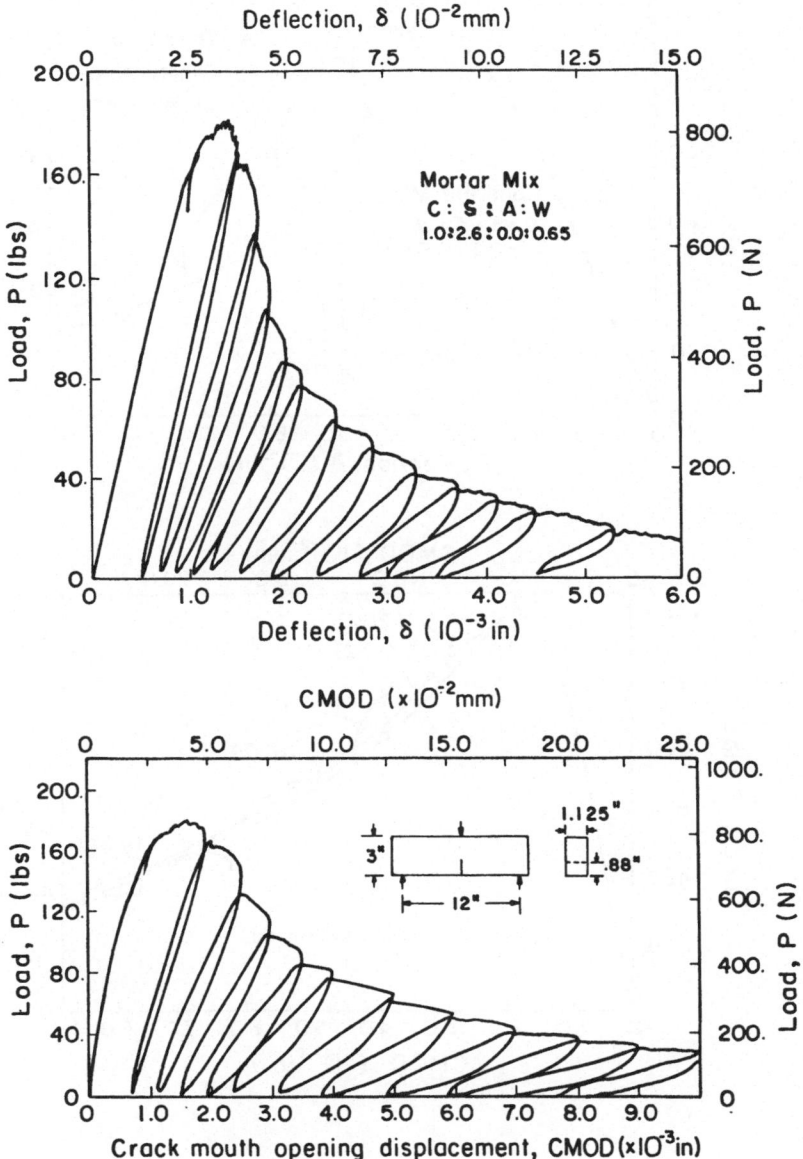

Fig. 6 - Typical Load-Displacement Curves

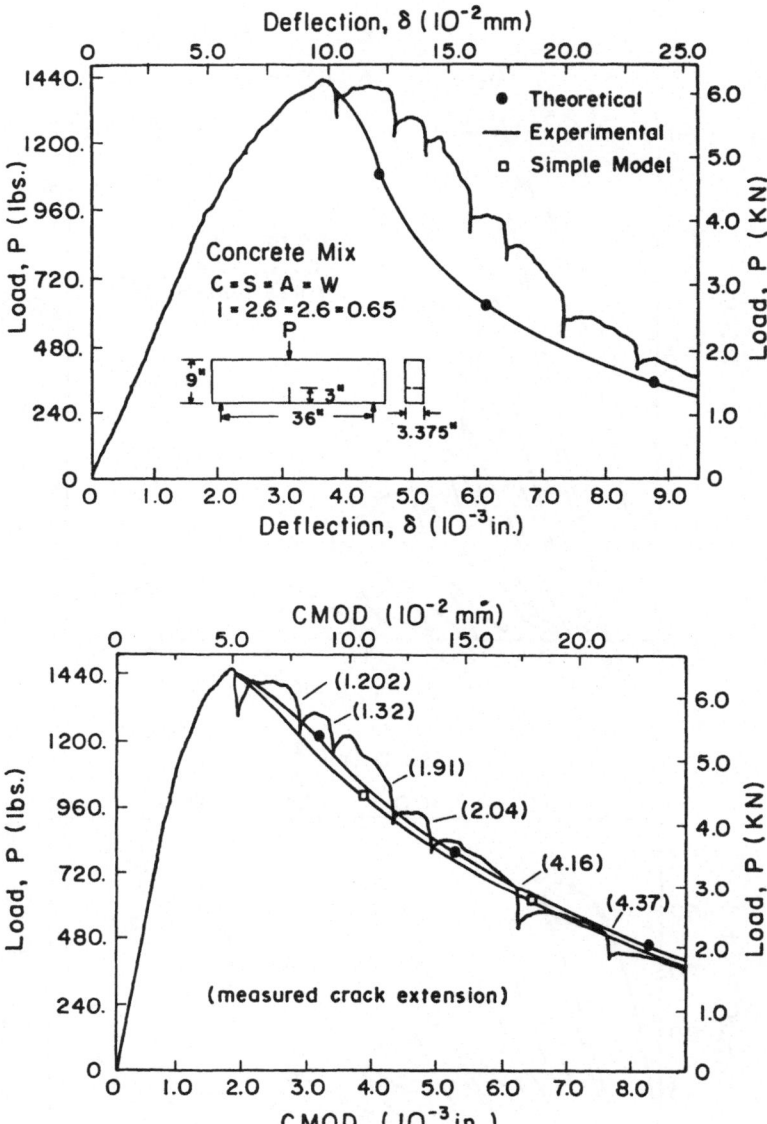

Fig. 7 - Theoretical Prediction and Experimental
Results - C1L1

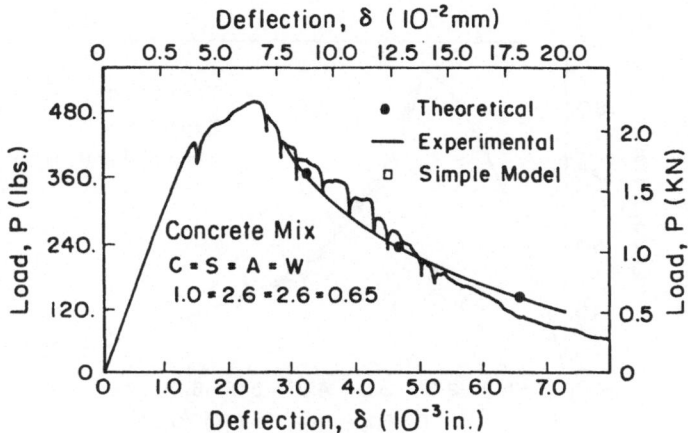

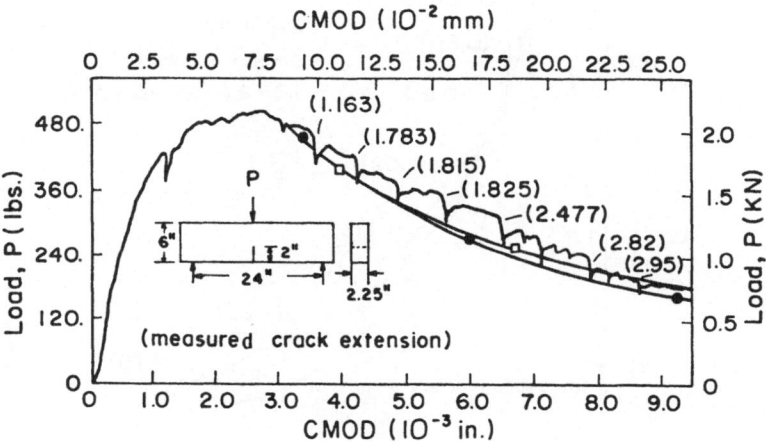

Fig. 8 - Theoretical Prediction and Experimental
Results - C1M2.

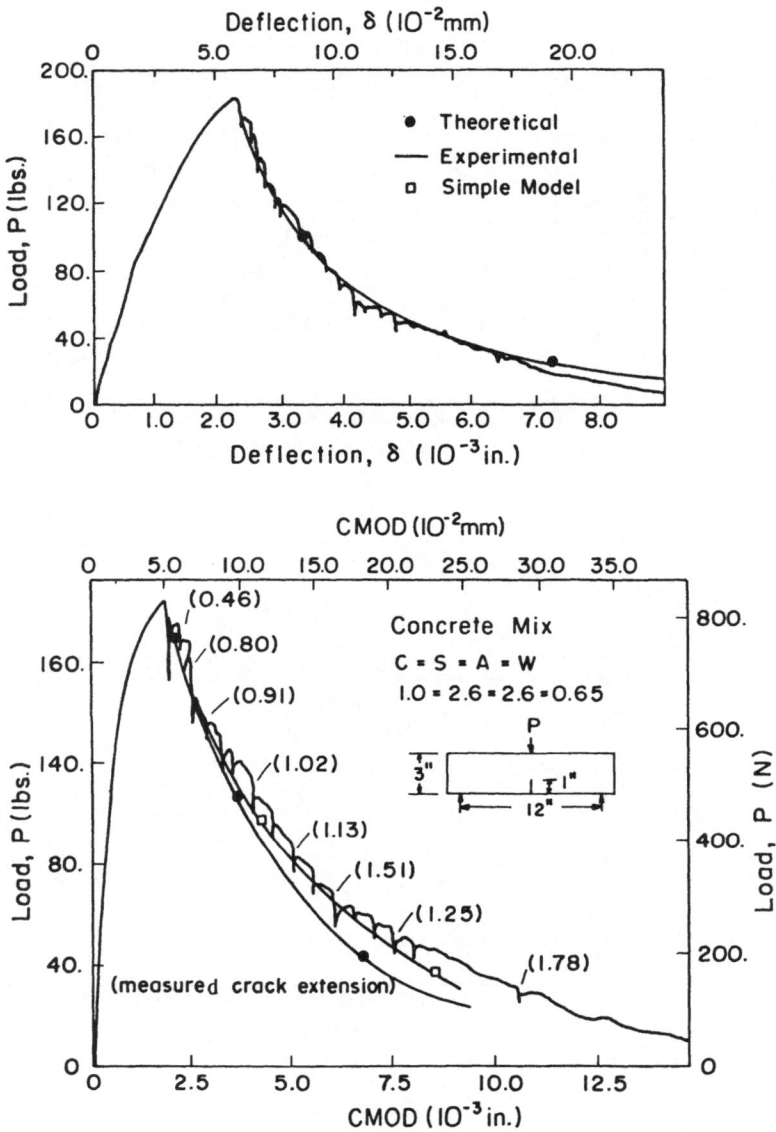

Fig. 9 - Theoretical Prediction and Experimental
Results - C1S3.

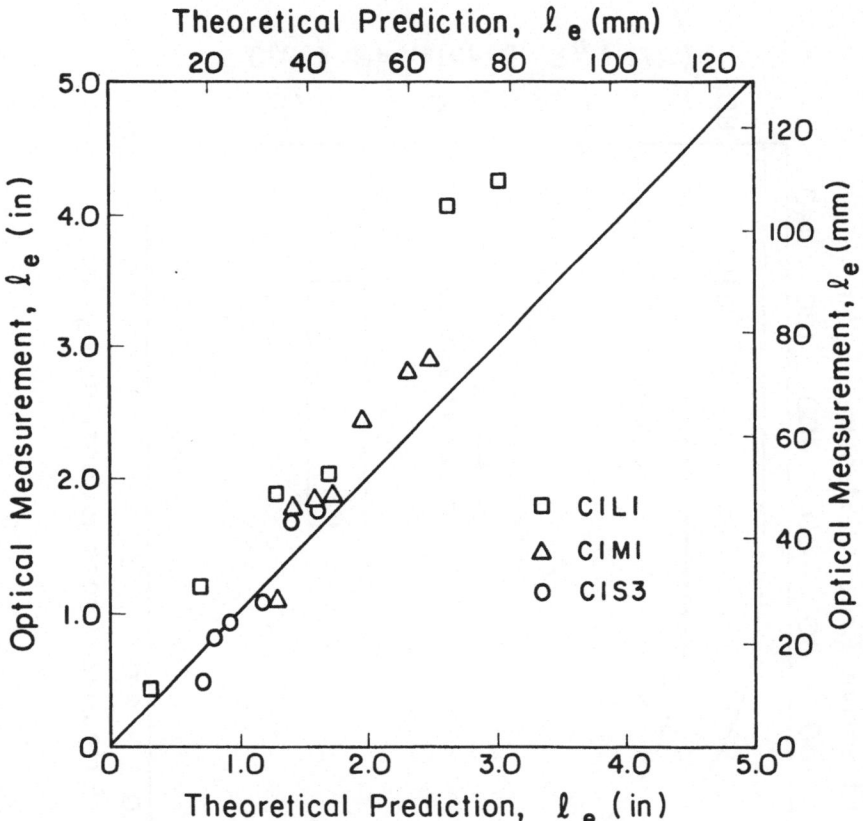

Fig. 10 - Comparison of Effective Crack Extension
(C1 Series).

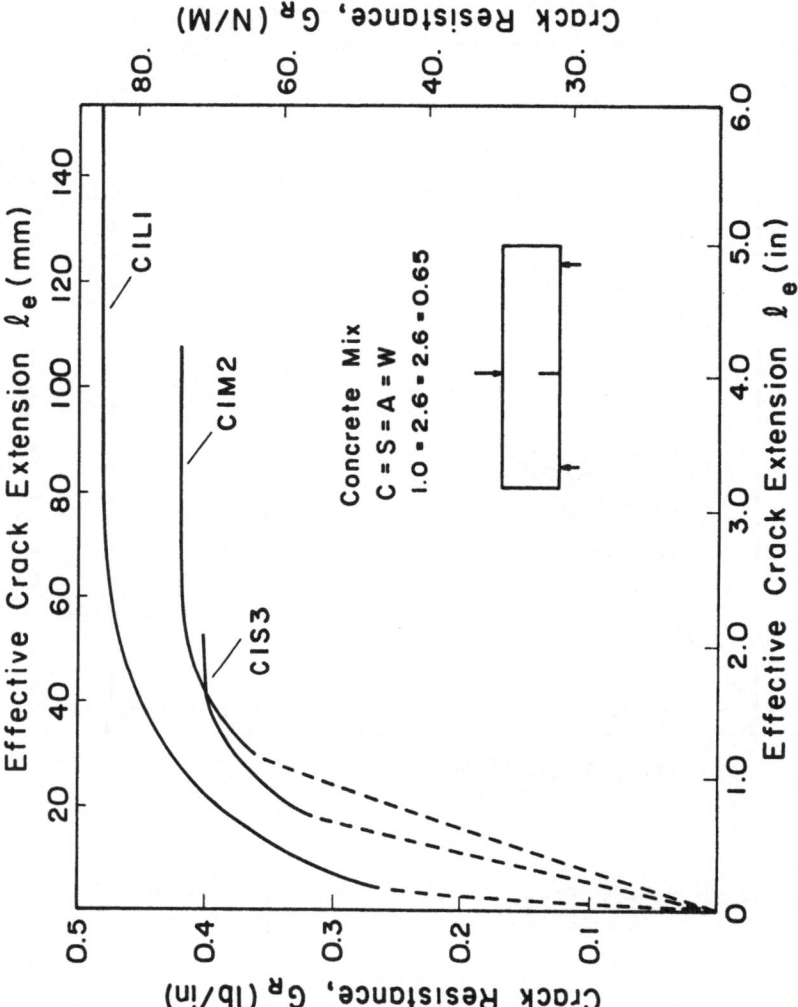

Fig. 11 – Crack Resistance Curves for Concrete (C1 Series).

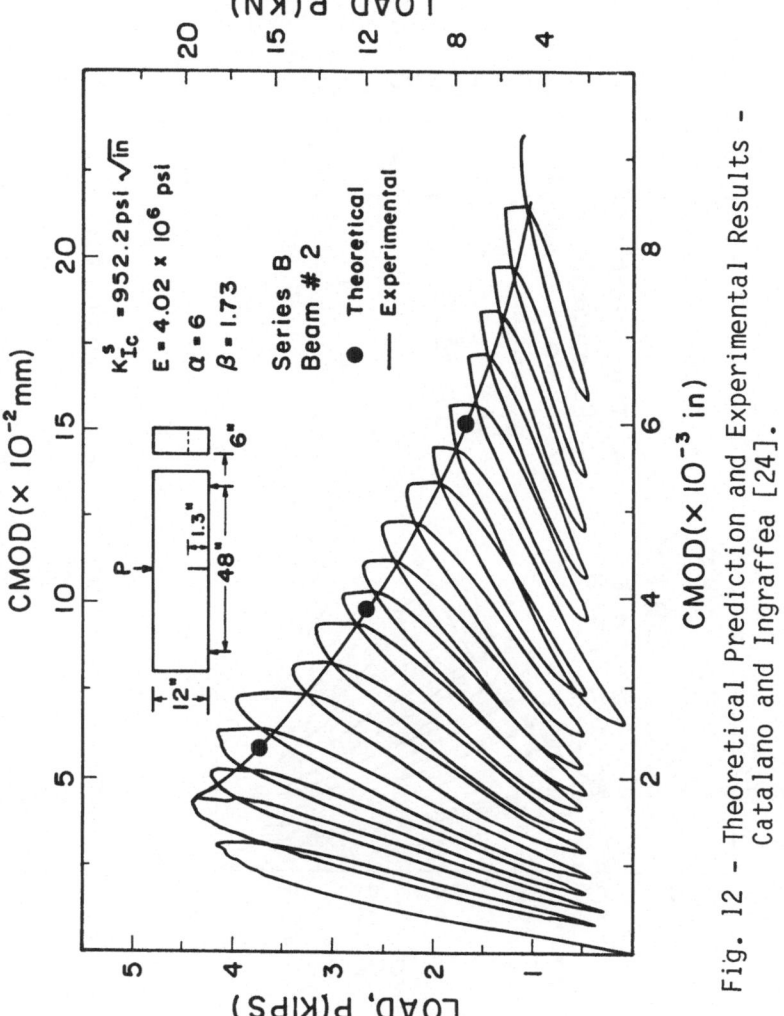

Fig. 12 - Theoretical Prediction and Experimental Results -
Catalano and Ingraffea [24].

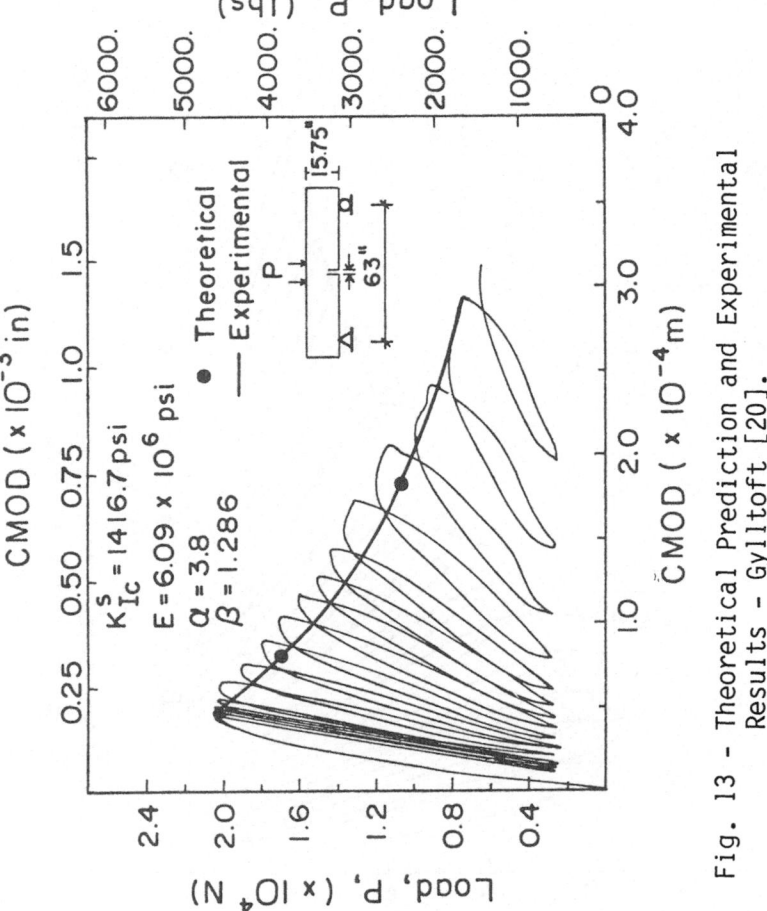

Fig. 13 – Theoretical Prediction and Experimental Results – Gylltoft [20].

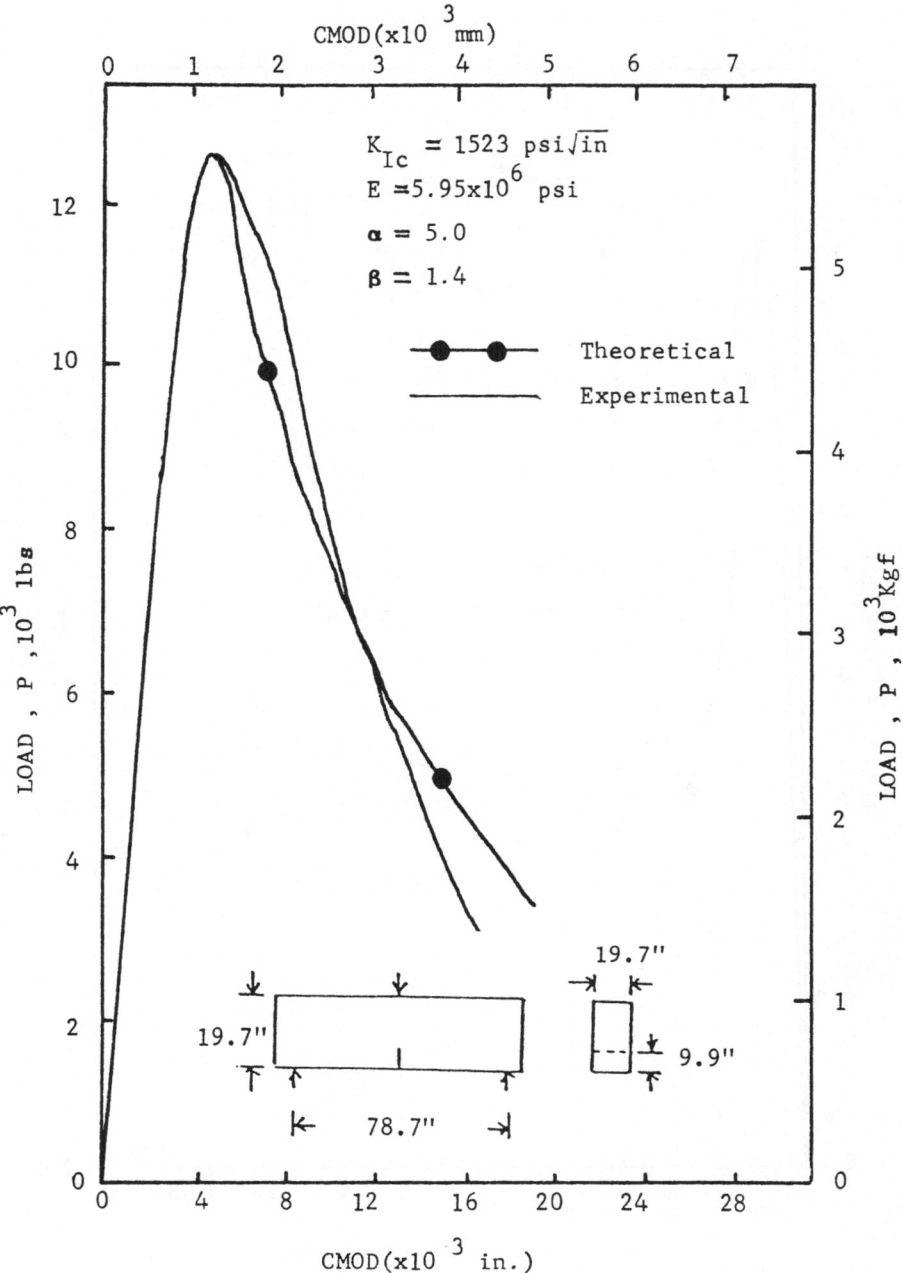

Fig. 14 - Theoretical Prediction and Experimental
Results - Imperato [32].

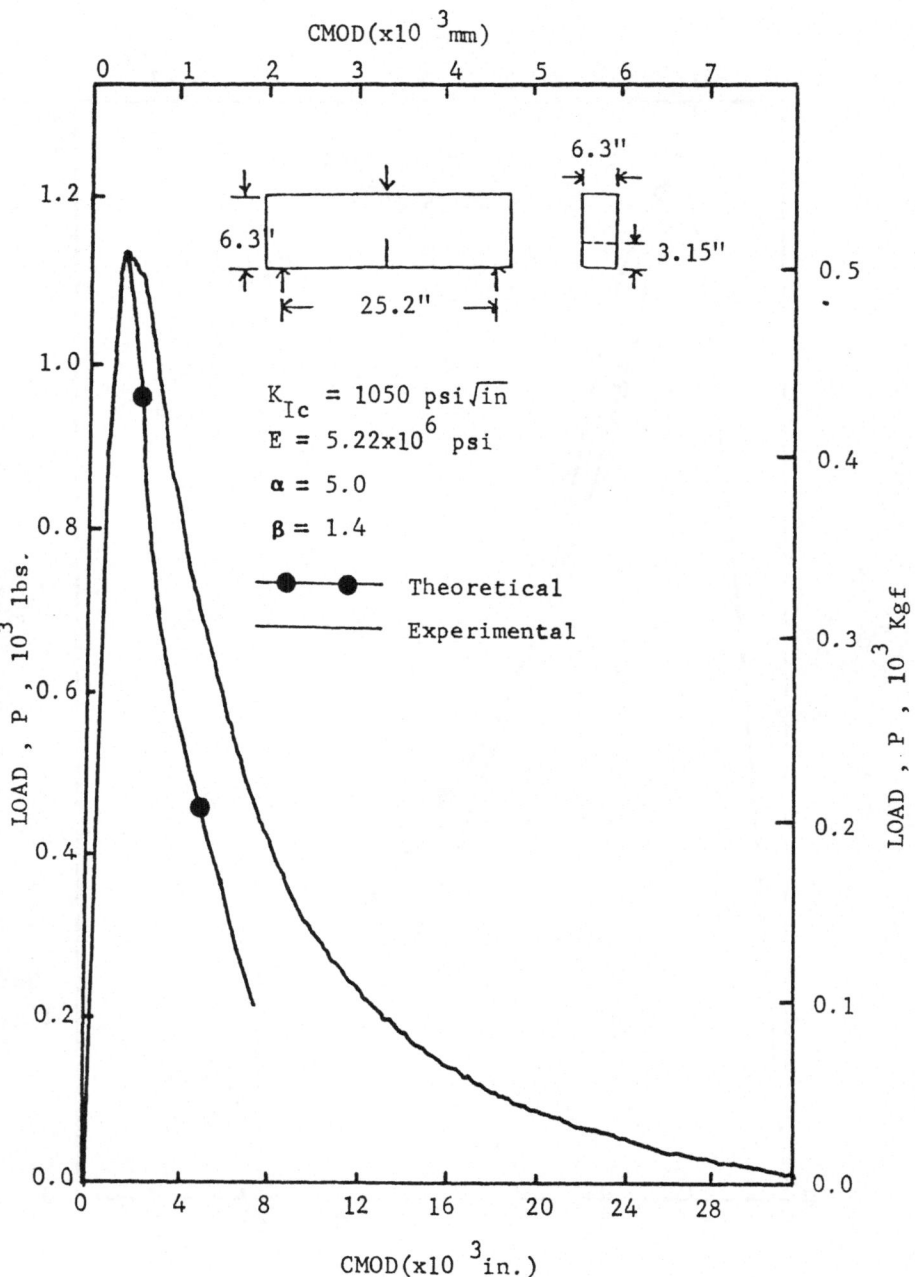

Fig. 15 - Theoretical Prediction and Experimental
Results - Imperato [32].

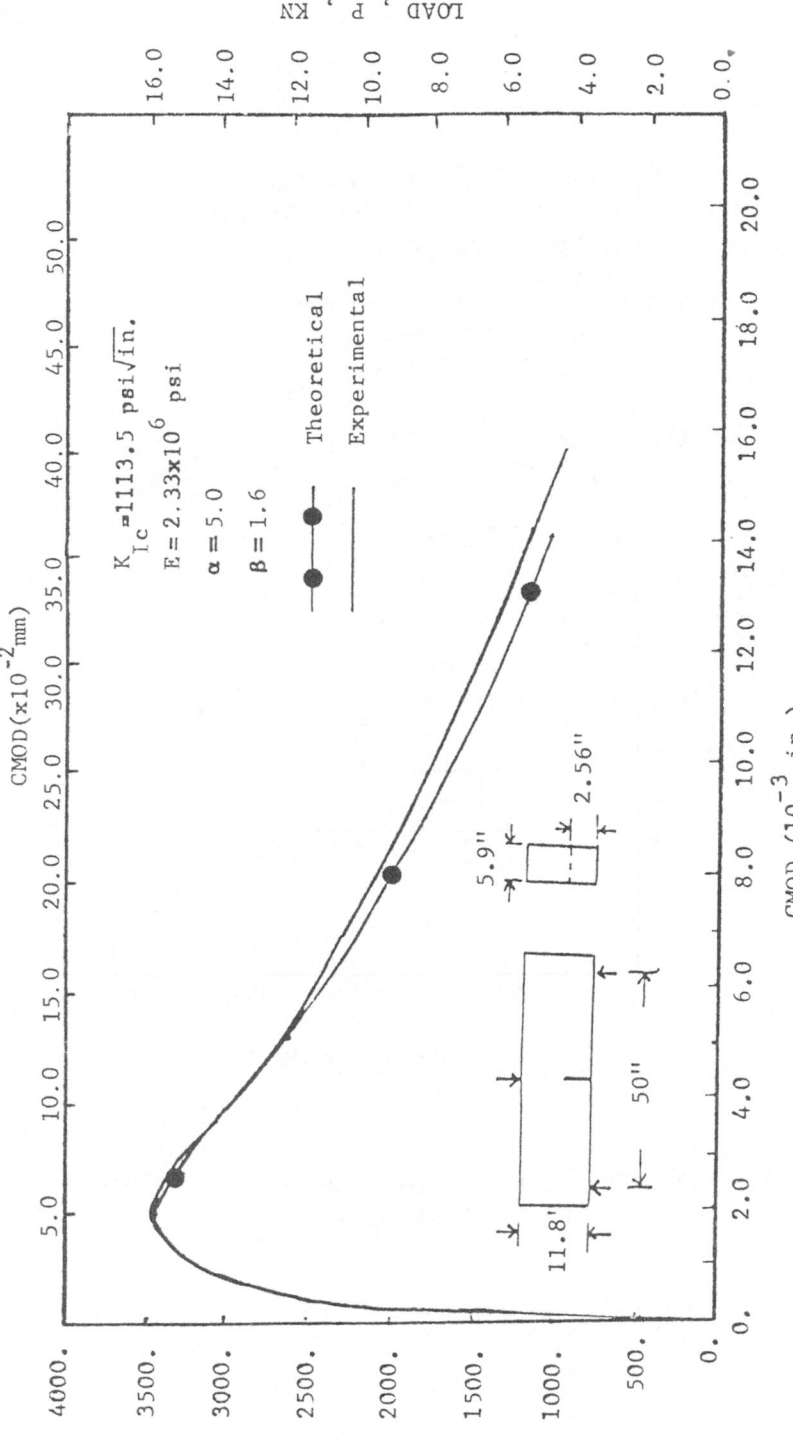

Fig. 16 – Theoretical prediction and Experimental Results – Eligehausen [33].

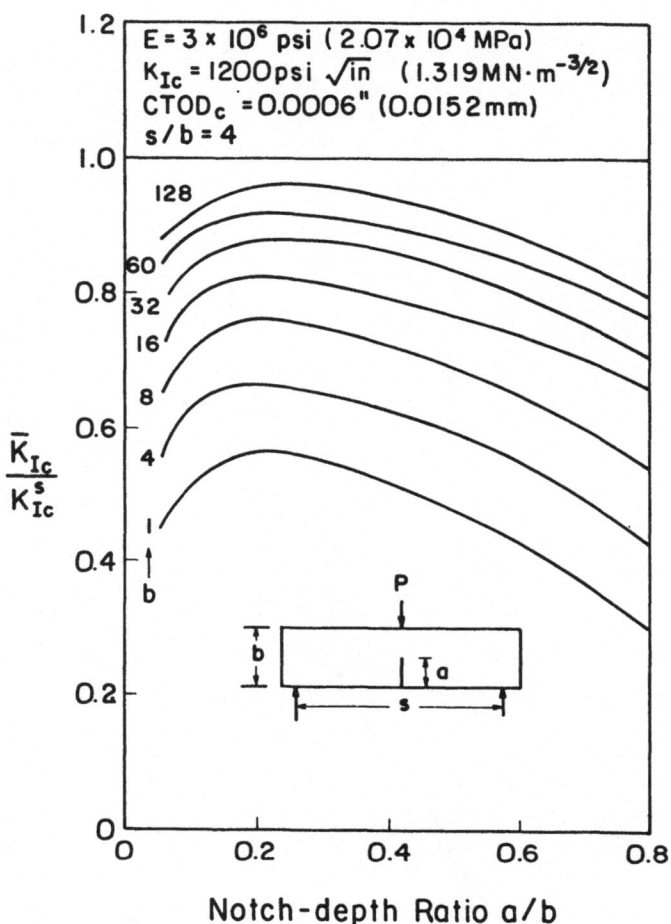

Fig. 17 - Theoretical Prediction of Size-Effect on Conventional $\bar{K}_{Ic}$.

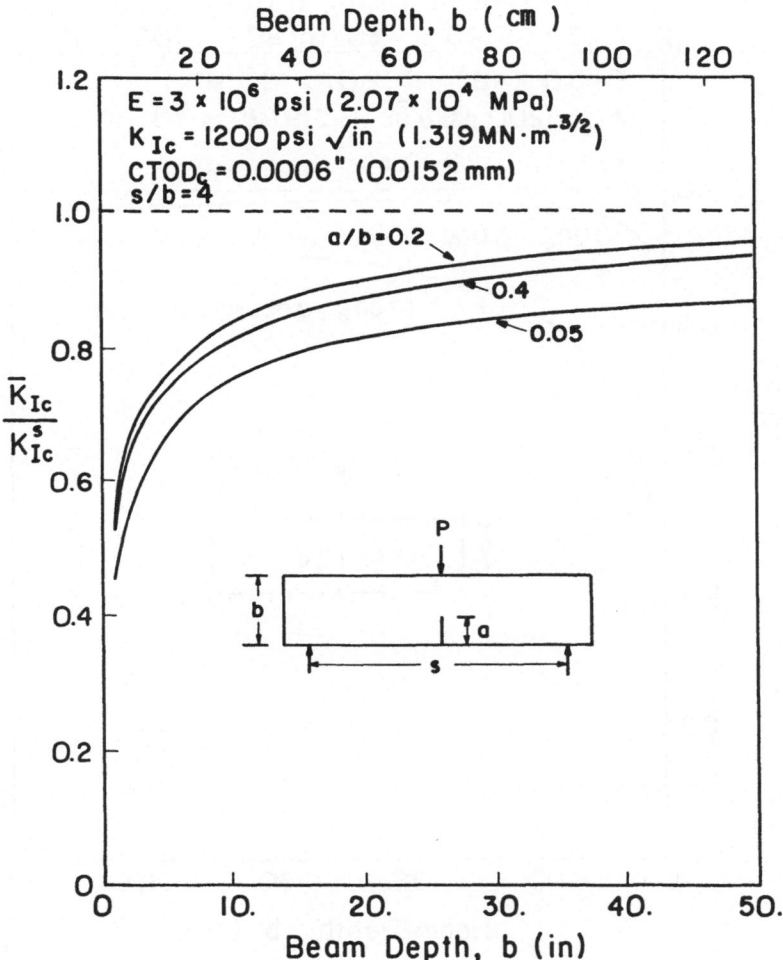

Fig. 18 - Theoretical Prediction of Size-Effect on Conventional $\bar{K}_{Ic}$ of a fixed notch-depth Ratio.

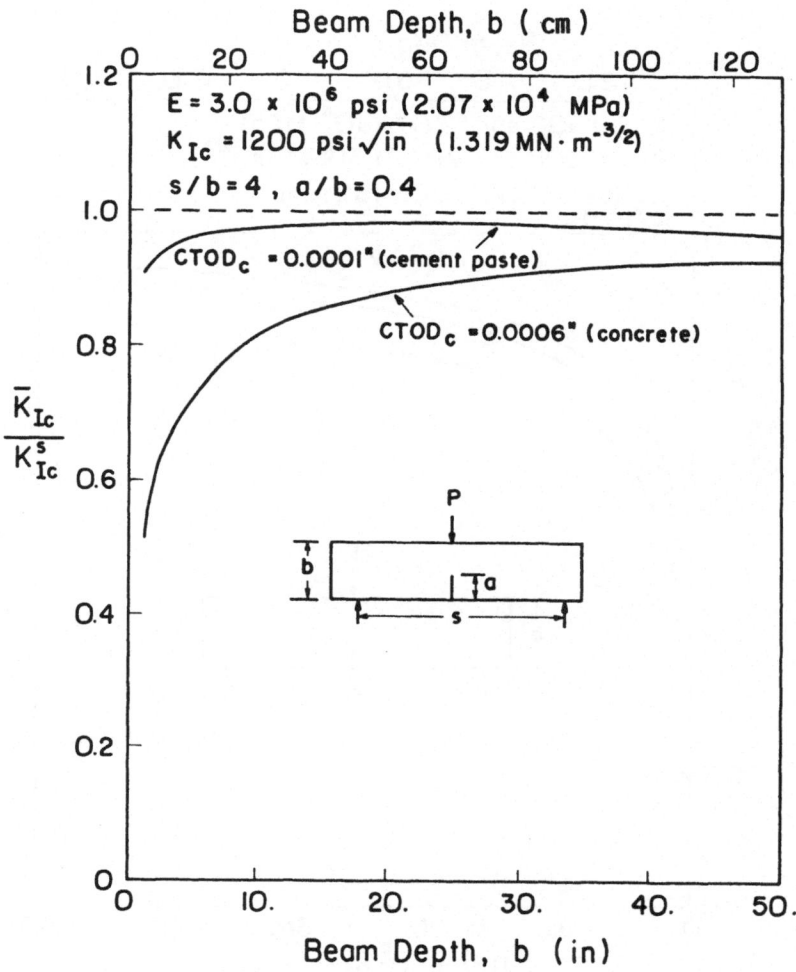

Fig. 19 - Theoretical Comparison of Size-Effect
on $\bar{K}_{Ic}$ of Concrete and Cement Paste.

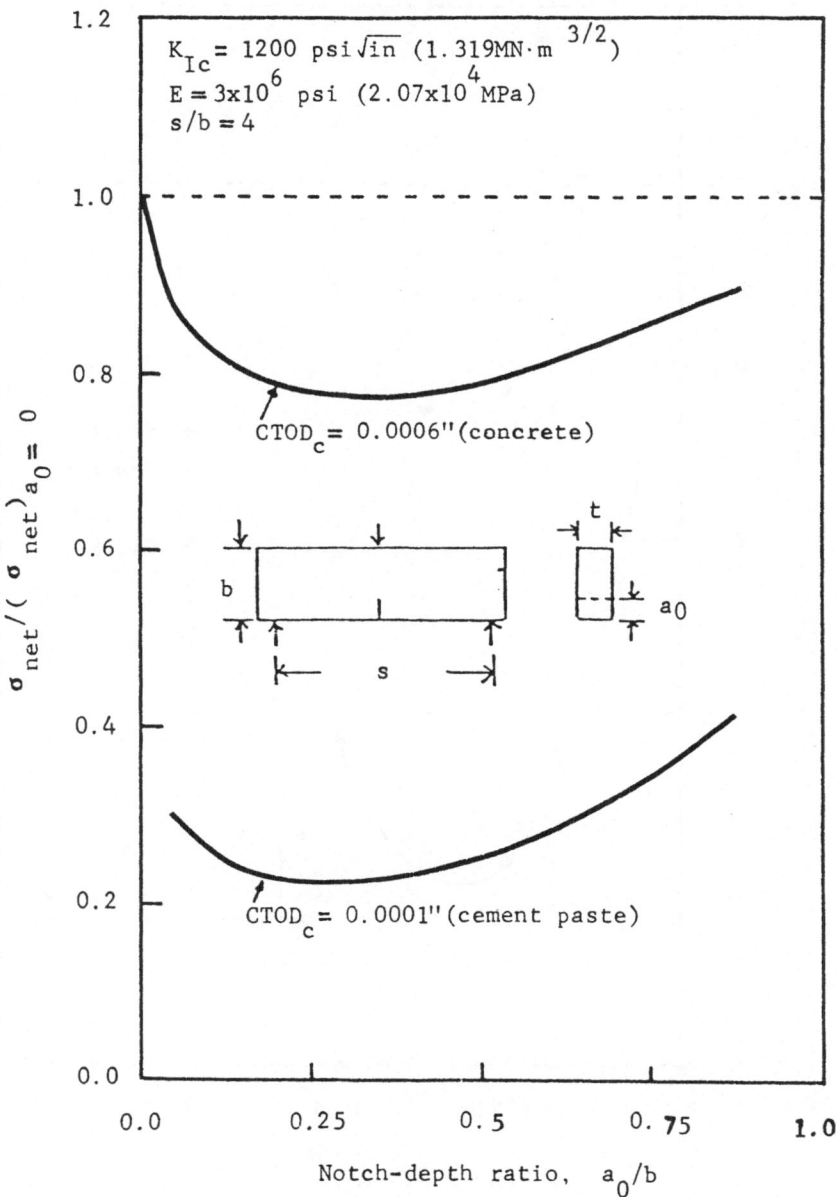

Fig. 20 - Theoretical Prediction of Notch-
           Sensitivity of Cement Paste and
           Concrete.

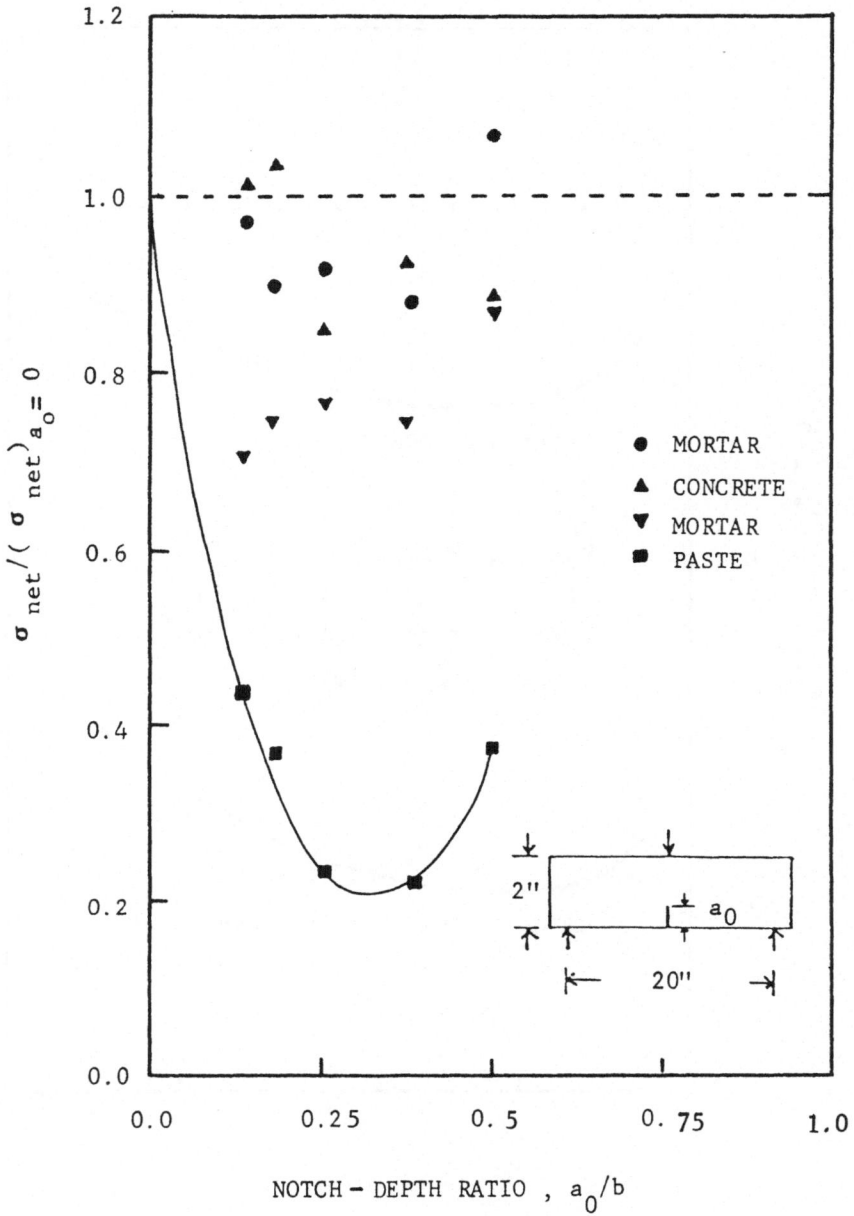

Fig. 21 - Experimental Results of Notch-
           Sensitivity of Cement Paste, Mortar,
           and Concrete - Shah and McGarry [5].

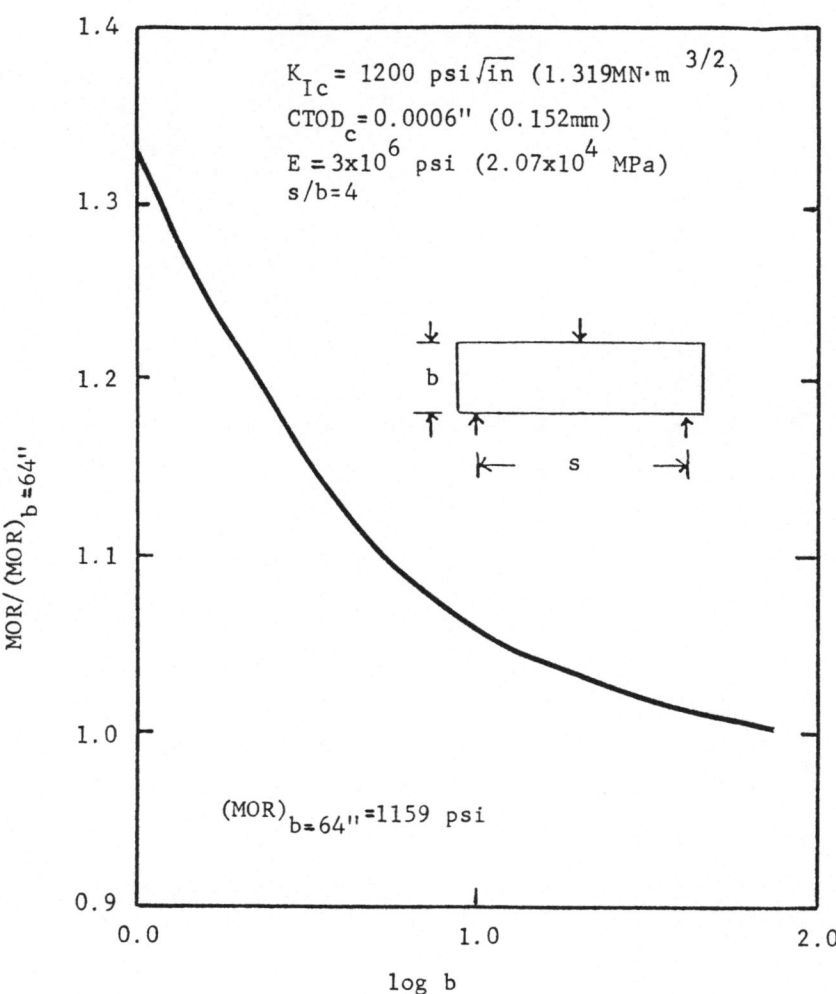

Fig. 22 - Theoretical Prediction of Size-
Effect on Modulus of Rupture of
Concrete.

**APPLICATION OF FRACTURE MECHANICS
TO CEMENTITIOUS COMPOSITES**
NATO-ARW - September 4-7, 1984
Northwestern Univeristy, U.S.A.
S. P. Shah, Editor

SIZE EFFECTS IN THE EXPERIMENTAL DETERMINATION OF FRACTURE
MECHANICS PARAMETERS

H.K. Hilsdorf, W. Brameshuber

Institut für Massivbau und Baustofftechnologie
University of Karlsruhe

1. Introduction

Because of the heterogeneity of concrete which causes the formation of a fracture process zone in front of a crack, conventional fracture parameters of concrete normally used in LEFM have
been found to be size dependent unless the dimensions of the
member under consideration exceed values which would make routine testing very difficult /1/.

Therefore, an experimental investigation is under way at the
Institut für Massivbau und Baustofftechnologie, Universität
Karlsruhe, to study the effect of member size on fracture characteristics. The objective of the study is to verify various
hypotheses regarding the dependence of fracture parameters on
size and to propose experimental procedures and evaluation methods which allow the reliable determination of fracture parameters of concrete in routine tests and on comparatively small
concrete specimens.

This paper is a preliminary presentation and evaluation of the
initial series of our experiments. The study will be continued.

2. Description of experiments

2.1 Test program

2.1.1 Materials, mix proportions and hardened concrete properties

Plain mortar (Mo) and concrete (C) beams have been cast using materials and mix proportions described in Table 2.1. A portland cement PZ 35 F (Typ I), Rhine River Sand and Rhine River Gravel have been used throughout.

Test Series	Amount of Cement [kg/m^3]	Max. Aggregate Size [mm]	Amount of Aggregate [kg/m^3]	w/c
concrete (C)	315	32	1877	0.54
mortar (Mo)	535	2	1372	0.54

Table 2.1: Materials and mix proportions

The compressive strength, f_{cu}, the tensile splitting strength, f_{tu}, and the modulus of elasticity, E_c of the concrete and of the mortar as determined on cylinders 150/300 mm are given in Table 2.2.

	f_{cu} [N/mm^2]	f_{tu} [N/mm^2]	E_c [N/mm^2]
C	31.0	2.7	32250
Mo	35.0	3.0	25650

Table 2.2: Properties of hardened concrete

2.1.2 Specimens

Table 2.3. summarizes the dimensions of specimens investigated:

Type	Depth [mm]	Width [mm]	Span [mm]	Total Length [mm]	Span Depth [-]	Weight [kg]	Number of specimens (C + Mo)
small (S)	100	100	500	550	5	13	30
medium(M)	400	200	2000	2300	5	450	20
large (L)	800	400	4000	4500	5	3200	9
halves of large beams (H)	800	400	2000	2250	2.5	1600	9

Table 2.3: Dimensions of specimens

For the small (S) and medium (M) size beams unnotched specimens as well as specimens with a notch at midspan and a relative notch depth, $a_0/d = 0.5$, have been cast. Since it became evident that no stable crack growth could be obtained for large, unnotched beams only notched specimens have been cast for series L.

2.2. Manufacturing of specimens

The S- and the L-beams have been manufactured in steel molds. Timber molds have been used for the M-beams.

With the exception of the H-beams the notches have been cast. For the S-beams a metal sheet, thickness 2 mm, has been attached to one side of the form work so that the direction of casting was perpendicular to the direction of testing. For the M-beams a wedge made of a bent steel plate has been attached to the top of the sides of the form. The average notch width was 3 mm. Prior to casting of the L-beams a steel plate with a thickness of 5 mm has been mounted to the upper half of the sides of the form. For the M- and the L-beams the tension side was the top side as cast. The steel forms as well as the metal plates have been greased prior to casting.

The halves of the L-beams have been used for further testing (H-beams). Notches with a thickness of 5 mm have been sawn into the halves after fracture of the L-beams.

For the S- and the M-beams the steel plates forming the notches have been removed prior to testing. This was not possible for the L-beams.

The concretes have been mixed in a continuous pan mixer and compacted by vibration. Whereas internal vibrators have been used for the S- and the M-beams, external vibrators have been used for the L-beams.

The specimens were covered with wet burlap up to the time of demolding which took place one day after casting for the S-beams and three days after casting for the M- and the L-beams. After demolding the specimens were wetted and sealed with plastic sheeting. Immediately prior to testing the plastic sheeting was removed at the supports and at the notch.

The age of testing was between 27 and 29 days for the S-, M- and L-beams and between 38 and 43 days for the H-beams.

2.3 Experimental procedures

2.3.1 Test set-up

All specimens have been subjected to three-point bending. However, because of the large dimensions and the weight of the beams different procedures have been used for the S-beams, and the M- and L-beams, respectively.

The test set-up used for the S-beams is shown in Fig. 2.1. For the M- and the L-beams a test set-up as shown in Fig. 2.2 has been used. The S-beams have been tested such that the notch was on the bottom side of the specimens whereas it was on the top side for the M- and the L-beams. Thus for the M- and L-beams the dead weight caused compressive stresses at the root of the notch. This facilitated handling of the specimens. It made the final evaluation of data more accurate as shown in Sect. 2.4. Furthermore, it appeared to be easier to obtain stable crack growth when employing this approach because the dead weight of the beam closes rather than opens the crack.

For the M- and the L-beams preparation of testing proceeded in the following way (Fig. 2.2): The specimen was placed on auxiliary supports (Au). At that time the load cylinder (C) was in its lowest position and not in contact with the specimen. Then, the frames at both ends of the beams have been placed in their proper position as shown in Sect. A-A of Fig. 2.2. There were hinges and rolers between beam and frame which allowed free rotation and displacement of the ends of the beam. Subsequently, the beam was prestressed against the top traverse of the frame with the rods (R) by an amount exceeding the dead weight of the beam. Thus the dead weight moment was positive at all stages of the experiment. Prior to loading the beam was lifted off the auxiliary supports by an adjustment of the columns of the load frames. Thus possible restraint of the beam during the experiment has been minimized.

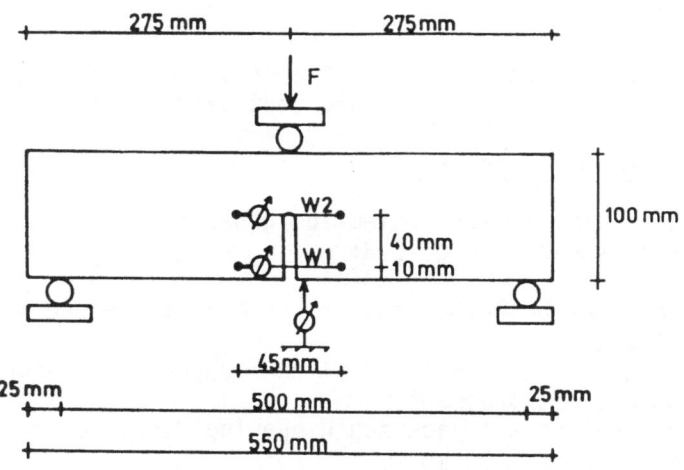

Fig. 2.1:   Test set-up and location of deflection and strain gages on S-beams

2.3.2 Test methods

Two types of experiments have been carried out. In the G_F-tests the specimens have been subjected to a constant rate of deflection (e.g. Fig. 3.1). In the compliance-tests (C-tests) the deflection has been increased until the load-deflection curve had a horizontal slope. Then the specimens were unloaded and again reloaded until the load-deflection curve became horizontal. Cyclic loading has been continued until the crack extended over the entire height of the beam (e.g. Fig. 3.8).

The S-beams have been tested in a universal testing machine such that a cross head speed of 0.1 mm/min has been kept constant. In testing the M- and the L-beams the load has been applied by means of a servo-controlled load cylinder such that a constant rate of deflection of 0.1 mm/min has been maintained. The rate of deflection has been controlled by two inductive strain gages mounted to both sides of the beam immediately adjacent to the notch.

2.3.3 Measurements of deflections and of strains

2.3.3.1 Small beams

The location of deflection and strain gages on S-beams is shown
in Fig. 2.1. Beam deflection has been measured at midspan. For
the S-beams no compensation has been made for displacements of
the supports since extreme care had been taken to minimize such
displacement. However, it has to be accepted that small errors
in the compliance and G_F calculations as described in sections
2.4.1 and 2.4.2 may be committed if support displacement is not
taken into account.

Strains have been measured at a distance of 10 mm from the
bottom of the beam (W 1) and at the root of the crack (W 2).

2.3.3.2 Medium and large beams

The location of strain and deflection gages for the M- and for
the L-beams is shown in Fig. 2.3:

W 1:   two gages on both sides of the beam to record midspan
        deflection
W 2:   two gages on both sides of the supports to record
        support displacement.
W 3:   measurement of crack mouth opening displacement CMOD
        (gage length 100 mm)
W 4-W 6: strain measurements in the ligament above the notch
        including crack tip opening displacement CTOD (gage
        length 45 mm)

In all cases the strain measurements were recorded continuously
using X-Y-recorders.

For a number of selected L-beams, 14 electrical resistance
strain gages with an active gage length of 60 mm have been used
to measure concrete strains across the ligament at midspan. The
strain data have been recorded on a digital recording system.

2.3.4 Observation of crack development

In order to observe crack growth during the experiments a method
originally proposed by Petersson /2/ has been used for the
S- and the M-beams. Prior to the experiment the surfaces of the
concrete beams above the notch were ground smoothly. During the
experiment the beams were repeatedly wetted with water and then
dried with compressed air. Since water is retained for a prolonged period of time in the cracked regions, the process zone
becomes clearly visible.

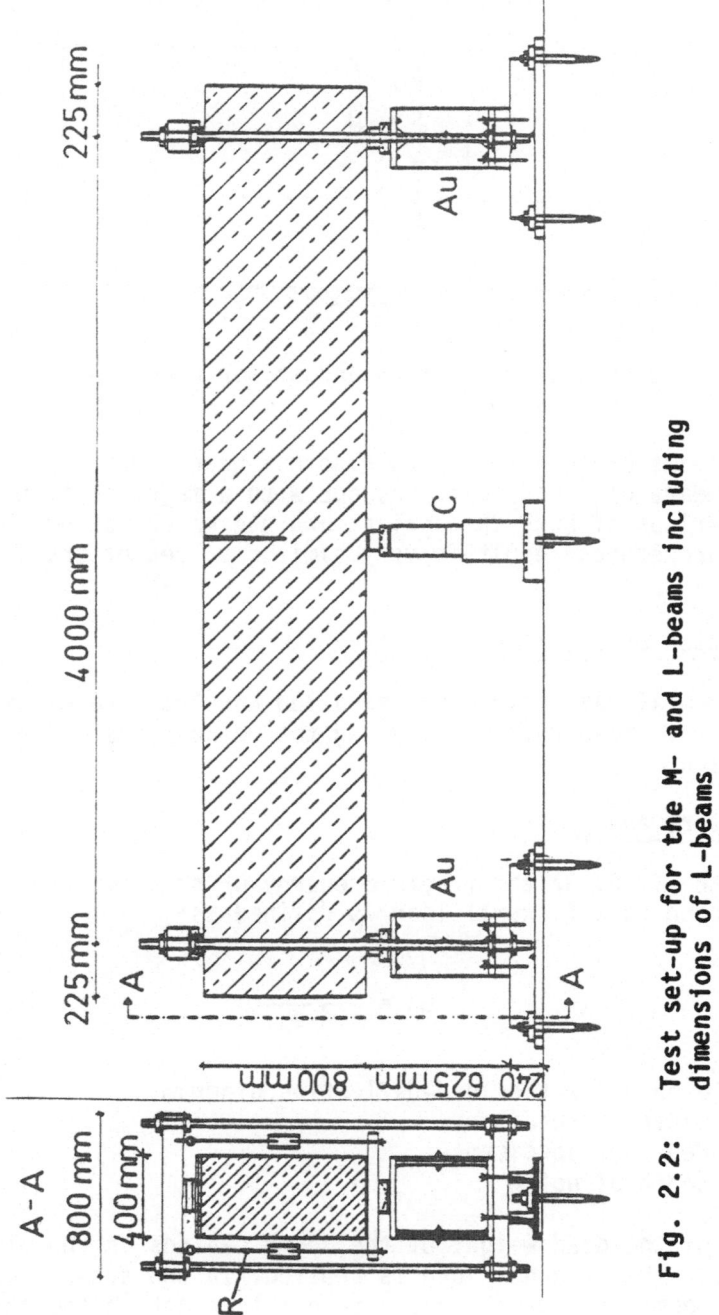

Fig. 2.2: Test set-up for the M- and L-beams including dimensions of L-beams

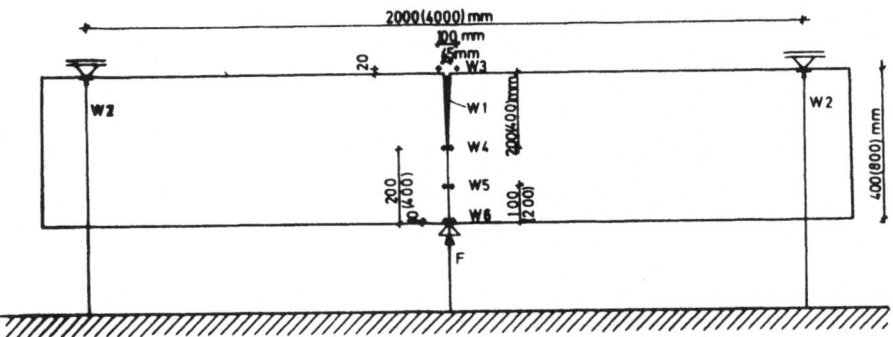

Fig. 2.3:   Location of deflection and strain gages on
             M- and L-beams

For the L-beams crack development also has been directly observed by means of a stereo-microscope with a magnification of
13 X. Comparison of both the optical and the water method showed
that both procedures result in very similar values of crack
length.

2.4 Evaluation of experimental data

The experimental data have been evaluated and fracture mechanics
parameters have been determined using the methods described in
the following.

2.4.1 Fracture energy G_F

According to /1, 2, 3/ the fracture energy G_F of a notched
beam subjected to a flexural load is defined as:

$$G_F = \frac{A}{b(d-a_0)}$$

where A = area under the load-deflection diagram
      b = width of specimen
      d = depth of specimen
      a_0 = depth of notch

The effect of the dead weight of the beams depends on the direction in which the external load is applied. In the set-up used to
test the S-beams dead weight and external load act in the same
direction. Then for the determination of G_F the load deflection
curve has to be corrected as shown in Fig. 2.4a. The area A_1
can be obtained from the load-deflection curve measured in the

experiment. The area A_2 represents the effect of the dead weight. According to Petersson /2/ it may be estimated as follows:

$$A_2 = 2 \cdot F_0 \cdot f_0$$

$$F_0 = G_0 \cdot g/2$$

where f_0 = deflection at fracture observed in the experiment
G_0 = mass of the beam (kg)
g = 9.81 m/sec.^{2}
$2F_0$ = dead weight of the beam (N)

Then the fracture energy of the beam taking into account its dead weight may be calculated from:

$$G_F = \frac{A_1 + G_0 \cdot g \cdot f_0}{b(d-a_0)} = \frac{A_1 + 2 \cdot F_0 \cdot f_0}{b(d-a_0)} \tag{1}$$

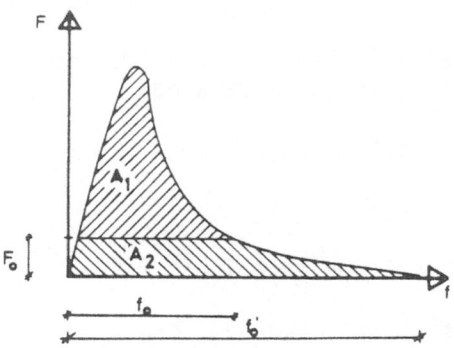

Fig. 2.4a: Determination of G_F on S-beams

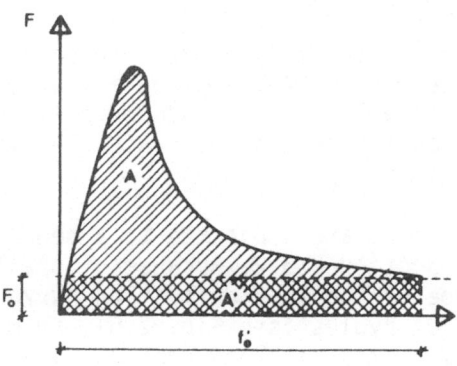

Fig. 2.4b: Determination of G_F on M- and L-beams

The M- and the L-beams have been tested such that dead weight
and external load act in opposite directions. After complete
fracture the specimen still rests on the load cylinder so that
an external load corresponding to half of the dead weight of the
beam is recorded. Therefore, for the determination of fracture
energy the load-deflection diagram as recorded in the experiment
has to be corrected by subtracting the effect of the dead
weight. This corresponds to a shifting of the f-axis upword by
an amount F_0 as shown in Fig. 2.4b. Then G_F may be calculated from:

$$G_F = \frac{A - A'}{b(d-a)} \tag{2}$$

Eq. (2) should give a more accurate result than eq. (1).

G_F-values have been calculated from the G_F-tests as described in sect. 2.3.2. In addition, values of A in eqs. (1) and
(2) also have been determined from the envelope to the load-
deflection curves obtained in the compliance tests (C-tests).

In addition the characteristic length l_{ch} as defined by
Hillerborg et al. /1, 2/ has been calculated:

$$l_{ch} = \frac{E_c G_F}{f_{tu}^2} \tag{3}$$

where E_c and f_{tu} have been taken from Table 2.2.

2.4.2 Determination of fracture toughness K_{Ic}

In linear elastic fracture mechanics, fracture toughness of a
material is calculated from the critical load, the corresponding
crack length and the calibration function for a particular type
of specimen.

In our study the critical crack length has been determined using
the reloading compliance method. This method is valid for an
elastic material, a specimen with a defined crack length and
without limited load transfer across the crack. The method is
described in detail e.g. in /4/.

In the C-tests the specimens are subjected to a sequence of load
cycles as described in section 2.3.2. Because of stable crack
growth the average slope of the load deflection curve decreases
with increasing number of cycles (see Fig. 2.5). From the reloading compliance of a particular load cycle C_i and the
initial loading compliance of an unnotched beam C_0 the crack

length for the preceding load cycle a_{i-1} can be estimated using the following calibration function, eq (4), which has been taken from /5/.

$$f(\tfrac{a}{d}) = 1.93 - 3.07 \, (\tfrac{a}{d}) + 14.53 \, (\tfrac{a}{d})^2 - 25.11 \, (\tfrac{a}{d})^3 + 25.80 \, (\tfrac{a}{d})^4 \quad (4)$$

$$0 \leq \tfrac{a}{d} \leq 0.6$$

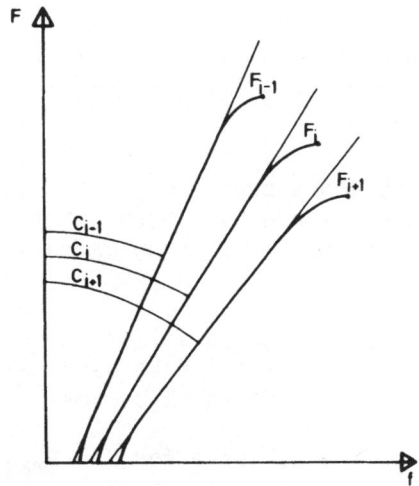

Fig. 2.5:   Definition of compliance for loading and reloading cycles

Then the fracture toughness has been calculated using eq. (5) from /5/:

$$K_{Ici} = \frac{1.5 \cdot F_i \cdot 1}{b \cdot d^2} \cdot \sqrt{a_{i+1}} \cdot f(\frac{a_{i+1}}{d}) \quad (5)$$

where K_{Ici} = fracture toughness for load cycle i
F_i = critical load for load cycle i
a_{i+1} = crack length for cycle (i + 1)
b = width of beam
d = depth of beam
1 = span
$f(a_{i+1}/d)$ = calibration function

2.4.3 K_R-curves and R-Curves

For the determination of K_R load-deflection curves have been chosen from the C-tests for which the relative crack depth a/d calculated according to 2.4.2 was close to 0.5. This is in accordance with the recommendations given in /6/.

As shown in Fig. 2.6 up to 10 intersects between the load-deflection curve for an initial crack length a_0 as determined according to section 2.4.2 and straight lines representing given values of C have been determined. For these intersects values of K_{Rn} have been calculated from eq. (6):

$$K_{Rn} = \frac{1.5 \cdot F_n \cdot 1}{b \cdot d^2} \cdot \sqrt{a_0 + \Delta a_n} \cdot f(\frac{a_0 + \Delta a_n}{d}) \qquad (6)$$

Then relations between K_R and a crack length $a_0 + \Delta a$ which corresponds to a given compliance have been plotted as shown in Fig. 2.7. There Δa_n is the crack extension calculated for a given load F_n.

Furthermore, relations between stress intensity factor K_I and crack length $a = a_0 + \Delta a$, for a given F have been calculated from eq. (6) using the proper indices (K-curves).

The K-curve is tangent to the K_R-curve for $F = F_{max}$. The intersect between the two relations gives the critical stress intensity factor K_c. Stable crack growth occurs for a crack length less than the critical value a_c whereas unstable crack growth has to be anticipated for crack lengths larger than a_c.

Also R-curves have been determined from:

$$G_R = \frac{K_R^2}{E} \cdot (1 - \nu^2) \qquad (7) \qquad \text{where } \nu = 0.2 = \text{Poisson's ratio}$$

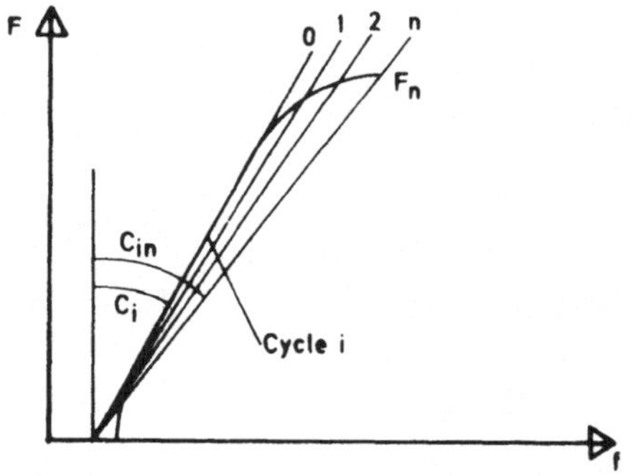

Fig. 2.6:   Determination of K_R-curves

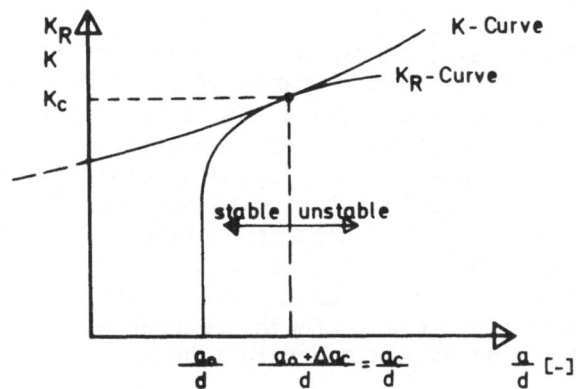

Fig. 2.7   Critical crack length a_c and fracture toughness K_c

2.4.4 Evaluation of strain measurements

Measurements of local strains at the critical cross-section of
the beam allow an estimate of the location of the neutral axis,
the length of the process zone and the amount of stress transfer
across parts of the crack. For the concrete (C) beams the gage
length of 45 mm used for the strain measurements is only
1.5 x maximum aggregate size and thus is in the range of the
width of the process zone w_c as proposed e.g. in /1, 7/.
Therefore, reasonably accurate results should be obtained when
estimating stress transfer from these local strain measurements.
However, the gage length is considerably larger than w_c for
the mortar (Mo) beams. This will limit the accuracy of estimates
obtained from these strain measurements.

For a reliable estimate of stress transfer relations between
stress and displacement for the process zone should be obtained
from measurements on concrete and mortar specimens loaded in
concentric tension including the strain softening branch such as
described in /2, 9, 10/. Particularly for mortar, rather elaborate experimental procedures are required in order to avoid
unstable crack growth in such tests. Therefore, in this investigation the behavior of mortar and concrete subjected to concentric tension has not as yet been studied. However, such experiments are planned for the future. For a preliminary data evaluation approximate relations as given e.g. in /2, 9, 10/ are
used.

Some typical results of strain measurements of a concrete L-beam
(specimen No. 5) are shown in Figs. 3.4 - 3.7. More detailed results and evaluations will be given at some later time.

3. Experimental results

3.1 Stability of crack growth

Depending on materials, specimen size and type of experiments - G_F-tests or C-tests - stable or unstable crack growth occurred. For some of the L-beams stability of crack growth has been checked by plots of deflection versus time. In Table 3.1 the results are summarized with regard to stability of crack growth. From this it follows that with the exception of the H-beams stable crack growth has been obtained for all of the notched specimens. However, the large mortar beams may have undergone a short period of rapid crack growth. In the case of unnotched beams stable crack growth occurred only in the C-tests on the S- and on one of the concrete M-beams.

Type of specimen		Unnotched Beams	Notched Beams	
Size	Material	C-Tests	G_F-Tests	C-Tests
S	C	S	S	S
	Mo	U	S	S
M	C	S, U	S	S
	Mo	U	S	S
L	C	-	S	S
	Mo	-	S(?)	S
H	C	-	U	S
	Mo	-	U	U

S = stable   U = unstable

Table 3.1: Stability of crack growth

These observations ascertain the criteria for stability of crack growth as proposed by Petersson /2/.

3.2 Load-deflection and load-strain diagrams

3.2.1 G_F-Tests

Figs. 3.1 - 3.3 show typical load-deflection diagrams for concrete S-, M- and L-beams. The deflections have been corrected for displacements of the supports wherever such data are available, however not for the effect of dead weight.

Figs. 3.4 - 3.7 show typical load-strain diagrams as obtained from gages W 3, 4, 5 and 6 of a concrete L-beam. The location of strain gages as well as the gage length are shown schematically in the corresponding figures.

3.2.2 Compliance-Tests

In Figs. 3.8, 3.9 and 3.10 typical load-deflection diagrams for concrete S-, M- and L-beams, respectively, are given. Fig. 3.10 also includes the load-deflection diagram as obtained from a G_F-test on a concrete L-beam.

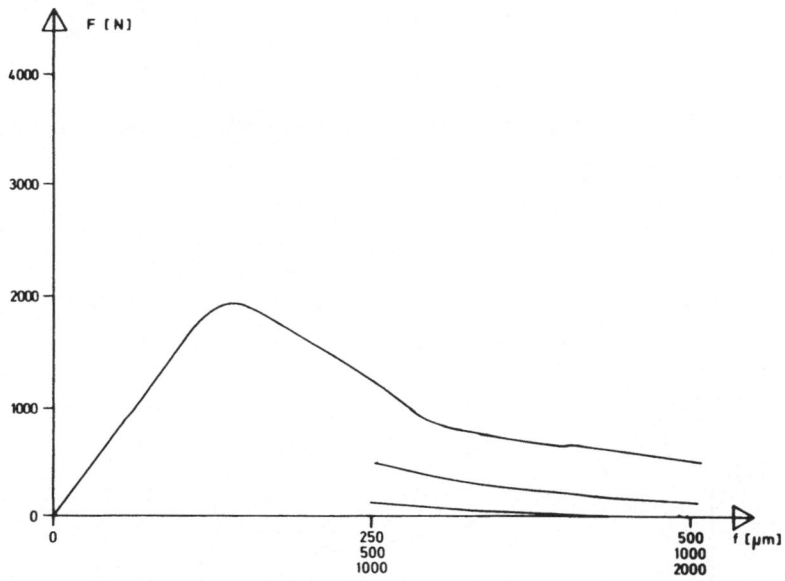

Fig. 3.1:   Load-deflection diagram, concrete S-beam, G_F-test

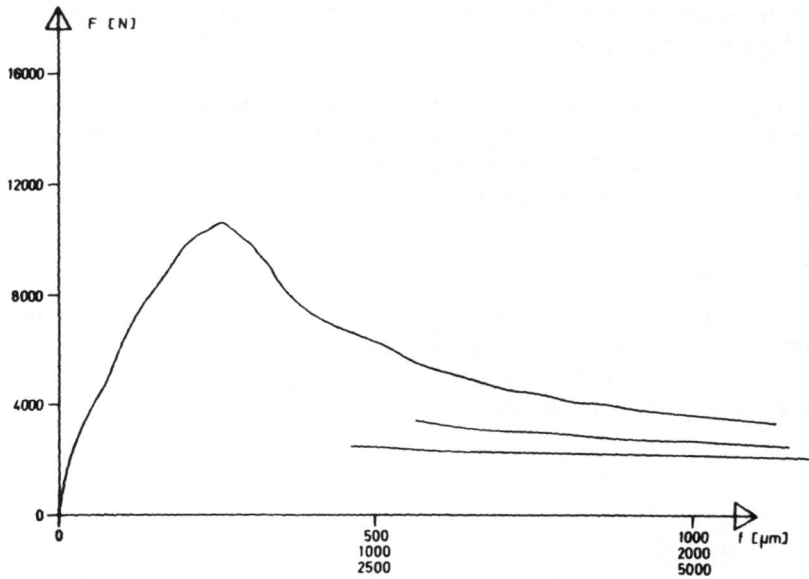

Fig. 3.2:   Load-deflection diagram, concrete M-beam, G_F-test

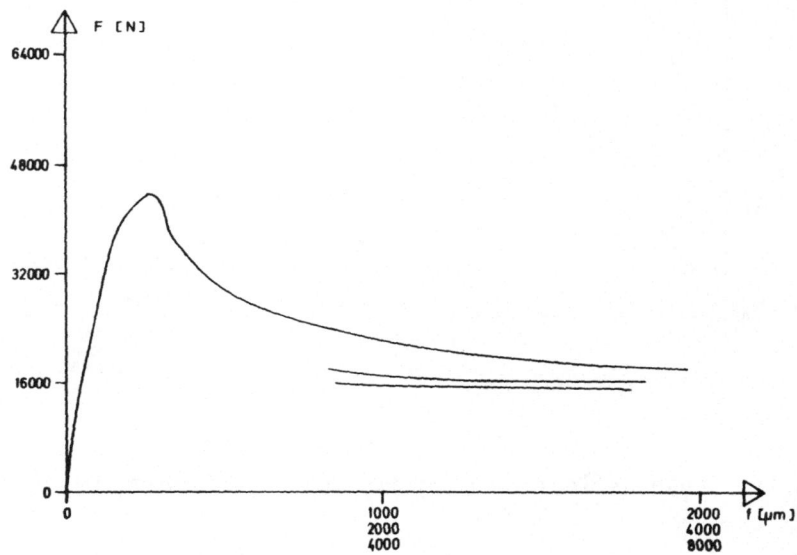

Fig. 3.3:   Load-deflection diagram, concrete L-beam, G_F-test

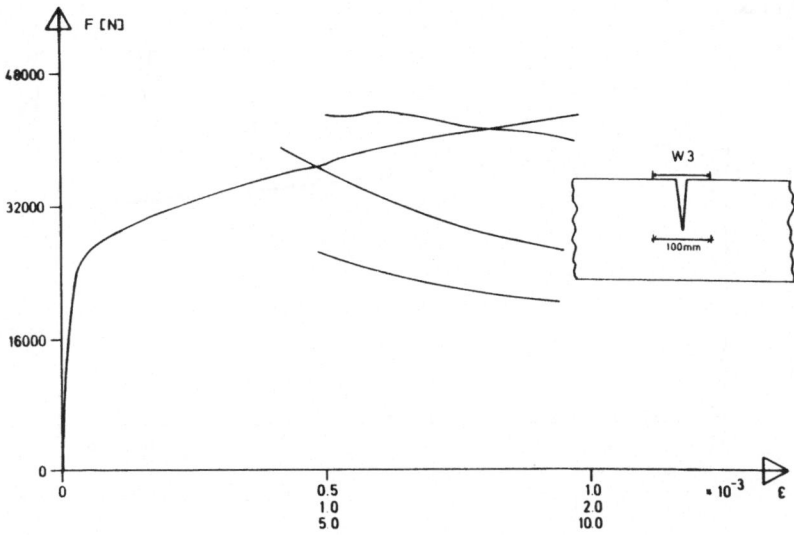

Fig. 3.4:   Load-strain diagram, strain gage W3, (CMOD),
            concrete L-beam, G_F-test

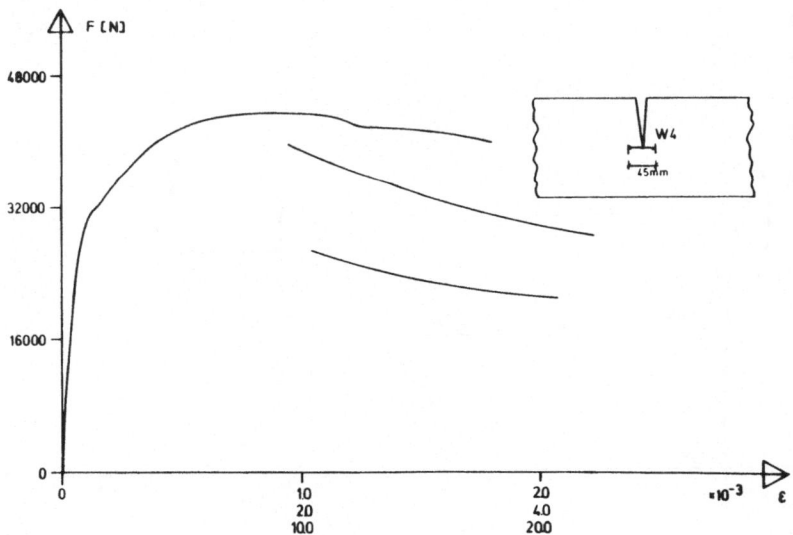

Fig. 3.5:   Load-strain diagram, strain gage W4, (CTOD),
            concrete L-beam, G_F-test

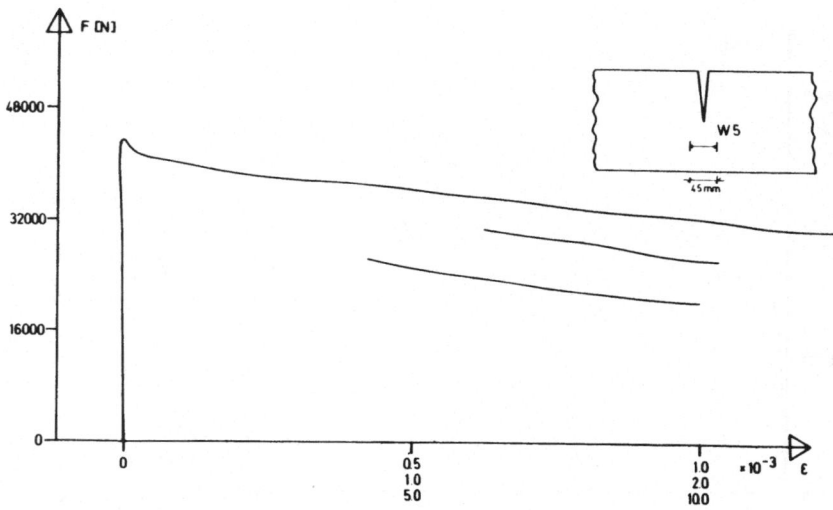

Fig. 3.6:   Load-strain diagram, strain gage W5,
             concrete L-beam, G_F-test

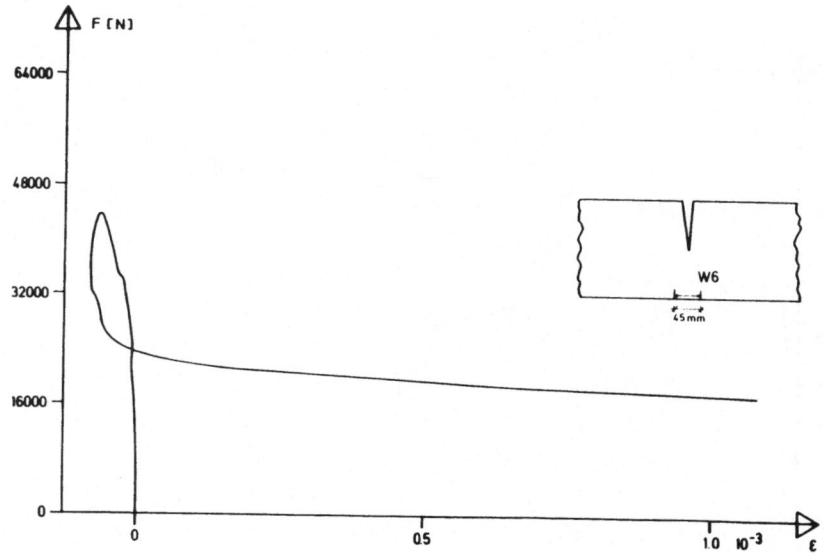

Fig. 3.7:   Load-strain diagram, strain gage W6,
             concrete L-beam, G_F-test

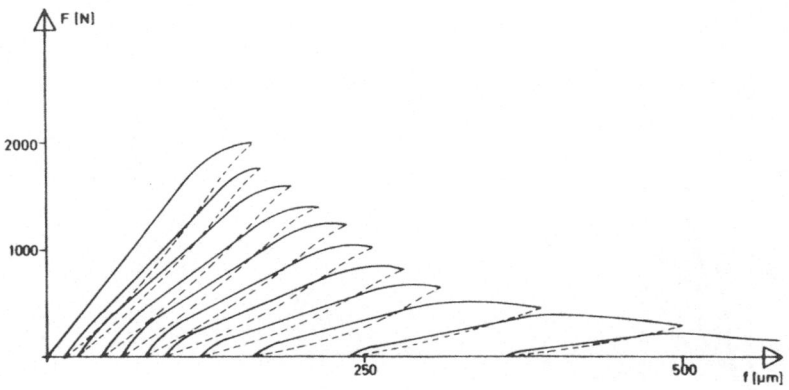

Fig. 3.8:   Load-deflection diagram, concrete S-beam, C-test

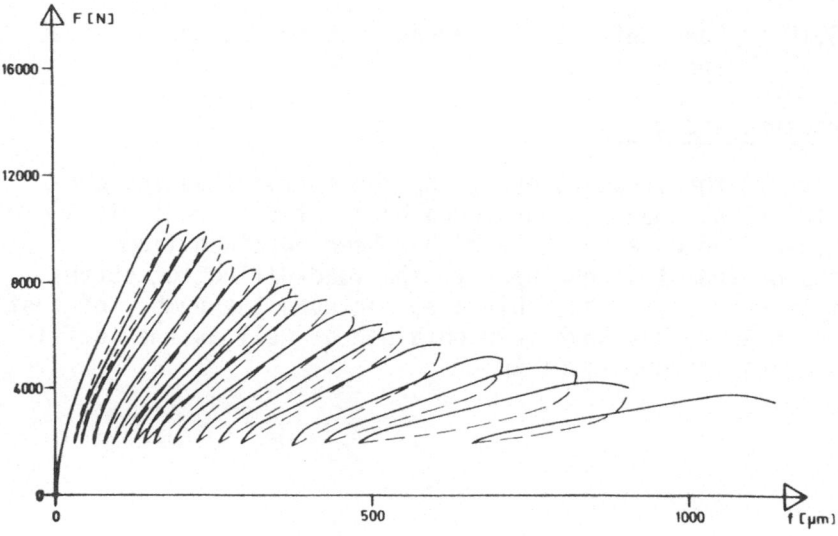

Fig. 3.9:   Load-deflection diagram, concrete M-beam, C-test

380

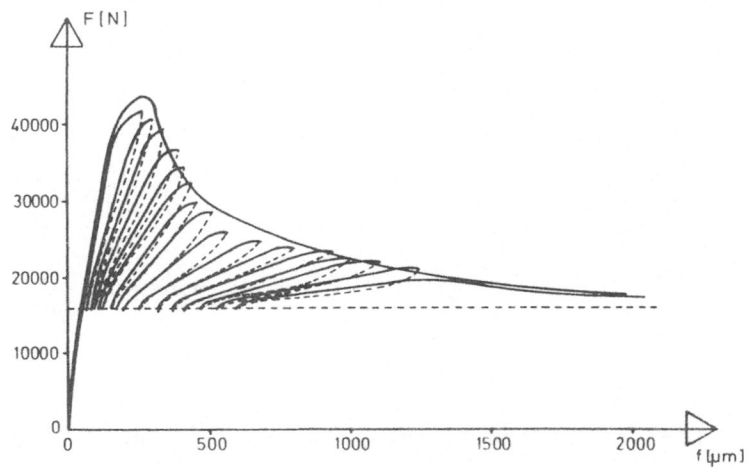

Fig. 3.10: Load-deflection diagram, concrete L-beam, C-test
and G_F-test

3.3 Fracture energy G_F

In Table 3.2 the fracture energy G_F for mortar (Mo) and concrete (C) of notched and unnotched beams of different sizes are
summarized. These values have either been obtained from
G_F-tests or from the envelopes to the load-deflection curves
of compliance tests. The table also includes the number of tests
from which G_F-values have been obtained as well as the coefficients of variation, v, of G_F.

Size and Material		G_F-Tests			C-Tests			Average		
		G_F [N/m]	v [%]	No.of Tests –	G_F [N/m]	v [%]	No.of Tests –	G_F [N/m]	v [%]	No.of Tests –
Notched beams										
S	C	141.3	22.3	12	144.4	18.4	4	142.1	20.8	16
	Mo	53.8	8.3	5	51.7	12.9	3	53.0	9.5	8
M	C	140.5	5.9	3	142.9	11.7	3	141.7	8.4	6
	Mo	48.4	14.3	4	50.5	5.7	3	49.3	10.8	7
L	C	175.6	7.1	2	168.0	8.5	3	169.9	7.2	5
	Mo	(34.5	5.8)	2	43.8	7.2	2	(37.6	14.8)	4
H	C	–	–	–	136.8	10.9	4	136.8	10.9	4
	Mo	–	–	–	–	–	–	–	–	–
Unnotched beams										
S	C	–	–	–	178.0	4.3	3	178.0	4.3	3
M	C	–	–	–	194.2	–	1	194.2	–	1

Table 3.2: Fracture energy G_F

From Table 3.2 it follows that G_F-values as determined in G_F-Tests or in C-Tests are almost identical. This is of significance because stable crack growth is easier to obtain in the C-tests than in the G_F-tests.

The fracture energy of the notched concrete S- and M-beams is almost identical, whereas the L-beams show a fracture energy which is approximately 20 percent larger than that of the S- and M-beams.

For the notched mortar specimens, there is a decrease of G_F up to 15 percent with increasing specimen size. The low value of G_F as obtained in the G_F-Tests on the mortar L-beams is uncertain because of the aforementioned possible period of rapid crack growth. Therefore, for the mortar L-beams only the G_F-values as obtained from C-tests will be used in the following discussion.

The G_F-values observed on unnotched M-beams in C-experiments are approximately 25 percent larger than the corresponding values of the notched beams. Only one value is available for M-beams.

3.4 Fracture toughness K_{Ic}

In Fig. 3.11 the dependence of fracture toughness on the depth of beam is given for concrete and mortar and for a crack depth $a/d = 0.5 - 0.6$ as calculated according to sect. 2.4.2. For this evaluation the second load cycle has been used in order to ascertain the existence of a sharp crack.

K_{Ic} has also been calculated for larger values of a/d and subsequent load cycles. However, these data are less reliable because of the limited range of validity of the calibration function, eq. (4).

With increasing depth of beams K_{Ic} increases by about 55 percent for the concrete and by about 45 percent for the mortar beams.

From fracture toughness the critical strain energy release rate G_{Ic} has been calculated using eq. (8):

$$G_{Ic} = \frac{K_{Ic}^2}{E} \cdot (1-\nu^2) \tag{8}$$

In Fig. 3.12 the ratio G_{Ic}/G_F is given as a function of the depth of the beams d. This ratio increases with increasing d. For the large mortar beams G_{Ic}/G_F approaches 1.0, whereas for the concrete beams values below 0.5 have been obtained.

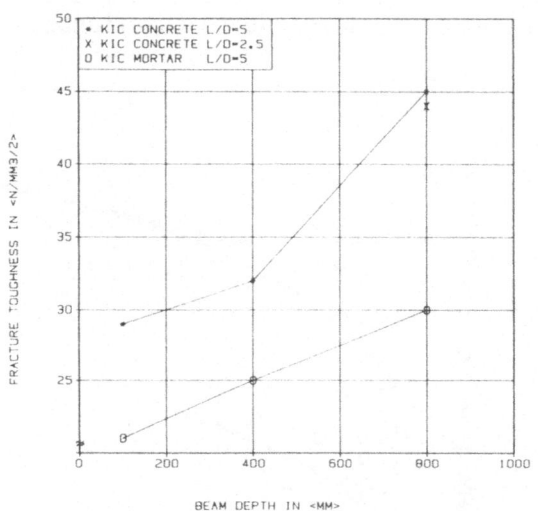

Fig. 3.11: Fracture toughness K_{Ic} for beams of different depth

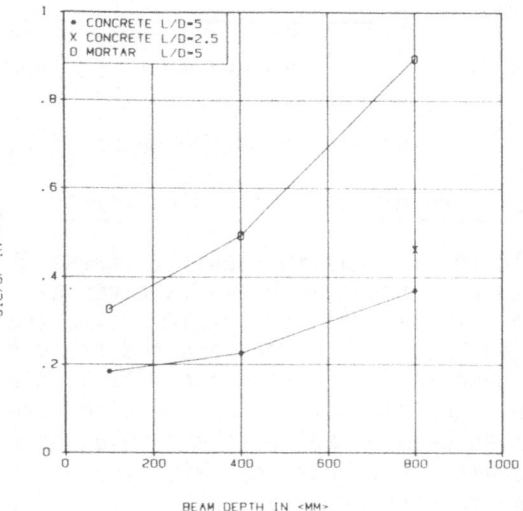

Fig. 3.12: G_{Ic}/G_F for beams of different depth

3.5 R-curves

Fig. 3.13 shows average R-curves as obtained for concrete beams of different sizes. They have been obtained from the second load cycle of C-tests. Apparently, the R-curves are size dependent when estimating Δa from reloading compliance.

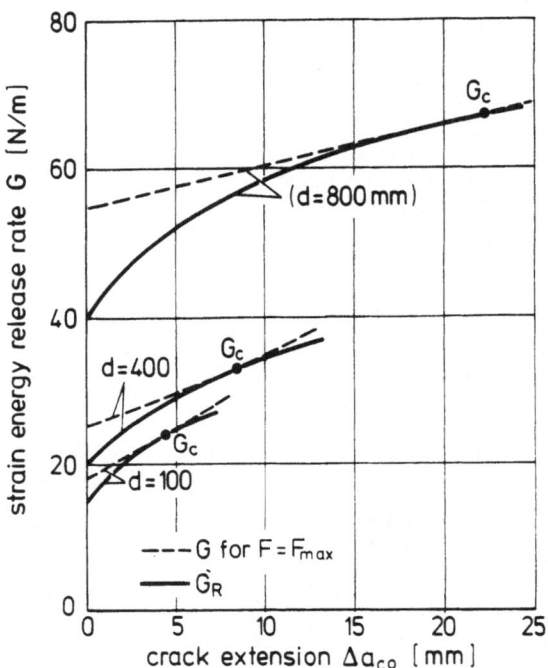

Fig. 3.13:   R-curves for concrete beams of different sizes

3.6 Flexural strength and notch sensitivity

In Fig. 3.14 the net maximum tensile stress σ_N as obtained from notched and unnotched beams is given as a function of specimen depth. With increasing specimen size the net maximum tensile stresses both for the notched and the unnotched beams decrease. As generally accepted (e.g. /11/), the notch sensitivity expressed by the ratio of the net tensile stress at maximum load of the notched beams, σ_N, to the modulus of rupture of the unnotched beams, f_{tu}, increases with increasing size both for concrete and for mortar. From Table 3.3 it follows that the mortar beams are considerably more notch sensitive than the concrete beams.

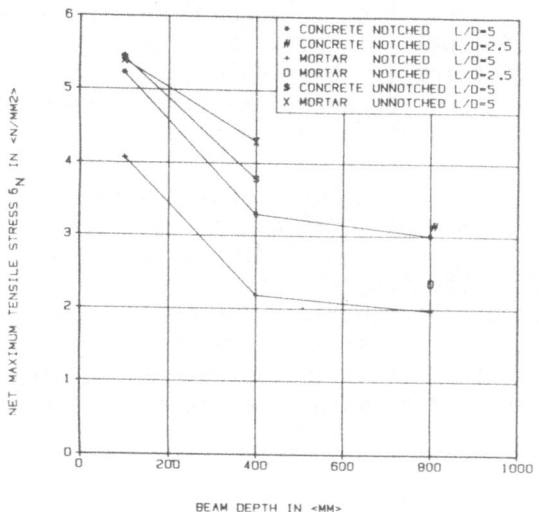

Fig. 3.14:   Net maximum tensile stress for beams
             of different depth

Size	Material	δ_N/f_{tu}
S	C	0.96
	Mo	0.75
M	C	0.87
	Mo	0.51

Table 3.3.: Notch sensitivity of concrete and mortar beams

3.7 Observation of crack growth

In Fig. 3.15 the relation between external load F and crack
length is given for a concrete L-beam (specimen No. 5). From
this it follows that crack growth becomes visible at about
80 percent of the maximum load.

In Table 3.4 the crack length cal a as deduced from a C-test is compared to the crack length directly observed on the specimen, obs a. Generally, the directly observed crack lengths are larger than the calculated values. However, the difference becomes smaller with increasing depth of the beam.

An estimate of the stresses in the compression zone of the beams under the assumption that the location of the neutral axis is given by the observed crack length indicates that for large values of crack length the concrete stresses would exceed the compressive strength of the concrete. Therefore, the directly observed crack length is too large or the crack length close to the surface is larger than it is in the interior of the cross-section. This is in agreement with observations reported in /12/.

S-beam		M-beam		L-beam		H-beam	
cal a/d [mm]	obs a/d [mm]	cal a/d [mm]	obs a/d [mm]	cal a/d [mm]	obs a/d [mm]	cal a/d [mm]	obs a/d [mm]
0.50	0.50	0.50	0.50	0.50	0.50	0.52	0.52
0.52	0.61	0.52	0.54	0.54	0.55	0.55	0.53
0.56	0.71	0.55	0.60	0.58	0.60	0.61	0.53
0.62	0.76	0.58	0.62	0.62	0.64	0.63	0.59
0.66	0.78	0.60	0.64	0.64	0.66	0.67	0.64
0.70	0.80	0.64	0.66	0.65	0.68	0.69	0.65
0.72	0.85	0.66	0.68	0.66	0.69	0.71	0.71
0.76	0.88	0.68	0.71	0.69	0.72	0.72	0.73
0.79	0.90	0.71	-	0.70	0.75	0.74	0.78
0.82	0.91	0.73	0.74	0.73	0.79	0.75	0.82
0.84	0.92	0.75	0.77	0.75	-	0.78	0.83
0.87	0.95	0.77	-	0.76	0.81	0.79	0.84
0.91	0.97	0.78	0.80	0.78	0.82	0.83	0.86
0.91	0.98	0.79	0.81	0.80	0.84	0.85	0.88
		0.82	0.84	0.82	0.85	0.87	0.89
		0.85	0.88	0.84	0.87	0.90	0.90
		0.88	0.90	0.86	0.89	0.92	0.92
		0.92	0.95	0.88	0.91	0.94	0.93
		0.95	-	0.91	0.91	0.97	0.95
				0.95	0.93		
				0.97	0.96		

Table 3.4: Comparison of calculated crack length cal a = $a_0 + \Delta a_{co}$ and observed crack length obs a

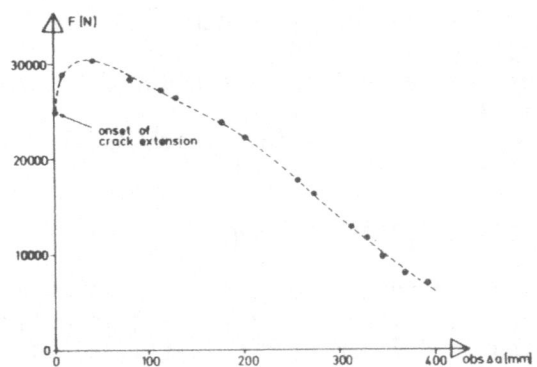

Fig. 3.15: Relation between external load F and change of crack
length obs△a, concrete L-beam (the load has been
corrected for the effect of dead weight)

4. Evaluation of experimental results

The primary objective of this investigation is to define fracture parameters of concrete

- which are suitable for numerical methods in which fracture
  mechanics concepts are applied
- which can be decuced from a limited number of comparatively
  simple experiments
- whose size and shape dependence does not exceed the size
  and shape dependence normally accepted for other concrete
  properties, and
- which may be estimated with some accuracy from other concrete properties known to the designer such as compressive
  or tensile strength and/or maximum aggregate size.

The major reasons that conventional fracture parameters such as
K_{Ic}, G_{Ic} or J_{Ic} do not satisfy these criteria are:

- the development of a process or fracture zone at the crack
  tip which in turn causes stress transfer across portions of
  the crack and the dissipation of energy due to microcracking
- difficulties in defining a true crack length, and
- dissipation of energy in parts of the specimen away from
  the fracture plane.

Several methods have been proposed in the past to define mostly
nonlinear fracture parameters which may satisfy the conditions
stated above. Some of these methods will be discussed in the

following and compared with the experimental results described in the preceding chapters. In particular we will refer to the

- concept of effective crack length
  (Shah et al. /8, 9, 14, 15/)
- R-curve and size effect law for blunt fracture
  (Bažant et al./7, 13, 16/)
- fracture energy and fictitious crack model
  (Hillerborg et al. /1, 2, 3, 17/)

4.1 Concept of effective crack length

In /8, 9, 14, 15/ a method is described in which a modified fracture toughness K_{Ic} is calculated using an effective crack length a_{eff} which is defined as the initial notch length plus an effective crack extension. For this CMOD has to be measured in cyclic three-point bend tests. The effective crack length may be estimated by means of compliance calculations and an iteration procedure such that CMOD at peak load, calculated from LEFM and assuming a value of a_{eff}, is equal to the measured CMOD. In the precise calculation also the effect of closing pressure in the process zone is taken into account. An approximate and simplified method has been suggested as a standard test procedure, in which the closing pressure is assumed to be zero /14/. In /15/ it has been shown that K_{Ic}^{S} calculated on the basis of the simplified method is independent of specimen size, so that it may be used in a LEFM-analysis. In our experiments accurate CMOD measurements have only been conducted on M- and L-beams whereas for the S-beams the location of the gage was 10 mm away from the mouth of the notch (Fig. 2.1). Since also the gage length used for the CMOD measurements was comparatively large (45 mm for the S-beam and 100 mm for the M- and L-beams) an evaluation has been carried out using the deflection rather than the CMOD measurements. The results are summarized in Table 4.1.

Fracture toughness	Material	Type of Specimen		
	-	S-beam	M-beam	L-beam
K_{Ic} compliance	concrete	26.1	30.7	42.0 N/mm$^{3/2}$
K_{Ic}^{S} deflection		23.1	28.6	41.9 N/mm$^{3/2}$
K_{Ic} compliance	mortar	19.4	22.6	26.4 N/mm$^{3/2}$
K_{Ic}^{S} deflection		17.2	21.5	27.7 N/mm$^{3/2}$

Table 4.1: Effective crack length concept:
calculation of K_{Ic}^{S} based on deflection measurements

From Table 4.1 it follows that deflection measurements are not suitable to calculate K_{Ic}^S since these values are highly size dependent and differ little from K_{Ic} compliance. Apparently it is mandatory for this approach to exclude inelastic effects away from the fracture plane which is possible by means of careful CMOD measurements.

4.2 R-curve and size effect law

In /7/ a finite element analysis to simulate fracture processes is described which is based on non-linear fracture mechanics. There the width of the process zone w_c and the fracture energy G_F in addition to E_c and f_{tu} are required as fracture parameters. Because of possible uncertainties in the experimental determination of G_F it is suggested in /13, 16/ to determine G_F from R-curves which in turn may be deduced from tests on notched beams of different sizes utilizing a "size effect law" which describes the relation between the net stress at failure of a notched beam, σ_N, and the characteristic size, d, of a specimen of given geometry (eq. 9):

$$\sigma_N = A \cdot f_{tu}(1 + d/B)^{-1/2} \tag{9}$$

where f_{tu} = tensile strength of unnotched specimen
A, B = coefficients

Subsequently $G = f(a_0 + \Delta a)$ relations e.g. according to eq. (6,7) for a given combination of σ_N and d, may be calculated which satisfy eq. (9). In /13, 16/ it is shown that the envelope to the family of G-curves corresponds to a unique R-curve. This curve yields a value of G_F for large values of d or crack extensions Δa. The approach is based upon the assumption that a unique R-curve independent of specimen size exists. According to Fig. 3.13 the R-curves for S-, M- and L-beams do not coincide if Δa is calculated from compliance measurements. The R-curve deduced from the size effect law, therefore, gives the relation between critical values of G and an equivalent crack extension similar to the approach given in sect. 4.1 though both approches may yield different values of Δa. The advantage of this R-curve method is, that G_F may be estimated e.g. from three-point bend tests on specimens of different sizes without strain or deflection measurements and without having to satisfy the condition of stable crack growth required for G_F-tests as described in sect. 2.3.2. In /13, 16/ the following relation (eq. (10)) is suggested to describe the R-curve:

$$G_c (\Delta a) = G_F \cdot \left[1 - \left(1 - \frac{\Delta a}{\Delta a_m}\right)^n \right] \quad \text{for} \Delta a \leq \Delta a_m \tag{10}$$

where   n   = exponent
        Δa_m = coefficient corresponding to the critical
               crack extension for large specimens

Eq. (9) and (10) have been evaluated using the experimental data obtained in this investigation. In Fig. 4.1 the size effect relations are shown for the mortar and the concrete beams respectively, whereas Fig. 4.2 gives a R-curve (solid line) for the concrete beams which has been calculated from the size-effect law as described in /13/. The values for the coefficients of eq. (10) are given in Table 4.2 together with the values of G_F which had been determined experimentally.

	G_F	Δa_m	n	G_F exp.
	[N/m]	[mm]	-	[N/m]
mortar	62	35	3.1	44-53
concrete	131.5	60	3.3	142-170

Table 4.2:   Results of evaluation of eq. (10)

Apparently, there is reasonably good agreement between G_F as deduced from the size effect law and the experimental values of G_F. To further verify the validity of this approach experimental values of G_c and Δa_{co} have been deduced for the concrete beams which had been subjected to a sequence of load cycles (C-tests). For each load cycle a value of Δa_{co} has been calculated from the reloading compliance. These individual data points are also given in Fig. 4.2. They are well represented by the R-curve eq. (10).

The size effect law will not be discussed in this paper in detail. Howewer, it has to be pointed out that as a minimum at least three sets of experiments on specimens of substantially different size would have to be carried out in order to deduce a reasonably reliable R-curve. Since for concrete this would include testing of beams of a depth of 400 mm or more, this can no longer be labeled "routine testing" as required in the introduction to sect. 4.

Nevertheless, the R-curve approach may be a very useful one if use is made of values of G_F obtained from load-deflection measurements: Evaluations of some experimental results showed that in eq. (10) the exponent $3 \leq n \leq 4$. Furthermore, Δa_m may be a linear function of the maximum aggregate size, though this would have to be further varified by experiments. With $n \approx 3$, $\Delta a_m = f$ (max. aggregate size) and G_F from experiments an R-curve may be deduced from tests on specimen of one size only. In Figs. 4.1 and 4.2 the dotted lines have been obtained from eq. (10) with the experimental value of G_F for concrete, $G_F = 151.5$ N/m, $\Delta a_m = 60$ mm and $n = 3$. They differ little from the relation obtained from the experiments on specimens of different sizes.

This approach may be further improved if the R-curve is described e.g. by the following relation (eq. (11)):

$$G_c(\Delta a) = G_F \cdot \left[1 - e^{\left(-\frac{\Delta a}{\Delta a_m} \right)^n} \right] \tag{11}$$

A preliminary evaluation showed that a good fit can be achieved with $n = 1.2$. The coefficient Δa_m in eq. (10) may be estimated from the net failure stress σ_N, so that both G_F and σ_N and thus $G_c(\Delta a)$ may be obtained from experiments on specimens of one comparatively small size without the need to estimate crack extension. Apparently, knowledge of a size independent (equivalent) R-curve would be a useful tool in fracture mechanics analysis.

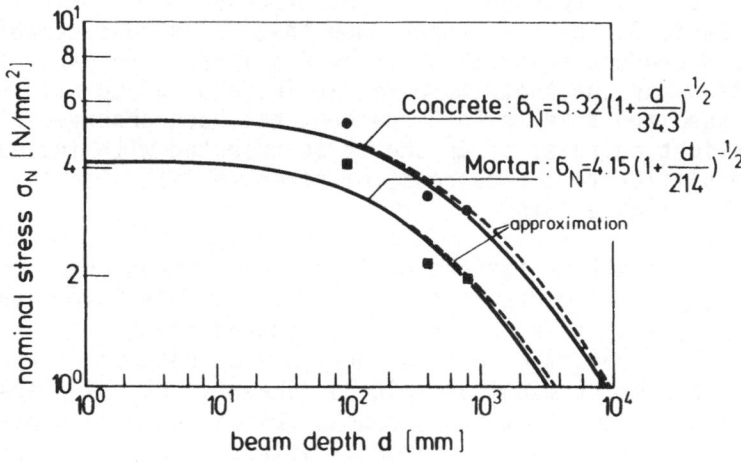

Fig. 4.1: Evaluation of experimental data: size effect law

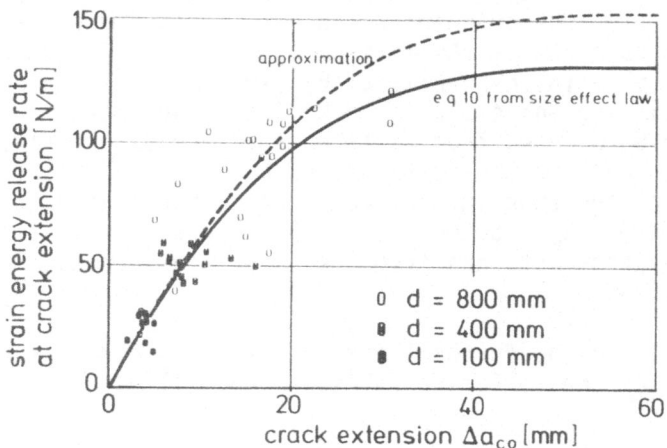

Fig. 4.2: R-curve evaluation and experimental data, concrete

4.3 Fracture energy and fictitious crack model

Hillerborg et al. developed and successfully applied the fictitious crack model to the fracture analysis of concrete members /e.g. 2, 17/. For the application of this method the characteristic length as defined in sect. 2.4.1 is needed which in turn requires knowledge of G_F. E.g. in /1, 2/ it has been suggested that G_F should depend little on the size of a specimen because when determining G_F from experimental data only the well defined total area of the ligament is needed, however, not the length of the crack and/or of the process zone. The results given in sect. 3.3 of this paper show that G_F is considerably less size dependent than other conventional fracture parameters. Nevertheless for the concrete beams an increase of G_F with increasing specimen size up to 20 percent has been observed. Instead a slight decrease of G_F should be expected with increasing size similar to the decrease of strength with size as observed for all materials.

Despite these small variations of G_F the fictitious crack model predicts well the experimental results in this investigation: In Fig. 4.3 values of G_{Ic}/G_F as deduced from experimental results are compared to an analysis by Petersson based on the fictitious crack model assuming a linear stress-displacement relationship for the fictitious crack. The values of G_{Ic} have been calculated from the experimental values of δ_N and a crack length $a = a_0 + \Delta a$ as estimated from reloading compliance for the second load cycle:

$$G_{Ic} = \delta_N^2 / E \cdot g(a_0 + \Delta a; d) \qquad (12)$$

The agreement between analysis and experimental results is very convincing underlining the validity of both the fictitious crack model and the G_F-concept.

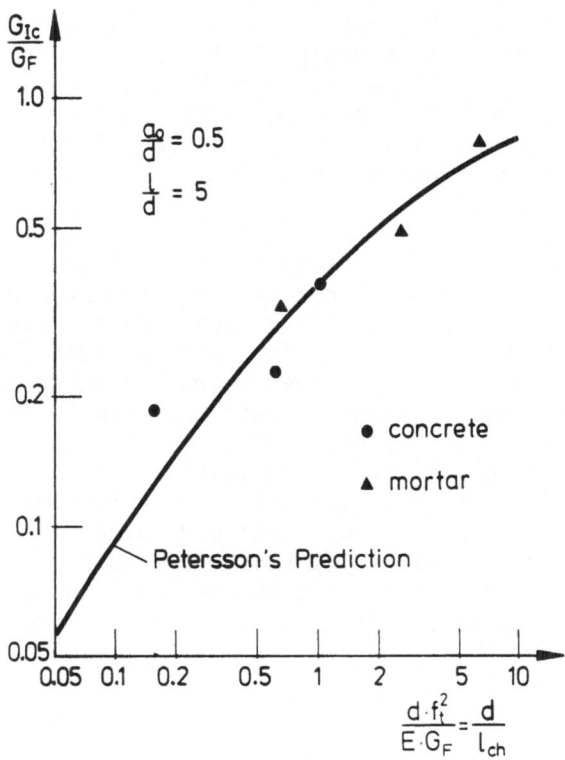

Fig. 4.3:   Comparison of experimental values G_{Ic}/G_F
with analytical values (Petersson /2/)

Nevertheless, several investigators have found G_F to be highly size dependent. In an interlaboratory study conducted by RILEM-Committee 50-FMC differences in G_F up to 80 percent for specimens of different sizes have been observed by some investigators. Such size dependence may be caused by several factors:

(1) the width of the process zone is specimen size dependent
(2) the length of the process zone varies within the strain softening branch
(3) insufficient stiffness of the testing machine
(4) inelastic deformation at the supports of the beam
(5) inelastic strains in the specimen away from the fracture plane

Though factors (1) and (2) cannot be excluded, they do not appear to be very probable. Factor (3) has been studied e.g. by Petersson /2/ who also postulated certain requirements regarding the experimental set-up to be satisfied in G_F-experiments. Inelastic deformations of the supports (4) may tend to increase G_F particularly for high loads i. e. large specimens and small ratios span/depth of beam. Neglecting inelastic support deformations of the L-beams of this investigations would have caused an apparent increase of G_F by 14 percent for the beams with $l/d = 5$ and of 104 percent for the beams with $l/d = 2.5$. Nevertheless such errors can be avoided by a precise definition of required experimental procedures and by suitable arrangement of deflection gages.

Inelastic strains away from the fracture plane (5) may be substantial. The difference in G_F of notched and unnotched beams shown in Table 3.2 may be attributed to such effects. Since the stresses in the unnotched beams away from the fracture plane are considerably larger than in the notched beams the occurrence of inelastic strains in these regions is much more likely for the unnotched beams than for the notched beams resulting in higher values of G_F. The possible effect of inelastic strains on G_F will be investigated in an analytical study. Nevertheless, experimental procedures will have to be developed which exclude such effects if they occur. A possible approach is a correction of G_F as determined from deflection measurements by means of CMOD-measurements. Such an approach is being investigated at this time.

5. Summary

5.1   The testing of large notched beams with the tension side up proved to be advantageous with regard to handling the specimens, stability of crack growth and evaluation of fracture data.

5.2 G_F-values obtained from experiments in which the rate of deflection has been kept constant throughout (G_F-tests) is identical with G_F-values deduced from experiments in which the beams have been subjected to a sequence of loading and unloading cycles (C-tests).

5.3   The fracture energy G_F both for the concrete and for the mortar specimens depends less on specimen size than other fracture parameters for beams ranging in depth between 100 mm and 800 mm.

5.4   The fracture toughness K_{IC} as determined from reloading compliance increases markedly with increasing depth of the specimen.

5.5 For the large mortar specimens the ratio G_{IC}/G_F approaches unity. For concrete specimens with a depth up to 800 mm the ratio of G_{IC}/G_F is less than 0.5.

5.6 In general, the crack length observed experimentally on the surfaces of the beams is larger than the crack length calculated from the compliance of the beams. From an estimate of concrete stresses in the compression zone it follows that the average crack length through the entire width of the beam is smaller than the crack length observed on the surface.

5.7 An evaluation of the experimental data applying an equivalent crack approach showed that the size dependence of K_{IC} cannot be eliminated if the calculations are based on deflection measurements.

5.8 The R-curves estimated by utilizing the size effect law as proposed by Bažant et al. agreed well with experimental values of G_F, G_C and Δa.

5.9 A method is proposed to estimate R-curves from experiments on specimens of one size making use of the size effect law.

5.10 Very good agreement has been obtained between the experimental results and analytical results based on the fictitous crack model.

5.11 Factors which may be responsible for a size dependence of G_F have been discussed and means to eliminate them have been proposed.

6. References

/1/ Hillerborg, A.
Analysis of one single crack,
Report to RILEM-Committee TC 50-FMC, January 1981

/2/ Petersson, Per-Erik
Crack growth and development of fracture zones in plain concrete and similar materials,
Division of Building Materials, Lund Institute of Technology
Report TVBM-1006, Lund, Sweden 1981

/3/ Proposed RILEM recommendation, 29th January, 1982, revised version June 1982,
Determination of the fracture energy of mortar and concrete by means of three point bend tests on notched beams,
Division of Building Materials, Lund Institute of Technology
Lund, Sweden, 1982

/4/ Brown, J. H.
Measuring the fracture toughness of cement paste and mortar,
Magazine of Concrete Research, Vol. 24, No. 81, Dec. 1972

/5/ Brown, W. F.; Srawley, J. E.
Plain strain crack toughness testing of high strength
metallic materials,
ASTM-STP No. 410, 1967

/6/ Tentative recommended practice for R-curve determination,
E 561 - 76, ASTM-STP No. 632, 1974

/7/ Bažant, Z. P.; Oh, B. H.
Crack band theory for fracture of concrete,
Materiaux et Constructions, May - June 1983, No. 93

/8/ Wecharantana, M.; Shah, S. P.
Predictions of nonlinear fracture process zone in concrete,
Technological Institute, Department of Civil Engineering,
Northwestern University, Evanston, Illinois, Jan. 1983

/9/ Gopalaratnam, V. S.; Shah, S. P
Softening response of plain concrete in direct tension,
Technological Institute, Department of Civil Engineering,
Northwestern University, Evanston, Illinois, June 1984

/10/ Heilmann, H. G.; Hilsdorf, H. K.; Finsterwalder, K.
Strength and strain characteristics of concrete subjected
to tensile stresses,
(Festigkeit und Verformung von Beton unter Zugspannungen)
DAfStb, Heft 203, 1969

/11/ Ziegeldorf, S.; Müller, H. S.; Hilsdorf, H. K.
A model law for the notch sensivity of brittle materials,
Cement and Concrete Research, Vol. 10, No. 5, 1980

/12/ Mc Clintock, F. A.; Irwin, G. R.
Plasticity aspects of fracture mechanics,
ASTM-STP No. 381, 1965

/13/ Bažant, Z. P.; Kim, J.-K.
Nonlinear fracture properties from size effect tests,
Center for Concrete and Geomaterials,
The Technological Institute, Northwestern University,
Evanston, Illinois, May 1984

/14/ Jenq, Y. S.; Shah, S. P
Determination of the fracture toughness of plain concrete
using three-point bend test,
Proposal for a standardized test method to be discussed in
RILEM-Committee TC 50-FMC, 1984

/15/ Jenq, Y. S.; Shah, S. P.
Nonlinear fracture parameters for cement based composites:
Theory and experiments,
Preprints, Proceedings NATO Advanced Research Workshop on
Application of Fracture Mechanics to Cementitious
Composites, Evanston, Illinois, 1984

/16/ Bažant, Z. P. ; Kim, J.K.
Determination of nonlinear fracture parameters from size
effect tests,
Preprints, Proceedings NATO Advanced Research Workshop on
Application of Fracture Mechanics to Cementitious
Composites, Evanston, Illinois, 1984

/17/ Gustafsson, P. J.; Hillerborg, A.
Improvements in concrete design achieved through the
application of fracture mechanics,
Preprints, Proceedings NATO Advanced Research Workshop on
Application of Fracture Mechanics to Cementitious
Composites, Evanston, Illinois, 1984

APPLICATION OF FRACTURE MECHANICS
TO CEMENTITIOUS COMPOSITES
NATO-ARW - September 4-7, 1984
Northwestern University, U.S.A.
S. P. Shah, Editor

FRACTURE MEASUREMENTS OF CEMENTITIOUS COMPOSITES

Y.W. Mai

Department of Mechanical Engineering
University of Sydney
Sydney, NSW 2006, Australia

ABSTRACT

A state-of-the-art review is given on the application of both
linear and non-linear elastic fracture mechanics parameters for
characterising the fracture behaviour of cementitious composites.
Experimental techniques for the measurement of these fracture parameters are discussed and critically assessed in relation to fibre
cements, mortars and concretes. It is shown that the fracture
behaviour is best described by the R-curve which is believed to be
a material property provided that certain size requirements are
met. Theoretical models for the prediction of R-curves are
considered. Some recent experimental results on fracture measurements of asbestos and cellulose fibre cements are also described.
In particular, the effects of specimen size, geometry and environment on the R-curve are given.

1   INTRODUCTION

Conventional design based on tensile strength alone is inadequate to guard against structural failure since all structures
contain some kind of defects. Catastrophic failure can occur if
the material does not have sufficient toughness or if the subcritical defects grow to a critical size under sustained or cyclic
loads. The assessment of the behaviour of cracks under load is
within the frame work of fracture mechanics. The development and
acceptance of this new scientific knowledge has proven to be a
valuable aid to design against fracture and in the selection of
candidate materials to be used in many high technology areas. It

400

is the success of fracture mechanics when applied to materials such
as metals, polymers and ceramics [1-3] that has led to its use in
analysing the fracture behaviour of cementitious matrices and their
fibre composites. However, there has been considerable debate as
to whether linear or non-linear elastic fracture mechanics techniques
should be used. Figure 1 shows the deformation zones ahead of a
highly strained crack tip. For many materials there is usually an
outer plastic zone whose shape is dependent on the loading/specimen
configuration and an inner process zone in which fracture takes
place, Figure 1(a). Only the work that goes into the fracture
process zone can be regarded as a material constant. In brittle
materials the fracture process zone is intimately associated with
the plastic zone so that linear elastic fracture mechanics (LEFM)
is valid and a one-parameter criterion such as the critical strain
energy release rate (G_c) or the critical stress intensity fractor
(K_c) suffices to describe the critical condition. For ductile
materials much of the plastic flow at the tip of the crack is not
directly involved in the fracture process and in laboratory size
specimens the total work of fracture is not a material constant.

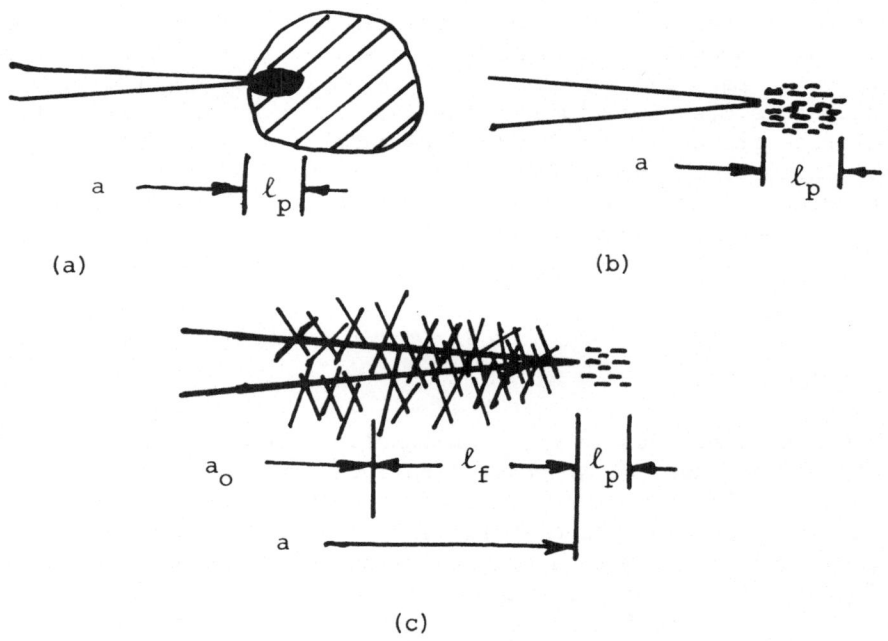

Fig. 1 Crack tip deformation zones in (a) metals
and polymers (b) cementitious matrices
and (c) fibre cement composites

Ductile fracture mechanics parameters such as the critical J-integral (J_c), critical crack tip opening displacement (δ_c) and the specific essential work of fracture (W_e) [4-6] have to be used. In cementitious matrices there is no outer plastic zone and only a narrow band of the fracture process zone exists at the crack tip, Figure 1(b). For both mortars and concretes this process zone consists of a region of interlocking aggregates and a region of microcracking. For cement pastes only the microcracking zone exists at the crack tip. If the length of the process zone (ℓ_p) is small relative to the overall dimensions of the specimen then LEFM para-meters, K_c and G_c, are sufficient to describe the critical fracture event. Otherwise, both K_c and G_c are size dependent. There are some direct experimental measurements of ℓ_p for cement pastes which give $\ell_p \approx 1$-2 mm [7,8]. However, direct measurements of ℓ_p for mortars and concretes are scarce. Petersson [9] suggested that a nominal value for mortar was about 40 mm and for concrete this was some 80 mm. Obviously, ℓ_p varies with the size of the aggregates $(\bar{d})$ used and for concretes Bazant and Oh [10] suggested that $\ell_p \doteq 12\ \bar{d}$. Based on these ℓ_p values it is easy to see that cement pastes can be analysed adequately with LEFM concepts. But for mortars and concretes LEFM cannot be applied directly without modifications. Two useful reviews on the application of fracture mechanics to cementitious matrices are recently given in [11,12]. It must, how-ever, be pointed out that despite many firmly held views that LEFM is inapplicable for concrete structures Saouma and Ingraffea [13] recently reanalysed the experimental data obtained by Kesler, Naus and Lott [14] using more accurate stress intensity factor (K) solutions and from these results they concluded that LEFM could in fact be applied to concretes. K_c was a material property and was independent of both notch length and specimen size. Swartz et al [15] also reported that there was no size effect on the fracture toughness of precracked concrete specimens. It is difficult to comment on these findings [13,15] until the fracture process zone sizes are accurately measured. Unless they are small relative to the size of the specimen LEFM parameters in theory cannot be used to describe the fracture behaviour.

For cementitious fibre composites, in addition to the matrix process zone (ℓ_p), there is a fibre bridging zone (ℓ_f) across the cracked faces of the matrix, Figure 1(c). ℓ_f is expected to vary directly with fibre length, specimen geometry and loading configu-ration. For example, in double-cantilever-beam (DCB) steel fibre reinforced concretes $\ell_f \doteq 500$ mm for 12.5 mm long fibres [16]. It is this large length of ℓ_f that has led many investigators to suggest that LEFM is inapplicable to fibre reinforced cementitious materials [e.g. 16-18]. It is shown in Section 3 that this should not be the single deciding factor whether LEFM can or cannot be applied because the fibre bridging zone can be suitably modelled by closure stresses acting across the crack opening. Provided the

matrix process zone size (ℓ_p) is small relative to the specimen
size [19,20] LEFM is certainly valid. Unfortunately, there are
very few direct measurements of ℓ_p for fibre cements and fibre
concretes.

2   FRACTURE MEASUREMENTS OF CEMENTITIOUS COMPOSITES

2.1   Linear and Non-Linear Elastic Fracture Parameters

Linear elastic fracture parameters include both K_c and G_c.
For cementitious materiasl the difference between plane strain and
plane stress is small [9]. Therefore, by neglecting the term
involving Poisson's ratio $K_c^2 = E\,G_c$. Detailed procedures for the
measurement of K_c are given in standard test methods [19,20] using
either bend or compact tension specimens. Other specimen geometries
and loading configurations can also be used provided that appropriate stress intensity factor solutions are available [e.g. 21-23]
and size requirements for valid testing are satisfied [19,20].
There are several methods to measure G_c the most common of which
is the compliance technique, i.e.

$$G_c = \frac{P_c^2}{2B} \frac{d}{da} \left(\frac{u}{P}\right) \tag{1}$$

where P_c is the fracture load, a, the crack length, B, the specimen
thickness and u, the displacement of the load point. The compliance
(u/P) can be predicted from elasticity theory or measured experimentally. G_c may also be calculated indireclty from K_c using the
relationship: $G_c = (K_c^2/E)$. These two fracture parameters are
valid only for linear elastic materials in which the crack tip
plastic zone size is small compared to that of the specimen [19,20].
Irwin's plastic zone correction factor can be used to correct for
small scale plasticity at the crack tip [24]. Dugdale's model [25]
also does the same thing. However, in Irwin's case there is still
a stress singularity at the effective crack tip, but in Dugdale's
case K is non-existent since the singularity is now removed.

Non-linear elastic fracture parameters include the critical
J-integral (J_c) and the critical crack tip opening displacement
(δ_c). Standard test methods have been established for evaluating
both J_c [26] and δ_c [27] using notched bend specimens. J_c is
strictly only valid for crack initiation. However, owing to difficulties of identifying the exact location of the crack tip in
cementitious materials J_c is usually evaluated at maximum load of
the test. The determination of δ_c in bend specimens uses a concept
of a plastic rotation constant (r) which for metals is 0.4 [27].
Exact r values may be estimated from slip-line theory [28] but it
is very doubtful that slip-line exists in cementitious materials
whose process zone geometry is a narrow crack band and there is no

outer plastic zone. It is, therefore, believed that δ_c cannot be
calculated using the rotation constant concept given in standard
test methods [27]. However, a critical δ_c fracture criterion can
still exist but it has to be measured using some other techniques.

Gurney and his co-workers have developed an elegant technique
for measuring the specific work of fracture (R) in both linear and
non-linear elastic materials [29-33]. Consider Figure 2 which shows
a typical load-deflection (P-u) record OABO for a linear elastic
solid. OA represents the compliance line for crack length a and OB
for crack length a + Δa; AB is the fracture locus. It can be easily
shown from energy balance consideration that area OABO represents the
fracture work for a crack extension BΔa. This is, therefore, a
simple method to determine R which for linear elastic materials is
equal to G_c. Only the irreversible work area OABO is required and
there is no need to measure compliances as required by equation (1).
Provided that stable cracking can be achieved in suitable specimen
geometries many R measurements can be obtained by repeatedly loading the sample after each crack extension. Conditions governing
the stability of cracking have already been discussed in [29,31,34].
If crack propagation is not interrupted and allowed to continue
until the whole sample breaks then the total work area under the

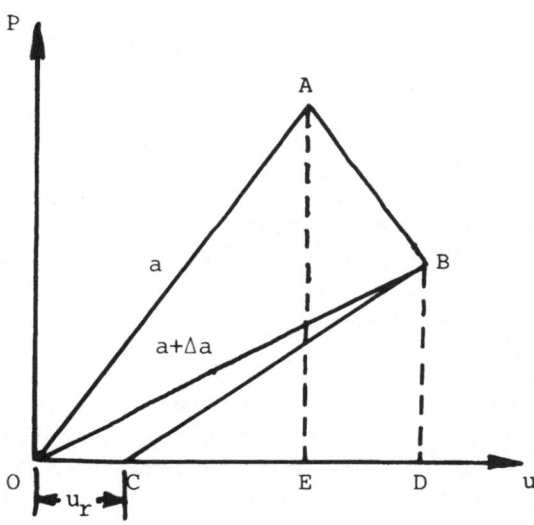

Fig. 2   Schematic load (P)-displacement (u)
diagram for crack propagation with and
without residual displacements (u_r)

P-u curve can still be used to calculate R. This method is now similar to the work of fracture technique given by Tattersall and Tappin [35] for ceramics and to the fracture energy test suggested by Petersson [9] for cementitious materials.

For non-linear elastic materials the total work of fracture after a crack extension Δa is given by OABCO in Figure 2. Unloading at B does not return to the origin O and there is a residual displacement u_r at zero load. It is important to determine if all the work area OABCO is for essential fracture work or if some of this is expended as plastic work not intimately associated with the fracture process. This problem may be analysed as follows. The strain energy release is the difference of the triangles BCD and AEO. For elastic deformation the strain energy is:

$$U = \frac{1}{2} Pu = \frac{1}{2} P^2(u/P) = \frac{1}{2} P^2 C^* \tag{2}$$

where C^* is the compliance measurement when $u_r \neq 0$. The strain energy release rate is given by:

$$\frac{dU}{da} = PC^* \frac{dP}{da} + \frac{P^2}{2} \frac{dC^*}{da} \tag{3}$$

The total external work is given by area ABDE and if it is assumed that α of this goes into plastic work and $(1-\alpha)$ goes to essential fracture work, then the total specific fracture work or the modified strain energy release rate [36,37] is given by:

$$R = G_c^* = \frac{(1-\alpha)}{B} P \frac{du}{da} - \frac{PC^*}{B} \frac{dP}{da} - \frac{P^2}{2B} \frac{dC^*}{da} . \tag{4}$$

For ductile metals and polymers $\alpha = 1$, i.e. all the external work has gone into the outer plastic zone, equation (4) is reduced to the familiar expression given by Hodgkinson and Williams [36]:

$$G_c^* = - \frac{P^2}{2B} \left[\frac{dC^*}{da} + \frac{2C^*}{P} \frac{dP}{da} \right] . \tag{5}$$

For cementitious materials it is justified, in view of the crack tip process zone geometry and the absence of an outer plastic zone, to assume that all the external work is used up as essential fracture work so that $\alpha = 0$. Equation (4) can be rewritten as:

$$G_c^* = \frac{P^2}{2B} \frac{dC^*}{da} + \frac{P}{B} \frac{du_r}{da} \tag{6}$$

using the relationship

$$u = u_r + PC^* . \tag{7}$$

Clearly, when $u_r = 0$, equation (1) is recovered. When $u_r \neq 0$ it can be shown that:

$$G_c^* = G_c + \frac{P}{2B} \frac{du_r}{da} \tag{8}$$

The measurement of G_c^* using equation (6) requires that the actual unloading/reloading compliance in the presence of residual displacement (C^*) and u_r be measured as functions of crack length (a). Equation (6) is also identical to the modified strain energy release rate expression obtained by Wecharatana and Shah [37].

2.2 Measurement of Fracture Parameters for Cementitious Composites

A good review on the application of linear and non-linear elastic fracture mechanics parameters to describe fractures in fibre cement composites has already been given by Mindess [18]. Thus, only a brief but critical discussion of previous investigations is given here.

As far as fibre reinforced cement pastes are concerned LEFM was first applied by Brown to glass cements [38] using four point notched bend (NB) test specimens (250x38x38 mm). It was shown that the single fracture parameter K_C could not be used to describe the fracture behaviour sufficiently. Slow crack growth (Δa) was predominant and K_C was observed to increase with Δa thus giving rise to a so-called R-curve as in metals. Lenain and Bunsell [39] also found that fracture in compact tension (CT) asbestos cements (HxW = 24x120 mm) was best described by the R-curve in which K_R was plotted against Δa. In both of these investigations [38,39] slow crack growth (Δa) was caused by the bridging fibres pulling out of the cracked cement paste and Δa was estimated from compliance calibrations using notched specimens. Although Lenain and Bunsell reported that the compliance curves were identical for notched and cracked samples it is difficult to see why the bridging fibres do not reduce the compliance of the latter. To avoid compliance measurements Patterson and Chan [40] in their work on CT glass reinforced cements (HxW = 102x51 and 127x63 mm) used Gurney's irreversible work area method to calculate G_C. However, they still had problems in defining exactly where the crack tip position was. They observed that the average G_C was relatively constant being independent of crack length and specimen size. It is very likely that the G_C values quoted correspond to the steady state when the whole fibre bridging zone is translated with crack growth. The early rising part of the $G_R - \Delta a$ curve would not have been noticed in the averaging procedure adopted by these authors to find G_C.

Residual displacements (u_r) upon unloading after incremental crack extension were noted in all these studies [38-40] but no attempt was made to analyse the cause of u_r. The size of the process zone was only measured in [39] using acoustic emission and this gave ℓ_p = 28 mm for asbestos cements.

Mindess and Bentur [41] recently investigated the fracture behaviour of wood fibre cements in both dry and wet conditions using NB specimens (400x100x15 ~ 19 mm). They concluded that LEFM was inappropriate. Fracture was primarily by fibre pull-out although there were also some fractured fibres. They observed that unlike glass cements [40] the mode of fracture was the propagation of a single crack through the specimen with little or no multiple cracking. Their work seems to suggest a relatively small matrix process zone size. It is thus possible to model the slow crack growth behaviour in wood fibre cements using LEFM concepts as discussed in Section 3.

Mai and co-workers studied the fracture behaviour of asbestos [42-44], cellulose [45-47] and cellulose/polypropylene [48] hybrid fibre cement mortars. For asbestos cement mortars K_c determined from the maximum load was not constant and depended on the size of the NB specimens [42,43]. It was suggested that the slow crack growth behaviour should be studied using the K_R - Δa curve. Mai [42] also determined J_c at maximum load for three point bend specimens using the analytical method described by Rice et al [49]. The results showed that J_c was sensitive to the amount of fibres in the cement mortar matrix. Unfortunately, the maximum load J_c was dependent on specimen size. Visalvanich and Naaman [50] also investigated the crack growth resistance curve for asbestos cement mortars using tapered DCB specimens. Four different techniques were used to measure G_R: (i) by equation (1) and compliance measurements, (ii) by direct measurement of K_R using known stress intensity factor solution for the DCB geometry and converting K_R to G_R, (iii) by Gurney's irreversible work area method when u_r = 0, and (iv) by Gurney's method when $u_r \neq 0$. All methods gave the same G_R - Δa curve and it was concluded that LEFM could be applied to asbestos cement mortars. This finding [50] seems to vindicate the LEFM approach used in [39,43,44] to calculate the K_R - Δa curve for asbestos cement mortars. Visalvanich and Naaman [50] further found that for steel fibre reinforced concretes method (iv), i.e. G_R^*, had to be used in order to characterise the fracture behaviour. Again, process zone size measurements were lacking in all these investigations on fibre cement mortars.

A lot more work were done on fibre concretes. With a few exceptions where glass and polypropylene fibres were used all previous investigations were on steel fibre concretes. For example, Harris et al [51], Mindess et al [52], Nishioka et al [53] and Velazco et al [54] measured K_c for steel fibre concretes in bending with beam depths varying from 10 to 100 mm. K_c was found to depend on specimen size [53] (when K_c was calculated based on maximum load) and to increase with crack growth [54] (when the current P and a values were used in the K-solutions). Maximum load K_c was also insensitive to steel fibre additions. The same general comments also apply to the glass and polypropylene fibre concretes investigated by Swamy [55].

Non-linear elastic fracture parameters such as J_c and δ_c had been applied to steel fibre concretes by a number of investigators. Mindess et al [52] and Halvorsen [56] found that J_c was a useful parameter for evaluating fibre addition effects on fracture behaviour. Since their J_c values were calculated at maximum load some crack growth had already taken place and any fibre reinforcement effects owing to varying amounts of fibre additions would be amplified. Had they determined the crack initiation J_c values these effects would have been less obvious [54]. The maximum load J_c values were also found to be affected by notch length and specimen size [54,56]. Indeed J_c can only be used to characterise the crack initiation toughness of steel fibre reinforced concretes [57].

As discussed in Section 1 the use of standard test method [27] to evaluate δ_c for fibre cements and concretes is problematical. Velazco et al [54] found that the rotation constant is 0.9 which cast serious doubts on whether useful δ_c values could be calculated. Nishioka et al [53] also attempted to measure the crack tip opening displacement as a fracture criterion for steel fibre concretes. Unfortunately, they only measured the crack mouth opening displacement and not δ_c. This perhaps explains why their experimental crack mouth displacements increase with specimen size.

In summary, the inability of both linear and non-linear elastic fracture mechanics parameters such as G_c, K_c, J_c and δ_c to describe fractures in fibre cementitious composites is due to two not unrelated factors: size effect and slow crack growth. Size effect is caused by the fact that enormously large specimens are required to accommodate the large fibre bridging and process zone sizes (i.e. $\ell_f + \ell_p$) in order to promote brittle fracture. Slow crack growth is largely a result of the fibre debonding and subsequent pull-out from the cracked matrix. This occurs irrespective of the specimen size. If the size is large enough full fibre pull-out is permitted and the maximum fracture toughness is developed. Otherwise, the maximum toughness is limited by the specimen size. Slow crack growth is best described by the R-curve. Velazco et al [54] have shown that the $K_R - \Delta a$ curve for steel fibre concretes is independent of notch length and sensitive to fibre additions. This is encouraging but further work needs to be done to establish if specimen size and geometry would have any effects on the R-curve.

2.3 Fracture Mechanisms in Cementitious Composites

To properly model the fracture behaviour of composites it is
important to understand the various failure mechanisms which contribute to the total fracture toughness. Harris [58] has recently
given an excellent review on resin-based fibre composites. Failure
mechanisms and their toughnesses such as fibre-matrix debonding,
stress redistribution owing to fibre failure, fibre pull-out,
fractures of matrix and fibre are discussed. For cementitious fibre
composites the matrix always cracks first and the major contribution
to fracture toughness is that of the work used in pulling the fibres
out of the matrix. When the length of the fibre (ℓ) is less than
the critical fibre length (ℓ_c) the fibres are simply pulled out but
not broken. This occurs in many high strength steel fibre reinforced
concretes. The total fracture work is, therefore, given by the sum
of the matrix fracture energy and the fibre pull-out work i.e.

$$R = \frac{v_f \ell^2 \tau}{6d} + \left[v_m + \frac{v_f \ell}{d} \right] R_m \qquad (9)$$

where v_f is the fibre volume fraction, v_m the matrix volume fraction,
d the fibre diameter, τ the fibre-matrix interfacial frictional
strength, and R_m the matrix fracture toughness. The second term in
equation (9) is usually negligible. For many glass and polymeric
fibre cements ℓ is usually longer than ℓ_c so that those fibres which
are embedded to a distance more than $\ell_c/2$ will be broken first before
they are pulled-out and those fibres embedded to a distance less than
$\ell_c/2$ are simply pulled out. The total fracture work is now given by
[42]:

$$R = \frac{v_f \ell_c^2 \tau}{6d} + \frac{4v_f \ell_c^3 \tau^{*2} \beta}{3E_f d^2} + \left[v_m + \frac{v_f \ell_c}{d} \right] R_m + \beta v_f R_f. \qquad (10)$$

where β is the fraction of fibres that are broken, τ^* is the fibre-
matrix bond strength, and R_f is the fracture toughness of the fibre.
The first and second terms correspond to fibre pull-out and stress
redistribution mechanisms, and the third and fourth terms are for
creation of new surfaces due to fibre and matrix fractures as well
as fibre-matrix debonding. The stress redistribution term is small
and can be ignored. Also R_f is negligible for high strength fibres
such as glass, asbestos, carbon, etc. However, ductile fibres such
as Kevlar and polypropylene, both of which have been used for reinforcing cement matrices, R_f can be quite substantial.

Fibre reinforcement in cementitious matrices is always randomly
orientated. This introduces a considerable effect on the fibre pull-
out toughness contribution to equations (9) and (10). The pull-out
term has to be multiplied by an orientation efficienty factor (η) of

either 0.41 or $2/\pi$ depending on the degree of randomness. While this has the adverse effect of reducing the fibre pull-out toughness the fact that the fibres are randomly orientated across a fracture plane may introduce some toughness enhancement in certain types of fibre cements such as steel fibre concretes $(\ell < \ell_c)$. This comes about because the steel fibres lying at an angle (θ) to the fracture plane are plastically sheared and the work dissipated is [51]:

$$R_s = v_f \tau_y \ell \theta \tag{11}$$

which should be added to equation (9). Here τ_y is the shear yield strength of the steel fibre. Helfet and Harris [59], Naaman and Shah [60] as well as Morton [61] are aware of this beneficial toughness effect owing to fibre orientation in steel fibre concretes. Apparently, there is no such toughness enhancement in other more flexible fibres such as glass, asbestos, cellulose and polypropylene.

Equation (9) has been used successfully by Mai and his co-workers to predict the total fracture work for both asbestos and wood fibre cement mortars [42,43,45]. They identify that fibre pull-out is the major failure mechanism in these cement composites although there is a small contribution due to fibre fracture. Harris et al [51] have also applied equations (9) and (11) to predict the fracture work for steel fibre reinforced concretes. It is noted that for mild steel fibres the fibre pull-out toughness is much larger than the plastic shearing work; but for high carbon steel fibres R_s becomes the major toughness contributor.

Although the study of failure mechanisms yields very useful and detailed information about fibre behaviour and fibre-matrix inter-actions, which in the end will be useful in improving the strength and toughness of fibre cementitious structures, there does not seem to be enough research efforts in this direction.

Finally, it should be noted that equations (9) and (10) are derived on the assumption that all the fracture energies due to fibre pull-out, stress redistribution, fibre and matrix failure are confined to a small volume of the material on either side of the matrix crack. This assumption seems justified since R measurements in laboratories always use pre-cracked specimens and the process zone is a line zone ahead of the crack tip. Hibbert and Hannant [62] have, however, drawn attention to the tensile fracture of cementi-tious composites in which multiple matrix cracking predominates over a substantial volume of the material. Final failure is not by fibre pull-out but by fibre fracture. This failure mode usually occurs in cement composites with long or continuous fibres and v_f is such that the fibres take up all the composite load when the matrix fails. Hibbert and Hannant also suggest that for these cementitious compo-sites it is appropriate to discuss about energy absorption per unit volume of the material under stress rather than R which is a measure of the fracture energy per unit area of crack extension.

3   MODELLING OF SLOW CRACK GROWTH IN CEMENTITIOUS COMPOSITES

It is suggested in Section 2.2 that the fracture behaviour of fibre cementitious composites is described adequately by the R-curve. The increasing resistance to crack growth is primarily due to the progressive failure processes taking place in the fibre bridging zone. These include fibre debond and pull-out or fracture as well as any other inelastic deformation of the fibres. For many practical fibre cements, mortars and concretes the fibre lengths are usually less than the critical fibre length so that fibre pull-out becomes the major failure mechanism. Consequently, all previous models are based on the fibre bridging concept [39,44,63-65]. When the plateau value of the R-curve is reached the fibres at the original crack tip are just completely pulled-out from the matrix and the fibre bridging zone has reached its maximum length. Further cracking causes the whole fibre bridging zone to translate forward. The plateau R value will be maintained if geometrical similarity exists.

Whether LEFM can be applied to calculate the R-curve depends on the process zone size of the reinforced matrix (ℓ_p) and does not depend on the length of the fibre bridging zone (ℓ_f). If ℓ_f is considered as an integral part of the fracture process zone then obviously LEFM is invalid unless extremely large specimens are available. However, if the effective crack length (a) is taken to be ($a_0 + \ell_f$) then the fibre bridging zone can be modelled by closure stresses acting across the crack faces, Figure 1(c). Provided ℓ_p is small relative to the size of the specimen [19,20] application of LEFM to calculate K_R should be valid. When ℓ_f/a is small the stress intensity factor due to the closure stresses (K_r) is independent of specimen geometry and the K_R - Δa curve determined should be a useful material property.

Lenain and Bunsell [39] as well as Mai and co-workers [43,44] independently modelled slow crack growth in fibre cement pastes and mortars using LEFM. These authors proposed that cracking occurred when the effective stress intensity factor at the tip of the cracked matrix reached a critical value K_i corresponding to the fracture initiation of the reinforced matrix. The stress intensity factor at the tip of the cracked matrix can be separated into two components:

$$K = K_i = K_R + K_r \tag{12}$$

where K_R is the stress intensity factor at the tip of the cracked matrix if there were no fibres bridging the crack and K_r is the stress intensity factor due to the bridging fibres closing the crack faces and is negative.

Lenain and Bunsell [39] used acoustic emission to measure the length of the matrix process zone in their compact tension specimens and obtained $\ell_p \approx 28$ mm. To account for this rather large process zone size they used an idea similar to the Irwin plastic zone correction factor and assumed that a fraction of this length $\xi \ell_p$, where $\xi \approx 1/3$, should be added to the fibre bridging zone ℓ_f. A uniform post-cracking stress was then applied over the length $(\ell_f + \xi \ell_p)$ and K_r was evaluated from Paris and Sih's analytical expression [66] for an internal crack in an infinite sheet under a region of uniform closing pressure on one side of the crack. K_R was then calculated from equation (12). The approach is reasonable and uses well-developed LEFM plasticity correction concepts in metals. Unfortunately, they used the wrong K-expression for K_r which is inappropriate for their compact tension test geometry. In addition, the fibre pull-out stresses cannot be constant and must decrease with crack face opening so that their predicted K_R values are only upper-bound solutions.

Foote et al [44] realised the deficiencies of Lenain and Bunsell's model and suggested that a size and geometry independent $K_R - \Delta a$ curve could be theoretically developed to compare with experimental data which might be affected by these variables. Such a model is based on a semi-infinite crack with closure stresses p(t) in the fibre bridging zone but not in the matrix process zone as they have assumed this to be small. The displacements of the crack faces due to p(t) is negative and can be calculated from:

$$\delta_r = \frac{4}{E} \text{ Im} \int_0^z \Phi(z)\,dz \tag{13}$$

and $\Phi(z)$ is the Muskhelishivilis̀ potential function given by [67]:

$$\Phi(z) = \frac{1}{2\pi i z^{\frac{1}{2}}} \int_0^\infty \frac{p(t)\,t^{\frac{1}{2}}dt}{(t-z)} \,. \tag{14}$$

The stress intensity function due to the bridging fibres is

$$K_r = -\left(\frac{2}{\pi}\right)^{\frac{1}{2}} \int_0^\infty \frac{p(t)\,dt}{t^{\frac{1}{2}}} \tag{15}$$

but this cannot be calculated since p(t) is dependent on the total displacement of the crack faces (δ_t) i.e.

$$\delta_t = \delta_R + \delta_r$$

where $\delta_R = \dfrac{2K_R}{E}\sqrt{\dfrac{2t}{\pi}}$ (16)

is the displacement due to the external load. A fifth order poly-nomial is assumed for p(t) and an iterative method is adopted to determine the K_R curve using the initiation criterion given by equation (12) for an asbestos cement mortar. Due to the lack of experimental data for the p(t) - δ_t relationship agreement with experimental results is poor. However, the data do suggest that for practical purposes a unique K_R - Δa curve may be considered to exist provided the specimen size is large enough for the complete fibre bridging zone to be developed and ℓ_f/a is small. For small samples and when $\ell_f/a > 0.60$ there are definite size and geometric effects on the K_R - Δa curve as discussed in Section 4.2.1.

Improvements on the semi-infinite crack model have been attempted. Finite element solutions based on a negative stiffness model and an influence coefficient model [65] were developed to account for the finite size of the specimen but these models also suffered from a lack of accurate data on the p(t) - δ_t relationship. The semi-infinite crack model solution [44] is bounded by the two solutions corresponding to a constant p(t) and a linear varying p(t) over the fibre bridging zone. These solutions do not require iterations and can be easily obtained for any finite size specimen geometries. A discussion on this bounding method for R-curves for metals and fibre cementitious composites is given elsewhere [68].

When inelastic deformations occur in the fibre bridging and matrix process zones K_R is not a suitable parameter. It is appro-priate then to consider the work done in the fibre bridging and matrix process zones as the crack extends. The fracture process may be modelled by removing the stress singularity at the crack tip and replacing a closure stress p(t) distribution over the length $(\ell_f + \ell_p)$. If inelastic work is confined to the fibre bridging zone such as in asbestos and cellulose cement mortars only p(t) over the region ℓ_f needs to be known. The fictitious crack model [63,64] has been developed in conjunction with finite element stress analysis to calculate the fracture toughness. Perhaps the most promising solu-tion is that given by Wecharatana and Shah [69]. They considered the incremental work done as the crack advanced an extension Δa by the modified strain energy release rate (G_R^*). This involved calcu-lating the compliance (C^*) and residual displacement (u_r). The length of the matrix process zone (ℓ_p) was obtained by an iterative procedure as outlined in [69] using a critical crack opening dis-placement (δ_c) at the end of the matrix process zone as a fracture criterion. For steel fibre concretes both u and u_r could be obtained theoretically thus allowing C^* to be determined from equation (7). The model also assumed a linear crack profile and p(t) over the region ℓ_p was neglected in the iterations. These assumptions are somewhat questionable particularly when ℓ_p is large and the closure stresses are close to the maximum fibre pull-out stress value. Furthermore, u_r is difficult to be theoretically calculated for other fibre cements.

Whatever models are to be used the major problem is to obtain an accurate estimate of the $p(t) - \delta_t$ relationship. Owing to the many factors, such as fibre orientation, fibre pull-out and debond, fibre inelastic deformation and fracture, etc., which may occur in one type of fibre cements but not in another, it is not possible to derive a theoretical relationship. Both Petersson and Hillerborg [63,64], Wecharatana and Shah [69] suggested that the $p(t) - \delta_t$ relationship should be obtained experimentally from uniaxial tensile tests conducted on fibre cement samples because this could be identified as the tensile stress-displacement relationship in the post cracking region. Provided experimental results are reproducible the $p(t) - \delta_t$ relationship determined in this way is a genuine representation of the failure processes that take place in the fibre bridging zone. In connection with Wecharatana and Shah's model [69] there is also a need to experimentally determine the matrix process zone size and compare this with the predicted value. Methods must also be developed to measure accurately the matrix crack tip opening displacement (δ_c).

Finally, in the construction of the R-curves, whether $K_R - \Delta a$ or $G_R^* - \Delta a$, the major difficulty is determining the crack length (a) and hence crack growth (Δa). Many methods have been suggested but the most common one is by the compliance technique. Figure 3 shows the experimental data obtained from cellulose fibre cement mortars using DCB specimens. The "natural crack" compliances are always smaller than the saw-cut notch compliances due to the closure stresses of the fibres acting across the crack faces. This means that crack length measurements based on saw-cut notch compliances are unreliable. Direct measurement techniques have to be developed. These include video camera, crack foil gauges and electrical conductive surface grids. These experimental techniques are presently being developed in the author's laboratory.

4   SOME EXPERIMENTAL RESULTS ON ASBESTOS AND WOOD FIBRE CEMENT
    MORTARS

The experimental results given in this section are concerned with asbestos and cellulose fibre cement mortars. All materials were prepared by James Hardie Research and Engineering Laboratory using either the commerical Hatschek process or castings made in a vacuum filter box. The castings did not have a laminated structure as the commercial sheets. The test samples were cured at 100% r.h. for 24 hours and subsequently autoclaved at 140°C and 0.4 MPa for

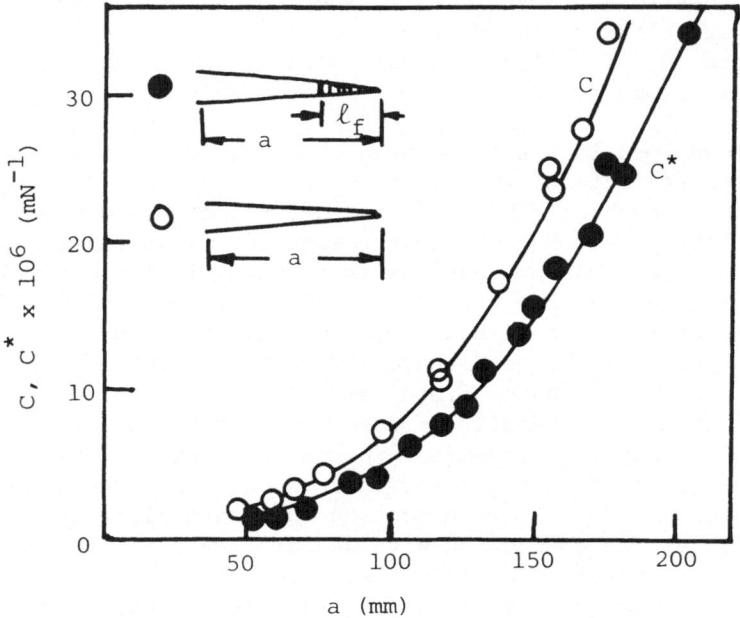

Fig. 3 Compliances curves for DCB cellulose cements
with saw-cut notches (C) and propagated cracks
(C*)

another 24 hours. Fracture energies and slow crack growth resistance curves were investigated for these fibre cement mortars.

4.1 Work of Fracture Measurements

Specific fracture energies (R) were measured using three-point
NB samples in an Instron testing machine. Crack propagation was
stable in all these experiments. Figure 4 shows the experimental
data for the cast cellulose and asbestos cement mortars as a function
of fibre mass fraction (m_f). The effect of increasing fibre additions is obvious from these results. Equation (9) can be used to
predict R for these fibre cement mortars using for cellulose/mortar
composites: $\ell/d = 135$, $d = 27.5$ µm, $\tau = 0.35$ MPa; and for asbestos/
mortar composites: $\ell/d = 160$, $d = 25$ µm, $\tau = 0.83$ MPa [42,45]. The
fibre orientation efficiency factor (η) is assumed equal to 0.41 and
R_m is 20 J/m². Clearly, the predicted results are in good agreement
with the experimental measurements. Although the fracture surfaces
do contain some broken fibres the major fracture mechanism in these
two types of fibre cementitious composites is by fibre pull-out.
Since these fracture measurements were made on beams with only 20–25
mm depth it became necessary to see if R was independent of beam
depth (W) and notch length (a). Figure 5 shows such results for

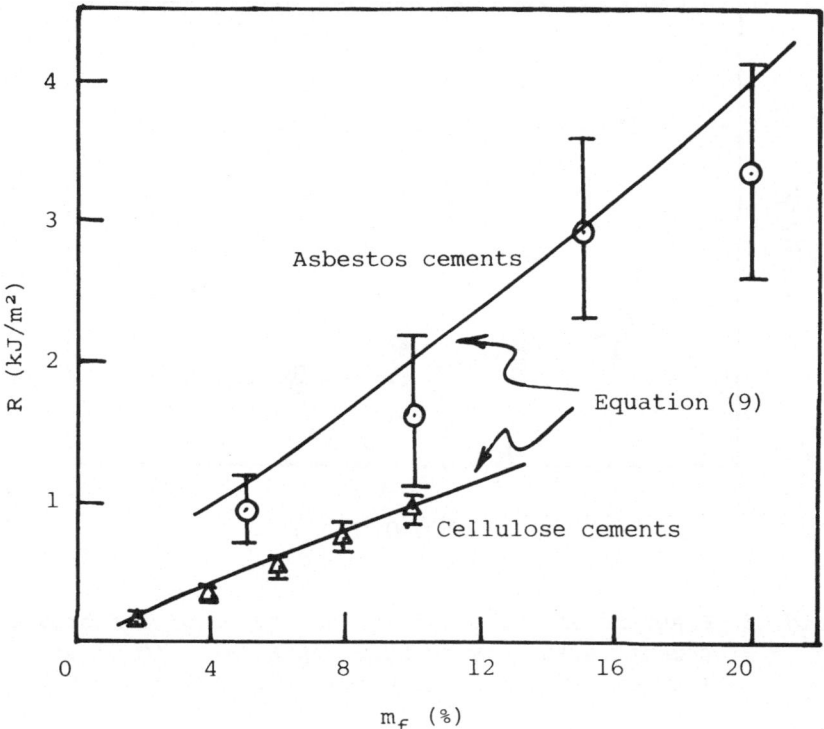

Fig. 4 Variations of fracture energies (R) with mass
fraction (m_f) of fibres for an asbestos and a
cellulose cement

two different commercial Hardiflex sheets which contain a mixture of
8% asbestos and 7% cellulose fibres. The crack was orientated in
the weak direction. These data suggest that R measured with this
technique is a material property. For the weaker asbestos cement
sheets, R decreases with a/W but tends to remain constant when a/W
$\geq$ 0.50. This effect can be explained in terms of [39], in which R
is found to increase with crack speed and small a/W favours rapid
uncontrolled crack propagation. There is no such dependence of R
on a/W for the stronger asbestos cements presumably due to the more
pronounced stabilizing effect of the fibre pull-out mechanism [43].
Once again, equation (9) gives a good estimate of the toughness of
the Hardiflex sheets and also confirms that fibre pull-out is the
main contributor to R.

This simple method of measuring fracture resistance in cementitious composites may be recommended as a standard procedure for
assessing candidate materials for structural design. For example,
it has been shown possible to use R to distinguish the effects of

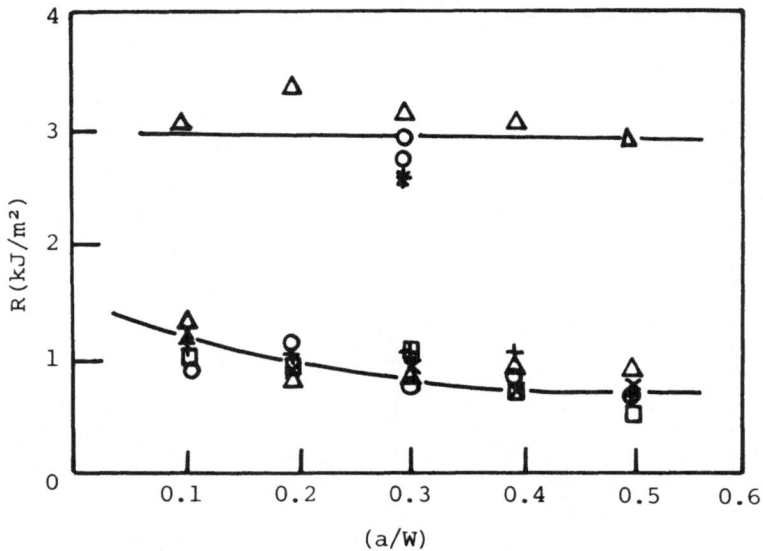

Fig. 5 Fracture energies (R) for two types of asbestos
cements with varying beam depth (W). (□ 25 mm,
+ 50 mm, ● 75 mm, x 100 mm, O 150 mm, △ 200 mm)

bleaching, water or moisture absorption and fibre orientation for
cellulose fibre cement sheets which have now completely replaced
asbestos building sheets [47]. In addition, as given in [9] this
fracture measurement also provides a simple indirect means of calculating the true K_C (=$\sqrt{ER}$) value had the specimen size been big
enough.

4.2 Slow Crack Growth Resistance Curves

Slow crack growth resistance curves were obtained for a
commercial Hardiflex asbestos cement mortar and an experimental
cellulose (West Hemlock pulp) cement mortar. Effects of specimen
size, fibre orientation and specimen geometry were studied for the
asbestos cement mortar composites. As for the cellulose cement
mortars the effect of water absorption was investigated. K_R was
used for characterising crack growth in the asbestos and dry cellulose cement mortar but G_R^* was required for the wet cellulose cement
mortar. Visalvanich and Naaman [50] also found that for asbestos
cement mortars LEFM was a valid approach. Accurate crack growth
(Δa) measurements are difficult to obtain and crack lengths evaluated
from calibrated compliances using saw-cut notches are always under-
estimated (see Fig. 3). In the present work the crack tip was
photographed at regular intervals until complete failure occurred.
The camera was equipped with a telephoto lens and close-up bellows.

A flash was used and synchronised with the Instron machine so that
each exposure was automatically recorded on the load-displacement
curve. The crack lengths were measured from negatives using enlarg-
ing projector. The accuracy of measurement was within ± 2 mm. It
was also found from these enlarged photographs that the failure mode
was a single crack through the specimen with very little multiple
cracking on either side of the crack. There was no evidence of a
large matrix process zone size ahead of the crack tip. For the
cellulose cement mortars residual displacements were also obtained
by unloading the sample after each crack extension.

4.2.1 Asbestos cement mortars. Figure 6 shows the experimental
K_R - Δa curves with crack propagation in the strong direction. The
results were obtained from NB specimens with beam depths varying
from 25 to 200 mm. A plateau K_R is reached when $\Delta a \geq 70$ mm. This
means that notched beams with $W \geq 200$ mm must be used to obtain the
complete K_R - Δa curve. For notched beams with $W \leq 100$ mm, some of
the data concave upwards and do not fall within the scatter band of
the R-curve. It was suggested that this was because the ratio of
the bridging zone (ℓ_f) to the effective crack length (a) was large
so that K_r was dependent on specimen size and geometry [43]. From

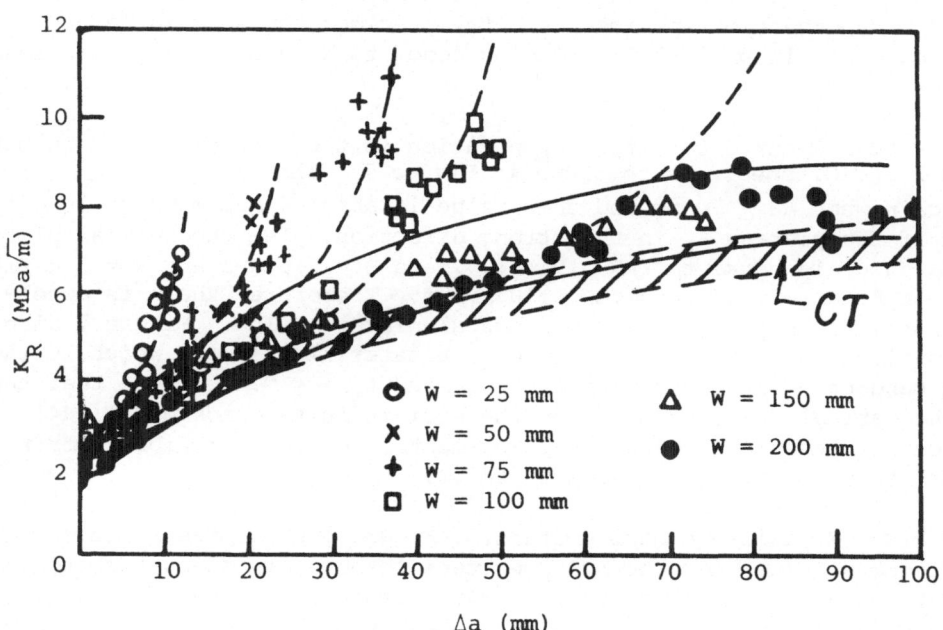

Fig. 6   R-curves for NB and CT asbestos cement speciments in
strong direction. Dashed curves for NB specimens when
fibre bridging zone is large compared to crack length

Figure 6 it can be shown that when $(\ell_f/a) \geq 0.60$ size effects exist. This size effect on K_r can be studied using the following K-solution given in [23] where

$$K = \frac{2P}{\sqrt{\pi a}} \, F\!\left(\frac{a_0}{a} \, , \, \frac{a}{W}\right) \, .$$

K_r is obtained by integrating the above equation over the fibre bridging zone of a given length with P replaced by the actual stress profile. Predictions based on a linear varying stress profile from 0 to p are superimposed in Figure 6. p is approximately 6.54 MPa and is the maximum mean fibre pull-out stress for the given fibres and orientation [44]. Although this is a lower bound solution for K_r [68] the agreement between predictions (represented by dashed curves) and experimental results is very good and the trend of the data is thus confirmed as purely size effects.

Experimental results were also obtained from CT specimens in this strong direction (H x W = 120 x 200) with $a_0 = 60$ mm and $\ell_f/a \leq 0.60$. The scatter band of these data is also plotted in Figure 6. Although it lies slightly below the scatter band of the NB results they are sufficiently close to suggest that for practical purposes a unique K_R - Δa curve exists for the asbestos cement mortar which is independent of specimen geometry provided the size is large enough to allow the full bridging zone to develop and ℓ_f/a is less than 0.60.

The R-curve for crack propagation in the weak direction in both DCB and CT specimens is shown in Figure 7. The plateau K_R is reached again when $\Delta a \geq 70$ mm and this value is about 4 MPa$\sqrt{m}$ which is only half the plateau K_R in the strong direction. The theoretical plateau K_R value is equal to $\sqrt{ER}$ [9,43], where E = 5.6 GPa and R = 3 kJ/m² from Figure 5, which gives 4.10 MPa$\sqrt{m}$. Therefore, there is excellent agreement with the experimental plateau value. Figure 7 also confirms that the K_R - Δa curve is a material property which is independent of specimen geometry provided $\ell_f/a \leq 0.60$. Note that the theoretical R-curves based on the semi-infinite crack model [44] compare favourably with the experimental results for fibre-matrix frictional strenghs between 2τ to 3τ.

4.2.2 Cellulose cement mortars. Figure 8 shows the residual displacements (u_r) for the DCB specimens in both dry and wet states. For the dry samples u_r is linearly dependent on a but its magnitude is so small that this is easily caused by broken or pull-out fibres protruding on the fracture surfaces [46]. For the wet samples u_r is non-linearly dependent on a and it is thought to be caused by inelastic work done in the fibre bridging and matrix process zones. Consequently, K_R is not valid and G_R^* is used. Figure 9 gives the G_R^* - Δa curves for both dry and wet cellulose cement mortars.

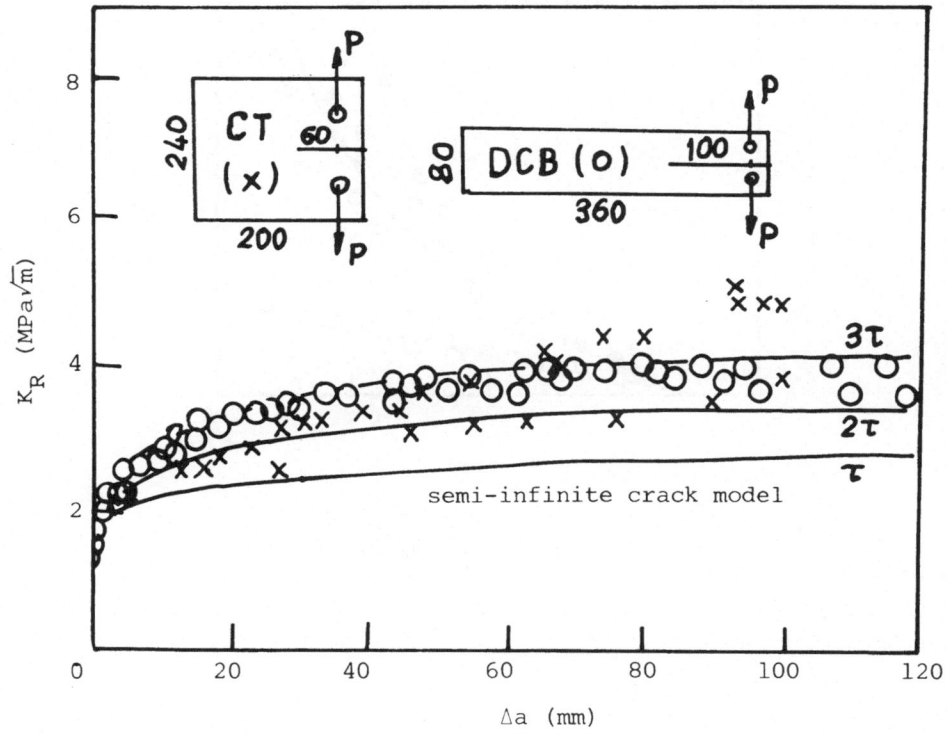

Fig. 7 R-curves for CT and DCB asbestos cement specimens
in weak direction

Clearly, the wet fibre mortar is much more resistant to fracture
than the dry fibre mortar. The toughening effect due to water
absorption comes about in that considerable inelastic work is needed
to stretch and pull-out the wet fibres. Unfortunately, this toughness enhancement is achieved at the expense of some considerable
loss in flexural strength and Young's modulus [47].

It is noted in Figure 9 that had equation (1) been used to
calculate G_R, i.e. excluding the effect of u_r, the G_R - Δa curve
would be very much under-estimated for the wet samples. There is
not much difference between the G_R^* - Δa and G_R - Δa curves for the
dry samples since u_r is not significant in this case.

4.3 Load Instability Prediction Using R-Curve

The usefulness of the R-curve is its ability to predict the
maximum load that can be withstood by a given structure. Mai and
Cotterell [70] have suggested that if the R-curve is both geometry
and size independent and is given by:

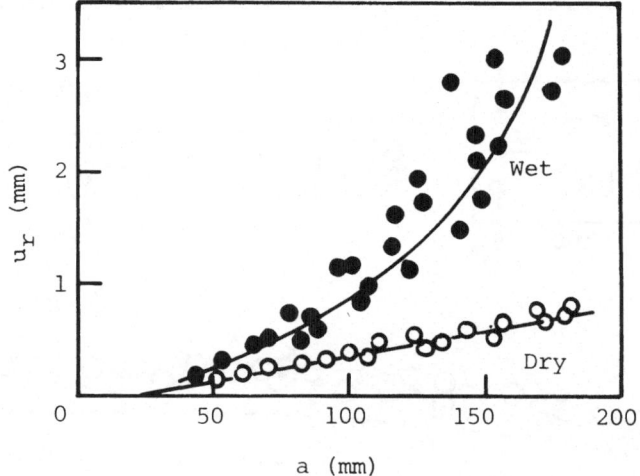

Fig. 8 Variation of residual displacement with
crack length for cellulose cements

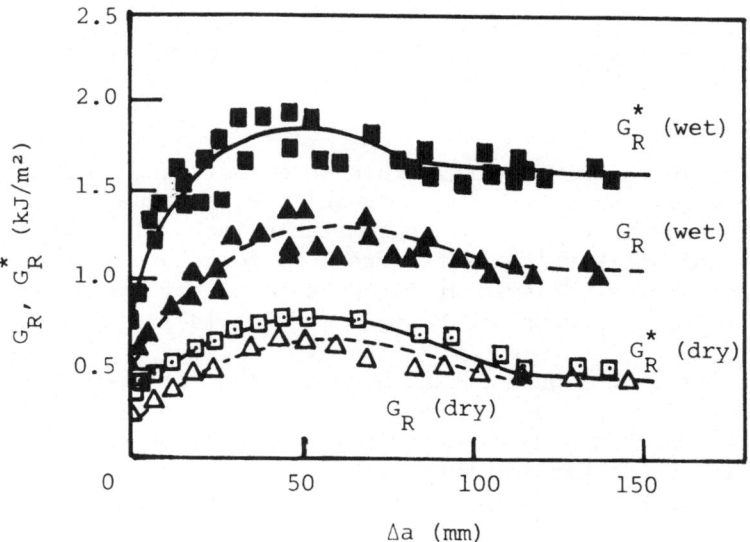

Fig. 9 R-curves for dry and wet cellulose cements

$$\Delta a = (a - a_0) = \gamma K_R^m \tag{17}$$

where γ and m are constants, then under load-controlled conditions, instability occurs when

$$K_a = K_R = K_c \quad \text{(at maximum load)}$$

and

$$\left(\frac{\partial K_a}{\partial a}\right)_P \geq \frac{dK_R}{da} . \tag{18}$$

K_a is the applied stress intensity factor and is given by $K_a = PF(a)$, where $F(a)$ is dependent on specimen geometry and crack length and is obtainable from many stress intensity factor handbooks [21-23]. The maximum load K_c value can be obtained from [70]:

$$K_c = \left(\frac{\Psi}{m\gamma}\right)^{1/m} \tag{19}$$

where $\Psi = F(a)/F'(a)$ is a geometric parameter. Consequently, the maximum load K_c increases with Ψ which is a geometric effect. The validity of equation (19) can be examined with the asbestos cement results obtained in [39,43,44] using a wide range of specimen geometries and sizes. The experimental K_R - Δa curve for this material is shown in Figure 10. Due to the slow crack growth that has occurred prior to the maximum load it is necessary to determine (P_{max}, a_{max}) in order to calculate the maximum load K_c and $\Psi(a_{max})$. Figure 11 shows a plot of $\ln K_c$ versus $\ln \Psi(a_{max})$ for all the asbestos cements [39,43,44] and there is a fairly good correlation between these two parameters. The coefficient of correlation is 0.85. γ and m can be obtained from Figure 11 and are respectively 0.26 and 3.70 so that the R-curve: $\Delta a = 0.26 K_R^{3.70}$ can be replotted in Figure 10. The agreement with experimental data is reasonably good. Much better agreement would have been obtained if it is realised that crack growth does not occur until K_R exceeds the initiation value K_i of about 1.3 MPa$\sqrt{m}$. However, the instability maximum load K_c has to be evaluated from a more complicated expression which is:

$$K_c (K_c - K_i)^{m-1} = (\Psi/m\gamma) . \tag{20}$$

The application of equation (19) is that if Ψ at instability is known K_c can be calculated which in turn allows the maximum load the structure can sustain to be obtained.

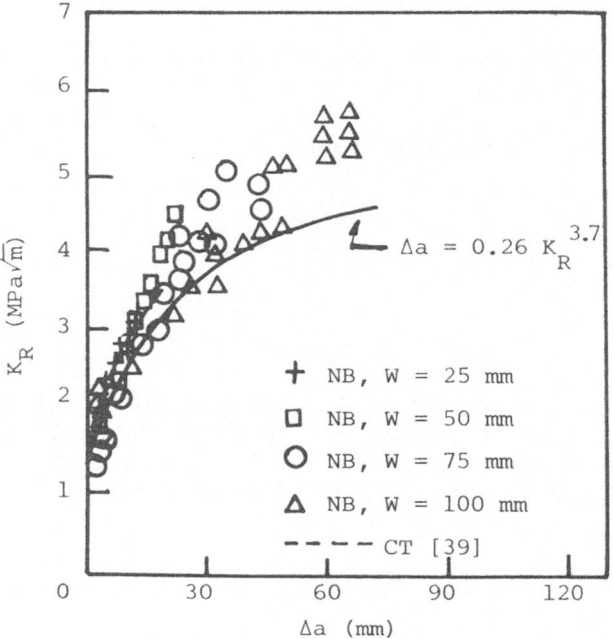

Fig. 10   R-curve for asbestos cement

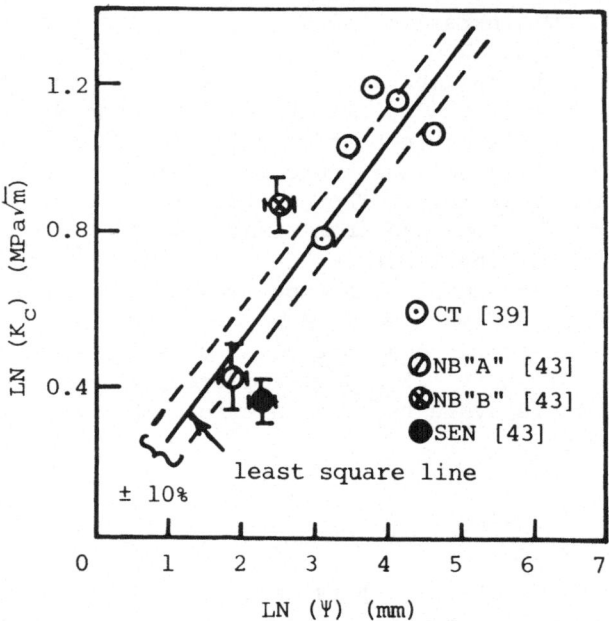

Fig. 11   A natural logarithmic plot of maximum load K_C against geometric parameter Ψ for asbestos cement

5   CONCLUDING REMARKS AND SUGGESTIONS FOR FUTURE WORK

A critical review of the application of both linear and non-linear elastic fracture mechanics to cementitious composites shows that no one-parameter criterion can be used to describe sufficiently the fracture behaviour owing to size effect and slow crack growth. The most promising approach seems to be the R-curve. For linear elastic materials such as asbestos and dry cellulose cement mortars a linear elastic fracture mechanics approach is sufficient and the fracture parameter K_R is valid to describe the fracture process. For the wet cellulose cement mortars considerable inelastic work is done in the fibre bridging zone and perhaps also in the matrix process zone the modified strain energy release rate G_R^* is used. The R-curve is useful for predicting fracture instabilities in given specimen geometries and loading configurations. It can also be used to evaluate the performance of candidate materials for structural applications although a more convenient and simple method, owing to Gurney [29] and Petersson [9], is to measure the specific work of fracture (R) in NB specimens.

Accurate modelling of the slow crack growth process is complex and much further work is required to provide detailed information for refining existing models. Future work should be carried out to: (a) obtain a better understanding of the various failure mechanisms which may vary between different cementitious composites, (b) develop better experimental techniques to measure crack length and to quantify the physical size of the matrix process zone so that direct comparisons can be made with theoretical predictions, and (c) investigate the fibre stress-crack face opening displacement relationships over the fibre bridging zone for various types of fibre cements. Further work should also be conducted to establish the conditions under which the R-curve is independent of specimen size and geometry. It is found that for asbestos cements K_r is independent of size and geometry if $\ell_f/a \leq 0.60$. This ratio seems too large. Also, the fully developed fibre bridging zone length must depend on specimen size, geometry and loading configuration. This may not affect the plateau R-curve value but may influence the shape of the rising R-curve. Another relatively unexplored area of fracture mechanics application to cementitious composites is fatigue crack propagation although there is already some work done on concretes [71].

ACKNOWLEDGEMENTS

The author wishes to thank James Hardie & Coy. Pty. Ltd. for a research grant and Dr. B. Cotterell for useful comments and discussions.

424

REFERENCES

1. Knott, J.F. Fundamentals of Fracture Mechanics. (London: Butterworths, 1973).

2. Kinloch A.J. and R.J. Young. Fracture Behaviour of Polymers. (London: Applied Science Publishers, 1983).

3. Lawn, B.R. and T.R. Wilshaw. Fracture of Brittle Solids. (London: Cambridge University Press, 1975).

4. Begley, J.A. and J.D. Landes. The J-integral as a Fracture Criterion. ASTM STP 514 (1972) 1-20.

5. Wells, A.A. Application of Fracture Mechanics at and Beyond General Yielding. British Welding Journal 10(1963) 563-570.

6. Cotterell, B. and Y.W. Mai. Plane Stress Ductile Fracture, in D. Francois, ed., Advances in Fracture Research (London: Pergamon Press, 1981) Vol. 4, pp. 1683-1694.

7. Pascoe, K.J. Private communication, November, 1979.

8. Higgins, D.D. and J.E. Bailey. A Microstructural Investigation of the Failure Behaviour of Cement Paste, in Proc. Conf. on Hydraulic Cement Pastes, Sheffield University (Cement and Concrete Association, U.K., 1976) pp. 283-296.

9. Petersson, P.E. Crack Growth and Development of Fracture Zones in Plain Concrete and Similar Materials. Doctoral Dissertation, December, 1981. Lund Institute of Technology, Sweden.

10. Bazant, Z.P. and B.H. Oh. Crack Band Theory for Fracture of Concrete. Materials and Structures 16(1983) 155-177.

11. Swamy, R.N. Fracture Mechanics Applied to Concrete, in F.D. Lydon, ed., Developments in Concrete Technology -1 (London: Applied Science Publishers, 1979), pp. 221-281.

12. Carpinteri, A. Application of Fracture Mechanics to Concrete Structures. Journal of the Structures Division, Proc. ASCE, 108(1982) 833-848.

13. Saouma, V.E. and A.R. Ingraffea. Fracture Toughness of Concrete: K_{IC} Revisited. Journal of the Engineering Mechanics Division, Proc. ASCE, 108(1982) 1152-1166.

14. Kesler, C.E., D.J. Naus and J.L. Lott. Fracture Mechanics - Its Applicability to Concrete. in Proc. Int. Conf. on Mechanical Behaviour of Materials, Japan, Vol. 4, 1972, pp. 113-124.

15. Swartz, S.E., K.K. Hu, M. Fartash and C.M.J. Huang. Stress Intensity Factor for Plain Concrete in Bending - Prenotched Versus Precracked Beams. Experimental Mechanics 22(1982) 412-417.

16. Naaman, A.E., A.S. Argon and F. Moavenzadeh. A Fracture Model of Fibre Reinforced Cementitious Materials. Cement and Concrete Research 3(1973) 397-411.

17. Argon, A.S. and W.J. Shack. Theories of Fibre Cement and Fibre Concrete, in RILEM Symp. Fibre Reinforced Cement and Concrete (Lancaster: The Construction Press Ltd., 1975) pp. 39-53.

18. Mindess, S. The Fracture of Fibre-Reinforced and Polymer Impregnated Concretes. International Journal of Cement Composites 2(1980) 3-11.

19. American Society for Testing and Materials. Standard Method of Test for Plane-Strain Fracture Toughness of Metallic Materials. E 399-74, Annual Book, Part 10, 1978.

20. British Standards Institution. Method for Plane Strain Fracture Toughness (K_{IC}) Testing. BS 5447, 1977.

21. Rooke, P.P. and D.V. Cartwright. Compendium of Stress Intensity Factors. (London: H.M. Stationery Office, 1976).

22. Sih, G.C. Handbook of Stress Intensity Factors (Bethlehem: Institute of Fracture and Solid Mechanics, Lehigh University, 1973).

23. Tada, H., P.C. Paris and G.R. Irwin. Stress Analysis of Cracks Handbook. (Hellertown, Pennsylvania: Del Research Corporation, 1973).

24. Irwin, G.R. Fracture, in S. Flügge, ed., Handbook of Physics (Berlin: Springer, 1958) Vol. VI, pp. 551-590.

25. Dugdale, D.S. Yielding of Steel Sheets Containing Slits. Journal Mechanics and Physics of Solids. 8(1960) 100-104.

26. American Society for Testing and Materials. Standard Method - The Determination of J_{IC}, a Measure of Fracture Toughness. E813-81, Annual Book, Part 10, 1983.

27. British Standard Institution. Methods for Crack Opening Displacement Testing. BS 5762, 1979.

28. Matsoukas, G., B. Cotterell and Y.W. Mai. A Note on the Plastic Rotation Constant Used in Standard COD Tests. Progress Report to the Australian Welding Research Association, Feb. 1984.

29. Gurney, C. and J. Hunt. Quasistatic Crack Propagation. Proc. Royal Society, London, A299 (1967) 508-524.

30. Gurney, C. and K.M. Ngan. Quasistatic Crack Propagation in Non-linear Structures. Proc. Royal Society, London, A325(1971) 207-222.

31. Gurney, C. and Y.W. Mai. Stability of Cracking. Engineering Fracture Mechanics 4(1972) 853-863.

32. Gurney, C., Y.W. Mai and R.C. Owen. Quasistatic Crack Propagation in Materials with High Toughness and Low Yield Stress. Proc. Royal Society, London, A340 (1974) 213-231.

33. Atkins, A.G. and Y.W. Mai. Elastic and Plastic Fracture. (Chichester: Ellis Horwood Publishers, in course of preparation).

34. Mai, Y.W. and A.G. Atkins. Crack Stability in Fracture Toughness Testing. Journal of Strain Analysis 15(1980) 63-74.

35. Tattersall, H.G. and G. Tappin. The Work of Fracture and Its Measurement in Metals, Ceramics and Other Materials. Journal of Materials Science 1(1966) 296-301.

36. Hodgkinson, J.M. and J.G. Williams. J and G_c Analysis of Tearing of a Highly Ductile Polymer. Journal of Materials Science 16(1981)50-56.

37. Wecharatana, M. and S.P. Shah. Double Torsion Tests for Studying Slow Crack Growth of Portland Cement Mortar. Cement and Concrete Research 10(1980) 833-844.

38. Brown, J.H.. The Failure of Glass Fibre Reinforced Notched Beams in Flexure. Magazine of Concrete Research 25(1973) 31-38.

39. Lenain, J.C. and A.R. Bunsell. The Resistance to Crack Growth of Asbestos Cement. Journal of Materials Science, 14(1979) 321-332.

40. Patterson, W.A. and H.C. Chan. Fracture Toughness of Glass Fibre-Reinforced Cement. Composites 6(1975) 102-104.

41. Mindess, S. and A. Bentur. The Fracture of Wood Fibre Reinforced Cement. International Journal of Cement Composites and Light Weight Concrete 4(1982) 245-249.

42. Mai, Y.W. Strength and Fracture Properties of Asbestos Cement Mortar Composites. Journal of Materials Science 14(1979) 2091-2102.

43. Mai, Y.W., R.M.L. Foote and B. Cotterell. Size Effects and Scaling Laws of Fracture in Asbestos Cement. International Journal of Cement Composites 2(1980) 23-34.

44. Foote, R.M.L., B. Cotterell and Y.W. Mai. Crack Growth Resistance Curve for a Cement Composite, in D.M. Roy, ed., Advances in Cement-Matrix Composites (Pennsylvania: Materials Research Society, 1980) pp. 135-144.

45. Andonian, R., Y.W. Mai and B. Cotterell. Strength and Fracture Properties of Cellulose Fibre Reinforced Cement Composites. International Journal of Cement Composites 1(1979) 151-158.

46. Mai, Y.W. and M.I. Hakeem. Slow Crack Growth in Cellulose Fibre Cements. Journal of Materials Science 19(1984) 501-508.

47. Mai, Y.W., M.I. Hakeem and B. Cotterell. Effects of Water and Bleaching on the Mechanical Properties of Cellulose Fibre Cements. Journal of Materials Science 18(1983) 2156-2162.

48. Mai, Y.W., R. Andonian and B. Cotterell. On Polypropylene-Cellulose Fibre-Cement Hybrid Composites, in A.R. Bunsell et al, eds., Advances in Composite Materials (Oxford: Pergamon Press, 1980) Vol. 2, pp. 1687-1699.

49. Rice, J.R., P.C. Paris and J.G. Merkle. Some Further Aspects of J-Integral Analysis and Estimates, in ASTM STP No. 536, 1973, pp. 231-245.

50. Visalvanich, K. and A.E. Naaman. Fracture Methods in Cement Composites. Journal of the Engineering Mechanics Division, Proc. ASCE, 107(1981) 1155-1171.

51. Harris, B., J. Varlow and C.D. Ellis. The Fracture Behaviour of Fibre Reinforced Concrete. Cement and Concrete Research 2(1972) 447-461.

52. Mindess, S., F.V. Lawrence and C.E. Kesler. The J-Integral as a Fracture Criterion for Fibre Reinforced Concrete. Cement and Concrete Research 7(1977) 731-742.

53. Nishioka, K., S. Yamakawa, S. Hirakawa and S. Akihama. Test Method for the Evaluation of the Fracture Toughness of Steel Fibre Reinforced Concrete, in RILEM Symp. Testing and Test Methods of Fibre Cement Composites (Lancaster: The Construction Press, Ltd., 1978) pp. 87-98.

54. Velazco, G., K. Visalvanich and S.P. Shah. Fracture Behaviour and Analysis of Fibre Reinforced Concrete Beams. Cement and Concrete Research 10(1980) 41-51.

55. Swamy, R.N. Influence of Slow Crack Growth on the Fracture Resistance of Fibre Cement Composites. International Journal of Cement Composites 2(1980) 43-53.

56. Halvorsen, G.T. J-Integral Study of Steel Fibre Reinforced Concrete. International Journal of Cement Composites 2(1980) 13-22.

57. Brandt, A.M. Crack Propagation Energy in Steel Fibre Reinforced Concrete. International Journal of Cement Composites 2(1980) 35-42.

58. Harris, B. Micromechanisms of Crack Extension in Composites. Metal Science 14(1980) 351-362.

59. Helfet, T.L. and B. Harris. Fracture Toughness of Composites Reinforced with Discontinuous Fibres. Journal of Materials Science 7(1972) 494-498.

60. Naaman, A.E. and S.P. Shah. Bond Studies on Oriented and Aligned Steel Fibres, in RILEM Symp. Fibre Reinforced Cement and Concrete (Lancaster: The Construction Press Ltd., 1975) pp. 171-178.

61. Morton, J. The Work of Fracture of Random Fibre Reinforced Cement. Materials and Structures 12(1979) 393-396.

62. Hibbert, A.P. and D.J. Hannant. Toughness of Fibre Cement Composites. Composites 13(1982) 105-111.

63. Petersson, P.E. Fracture Mechanical Calculations and Tests for Fibre-Reinforced Cementitious Materials, in D.M. Roy, ed., Advances in Cement-Matrix Composites (Pennsylvania: Materials Research Society, 1980) pp. 95-106.

64. Hillerborg, A. Analysis of Fracture by Means of the Fictitious Crack Model, Particularly for Fibre Reinforced Concrete. International Journal of Cement Composites 2(1980) 177-184.

65. Foote, R.M.L. and G.P. Steven. Modelling of Crack Growth Resistance Curves, in G.C. Sih, ed., Fracture Mechanics Technology Applied to Material Evaluation and Structure Design (The Hague: Martinus Nijhoff Publishers, 1983) pp. 295-304.

66. Paris, P.C. and G.C. Sih. Stress Analysis of Cracks, in Fracture Toughness Testing and its Application, ASTM STP 381, 1964, pp. 30-83.

67. Muskhelishivili, N.I. Some Basic Problems of the Mathematical Theory of Elasticity. (Leyden: Noordhoff International, 1976).

68. Mai, Y.W., B. Cotterell and R.M.L. Foote. To be published.

69. Wecharatana, M. and S.P. Shah. A Model for Predicting Fracture Resistance of Fibre Reinforced Concrete. Cement and Concrete Research 13(1983) 819-829.

70. Mai, Y.W. and B. Cotterell. Slow Crack Growth and Fracture Instability of Cement Composites. International Journal of Cement Composites and Lightweight Structures 4(1982) 33-37.

71. Gylltoft, K. Fracture Mechanics Models for Fatigue in Concrete Structures. Doctoral Thesis. Luleå University of Technology, Sweden, 1983.

**APPLICATION OF FRACTURE MECHANICS
TO CEMENTITIOUS COMPOSITES**
NATO-ARW - September 4-7, 1984
Northwestern Univeristy, U.S.A.
S. P. Shah, Editor

FRACTURE RESISTANCE PARAMETERS FOR CEMENTITIOUS MATERIALS AND
THEIR EXPERIMENTAL DETERMINATIONS

V. C. Li
Department of Civil Engineering
Massachusetts Institute of Technology
Cambridge, Massachusetts   02139

1   INTRODUCTION

The need for characterization of fracture resistance in
metals has long been recognized. Spurred on by the development
of turbine engines and the aviation industry, advances in linear
and non-linear fracture mechanics have rapidly established
standardized testing techniques to rank materials and to aid in
the design against fracture failure in many metal structures. Unfortunately, the same has not happened in the concrete industry.
The ACI code, for example, does not embody concepts from fracture
mechanics. It has been said that the code is a "low-tech" one
and at least part of the reason may be attributed to our lack of
understanding of the fracture behavior in concrete and the proper
application of fracture mechanics in concrete structure design.
Also, our inability to characterize fracture resistance in concrete almost certainly has an effect on prohibiting the rapid
development of new cementitious composites with improved strength
and ductility. The need for a rational basis of concrete structure design with regard to public safety and economy, and the
increasing demand in load carrying capability of concrete structures under severe environments are forcing us to reconsider our
past strategy. Adequate effort must be invested in the research
on the mechanical behavior of advanced cementitious composites
and their application to concrete structures. Fracture mechanics
provides a convenient tool to describe material and structural
behavior, particularly when cracking or severe localization of
deformation is involved.

A very basic need is defining what we mean by fracture resistance. This appears to be easier than it really is. If fracture

resistance is a material property, it should not depend on the details of how we perform a test, i.e. it must not depend on the size
of the specimen, the geometry of the specimen, or how the load is
applied. In addition to theoretical considerations, there are also
practical constraints. The parameter we choose to describe fracture resistance must be readily measured in the laboratory. This
implies that the testing procedure must be simple enough to be
standardized and repeated in different laboratories and the specimen must have size small enough to be easily handled and without
exceeding the loading capacity of regular testing machines. These
considerations expediate the general adoption of the testing method
which benefits the users of the material being tested. In this
report, we shall focus particularly on the theoretical aspects of
specimen size and geometry effects of some of the fracture resistance parameters discussed in this conference (Application of
Fracture Mechanics to Cementitious Composites) and particularly
in the session on Experimental Methods of Determining Fracture
Parameters.

In discussing the fracture resistance, it is useful to have
some understanding of the physical processes leading to fracture
failure of cementitious composites. In particular, concrete is a
brittle material, but is quite different from glass in its fracture
behavior because of its heterogeneity on the microstructural scale.
The brittleness in concrete makes it a much weaker material under
tension than in compression, because of the formation of microcracks and their extension and coalescence to form macrocracks.
It has been observed that for a given specimen geometry and loading
configuration, the tensile deformation localizes onto an eventual
fracture plane. The mechanics of such localization is still not
fully understood, but it is probably related to the presence of
local defects such as concentration of voids. Research in damage
mechanics should help in acquiring a handle on this aspect of concrete failure and in explaining the non-linear inelastic behavior
prior to peak load in a tensile stress-strain curve (see, e.g.
the session on "Damage and Continuum Modeling" of this volume).
Once a macroscopic crack has formed, the crack tip can experience
high tensile stress. Microcracking in the cement paste and in
the cement-aggregate interface as well as pull-out of the aggregate
from the cement matrix dissipate the energy which drives the propagation of the macrocrack. Observation under the microscope (see
e.g., the session on "Fracture Processes in Cement Composites:
Experimental Observations" of this volume) reveals interaction of
microcracks and voids in the region ahead of the macrocrack tip.
This zone of inelastic deformation, often known as the process
zone in concrete, appears to be rather planar on the macroscopic
scale. This is in contrast to the more ductile behavior exhibited
by metal, where inelastic deformation occurs by generation of
dislocations, and where the plastic zone does not in general lie
on the plane of fracture.

If the inelastic deformation at the crack tip is confined to a small region (compared to crack length, distance to boundary of structure, distance to nearest rebar etc.), then it is known as small scale yielding (ssy). This terminology is originally intended to describe the plastic yielding zone in metal, but we should have no problem in using it to describe concrete if it is understood that "yielding" here really means the inelastic processes as described in the previous paragraph. Further, if the material outside the process zone is behaving elastically, then the application of linear elastic fracture mechanics (LEFM) is suitable. In particular, Irwin's stress intensity factor K would be useful to describe the intensity of stress and the fracture toughness K_{IC} could be used to rank such material as to its resistance to tensile fracture. However, it is very doubtful whether cracking in concrete observes the assumptions used in LEFM. In particular the process zone size has been observed to be large relative to normal specimen sizes so that the assumption of ssy is not valid in most cases. For example, Hillerborg (1) shows that for 3-point bend specimens, beam depths of less than 1-2m would give invalid K_{IC} values. Ingraffea and Gerstle (2) also give a detailed discussion of this problem, with special reference to the length and development of the process zone. Hence it is not surprising that most measurements of K_{IC} for concrete and mortar reported in the literature show some kind of size dependence, with larger values of K_{IC} obtained for large size specimens. Except for K_{IC} measured with very large specimens (see e.g. Sok (3)) most reported K_{IC} values for concrete are therefore not true material property. It should be pointed out, however, that although K_{IC}-testing is probably not valid for laboratory specimens, it does not necessarily imply that LEFM cannot be used to describe fracture behavior in real structures. That is, even if the process zone is on the order of a meter, it may still be a relatively small size in a large concrete structure so that ssy is still observed. In the following paragraphs, we shall discuss a modified K_{IC} procedure as proposed by Jeng and Shah (4), a G_F-test method recommended by a RILEM Committee (5) based on work by Petersson (6), R-curves for concrete, and also briefly discuss a test technique under development at MIT, which allows one to obtain the tension softening curve and the energy release rate. In the discussion, we shall take the approach of asking two very basic questions: Is what one measures truly a material parameter that describes fracture resistance? If so, how could it be used in the analysis and design of concrete structures?

2 A MODIFIED K_{IC} PROPOSED BY JENQ AND SHAH

Motivated by the lack of constancy of K_{IC} measured values published in the literature, Jenq and Shah (4) proposed the measurement of a modified K_{IC}, which they claim to be a material constant. The modification comes about by explicitly accounting for the inelastic process zone. Thus the energy release rate is partitioned into energy absorbed into the linear elastic K-field and energy absorbed into the planar process zone:

$$G_c = K_{IC}^2/E' + \int_0^{w_c} \sigma(w)\,dw \qquad (1)$$

where σ is the tensile stress acting across the process zone and w is the material separation in the process zone. Their procedure of determining K_{IC} involves the use of a predetermined tension softening relation $\sigma(w)$, and the length of the process zone was determined by matching the numerically predicted and experimentally measured crack mouth opening displacement in a 3-point bending test. They found that the K_{IC} computed in this manner remains a constant for different specimen size and for different crack depths. The K_{IC} value they obtained for concrete is on the order of 1000 psi$\sqrt{in}$..

Is the partitioning of the energy release rate physically meaningful? Afterall, when Dugdale (7) and Barenblatt (8) introduced their strip yield model, their idea was to introduce a cohesive zone to cancel the stress singularity which would otherwise exist. So why would one still insist on the presence of the singular K-field ahead of the process zone? To answer this question, one should recall that concrete is a much more heterogeneous material than metal on the size scale of mm. The fracture process in concrete involves microcracking and aggregate pullout as explained earlier. It is possible that the aggregate pull out occurring in the process zone acts as a separate mechanism from the microcracking at the tip of the process zone. Of course in order that the microcracking maintains a K-field, it has to satisfy the ssy condition, i.e. the size of the microcracking zone must be small in relation to the crack length and the ligament size of the specimen. Furthermore, since the stress in this zone can be larger than the tensile strength, the size of this zone should be smaller than the continuum scale. These ideas are illustrated in Figure 1.

Figure 2 shows pictorially the partitioning of the energy release rate as in equation 1. In general a complete tension softening curve as illustrated in Figure 2c would adequately describe everything about the fracture resistance, but the delta-function

FRACTURE MECHANISMS IN CONCRETE

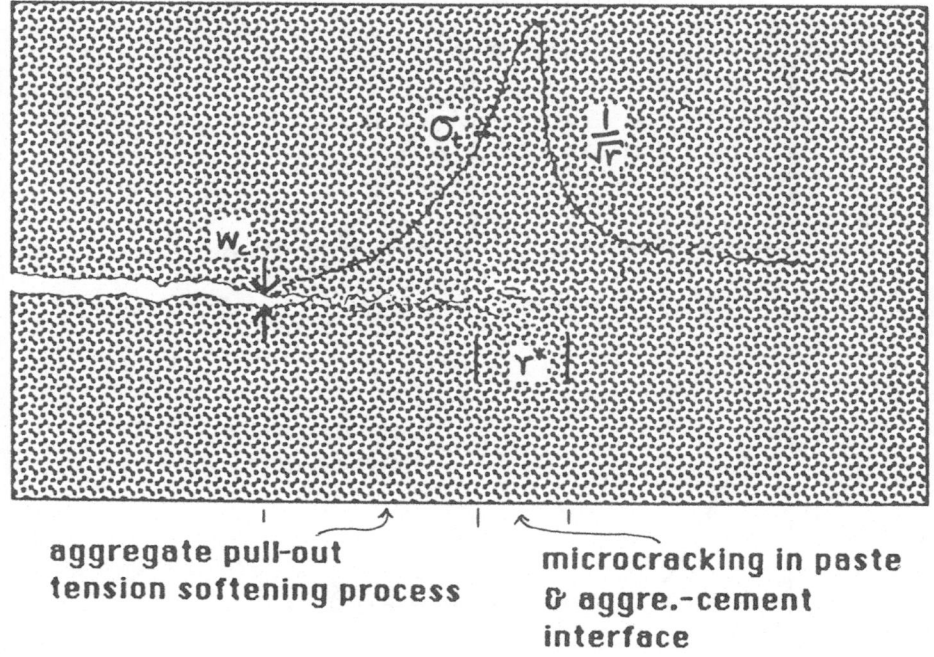

aggregate pull-out
tension softening process

microcracking in paste
& aggre.-cement
interface

Figure 1: Physical Interpretation of Jenq and Shah's Model

like stress jump may occur with very small material separation w, making it impractical to be measured within experimental accuracy.

As a rough estimate, the size of the microcracking zone may be obtained by using the asymptotic stress field and r* may be calculated as follows:

$$\sigma = \frac{K_{IC}}{\sqrt{2\pi r}} \quad \rightarrow \quad r^* = \frac{1}{2\pi} (K_{IC}/\sigma_t)^2$$

Typically, $K_{IC} \simeq 1000 \text{ psi}\sqrt{in}$

$\sigma_t \simeq 500 \text{ psi}$

$\rightarrow r^* \simeq 0.6 \text{ in}$

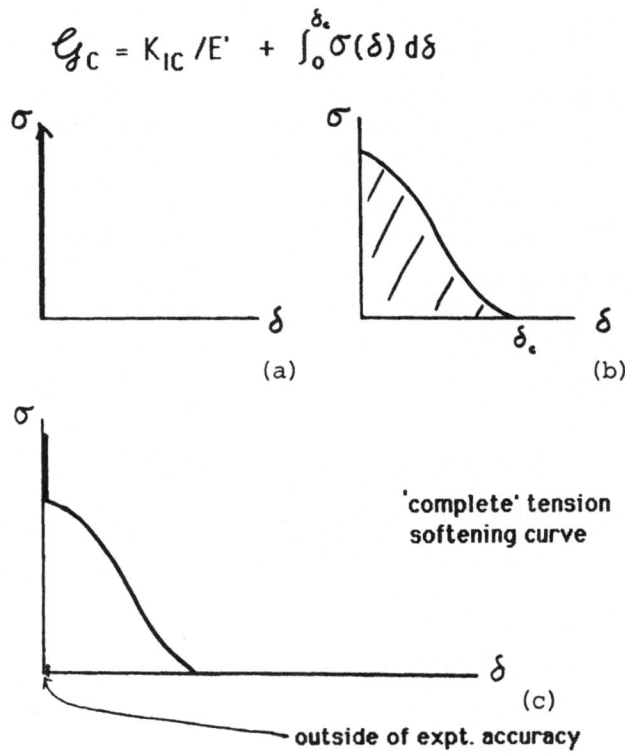

$$\mathcal{G}_C = K_{IC}/E' \, + \, \int_0^{\delta_c} \sigma(\delta)\,d\delta$$

(a)

(b)

'complete' tension
softening curve

(c)

outside of expt. accuracy

Figure 2:   Partitioning of Energy Release Rate

If we assume the continuum scale to be on the order of several times the maximum aggregate size (0.75 in. used in the concrete mixture in the specimens of Jenq and Shah), then r* as calculated is certainly less than the continuum scale. Hence the model proposed by Jenq and Shah does have a physical basis and their successful application to experimental results by several researchers makes it all the more impressive. It would be of interest to check if the same procedure would derive the same K_{IC} values for specimens of geometries other than notched beams.

It should be pointed out, however, that the K_{IC} as obtained in this manner does not carry the same meaning as Irwin's K_{IC}. First the energy absorption is not fully described by this modified K_{IC}, and hence it is not a true fracture resistance parameter. Second, when the stress at the crack tip rises in response to increase loading, and when K_I reaches K_{IC}, unstable crack propagation do not necessarily occur, as would be the case in the ideally brittle material. These differences of the modified K_{IC} from Irwin's K_{IC} are of course a result of the presence of the process zone. Further, it should be mentioned that the computation of

K_{IC} requires some knowledge of the tension softening curve. These considerations unfortunately make the modified K_{IC} less useful as a material parameter characterizing fracture resistance of concrete.

3 G_F-TEST AND SPECIMEN SIZE DEPENDENCE

The G_F-test has been under study by a RILEM-committee on Fracture of Concrete as a standard test for measuring fracture resistance of concrete (5). The test involves a 3-point bend test on a concrete beam loaded to complete failure. G_F is measured as the area under the load-deformation curve (with a correction for the weight of the beam) divided by the area of the fracture plane. The obvious advantage of this test method is the simplicity of the test procedure.

An assumption of this testing technique is that the work provided by the applied load all goes into the creation of the fracture plane. Care is used in ensuring that no energy is lost in crushing at the loading point and at the supports. However, it appears that G_F is still a specimen size-dependent parameter. This is illustrated in Figure 3 where G_F as measured by seven different laboratories are plotted as a function of ligament size. These data are based on reports (11), (6), (4) and (10). In each set of data, the w:c ratio, the maximum aggregate size, the span to depth ratio, and the concrete curing time have been kept constant. The data suggests a general increase of G_F with the ligament size. It is possible that prior to localization of deformation onto the eventual fracture plane, diffuse inelastic deformation or damage is occurring in the concrete beam. Thus part of the area under the load vs. load point displacement curve should have been attributed to this inelastic energy dissipation. Without properly accounting for this, the G_F value may be expected to be inflated and could only serve as an upperbound to the true fracture energy. The increase in G_F with beam size is probably associated with the large volume of material damaged during the loading process. This confirms earlier findings of G_F size-dependence by other authors (4), (13), (12).

The above discussion suggests that smaller beams may produce more accurate results. Indeed Figure 3 shows that the smaller beams in general seem to have less size dependence. More recently, the RILEM committee has initiated a second round-robin test involving 7 laboratories, with special attention paid to size effect (13). These tests are all performed with small size beams with ligament sizes of .0005m, .001m and .0015m. Even these small beams show a definite size effect. For example, the .001m beams show a 20% increase in G_F and the .0015m beams show a 30% increase in G_F over that for the small .0005m beams. Even

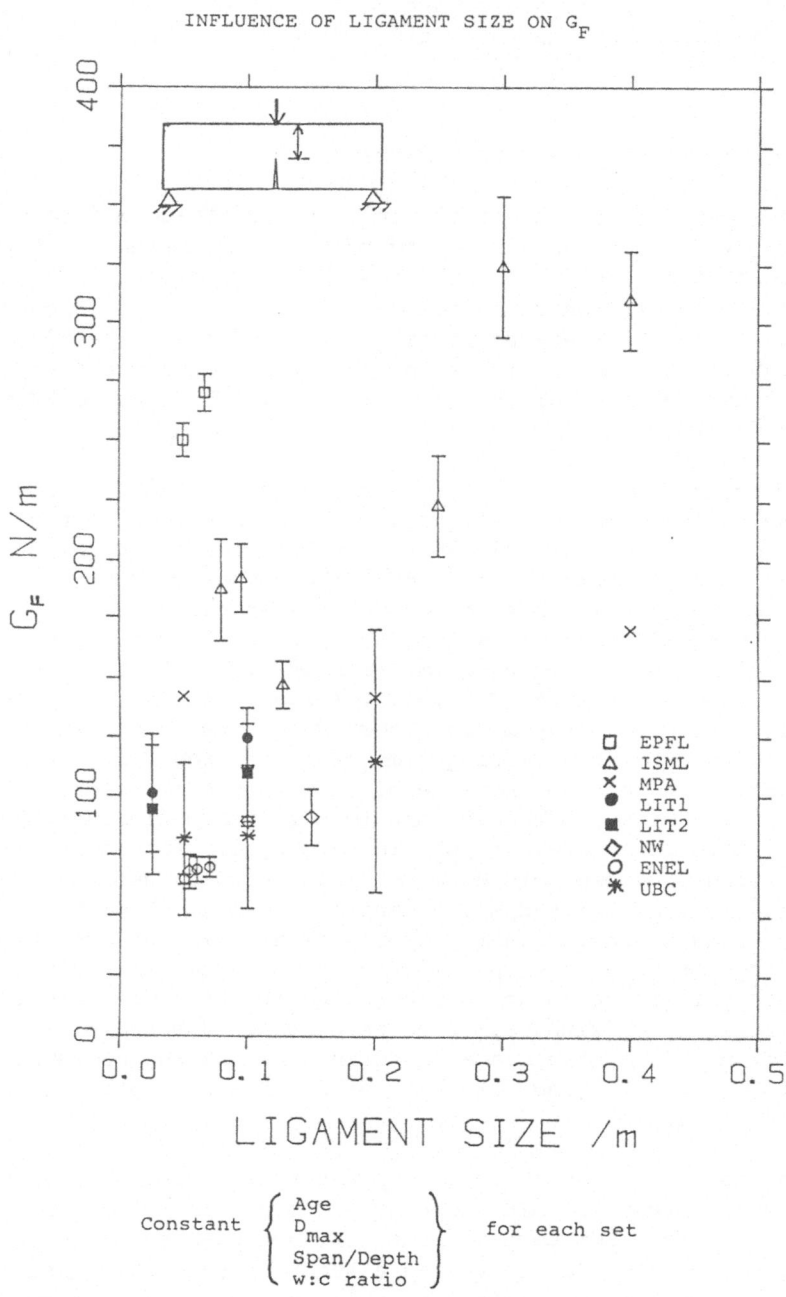

Figure 3: Influence of Ligament Size on G_F

though these deviations may not be large when compared to size effects of other parameters such as cylinder compressive strength commonly adopted for concrete, it should be pointed out that the 20% and 30% deviations are that above the .0005m beams. It is not clear how much the G_F value measured from the .0005m beams differ from the true value such as that measured as the area under an accurate tension softening curve. It is also worrisome that the G_F as measured, being an upper bound value, is unconservative for engineering design purposes.

4   R-CURVE FOR CONCRETE

Many measurements of resistance curves for concrete and for fibre-reinforced concrete (see, e.g. Wecharatana & Shah (14), Mai (15)) have been published in the literature. In the following we discuss whether an R-curve calculated from laboratory experiments could be used to characterize fracture resistance uniquely.

The R-curve concept is originally used in metal, where it is found that the thickness of specimens plays a strong influence on the fracture resistance. This is due to the development of the size of the plastic zone which is sensitive to the stress state at the crack tip. In particular, the plane stress state allows growth of a large plastic zone in comparison to the plane strain situation where material constraints inhibit plastic zone growth, resulting in a lower fracture toughness. Thus in moderately thin metal sheets, fracture resistance increases with stable crack growth accompanied by plastic zone growth. This also happens under conditions approaching plane strain in highly ductile materials. In concrete, it has been found that plane stress or plain strain does not play an important role in determining the fracture resistance. In contrast to metal, where the inelastic behavior at the crack tip owes itself to true plasticity (dislocations), the inelastic behavior in concrete is due to microcracking and aggregate pull-out, which is most sensitive to straining normal to the crack plane (in mode I) but not so much to the stress in the perpendicular direction. Rather, the increase in fracture resistance in concrete with crack growth is associated with the development of the process zone. Ingraffea and Gerstle (2), e.g., shows that the process zone increases in length with applied load before reaching a steady value by means of finite element calculation and the assumption of tension softening on the crack line. They also show that the length of the process zone is not a material parameter, but rather depends on the specimen geometry and loading configuration. These considerations imply that the R-curve in general could not possibly be a unique material property. It is possible, however, to produce R-curves which look rather similar (e.g. Wecharatana and Shah (14), Bazant et al (15)). The similarity may be attributed to similar loading conditions producing similar stress fields which

control the crack tip processes. For example the R-curves calculated by Bazant et al (15) were based on two 3-point bend tests
and a compact tension test. Both test configurations produce a
compressive zone at the outer fibres of the ligament and the resulting stress field may be expected to be rather similar.

Indeed, the applicability of the R-curve to describe fracture
resistance in a real structure made of the same material (as the
test specimen) depends on the presence of a dominant stress field
in the test specimen and in the real structure. As an example,
if the crack in a real structure experiences a dominant K-field,
then the laboratory calculated K_R-curve is applicable provided
the test has been performed under a dominant K-field. Such requirements are probably difficult to achieve for concrete.

Suppose one somehow obtains a unique R-curve, another concern
is how one might use it. In general the R-curve is useful in
determining stability of crack growth if one knows the increase of
energy release rate with crack length. For concrete, if the process zone is large, this last relation is not easily determined
so that the usefulness of the R-curve seems to be rather limited
in practice.

5 INDIRECT MEASUREMENT OF TENSION SOFTENING CURVE

It appears that one of the most basic properties of concrete
fracture is the relation between the tensile stress and the separation distance of material in the process zone, generally known as
the tension softening curve. The tension softening curve has been
used successfully as a constitutive relation on the crack line in
several numerical schemes to describe behavior of structures containing cracks (see, e.g., Hillerborg et al (17), Ingraffea and
Gerstle (2), Bazant and Oh, (18). Assuming no singular field
exists at the crack tip, the area under the tension softening curve
can be shown to be the critical energy release rate G_c. Several
attempts at measuring the tension softening curve have been made
(see e.g. Evans and Marathe (19), Petersson (6), **Gopalaratnam** and
Shah (21), Reinhardt (22)). An inherent difficulty of this kind
of testing is that the deformation is unstable and some kind of
stiffening of the testing machine is necessary to perform a valid
test. Various kinds of stiffening mechanisms have been used,
including parallel steel bars in the direction of loading and
closed loop feedback systems. Some of these tests have performed
quite well, although the testing procedure can hardly become a
standard one because of the need of rather intricate modifications
on the loading machines. In the following paragraphs, we shall
describe an indirect technique which has the following characteristics:

a) simple testing procedure
b) simple testing machine
c) small specimens

These characteristics are important if the technique is going to be widely adopted. They are also consistent with economy of testing, especially when a large number of specimens are to be tested.

We first describe the theoretical basis of this testing technique. For a cohesive type model of fracture process, it may be shown that the J-integral (see, e.g. Rice (23),(24)) when taken on a contour surrounding the process zone, may be written in the form

$$J = \int_{0}^{w_c} \sigma(w) \, dw \qquad (2)$$

Taking the derivative with respect to w of equation (2) gives

$$\partial J / \partial w = \sigma(w) \qquad (3)$$

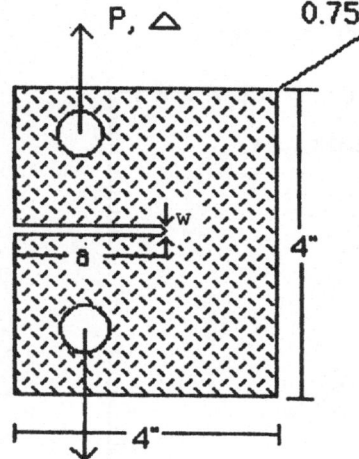

a=1.55 and 1.75 in.
cement:water:mortar sand:no. 8 sand= 2:1:2:2
curing: 7 days under water
cross head displ. rate= 0.002 in/min.

CONCRETE SPECIMEN

Figure 4: Specifications of Compact Tension Specimen

Hence the tension softening curve may be obtained from a knowledge of the change of J with respect to the separation w. For the calculation of J, we choose the compliance test on compact tension specimens. To fix ideas, two identical single edge notched specimens as shown in Figure 4 are loaded in tension. The only difference in the specimens are the crack lengths, which are cut by

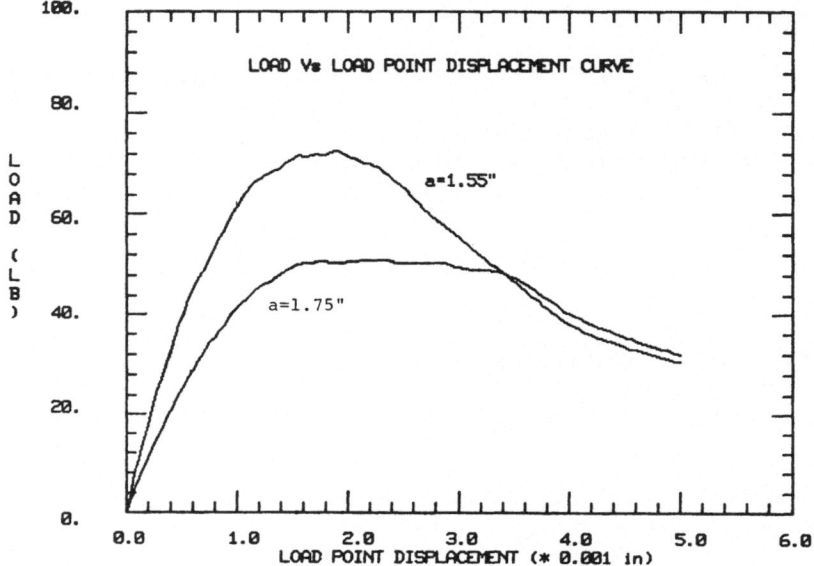

Figure 5a: Load vs. Load Point Displacement Curves from Experiment

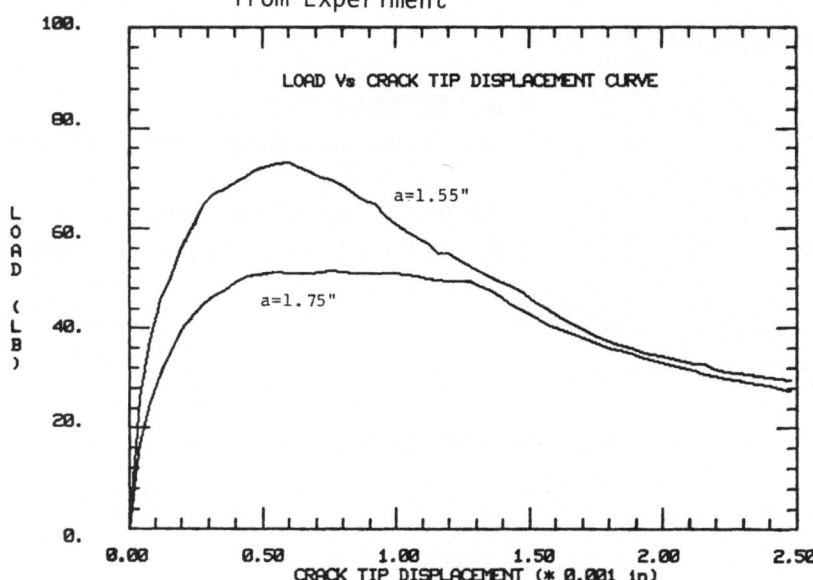

Figure 5b: Load vs. Crack Tip Displacement Curves from Experiment

diamond saws. During the test the load vs. load point displacement
and the load vs. crack tip opening is continuously recorded by
means of extensometers, as shown in Figures 5a,b. The area between
the load vs. load point displacement curves give the value of
$J(a_1-a_2)B$ where B is the thickness of the specimens, since an
alternative definition of J is

$$J(w) = \int_O^W (\partial P/\partial a)_W \, dw \tag{4}$$

where P is the load magnitude.

As the crack length difference (a_1-a_2) is known a priori, J
may be readily calculated. Once $J(w)$ is known, its slope then
gives the desired tension softening curve.

To illustrate, we show some preliminary results based on testing mortar specimens, with material and preparation specifications
as shown in Figure 4. Figures 6a and 6b show the J vs. w and the
$\partial J/\partial w$ vs. w results obtained from the test data shown in Figures
5a,b already mentioned. The $\partial J/\partial w$ curve shows an initial rise to
a peak before descending down to zero. We attribute this behavior
to diffuse inelastic damage experienced by the specimen prior to
localization of damage onto the plane of the eventual fracture.
This observation is consistent with what we had suggested earlier
for the beams of the G_F tests. Prior to localization the measurement of w reflects the diffuse inelastic deformation and do not
represent any real material separation of the crack tip. The corrected tension softening curve is shown as the dashed curve in
Figure 6b. Following the suggestion of Hillerborg we have reproduced the tension softening curve in normalized form, shown in
Figure 7. The separation distance w has been normalized by a
characteristic length $\ell_{ch} \equiv EG_c/f_t^2$, where the critical energy
release rate G_c has been measured at $0.28\ell b/in.$, the tensile
strength f_t at 380 psi. With an assumed value of 4.5×10^6 psi
for the Young's modulus E, ℓ_{ch} is equal to 8.7 in. Our preliminary
results reveal a tension softening curve which agrees qualitatively
with that measured by direct tensile tests by Petersson (6), by
Gopalaratnam and Shah (21), and by Reinhardt (22). As a rough
check, we compare their experimentally determined values of f_t,
G_c and the critical separation w_c with these obtained from our
experiments, shown in table 1. Information concerning concrete
mix and age, and specimen dimensions are given in table 2. Table 1
shows that our values are generally on the low side, perhaps due
to the early age at testing and the small aggregate size used for
our specimens. This is particularly true for w_c. Results from
Petersson (6) and Gopalaratnam and Shah (21) showed a long tail
and w_c in fact could not be determined in their experiments.

Table 1. Comparison of preliminary test results with
test results from (6), (21), (22).

Reference	f_t (psi)	w_c (10^{-3} in.)	G_c (lb/in)
Preliminary Test (this study)	380	1.29	0.28
Petersson [6]	479	>4.3*	0.58
Gopalaratnam & Shah (21)	405	>2.4*	0.42
Reinhardt [22]	464	6.9	0.77

* the w_c values given by Petersson, and by Gopalaratnam and Shah
are subjective numbers. Their G_c-values were calculated to the
indicated w_c values. Beyond that limit contribution to G_c
was neglected.

Table 2: Specimen and Material Compositions

Reference	Type of Test	Overall Dimension	Mix Proportion (c:s:A:w	Max. size of sand/ aggregate	Age at Testing (date)
Preliminary Test	Compact Tension	4"x4"x0.75"	1:1:1:0.5	sand:0.056" agg:0.093"	7
Petersson	Direct Tension	2"x2"x1.2"	1:0.6:2.3: 2.3	sand:0.157"	28
Gopalaratnam & Shah	Direct Tension	12'x3'x0.75"	1:0:2:0.5	agg:0.19"	28
Reinhardt	Direct Tension	9.8"x2.4"x 2.5"	1:0:3.4:0.5	agg:0.31"	55

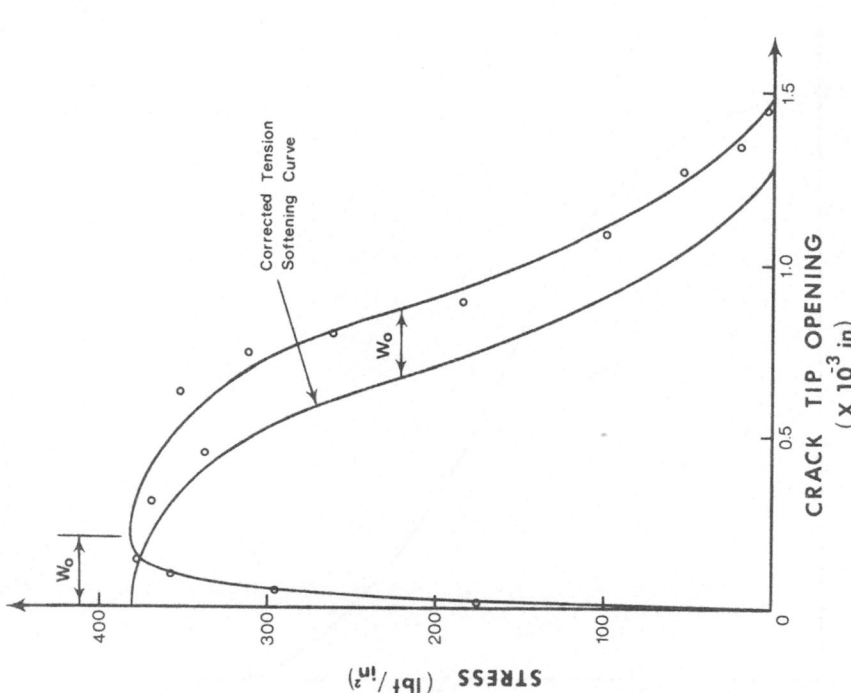

Figure 6b: Stress vs. Crack Tip Opening Deduced
from Fig. 6a based on Equation (3)

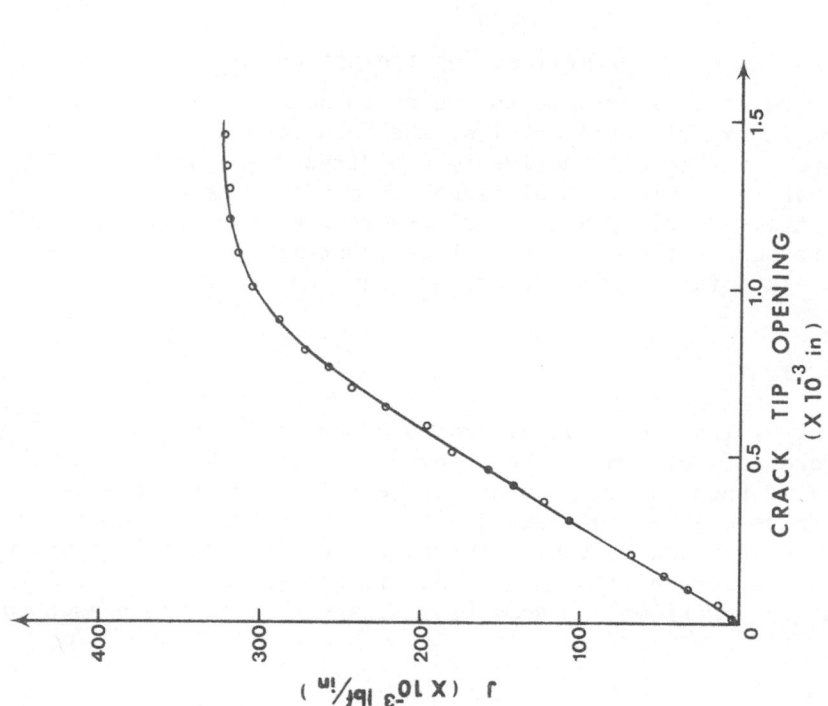

Figure 6a: J vs. Crack Tip Opening Displacement
Deduced from Fig. 5a,b.

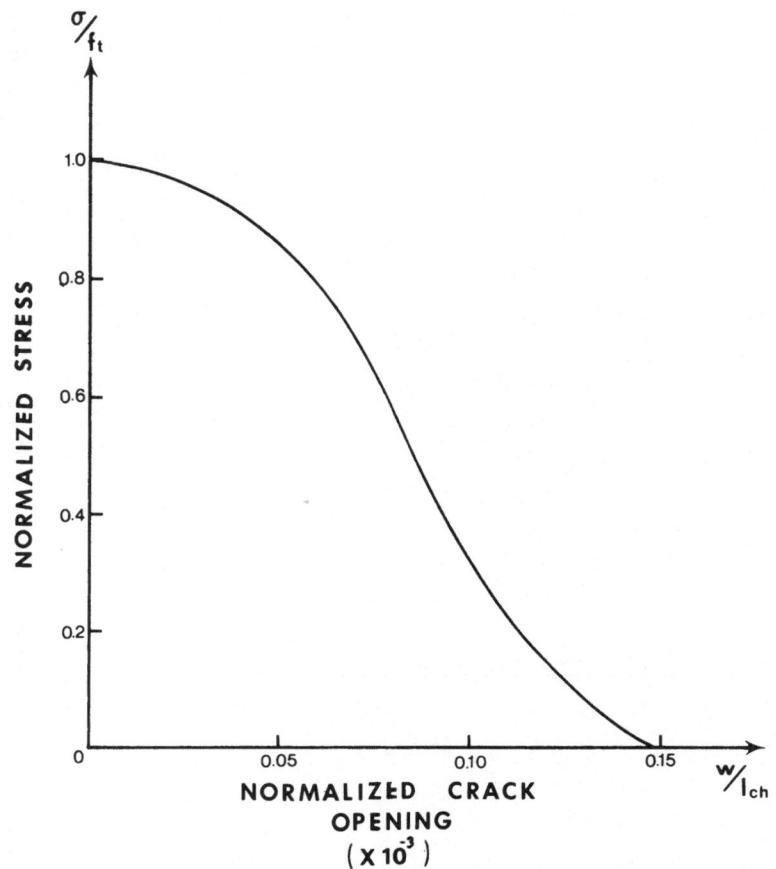

Figure 7: Normalized Tension-softening Curve

The qualitative agreements are encouraging. With further refinements in experimental details, the technique presented here may become a viable alternative to the direct tensile test. We expect that this experimental technique may also be applicable to other brittle materials with tension-softening behavior. Details of this testing technique and further experimental results are contained in a forthcoming report by Chan and Li (25).

6 CONCLUSIONS

We have discussed several fracture resistance parameters from the approach of getting at their physical basis. We have shown that some of these parameters are at best limited in their usefulness in describing concrete fracture resistance. A new technique of measuring tension-softening curves for brittle materials which overcomes problems of instability and is simple and economical to carry out is described and some preliminary results are presented.

7 ACKNOWLEDGEMENT

The author would like to thank the following individuals:
A. Hillerborg, for numerous stimulating discussions and for his
many insightful comments; J. R. Rice for his introducing the author
into the exciting world of fracture mechanics, and for his suggestion of the theoretical basis of the extraction of the tension-
softening curve, and C. M. Chan, for conducting the experimental
analysis of tension-softening behavior described in the last section of this paper. A grant from the Sloan Foundation at M.I.T.
supports this work.

8 REFERENCES

1. Hillerborg, A., Analysis of One Single Crack, Fracture Mechanics
   of Concrete, ed. F. H. Whitmann, (Elsevier, 1983), 223-250.

2. Ingraffea, A.R. and W.H. Gerstle, Non-linear Fracture Models for
   Discrete Crack Propagation, Application of Fracture Mechanics
   to Cementitious Composites, this volume, 1984.

3. Sok, C., Etude de la Propagation d'une Fissure dans un Beton now
   Arme, Bull. Liaison Lab. Ponts Chaussees, 98 (1978), 73-84.

4. Jenq, Y.S. and S.P. Shah, Nonlinear Fracture Parameters for Cement Based Composites: Theory and Experiments, Application of
   Fracture Mechanics to Cementitious Composites, this volume, 1984.

5. Determination of the Fracture Energy of Mortar and Concrete by
   means of Three-point Bending Tests on Notched Beams, Proposed
   RILEM recommendation, January 1982, revised June, 1982. Lund
   Institute of Technology, Division of Building Materials.

6. Petersson, P-E., Crack Growth and Development of Fracture Zones
   In Plain Concrete and Similar Materials, Lund Institute of Technology, Division of Building Materials, Report TVBM-1006, 1981.

7. Dugdale, D.S., Yielding of Steel Plates Containing Slits, J.
   Mechanics Phys. Solids, vol. 8 (1960), 100-108.

8. Barenblatt, G. I., The Mathematical Theory of Equilibrium Crack
   in the Brittle Fracture, Advances in Applied Mechanics, vol. 7
   (1962), 55-125.

9. Hillerborg, A., Additional Concrete Fracture Energy Tests Performed by 6 Laboratories according to a Draft RILEM Recommendation, Report to RILEM TC50-FMD, 1984.

10. Hilsdorf, H.K. and W. Brameshuber, Size Effects in the Experi-
    mental Determination of Fracture Mechanics Parameters, Application
    of Fracture Mechanics to Cementitious Composites, this volume 1984.

11. Hillerborg, A., Concrete Fracture Energy Tests Performed by 9 Laboratories According to a Draft RILEM Recommendation, Report to RILEM TC50-FMC, Report TVBM-3015, Lund Institute of Tech., 1983.

12. Nallathambi, P., Karihaloo, B.L., and B.S. Heaton, Various Size Effects in Fracture of Concrete.

13. Hillerborg, A., Summary of Test Results from 7 Laboratories on the Influence of the Specimen Size on the Measured Value of the Fracture Energy. Preliminary Version, 1984.

14. Wecharatana, M. and S.P. Shah, "Slow Crack Growth in Cement Composites", J. Structural Division, ASCE, vol. 108, No. ST6 (1982), 1401-1413.

15. Mai, Y.W., Fracture Measurements of Cementitious Composites, Application of Fracture Mechanics to Cementitious Composites, this volume 1984.

16. Bazant, Z.P., Kim, J.K. and P. Pfeiffer, Determination of Non-linear Fracture Parameters from Size Effect Tests, Application of Fracture Mechanics to Cementious Composites, this volume 1984.

17. Hillerborg, A., Modeer, M. and P-E Petersson, Analysis of Crack Formation and Crack Growth in Concrete by Means of Fracture Mechanics and Finite Elements, Cement and Concrete Research, vol. 6 (1976) 773-782.

18. Bazant, Z.P. and B.H. Oh, Crack Band Theory for Fracture of Concrete, Materials and Structures, (RILEM, Paris), vol. 16 (1983) 155-177.

19. Evans, R.H. and M.S. Marathe, Microcracking and Stress-Strain Curves for Concrete in Tension, Materiaux et Constructious, 1 (1968) 61-64.

20. Modeer, M., A Fracture Mechanics Approach to Failure Analysis of Concrete Materials, University of Lund, Report TVBM-1001, 1979.

21. Gopalaratnam, V.S. and S.P. Shah, Softening Response of Concrete in Direct Tension, submitted to American Concrete Institute Journal, 1984.

22. Reinhardt, H.W., Fracture Mechanics of an Elastic Softening Material like Concrete, HERON. Delft Univ. of Technology, vol. 29, No. 2, 1984.

23. Rice, J.R., Mathematical Analysis in the Mechanics of Fracture, in Fracture: An Advanced Treatise, Vol. 2 (Academic Press, 1968), 191-311.

24. Rice, J.R., A Path Independent Integral and the Approximate Analysis of Strain Concentration by Notches and Cracks, J. Appl. Mech., vol. 35, 379-386, 1968.

25. Chan, C.M. and V.C. Li, An Indirect Procedure of Measuring Tension Softening Curve for Concrete, in preparation, 1984.

SECTION V

DAMAGE AND CONTINUUM MODELING

APPLICATION OF FRACTURE MECHANICS
TO CEMENTITIOUS COMPOSITES
NATO-ARW - September 4-7, 1984
Northwestern Univeristy, U.S.A.
S. P. Shah, Editor

MECHANICS OF SOLIDS WITH A PROGRESSIVELY
DETERIORATING STRUCTURE

Dusan Krajcinovic
Professor, CEMM Department
University of Illinois at Chicago
Chicago, Illinois, USA

ABSTRACT

This report summarizes the applications of the continuum
damage theory to problems of mechanics of rocks and concrete. The
report discusses the basic aspects of the theory such as the
selection of the internal variable describing the distribution of
microdefects, establishment of the 'damage laws', the essential
structure of the governing equations etc. Finally, the discussion
focuses on a number of as yet unresolved problems of some
theoretical and practical significance.

1. INTRODUCTION

The spectacular and somewhat bewildering progress of the
computational capabilities enables us to tackle an ever widening
spectrum of problems which were only recently beyond our fondest
dreams. Somewhat belatedly, but just as painfully, it became
increasingly apparent that the computer itself does not and will
not offer a better and more accurate insight into the behavior of
solids until and unless a better understanding is developed into
the manner in which the microstructural changes reflect on the
macrostructural response.

This is not to say that the purely phenomenological theories
which served us so faithfully in the past should be discarded and
forgotten. Quite to the contrary, they should be used and even
developed further being aware of their limitations inherent to the
assumptions upon which they are based. It is both wrong and

dangerous, however, to use these phenomenological theories indiscriminately especially in cases when the precise mode of the microstructural kinetics has a profound effect on the macrostructural response.

In general, when subjected to changes in loads and ambient temperature the microstructure of the solid changes. Some of these changes are irreversible and they are to a large extent responsible for the nonlinear component of the material response in a purely phenomenological sense.

The two most prominent modes of the irreversible changes of the microstructure are:

- slip on the preferred crystallographic planes, and
- nucleation and growth of microcracks and microvoids.

The slip is promoted by the shear stresses available for moving and stacking dislocations (line defects) into a preferential configuration. As a results, slip occurs typically in a plane and direction in which the shear stress exceeds some threshold value which depends on the crystal structure, homogeneity of the crystalline lattice, second phase particles, alignement of grain boundaries as well as the changes in stress resulting from the slip in the neighboring planes. Since the material slips 'through' the crystalline lattice the number of bonds between particles remains practically unchanged. Consequently, the elastic properties of the material are relatively insensitive to the amount of slip already recorded in the solid. The plastic deformation of the solid is a phenomenological result of the slips on all active slip systems in the solid. As such the plastic strain is a reasonably good if not the best record of the dissipative process. This is not to say, though, that every deformation process which results in residual strains can be directly related to the slips in preferred crystallographic planes. Specifically, the rocks and concrete lack the crystalline lattice necessary for the sustained slip deformation.

In contrast, microcracking in the cleavage mode occurs in planes perpendicular to the direction in which the direct tensile strain exceeds some threshold value reflecting the cohesive and/or adhesive strength of the solid locally. Since the microcracking involves progressive loss of bonds between adjacent particles (grains) the elastic properties of the solid are affected as well. In other words, the solid will not unload along a path parallel to the initial (elastic) segment of the original stress-strain curve in loading.

The phenomena dominated by the slip are studied within the context of the theory of plasticity (or the slip theory). It must be pointed out, though, that the nonlinear behavior of the rock (see, for example, Wong 1982 and Evans and Wong 1983) depends also on the mineralogy and the grain orientation. The dilatancy and the pressure sensitivity of the rock prompted the appearance of a host of modifications of the plasticity theory even though the scanning microscope studies of Tapponnier and Brace (1976) indicate that the dilatancy is a result of cracking. A precis of these theories is beyond the scope of this report.

It is quite obvious that both the cause and the consequence of these two dominant modes of the irreversible changes in the microstructure are distinctly different. It is, therefore, unrealistic to expect that the theory of plasticity alone, irrespective of its incarnation, could account for the behavior caused by the interaction of both modes of the microstructural change.

Just as the theory of plasticity and the slip theory were developed to deal with the phenomena characterized by slip a recent surge in the interest in the continuum damage mechanics raises a hope that it will be possible to analyze the behavior of solids dominated by the nucleation and growth of microcracks. Owing to a better familiarity of the mechanics community with the theory of plasticity and the slip theory, it seems appropriate to focus on the continuum damage theory with a special emphasis on its application to the mechanics of rocks and concrete.

In this context it is important to underline that a particular solid is per se neither brittle nor ductile. Every solid, depending on a specific set of circumstances (such as temperature, strain rate, confinement, homogeneity etc), can behave as either brittle or ductile. In other words, a response dominated by slip in shear planes will be perceived as ductile. Conversely, the response characterized by microcracking in cleavage mode is typically classified as brittle. For example, a concrete or rock specimen in uniaxial compression will behave as brittle if unconfined and ductile if partially or totally confined. Naturally, it is very seldom that a response is either perfectly brittle or perfectly ductile. Different materials under different conditions cover the entire spectrum between these two extremes.

2. PRELIMINARIES

The utility of the continuum damage mechanics and its applicability over a wide range of diverse problems (see, for an example, Krajcinovic 1984) was first demonstrated by means of the simple, one-dimensional models. The ease with which some of these simple, design oriented models, based directly on the original Kachanov's idea (1958) furnished explanation of some rather complex aspects of the response of brittle materials provided a strong impetus for the further development of the continuum damage mechanics. As these simple models ran their course exhausting the supply of problems amenable to simple, algebraic solution on the back of the proverbial envelope, the attention shifted towards a more ambitious and difficult task of the development of a comprehensive theory based on the firm thermodynamical foundation.

Without pretense of generality the objective of this report is to summarize the part of the general effort which is directly related to the mechanics of rocks and plain concrete. Needless to say, the narative will be biased by the opinions of its author brought upon by his own work in this field.

3. GENERAL THEORY

In general, any continuum theory can be developed either as a memory or as a state type of a theory. From the purely practical viewpoint the latter approach has definite advantages and will, consequently, be the only one discussed in the sequel. In order to construct an analytical model which will be as simple as possible while being simultaneously able to duplicate the macroscopically observable trends it is necessary to assume 'that it is always possible to determine a finite set of internal variables...in such a way that their number is adequate to render the nonequilibrium state under consideration sufficiently close to a constrained state of thermodynamic equilibrium' (Kestin and Bataille 1977). The set of internal variables characterizing dissipative mechanisms (interior to the system) associated with the irreversible mechanical changes in the microstructure can be selected in a wide variety of ways reflecting just as wide a variety of reasons, speculations and often tenuous arguments. Nevertheless, it appears obvious that the accuracy with which a model can be expected to describe a specific process is proportional to the degree with which the set of selected internal variables can depict the kinematics of the irreversible changes in the structure of the material on the microscale.

3.1 Internal Variables

Assuming it to be possible, or even advisable, every local rearrangement of the structure of the solid can be characterized by a specific internal variable. From a detailed knowledge of the microstructural kinetics it should be, at least in principle, possible to reconstruct the macroscopic response of the specimen. However, since the number of sites in which the irreversible changes in the microstructure take place is very large, dependent on the size of the specimen and since most importantly the details of the process will change from one experiment to the other a strategy incorporating 'exact' kinetics of every microdefect would not be a reasonable one. In other words a rational transition from microscopic to a macroscopic theory by necessity involves averaging (Kroner, 1962).

For example, assuming that the inelastic strain results from the crystallographic slip alone Hill and Rice (1972) suggested the slipping rate of a particular slip system relative to the reference lattice as an appropriate choice for the internal variable. Incidentally, even crystalline slip already implies volume averaging of dislocation glides (Rice 1975) over all operative slip systems of a crystalline subelement.

The same strategy can be applied for the description of microdefects related to the brittle behavior of solids. A spherical microvoid should be characterized by a scalar delta function $a = a_0 \delta(x - x_0)$ where a_0 is the volume of the microcavity and x_0 its location. Similarly an axial vector $\underline{b} = \underline{b}_0 \delta(x - x_0)$ is a geometrically consistent representation of a penny-shaped flat microcrack (with b_0 being its area). Departure from the spherical shape in first and circular shape in the latter case would involve higher momenta of the selected variables and consequently increase in the complexity of the model. In view of the results computed by Budiansky and O'Connell (1976), according to which the compliance of an elastic body is relatively insensitive to the planform of the crack, this should usually not be necessary.

The selection of a rational procedure averaging these delta functions describing the microdefects of the solid into a continuous function of coordinates which performs the same task omitting the unnecessary details is one of the most contentious aspects of CDM. In case of spheroidal microcavities the volume density of voids (i.e. the porosity) appears to be a reasonable choice for a rational description of the void distribution. A comprehensive analytical model based on the porosity as a measure of 'damage' has been proposed in the past by Davison et al (1977). In fact the models of porous solids such as those suggested by Goodman and Cowin (1972) can be relatively easily recast into a CDM form.

The situation is not as straightforward in case when a solid is weakened by a multitude of flat penny-shaped microcracks. Fortunately, the original Kachanov's idea that the void area density in a cross section is a reasonably good measure of its integrity and load carrying capacity is not only almost universally accepted but also has an undeniable intuitive appeal. Nevertheless, this proposition must be carefully examined within the context of a general formulation. Since the number of microcracks while large is still finite and the number of cross sections in a solid through a point infinite the probability of a cross section to contain a microcrack is zero. As a result a great majority of cross sections will be 'undamaged' regardless of how many cracks permeate the volume around the observed point. Guided by the angle of the diffusion of stresses through solids and the experiments by Litewka and Sawczuk (1981), Krajcinovic (t.b.p.) resolved this problem introducing a characteristic length and an appropriate weighting (or influence) function. As a result it can be demonstrated that the distribution of microcracks in the observed material point can be properly described by a set of axial vectors $\underline{\omega}(x) = \omega_0(x) \underline{N}$ where ω_0 is the local density of microcracks with normal N in the observed material point.

It is important to note that the vectors $\underline{\omega}$ do not represent the damage in the plane with normal N through the material point $x = x_0$. According to the adopted definition damage in a cross section through the observed material point x_0 is obtained simply as a projection of all active and operative microcrack systems on the plane of the cross section,

$$D(x_0 ; \hat{\underline{N}}) = \sum_j \underline{\omega}^{(j)} \cdot \hat{\underline{N}} \tag{1}$$

with a normal $\hat{\underline{N}}$. The microcrack system is considered operative if its density still increases ($\dot{\omega} > 0$, $\omega > 0$), active if its magnitude remains constant but the area for the transfer of stresses is still reduced ($\dot{\omega} = 0$, $\omega > 0$) and passive if it behaves as if the microcracks do not exist at all. The criteria for the status of the microcrack system should not differ substantially from those discussed by Horii and NematNasser (1983) and Margolin (1983).

According to the proposed model the damage is, in a complete accord with the Kachanov's original idea, a nonnegative scalar (defining the loss of the load bearing area) associated with each cross section through the material point x_0. While this distinction between the microcrack distribution $\underline{\omega}$ and damage D appears to be purely formal, and is indeed lost or at least blurred in a variety of simple cases characterized by a monotonicaly increasing proportional loading, the associated implications are far from being trivial. For example, if a specimen weakened by a system of microcracks in x-y plane is subjected to a tensile stress in the direction of z-axis the the specimen responds as if it is damaged. If the stress along the z-axis is compressive the same specimen behaves as if the damage does not exist. In contrast, while the status of the microcrack system changes from active to passive they never cease to exist. Since the magnitude of the internal variables should not depend on the state of stress it is inappropriate to use the damage D instead of ω as internal variable. Damage is nothing but the consequence of the existing defects in the material and as such it can definitively depend on the manner in which these defects affect the response.

An additional important advantage of the suggested model is that it enables transition from the volume to the area density and as a result establishes a link between CDM and the theories dealing with estimates of the elastic compliance of solids permeated by a large number of defects (Budiansky and O'Connell 1976, Horii and Nemat-Nasser 1983 etc). The major difference between the two theories is that the latter one is based on the linear fracture mechanics and, consequently, predicts that a specimen with cracks in (x-z) and (y-z) planes behaves as if the cracks do not exist if subjected to a compressive stress in z direction. This is, of course, in contradiction with the experimental evidence obtained from uniaxial compression of unconfined rock or concrete specimens.

It is important to notice that both the slip γ and the density of microcracks ω in the plane with normal $\underline{N}$ are actually averaged variables in sense of Rice (1971). As such they entail the question of scale. The elementary volume must be large enough to contain a sufficient number of microcracks (i.e. achieve macro-uniformity) capturing the physics of the phenomenon and yet be small enough to be mapped on a material point. This is an especially troubling point when concrete is in question since the size of the aggregate and accordingly the cracks is typically rather large. This in effect means that in case of concrete and probably rock as well the criterion separating microcracks from macrocracks should attract more attention. Nevertheless, there is at least some experimental evidence that the fracture mechanics leads to unreliable results for rocks (Hoagland et al 1973) and concrete (Pak and Trapeznikov 1981) as well. The indications are that the criterion is not absolute but is related to the size of the grain (i.e. uniformity scale).

3.1.1. Phenomenological Models. As already stated above in case of monotonically increasing, proportional (in-phase) loadings it is advisable to resort to purely phenomenological models (such as these proposed by Lemaitre and Chaboche 1978, Chaboche 1978 and

1979 etc). Despite the fact that the theory proposed above and based on the selection of the slip $\gamma(x)$ and the microcrack density $\omega(x)$ in every operative plane through every sample point already involves averaging its application in case of nonuniform stress fields and complex geometry will require extensive bookkeeping i.e. large and powerful computer codes. Instead of keeping track of both variables in each plane through each material point, guided by the success of the conventional plasticity theory, it appears reasonable to search for a single internal variable (being a continuous and rather smooth function of coordinates) which will adequately approximate a set of variables $\underline{\omega}$ and define the state of material weakened by an arbitrary microcrack field. An attempt to formulate such a theory can proceed in many ways, three of which have been used in the past.

Firstly, the 'damage' variable can be selected on the basis of rather general considerations regarding the nature of the macroscopic stress-strain relationship in view of the anisotropy induced by microcrack field (Chaboche 1979). Secondly, the 'damage' variable may be derived on the basis of geometrical speculations (Murakami and Ohno 1981). Finally, guided by the relation existing between the plastic strain tensor and slips over all operative slip systems a similar relation was established between a continuous tensor variable $\underline{D}$ and a set of axial vectors $\underline{\omega}$ (Krajcinovic t.b.p.).

Recently, several authors such as Dougill (1976) and Nicholson (1981) suggested phenomenological models incorporating effect of the microcracking through an additional component of the strain tensor. According to this model the total strain comprises of elastic, plastic and damage component. The application of such a model for concrete was suggested by Bazant and Kim (1979). It appears, though, that such an approach needlessly obfuscates the issue without any tangible advantage. Not only that the physical insight grows fuzzier but the fact remains that a purely brittle solid weakened by microfractures should not exhibit any residual strains upon unloading. In other words, to record the history of an event characterized by a gradual deterioration of the solid it is necessary to use progressive loss of bonds rather than the accumulated strain.

4. GOVERNING EQUATIONS

The essential structure of the governing equations for the considered class of solids can be effectively and conveniently established along the lines suggested by Rice (1971). Writing the expression for the change in the internal energy density in transition from one constrained equilibrium state to the neighboring one the thermodynamic forces can be expressed as the derivative of the Helmholtz free energy with respect to its conjugate kinematic variable

$$
\underset{\sim}{S} = \frac{\partial \psi}{\partial \underset{\sim}{E}} \qquad \tau = \rho \frac{\partial \psi}{\partial \gamma} \tag{2}
$$
$$
R = - \rho \frac{\partial \psi}{\partial \omega}
$$

where $\underset{\sim}{S}$ is the Kirchoff stress, $\underset{\sim}{E}$ the Lagrangian strain, while τ and R are the thermodynamic forces conjugate to the slip and the microcrack density γ, ω respectively. In case of the infinitesimal strains τ is the Schmid shear stress (Rice 1971). As noticed by Chaboche (1978) and Krajcinovic and Fonseka (1981) the affinity R is in fact closely related to the fracture mechanics 'crack resistance force'. It should be mentioned that in general the increment $\dot{\omega}(x)$ involves both cleavage and change of orientation (Krajcinovic 1983). For simplicity, at this point of the development the change of the orientation can be neglected.

The fluxes and their affinities must satisfy the entropy production rate (or Clausius-Duhem) inequality

$$
\rho \dot{\mathcal{V}} = \sum_i \tau_i \dot{\gamma}_i + \sum_j R_j \dot{\omega}_j \geq 0 \tag{3}
$$

where the equality sign is associated with processes during which no energy dissipation takes place. As in (2) ρ is the mass density of the material.

4.1 Kinetic Equations for Internal Variables

One of the major problem in establishing a rational theory of practical value consists in formulating simple constitutive equations for internal variables with as few material parameters as possible. In general, assuming that the flux is not an explicit function of stresses, i.e. that the rate of stress influences the of change of internal variables only indirectly through affinities, it follows that

$$
R_i = f(R_j, \underset{\sim}{\omega}, \underset{\sim}{E}, \theta) \tag{4}
$$

where Θ is the absolute temperature. The specific form of the constitutive equation (4) must be reasonably simple and simultaneously sufficiently accurate in duplicating (with a moderate computational effort) the experimentally observed salient trends in the response of solids weakened by microcracks.

In principle, the constitutive equations can be always formulated in form of the evolution equations. However, in view of the number of unknowns and the necessity to consider the arbitrary case of loading (cyclic, out-of-phase etc) it appears somewhat simpler to try to define a potential surface in the field of thermodynamic forces from which the fluxes can be subsequently determined on the basis of the normality principle. This approach, familiar from the conventional plasticity theory, appeals to the intuitive feeling that in the neighborhood of a constrained equilibrium state exist other equilibrium states which can be reached at no loss of energy (unloading). This appears to be an especially appropriate assumption in case of microcracking which is intuitively associated with the increase of the tensile strain normal to the crack surface.

The existence of such a flow potential was proven in cases when:

- the flux-affinity relations are instantaneously linear (Kestin and Bataille 1977),
- the flux is fully determined from the thermodynamic force conjugate to it (Rice 1971),
- the conditions derived by Nemat-Nasser (1975) are satisfied, or finally when
- the generalized Drucker's postulate is assumed to hold.

Once any one of these conditions is satisfied a potential surface $F(\tau, R)$ exists in the space of affinities. Each flux is obtained as a derivative of such a potential with respect to its conjugate thermodynamic force and is geometrically represented by a vector normal to the surface of the constant flow potential.

In principle, the flow potential F is not in every case directly derivable or related to the dissipation power density. Therefore, assuming that the inequality (3) is always satisfied it is by no means necessary to define F as a function of affinities τ and R (defined by (2)). Nevertheless, in light of the definite physical meaning attributable to both τ and R and in absence to compelling reasons to the contrary it appears logical to define F as a function of affinities (2).

The exact form of the flow potential is still far from being settled. With little to go on but the rudimentary understanding of the microstructural kinetics and a host of macroscopic observations it is not always simple to determine the flow potential even for the two limiting cases: perfectly brittle (γ = 0) and perfectly ductile (ω = 0) solid.

The most difficult task in specifying the flow potential in case of perfectly ductile solids is to assess the degree to which the neighboring fields interact. It is well known that the slip in one system has a hardening effect on all other systems. A rule according to which the resistance to slip in a specific slip system changes as a result of the slip interaction is still a matter of some disagreement. The rules suggested by Hutchinson (1976), Rusinko (1981), Peirce et al (1982) are just few of the more recent ones.

The similar considerations for a purely brittle solid are at best in the rudimentary stage of the development. In lieu of a better choice, if it is assumed that the plain concrete is representative of the class of perfectly brittle solids the existing experimental data related to the strength of concrete in uniaxial and biaxial tension and compression offer a base for speculations leading to a rational assessment of the degree to which the increase in the microcrack density in one plane inhibits the growth of microcrack densities in the adjacent planes.

The increase in the microcrack density should, in all probability, decrease resistance to the slip in the same plane in proportion to the loss of load bearing surface. However, the understanding of the total interaction picture is, at least at this point, a task requireing a rather comprehensive experimental and analytical effort.

According to the proposed scheme the establishment of the kinetic equations for fluxes is reduced to the determination of the flow potential (yield–damage) surface $F(\tau ,R)$ for each loading case and ambient temperature. Initially the surface F is for a solid with random structure a regular polyhedral surface which in the limit (for an infinite number of operative systems) becomes a smooth convex ellipsoid. In fact, a typical macroscopic yield surface is the inner envelope of very large number of planar yield surfaces corresponding to individual slip planes. As the slip and microcracking are developing in preffered planes the innitially smooth surface develops vertices (Hill 1967 etc) in the intersection of latent assymptotes. These vertices act as the stress attractors and their introduction allows more flexibility in the analyses due to the absence of a unique normal in the intersection of two hyperplanes.

4.2 Helmholtz Free Energy

Since the affinities (2) are derived directly from the Helmholtz free energy (HFE) the determination of its most appropriate form is a very delicate and just as important task. The HFE is an objective scalar valued function of all state and internal variables (i.e. strain tensors and internal variables defining the distribution of all active and operative microcracks). If the model is formulated as an extension of the original Kachanov's model, in case of infinitesimal strains HFE must be quadratic in strains and linear in variables defining the 'damage'. Using the theory of invariants (see Spencer 1971 for example) HFE can be readily written in its objective form for the full group of orthogonal transformations combining the polynomial terms of the irreducible integrity basis for the involved second order tensors and vectors (polar or axial). At this point it is important to underline the distinction between the axial and polar vectors since their basic invariants are nor identical for a full group of orthogonal transformations.

A specific form for the HFE for a perfectly brittle solid was suggested by Krajcinovic and Selvaraj (1983). The current investigations indicate some shortcomings of this initial effort implying need for a more complex form in order to be able to duplicate the strong anisotropy induced by the presence of the microcracks. In a later development (Krajcinovic t.b.p.) demonstrated that in the limit as the number of the microcrack fields tends to infinity the dyadic products of the axial vectors figuring in the expressions for the free energy, material property (stiffness) matrix, thermodynamic forces and the energy dissipation rate density become under certain conditions tensors the order of which depends on the selected form of the HFE. This is not to say that the corresponding tensor and vector representations are identical. Knowing the microcrack density in each operative and active microcrack field it is always possible to compute all components of their dyadic product. In contrast, knowing the dyadic product it is, in general, not always possible to determine the microcrack density in all microcrack fields (unless the number of the microcrack fields is less or equal to the number of the components of the tensor). The same situation, naturally exists between the slips and the plastic strain tensor.

4.3 Stress-Strain Law

Consider for simplicity the case of the infinitesimal strains of perfectly brittle solids ($E \simeq \varepsilon$). In absence of the slips only elastic strains take place while the microcrack density vectors $\underline{\omega}$ are the only internal variables. Using the Voigt's notation from (2a)

$$\{\sigma\} = \frac{\partial \psi}{\partial \varepsilon} = [K(\omega_i)]\{\varepsilon\} \tag{5}$$

where ω_i are the microcrack densities in all operative and active fields. Differentiating the stress-strain relation above it follows that

$$\{\dot{\sigma}\} = [K(\omega_i)]\{\dot{\varepsilon}\} + \sum_j [K']\{\varepsilon\}\,\dot{\omega}_j \tag{6}$$

where the summation is over all $j < i$ active microcrack fields while the components of the matrix $[K']$ are the derivatives of the components of the original matrix $[K]$ with respect to ω_j. All square matrices and vectors are naturally of the sixth order. From the consistency equation written for every active hyperplane $F(R_j, \omega_j)$ obtained is a scalar equation in form of

$$\dot{\omega}_j = \lambda_j\, G_j(R,\omega)\dot{R}_j$$

where $\lambda_j = 1$ iff $(\partial F/\partial R_j)\dot{R}_j > 0$ (loading)
and $\lambda_j = 0$ otherwise (unloading).

Since every affinity R (2c) is, in general, a function of all axial vectors $\underline{\omega}$ and the components of the strain tensor after some manipulations it follows that

$$\dot{\omega}_j = \lambda_j\, \{L(\omega,\varepsilon)\}^*\,\{\dot{\varepsilon}\} \tag{7}$$

where the asterisk denotes transposed matrix. Substitution of the relation (7) into the incremental stress-strain law (6) leads finally to

$$\{\dot{\sigma}\} = [C(\omega)]\,\{\dot{\varepsilon}\} \tag{8}$$

where the material parameter matrix $[C]$ is

$$[C] = [K] + \sum_j \lambda_j\, [K']\{\varepsilon\}\{L\}^* \tag{9}$$

It is important to note that the components of the material parameters matrix depend on the already accumulated microcracks in all active and operative fields regardless of whether the point is in loading or unloading.

While the general procedure remains essentially the same the derivation is somewhat more complex and cumbersome in the case when the slips occur as well. An algorithm similar to one suggested by Halphen and Son (1977), involving the partitioning of the governing matrices, appears to be quite suitable for the purposes at hand.

5. APPLICATIONS TO CONCRETE AND ROCKS

The early applications of the continuous damage theory to concrete and rock problems involved, as expected, in most cases simple one dimensional models. Studying pure bending of a concrete beam with a rectangular cross section Krajcinovic (1979) offered a simple and yet convincing explanation of the difference existing between the strength in direct tension and flexure. It is important to notice that the proposed model involves a single additional constant (directly related to the stress at the apex of the stress-strain curve) and yet yields results which are in a surprisingly good agreement with the experimental measurements. This model was later extended to the beams with circular cross section in eccentric compression by Ouchterlony (1983) in order to study the behavior of rock specimens. Mazars (1981) and later Lemaitre and Mazars (1982) studied the cyclic tests of a double cantilever concrete beam as well as the three and four point bending of a plain and reinforced concrete beam. Their results again show a remarkably good fit of the experimental data. The main problem in the application of the damage theory to the bending of concrete beams is the fact that it is not possible to derive a simple expression for the change of the curvature of the neutral axis as a function of the bending moment. Making use of the fact that the 'exact' relationship for a rectangular cross section is approximated, to a maximum error of 6%, by the arc sine function Krajcinovic and Sestan (1983) suggested a simple design oriented method for the bending of beams with different boundary conditions. Same paper discusses the influence of the shear stresses which is in all probability important due to large thickness to span ratio characteristic for beams manufactured from brittle materials. This speculation is fortified by the fact that the actual thermodynamic force derived from the three-dimensional theory depends in addition of the direct tensile strains on the shear strains as well. It is further assumed that this dependence of the thermodynamic force on the shear strain explains the difference in the ultimate (rupture) bending moment in three- and four-point bending of concrete beams.

A more voluminous studies of concrete using simplified damage theories were published by Loland (1981) and Mazars (1984) who summarized their research on many aspects of the behavior of this important material which is not only often used but just as often misunderstood.

The problems of the rock mechanics were studied using the simple one-dimensional damage models by Passman et al (1980) and Grady and Kipp (1980). Their study of the rock fragmentation due to the impulsive loads was motivated by the model suggested earlier by Ericksen for the analysis of the flow of anisotropic fluids. Introduction of the 'crack inertia', extrinsic, intrinsic and inhomogeneity forces presents a new and worthwhile development of the basic theory. A carefull reexamination of all of these newly added terms as well as the determination of their physical meaning is a challenging task which has as yet to be tackled. A rate dependent degradation model for both concrete and rock (creep damage) was suggested by Mroz and Angelillo (1982) while Costin (1983) studied the uniaxial and triaxial tests on rock specimens. Finally, a computational fracture model for the rock describing nucleation, growth and coalescence of the brittle microcracks and production of rock fragments was formulated by McHugh et al (1980).

However, even in the uniaxial compression the response is not nearly one-dimensional since the normal to the microcrack field is perpendicular to the direction of the load vector. Consequently the need for a general three-dimensional theory became rather urgent even at the early stages of the development. The first two three-dimensional theories were formulated by Dougill et al (1977) and Dragon and Mroz (1979). In the first case the damage variable was assumed to be a scalar while the latter paper is an extension of an earlier effort by Vakulenko and M. Kachanov (1971) according to which the damage variable is a second order tensor formed as a dyadic product of the normal to the crack surface and the direction of the relative displacement of the microcrack faces. Such a model is in violation of the basic assumption that the response of the material depends only on the current state (i.e. current distribution and density of the microcracks) rather than the entire history (i.e. the manner in which the faces were displaced during the earlier stages of the deformation process). This model was later reformulated by M. Kachanov (1980). An entirely different model in which the inelastic strain is associated with the sliding, kinking and propagation of the cracks in a compressed rock was formulated by M. Kachanov (1982) (so called 'sliding crack' model).

A rather elaborate three-dimensional model for concrete based on the vectorial representation of the microcrack field was

developed by Krajcinovic and Fonseka (1981) and later refined by
Krajcinovic and Selvaraj (1983). A surprisingly good correlation
with the experimental results was obtained, despite a modest
number of additional material parameters (three), for uniaxial
tension and compression not only for the axial stress-strain curve
but also for the lateral and volumetric strain as a function of
the axial stress as well. The essentially same type of a theory
was extended by Suaris and Shah (1984) to the strain rate type of
the problem using the already mentioned ideas exploited by
Passmann et al (1980). The obtained results show again a very
good fit with the experimental results over a rather wide range
of the strain rates. A somewhat less ambitious scheme for the
spalling of concrete specimens was suggested by Krajcinovic and
Srinivasan (1983) who approximated the general threedimensional
wave propagation problem by a one-dimensional one neglecting the
lateral inertia and the influence of the lateral surfaces. The
ensuing model is amenable to a rather simple numerical solution
using the method of characteristics. The same model was
successfully used by Krajcinovic and Silva (t.b.p.) to explain the
experimental data on the large concrete breakwater armor blocks
subjected to impact and repeated impact loadings.

However, it was not long before it was realized that the
predictive capabilities of the models and theories based on a
single 'damage' variable regardless of its mathematical nature
fail in cases characterized by the interaction of many microcrack
fields or in cases when the microcrack fields change their status
from active to passive or vice versa.

Consider, for example, the uniaxial and biaxial tests on
plain concrete nad rock specimens. The rupture curves
according to several authors are summarized in Mazars (1981,
1984). The scatter, especially in the case of the biaxial
compression, reflects the sensitivity of the results to the
selection of the test procedure, precision of the measurements
as well as the composition of the concrete. Table 1 below
contains the directions of the principal microcrack fields for the
four characteristic points on the rupture curve.

Table 1: Damage fields in uniaxial and biaxial tension
and compression in (x,y) plane.

σ_x/σ_c	σ_y/σ_c	Damage in planes
0.1	0.1	$D(y,z) = D(x,z) > 0$
0.1	0	$D(y,z) > 0$
-1.	0	$D(x,z) = D(x,y) > 0$
-1.	-1.	$D(x,y) > 0$

where σ_c is the compressive strength of the cube. Also $\sigma_z = 0$.

Naturally, every principal microcrack field is surrounded by a bundle of fields of a somewhat lesser microcrack density. For example, in case of the uniaxial tension (second row of the Table 1) the microcracks will start nucleating in the (y,z) plane since the thermodynamic force R has a maximum value in the direction of the x-axis. The microcracking will commence as soon as the maximum R exceeds a certain threshold value R_0 depending on the adhesive strength of the mortar-aggregate interface or the cohesion of the mortar itself. If the external tensile force, i.e. the axial tensile strain, is further increased the affinities R will start exceeding R_0 in planes subtending an ever increasing angle with the the 'original damage plane' (y,z). In absence of carefull observations during the stages preceeding the final rupture it is difficult to speculate on the angle of the 'damage cone'. According to the experimental data by Halbauer et al (1973) and Evans and Wong (1983) that angle should not exceed 15 degrees.

In uniaxial compression the microfracturing takes place first in the planes containing the x-axis as a result of the tensile strain in the lateral direction. Therefore, the thermodynamic force associated with cleavage must be maximum in the directions perpendicular to the x-axis for compression and in the direction of the x-axis for tension. Moreover, the compressive strength must be approximatelly ten times larger than the tensile strength.

It might turn out that the combinations of the basic invariants containing products of two axial vectors $\underline{\omega}$ will not satisfy all of these conditions although the thermodynamic force R always has a maximum in the direction of the principal axes of the strain tensor. Thus, it will become necessary to use the invariant terms combining products of four axial microcrack distribution vectors. In the limit, as the number of active and operative microcrack fields tends to infinity, the dyadic products of four axial vectors $\underline{\omega}$ will form fourth order tensors. This conclusion based on a different, but equally valid, argument was reached by Chaboche (1979) who considered the damage induced anisotropy.

Consider next the uniaxial and biaxial compression tests. If the rate of the microcracking depends primarily on the lateral strain then the strength in biaxial compression should be approximately a half of the strength in uniaxial compression (as computed by Mazars 1984). However, the tests suggest that the strengths in uni- and biaxial compression are approximately equal. This can be explained only by the interaction between the microcrack fields which is substantially different in the two cases (see Table 1). The coupling between the microcrack fields can be introduced either through the expression for the free energy or through the flow potential (as in the slip theory). The first (and yet unpublished) computations using a model based on the representation of the microcrack field by 500 axial vectors demonstrate a very satisfactory fit of the experimental data. All four characteristic points from Table 1 are readily matched and the rupture curve ressembles the results obtained the three-parameter failure criterion suggested by Lade (1982). The results also suggest that the interaction through the flow potential is dominant and that the terms in HFE representing the interaction of axial vectors can be safely neglected. It is especially important to notice that the model captures the three-dimensional nature of the phenomenon and show proper trends with regard to the lateral and volumetric strain as well.

An essentially similar behavior in tension and compression is characteristic of rock specimens (see, for example, Jaeger and Cook 1979, p. 350). In its initial stages the stress-strain curve (Halbauer et al 1973, Jaeger and Cook 1979 etc.) reflects the influence of the porosity. This can be readily incorporated into the model along the lines suggested by Davison et al (1977) according to which the elastic properties depend on the volume density of the spheroidal cavities. In addition, the crystalline texture of the rocks, banding or random orientation of the constituent minerals, stability of the cementation, size and orientation of grains, compactness, pore pressure etc have a strong influence not only on the elastic parametrs of the rock but also on the nucleation and growth of the microcracks as well. Naturally, large discontinuities, such as faults or joints in large rock masses will typically dominate their response regardless of the material properties of the virgin material. As already stated above the behavior of rocks in a compression field is a substantially more complex phenomenon.

6. SUMMARY AND CONCLUSIONS

The presented abbreviated review of the applications of the continuum damage theory to the problems of the mechanics of rocks and concrete is only an attempt to summarize some of the developments occuring mainly in the last five to ten years. As a theory which closely mirrors the salient features of the microstructural kinematics, has a sound physical base and a rigorous mathematical and thermodynamical structure it will, hopefully, continue to attract the attention of the theoreticians and practitioners in the future as well. It is not often that a newly developing branch of the oldest discipline of physics contains such a cornucopia of problems of theoretical and practical importance. The theory may also lead to a rational resolution of many hitherto unresolved problems of material behavior.

Yet, a perhaps well deserved sense of euphoria should be tempered in view of many aspects of the theory deserving of a better, more deliberate and critical reappraisal. Among those important aspects which merit another look are the following ones:

(a) A rational definition of the exact domain within which the continuum damage theory could be expected to be valid. This question will center on the question of scale, i.e. the size of the defect which will separate a microdefect from the macrocrack. In effect the problem reduces to the determination of the onset of the localization, i.e. the point after which the fracture mechanics model becomes descriptive of the response of the solid.

(b) While the proposed approach, using the objective form of the Helmholtz free energy as a starting point, effectivelly reduces the number of material parameters a strong effort to identify these additional constants with the experimentally observable and measurable aspects of the state of the material is clearly necessary. The damage growth models suggested for metals by Curran et al (1977) or Leckie (1978) based on the metallurgical studies available in the literature, are perhaps one of the most appropriate avenues to take in formulating the 'damage laws' despite some shortcomings. A model like that has as yet to be formulated for concrete or various rocks. It appears almost certain that the fracture mechanics without proper modifications does not offer a direct insight into the laws of nucleation and growth of cracks of the size of a grain. Despite an apparent wealth of various measurements focused unfortunately on the ultimate rupture of a rock or concrete specimen there is a relative dearth of observations related to the details of the microcrack states at various points of the stress-strain curve (one of the stellar exceptions being the already mentioned paper by Halbauer et al 1973 as well as the microscopic studies summarized in Evans and Wong (1983)). Many other important points are in dire need of experimental confirmation. For example, is the microcrack density at the apex of the stress-strain curve (or at the onset of the localization) a material parameter or does it depend on the state of the stress as well? The manner in which the rate of the microcrack growth depends on the size and orientation of the grains, the texture of the rock or the casting direction of concrete, the type of curing etc. are just some of the problems that readily come to the mind without going into finer and more delicate problems of the strain rate sensitivity, size effect, time dependent deformations (creep rates), fatigue endurance limits to name just a few.

(c)   One of the most difficult problems centers on the nucleation and growth of the microcracks in the compressed planes. Along with it comes the question of the criteria separating active from operative from passive microcracks.   It is likely that the proposed algorithm will have to be revised and properly modified since the microcrack density growth might depend on the stresses explicitly and not only implicitly through the affinities. Additionally, the cleavage may cease to be the dominant mode of the microcrack growth when compared to sliding, kinking, branching etc.   The introduction of the friction will add to the complexity of the problem since the vector of the microcrack density rate will not be normal to the flow potential in the space of affinities.

(d)   Despite our sometimes boundless trust into the almost mystical powers of the conventional plasticity theory the question is whether the 'plastic' behavior of concrete and rocks is indeed related to the slips in the preferred crystallographic planes. The slip rupture lines showing in their utmost clarity in the photographs of the triaxial compression tests on the Carrara marble (Kovari and Tisa 1974) are a strong indication that at least under these conditions the theory of plasticity (or the slip theory) is a reasonable choice for the description of the phenomenon.   In contrast, the cyclic compressive loading of an unconfined rock or concrete specimen presents the different side of the problem.   For example, experiments on norite (Jaeger and Cook 1979, p. 82) show no residual strain but exhibit simultaneously rather large hysteresis loop with no flat (linear) segments to speak of.   Similar experiments on concrete clearly demonstrate that the 'unloading' paths are neither straight nor parallel to the initial (elastic) part of the loading segment of the stress-strain curve.   The secant modulus of concrete for the 'unloading' portion of the stress-strain curve decreases by almost 40% (Holmen 1975) prior to the fatigue failure.   However, the hysteresis loop is of significant girth and some residual strain is a matter of experimental record.   Consequently, the deformation process is not purely brittle.   The 'unloading' in the vernacular sense is not actually unloading since the curvature of the 'unloading' path suggests some energy dissipation.   However, instead of the classical plasticity in sense of some volume average of slips one is lead to believe that some kind of the small scale plasticity is the cause of the residual deformation. This speculation is fortified by the gradual increase in the volumetric strains measured by Shah and Chandra (1968) in case of a cyclic compression of an unconfined concrete specimen.   The gradual transition from a purely brittle (cleavage) rupture in planes perpendicular to the maximum tensile strain to the ductile rupture in the slip planes (Jaeger and Cook 1979, Kovari and Tisa 1974 etc) and the concomitant increase in the maximum stress with the increase in the lateral confinement is but another aspect of the overall problem waiting for an adequate and rational analytical description within the framework of the continuum mechanics.

In conclusion, this report should be understood as the author's assessment of the successes and shortcomings characterizing the present state of a rapidly developing discipline of the continuum mechanics. More importantly, it is an appeal for the initiation of a well designed and comprehensive experimental program aimed at the observation of the salient trends and modes of the microstructural kinetics under a host of different conditions and circumstances.

7. ACKNOWLEDGEMENTS

The author gratefully acknowledges the NSF (Geotechnical Program) grant to the University of Illinois at Chicago which supported the work on the presented research. The author would also like to express the gratitude to NATO (Scientific Affairs Division) for the travel grant which made possible numerous helpfull consultations with his French colleagues J. Lemaitre nad J. Mazars (ENSET, Cachan), J. L. Chaboche (ONERA, Chatillon) and G. Cailletaud (Ecole des Mines, Evry) clarifying many of the issues discussed above.

8. REFERENCES

1. Bazant Z.P., Kim S.S. (1979), Plastic-Fracturing Theory for Concrete, Journal of Engineering Mechanics Division, ASCE, 105, p. 429-446.

2. Budiansky B., O'Connell R.J. (1976), Elastic Moduli of a Cracked Solid, International Journal of Solids and Structures, 12, p. 81-97.

3. Chaboche J. L. (1978), Description Thermodynamique et Phenomenologique de la Viscoplasticite Cyclique avec Endommagement, ONERA, Publ. No. 1978-3.

4. Chaboche J. L. (1979), Concept of Effective Stress Applied to Problems of Elasticity and Viscoplasticity Combined with Anisotropic Damage (Le Concept de Contrainte Effective Applique a l'Elasticite et la Viscoplasticite en Presence d'un Endommagement Anisotrope), ONERA, T.P. No. 1979-77.

474

5. Costin L. S. (1983), A Microcrack Model for the Deformation and Failure of Brittle Rock, Journal of Geophysical Research, 88, p. 9485–9492.

6. Curran D. R., Seaman L., Shockey D. A., (1977) Dynamic Failure in Solids, Physics Today, 30, p. 46–55.

7. Davison L., Stevens A. L., Kipp M. E., (1977), Theory of Spall Damage Accumulation in Ductile Metals, Journal of Mechanics and Physics of Solids, 25, p. 11–28.

8. Dougill J. W., (1976), On Stable Progressively Fracturing Solids, ZAMP, 27, p. 423–437.

9. Dougill J. W., Lau J. C., Burt N. J., (1977), Toward a Theoretical Model for Progressive Failure and Softening in Rock, Concrete and Similar Material, ASCE–EMD 1976, University of Waterloo Press, p. 335–355.

9. Dragon A., Mroz Z., (1979), A Continuum Model for Plastic-Brittle Behavior of Rock and Concrete, International Journal of Engineering Science, 17, p. 121–137.

10. Evans B., Wong T. F., (1983), Shear Localization in Rocks Induced by Tectonic Deformation, in W. Prager Symposium on Mechanics of Geomaterials: Rocks, Concretes, Soils, Ed. Z.P. Bazant.

11. Goodman M. A., Cowin S. C., (1972), A Continuum Theory for Granular Material, Archive for Rational Mechanics and Analysis, 44, p. 249–266.

12. Grady D. E., Kipp M. E., (1980), Continuum Modelling of Explosive Fracture in Oil Shale, International Journal for Rock Mechanics, Mineral Sciences and Geomechanics, 17, p. 147–157.

13. Halbauer D. K., Wagner H., Cook N. G. W., (1973), Some Observations Concerning The Microscopic and Mechanical Behavior of Quartzite Specimens in Stiff, Triaxial Compression Tests, International Journal for Rock Mechanics, Mineral Sciences and Geomechanics, 10, p.713–726.

14. Halphen B., Son N. Q., (1975), On the Generalized Standard Materials, (Sur les Materiaux Standards Generalises), Journal de Mecanique, 14, p. 39–63.

15. Hill R., (1967), The Essential Structure of Constitutive Laws for Metal Composites and Polycrystals, Journal of Mechanics and Physics of Solids, 15, p. 79–95.

16. Hill R., Rice J. R., (1972), Constitutive Analysis of Elastic - Plastic Crystals at Arbitrary Strain, Journal of Mechanics and Physics of Solids, 20, p. 401-413.

17. Hoagland R. G., Hahn G. T., Rosenfeld A.R., (1973), Influence of Microstructure on Fracture Propagation in Rock, Rock Mechanics, 5, p. 77-106.

18. Holmen J. O., (1982), Fatigue of Concrete by Constant and Variable Amplitude Loading, in Fatigue of Concrete Structures, ed. S. P.Shah, ACI Publication SP-75.

19. Horii H., Nemat-Nasser S., (1983), Overall Moduli of Solids with Microcracks: Load Induced Anisotropy, Journal of Mechanics and Physics of Solids, 31, p. 155-171.

20. Hutchinson J. W., (1976), Elastic-Plastic Behavior of Polycrystalline Metals and Composites, Proceedings of the Royal Society, Series A, 319, p. 247-272.

21. Jaeger J. C., Cook N.G.W., (1979), Fundamentals of Rock Mechanics, third edition, Chapman and Hall, London.

22. Kachanov L. M., (1958), On the Creep Rupture Time, Izvestia AN SSSR, OTN, No. 8, p. 26-31 (in Russian).

23. Kachanov M. (1980), Continuum Model of Medium with Cracks, Journal of the Engineering Mechanics Division, ASCE 106, EM5, p. 1039-1051.

24. Kachanov M. L., (1982), A Microcrack Model of Rock Inelasticity, Parts I-III, Mechanics of Materials, 1, 19-41, 123-129.

25. Kestin J., Bataille J., (1978), Irreversible Thermodynamics of Continua and Internal Variables, in Continuum Models of Discrete systems, ed. Provan J.W., University of Waterloo Press, Study No. 12.

26. Kovari K., Tisa A., (1974), Upper and Lower Strength of Rocks in Triaxial Testing, (Hoechstfestigkeit und Restfestigkeit von Gesteinen im Triaxialversuch), ETH, Mitteilung Nr.26.

27. Krajcinovic D., (1979), A Distributed Damage Theory of Beams in Pure Bending, Journal of Applied Mechanics, 46, p.592-596.

28. Krajcinovic D., (1983), Constitutive Equations for Damaging Materials, Journal of Applied Mechanics, 50, p. 355-360.

29. Krajcinovic D., (t.b.p.), Continuous Damage Mechanics Revisited, submitted for publication.

30. Krajcinovic D., Fonseka G. U., (1981), The Continuous Damage Theory of Brittle Materials, Parts I,II, Journal of Applied Mechanics, 48, p. 809-824.

31. Krajcinovic D., Selvaraj S., (1983), Constitutive Equations for Concrete, in Proceedings of International Conference on Constitutive Laws for Engineering Materials, ed. Desai C.S., Gallagher R.H., Tucson Az.

32. Krajcinovic D., Sestan J. Z., (1983), Design Methods for Analysis of Damage Accumulating Structures, Proceedings of 7th International SMIRT Conference, Vol.L, 8/4, Chicago, IL.

33. Krajcinovic D., Silva M. A. G., (t.b.p.), Damage to Colliding Concrete Cubes, to appear in Journal of Structural Mechanics Division, ASCE.

34. Krajcinovic D., Srinivasan M. G., (1983), Dynamic Fracture of Concrete, in Time-Dependent Failure Mechanisms and Assessment Methodologies, ed. Early J.G., Cambridge University Press.

35. Kroener E., (1962), Dislocation: A New Concept in the Continuum Theory of Plasticity, Journal of Mathematics and Physics, 42, p. 27-37.

36. Lade P. W., (1982), Three-Parameter Failure Criterion for Concrete, Journal of the Engineering Mechanics Division, ASCE, 108, p.850-863.

37. Leckie F. A., (1978), The Constitutive Equations of Continuum Creep Damage Mechanics, Philosophical Transaction of the Royal Society, Series A, 288, p. 27-47.

38. Lemaitre J., Chaboche J. L., (1978), Phenomenological Aspects of Rupture through Damage, (Aspect Phenomenologique de la Rupture par Endommagement), Journal de Mecanique Appliquee, 2, p. 317-365.

39. Lemaitre J., Mazars J., (1982), Application of the Damage Theory in Nonlinear Response and Rupture of Concrete

Structures, (Application de la theorie de l'endommagement au comportement non lineare et a la rupture du beton de structure), Annales de l'ITBTP, No. 401.

40. Litewka A., Sawczuk A., (1981), A Yield Criterion for Perforated Sheets, Ingenieur-Archiv, 50, p. 393-400.

41. Loland K. E., (1981), Mathematical Modelling of Deformational and Fracture Properties of Concrete Band on Principles of Damage, (Matematisk Modellering av Betongens Deformasjons-og Bruddegenskaper Basert Pa Skademekaniske Prinsipper), University of Trondheim, Rapport BML 81.101.

42. Margolin L. G., (1983), Elastic Moduli of a Cracked Body, International Journal of Fracture, 22, p. 65-79.

43. Mazars J., (1981), Mechanical Damage and Fracture of Concrete Structures, in Advances in Fracture Research (Fracture 81) ed. Francois D., Vol. 4, p. 1499-1506.

44. Mazars J., (1984), Application of the Damage Mechanics in Non-linear Response and Rupture of Concrete Structures, (Application de la Mecanique de l'Endommagement au Comportement Non Lineare te a la Rupture du Beton de Structure), These de Doctorat d'Etat, Universite Pierre et Marie Curie, Paris 6.

45. McHugh S. L., Curan D. R., Seaman L., (1980), The NAG-FRAG Computational Fracture Model and its Use for Simulating Fragmentation and Fracture, SESA meeting, Fort Lauerdale FL.

46. Mroz Z., Angellilo M., (1982), Rate-Dependent Degradation Model for Concrete and Rock, in Numerical Models in Geomechanics, ed. Dungar R., A.A. Balkema, Roterdam.

47. Murakami S., Ohno N., (1981), A Continuum Theory of Creep and Creep Damage, in Creep in Structures, ed. Ponter A.R.S. Ponter, Springer-Verlag.

48. Nemat-Nasser S., (1975), On Nonequilibrium Thermo-Dynamics of Viscoelasticity: Inelastic Potentials and Normality Conditions, in Mechanics of Visco-Elastic Media and Bodies, ed. Hult J., Springer-Verlag.

49. Nicholson D. W., (1981), Constitutive Model for Rapidly Damaging Structural Materials, Acta Mechanica, 39, p. 195-205.

478

50. Ouchterlony F., (1983), A Distributed Damage Approach to Combined Bending and Axial Loading of Rock Beams, in Mechanical Behavior of Materials, ed. Carlsson J., Ohlson N.G., Volume 2, Pergamon Press.

51. Pak A. P., Trapeznikov L. P., (1981), Experimental Investigations Based on the Griffith-Irwin Theory Processes of the Crack Development in Concrete, in Advances in Fracture Research (Fracture 81), Vol. 4, ed. Francois D., p. 1531-1539.

52. Passman S. L., Grady D. E., Rundle J. B., (1980), The Role of Inertia in the Fracture of Rock, Journal of Applied Physics, 51, p. 4070-4075.

53. Peirce D., Asaro R. J., Needleman A., (1982), An Analysis of Nonuniform and Localized Deformation in Ductile Single Crystals, Acta Metallurgica, 30, p. 1087-1119.

54. Rice J. R., (1971), Inelastic Constitutive Relations for Solids: An Internal Variable Theory and its Application to Metal Plasticity, Journal of Mechanics and Physics of Solids, 19, p. 433-455.

55. Rice J. R., (1975), Continuum Mechanics and Thermodynamics of Plasticity in Relation to Microscale Deformation Mechanisms, in Constitutive Equations in Plasticity, ed. Argon A.S., the MIT Press.

56. Rusinko K. N., (1981), Theory of Plasticity and Nonstationary Creep, Lvov (in Russian).

57. Shah S. P., Chandra S., (1968), Critical Stress, Volume Change and Microcracking of Concrete, Journal of the American Concrete Institute, 65, p. 770-781.

58. Spencer A. J. M., (1971), Theory of Invariants, in Continuum Physics, Vol. I, ed. Eringen A.C., Academic Press.

59. Suaris W., Shah S. P., (1984), Rate-Sensitive Damage Theory for Brittle Solids, Journal of Engineering Mechanics, ASCE, 110, p. 985-997.

60. Tapponier P., Brace W. F., (1976), Development of Stress-Induced Microcracks in Westerly Granite, International Journal of Rock Mechanics and Mineral Science, 13, p. 103-112.

61. Vakulenko A. A., Kachanov M. L., (1971), Continuum Theory of Cracked Media, Mekhanika Tverdogo Tela, 6, p. 159-166.

62. Wong T. F., (1982), Micromechanics of Faulting in Westerly Granite, International Journal of Rock Mechanics and Mineral Sciences, 19, p. 49-64.

APPLICATION OF FRACTURE MECHANICS
TO CEMENTITIOUS COMPOSITES
NATO-ARW - September 4-7, 1984
Northwestern University, U.S.A.
S. P. Shah, Editor

A MICROSTRUCTURAL APPROACH TO FRACTURE OF CONCRETE

L.Seaman, J.Gran and D.R.Curran

SRI International
333 Ravenswood Avenue
Menlo Park,CA - 94025.

ABSTRACT

A mathematical model for concrete has been developed by combining a multiple-plane plasticity and fracture model with a cap model for compaction. The model provides for the anisotropy of tensile and shear behavior and also for nonlinear compaction. The model is primarily intended to represent the response of concrete structures to impact loading.

The model shows the usual differences in loading and unloading pressure-volume paths for compaction, and also shear stresses in the model enhance the compaction process. Shear stresses on each of seven planes are computed using a Mohr-Coulomb yield curve. Shear damage on a plane causes loss of shearing and tensile strength specifically on that plane and not on adjacent planes. Provision is made for rate-dependence in the shearing strength. Initially, the model has a macro character for ease in matching experimental data, but is extendable later to have micromechanical features for generality and for guidance in scaling calculations.

INTRODUCTION

A new model with the advantages of both multiple-plane and cap models is presented for use in simulating concrete behavior. The model provides for the anisotropy of tensile and shear behavior and also for nonlinear compaction. Although our original goal was to develop a micro model, the present model largely emphasizes macro features. However, the model is written in a manner that will allow the addition of micro features later.

The objective was to develop a model to represent the response of concrete structures to impact loading. Therefore, we needed a model for the high rate, high stress processes, one that could handle compaction and damage under shear and tensile loadings in a realistic way. We chose to disregard minor nonlinearities and the complexity of cyclic loadings that might be important in other applications.

With this objective in mind we planned the following features for the MPCap model:

(1) Shear-enhanced compaction process. As with any porous material, there must be different loading and unloading paths for the pressure-volume curve. The compaction must be shear-enhanced because shear stress induces compaction more readily than pure hydrostatic stress states do.

(2) Mohr-Coulomb shear strength behavior. Shear strength is a function of the normal stress on the shear plane. Strength is not simply related to the average pressure as it is in a Drucker-Prager [1] model.

(3) Anisotropic shear damage. Shear damage on a plane causes loss of shearing and tensile strength specifically on that plane and not on adjacent planes.

(4) Rate-dependent compaction strength, shearing strength, and tensile strength. These types of strength are all strongly rate dependent, especially in the strain rate range above 1 per second.

(5) A macro model extendable to micro features. The model should initially have a macro character for ease in matching experimental data, but be extendable later to have micromechanical features for generality and for guidance in scaling problems.

In the following discussion, we first describe relevant concrete data that guided us in the choice of the model features. Then several models are mentioned that have been used for concrete and other materials and that contain features of the present model. Finally, we outline the model and derive some of the major features, including the solution procedure.

BACKGROUND

A brief review of concrete data is given as a basis for the development of the present model. Then several multiple-plane and cap models are described, since the present model is an outgrowth of models of these types. The emphasis here is on macro models, that is, continuum models in which the material is considered to be homogeneous; micro structural aspects are not treated explicitly. However, some of the models mentioned below are micro models. This selection reflects the position of the MPCap model—it is a macro model in its initial formulation, but micro features can be easily inserted.

A. Experimental Concrete Properties

The following discussion of concrete behavior is taken largely from a text and a paper by W. F. Chen on plasticity in concrete [2,3]. At several points we have added our comments or disagreements with his descriptions.

Concrete exhibits nonlinear elastic behavior up to some stress level at which cracking, plastic flow and crushing are observed. The plastic flow, and possibly crushing, can be treated by plasticity models. Chen suggests using an associated flow rule, but there are no data to substantiate this choice. Most models of concrete show a path-dependence, although the data are from proportional loading tests that cannot distinguish whether there is path-dependence.

In describing a choice of yield models, Chen says that the Mises and Tresca yield models give about the same results. This conclusion may be valid for beams and columns, but we doubt that it is true for massive concrete, and for dynamic conditions in which significant confinement is present. Under confined conditions, the pressures and normal stresses can be much larger than the usual values. Lindberg and Schwer [4] have studied three different models to learn how each would predict strength and stress states around tunnels in rock, a high-confinement situation (see Figure 1). These include a Mohr-Coulomb model (strength or stress difference is proportional to the normal stress on each plane), Mises-Schleicher [5] (like Drucker-Prager [1]) model (in which strength or equivalent stress is proportional to pressure), and a Tresca model. They have shown that the Mohr-Coulomb model gives very different results from those of the Mises model. In addition, the Mohr-Coulomb model represents the behavior of tunnels much better.

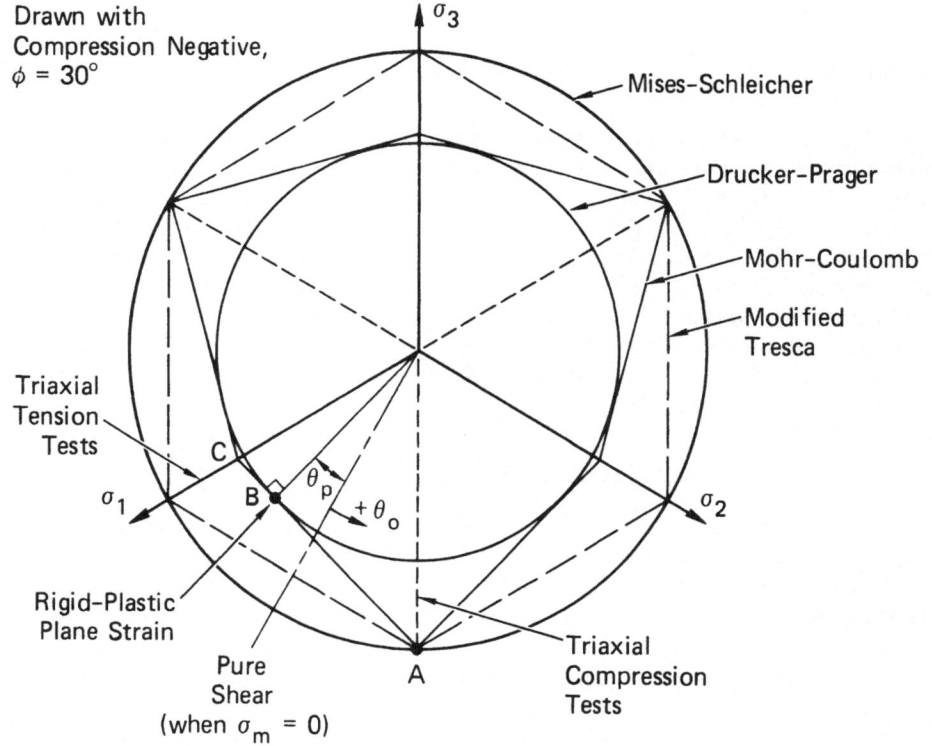

MA-317522-93

FIGURE 1   DISPLAY OF SEVERAL YIELD SURFACES ON A PLANE OF CONSTANT MEAN STRESS IN PRINCIPAL STRESS SPACE

Chen states that the hydrostatic compression process is important, although it is not considered in the traditional analysis of beams and columns. We note that the nonlinear crush (pressure-volume) curve that represents the hydrostat is similar to those we have seen for most porous materials, such as soils [6], rocks [7], aluminum [8], and woven composites.

Poisson's ratio varies, starting at 0.1 to 0.2 for small stresses and approaching 0.5 near failure. (Failure in this context probably means fragmentation.)

Although fracture in concrete occurs by a crack-growth process, concrete cannot simply be represented by an elastic-brittle

model. At least under compressive loadings, large plastic flows
(up to 6% according to Chen) may occur before fracture. In our
own work with the rapid collapse or expansion of concrete cylinders or spheres under explosive loading, we have qualitative
observations of 50% to 100% strain. The mean stress was several
GPa during the period of straining. In some of our experiments,
no fracture occurred even at these large strains. Chen noted that
the character of the fracture depends on the state of stress, and
that observation would agree with our results. Bazant [9] has
noted that fracture often occurs by the propagation of bands of
cracks, rather than by growth of a single crack. For this reason,
it is unlikely that the traditional fracture mechanics approach,
based on the growth of a single sharp crack, would be able to
treat fracture adequately. This observation agrees with our own
examinations of fracture is steels [10], shale [11], concrete
[12], and polycarbonate [13].

Isotropic models are generally used to describe all aspects
of the constitutive response of concrete, even for fracture, although tensile and shear damage are clearly orientation-dependent.
With such models the failure process can be treated in stress
invariant space. By using invariants, there is a natural transition from the usual one-dimensional test data base to the three-
dimensional state needed for calculations. However, it is not
clear that the three-dimensional predictions of these models are
correct.

Our conclusions from the concrete observations are that we
have insufficient data to generate a satisfactory model for three-
dimensional phenomena. Most of the data are from proportional
loading tests, or from tests in which the stress paths are only
slightly different from proportional loading (such as uniaxial
strain). Such data sample only a small portion of the stress
paths that are actually exercised in structural problems. Because
of this limitation of the data, all models can only represent fits
to standard experimental data and are extrapolations to other
loading paths. The differences between various models must lie in
the degree to which they are able to represent the available data
and the manner in which the extrapolation is made. For our MPCap
model we have chosen to make the extrapolation in the way that
seemed most physically reasonable to us. We have chosen to
emphasize direction-dependence, rate-dependence, and compaction
processes, and to simplify some of the other possible features,
such as nonlinear elastic behavior and hysteresis loops.

B.   Multiple-Plane Models

A multiple-plane model, as used here, provides a number of
specific planes or orientations on which yielding can occur. Such
models have often been used for single crystal studies in which

the planes of interest were known initially. Other models have treated polycrystalline materials. In these, a large number of planes were provided to represent approximately the continuum of possible orientations for slip. In either case (single or multiple crystals), calculations allowed slip only on the specific planes.

An early model of the multiple-plane type was given by Batdorf and Budiansky [14] for use in metals. They were successful in simulating yielding in a polycrystalline material. Como, Grimaldi, and D'Agostino [15,16] developed a multiple-plane model and simulated the work-hardening, anisotropic yielding, and Bauschinger effect observed in experiments of Nagdhi et al. [17]. Bazant has recently begun formulation of a multiple-plane model [18] to represent anisotropic plastic flow and fracture in concrete. For this model he has initially dealt with the accuracies to be expected for different numbers and orientations of planes.

We have used multiple-plane models to represent high-rate brittle fracture [19] and shear banding [20,21]. The model for brittle fracture (BFRACT) uses one or more planes to represent crack orientations. These discrete planes provide for the strong anisotropy that develops during brittle fracture. Our shear band model (SHEAR4) provides for seven discrete planes on which plastic flow, shear banding or shear cracking, and tensile fracture can occur. As with BFRACT, the shearing model permits a very strong anisotropy to develop as damage occurs on the planes. This anisotropy allows fracture to take place in one direction, but maintains full strength in orthogonal directions. Both these models are micromechanical because they provide explicitly for the nucleation and growth of microfractures and for the effect of these microprocesses on the macro stress-strain relations. The SHEAR4 model is used as one component of the new MPCap model for concrete.

C. Cap Models

A cap model has a second yield surface that primarily governs the pressure, rather than the shear stress. A typical pair of yield curves for a cap model are shown in Figure 2. The cap curve is allowed to move to the right as loading occurs.

Drucker [22] introduced the cap model idea initially for soils. The movable cap yield curve provided a way to represent compaction of porous materials as a type of work-hardening behavior. The cap also permits a shear-enhanced compaction that is observed in porous materials.

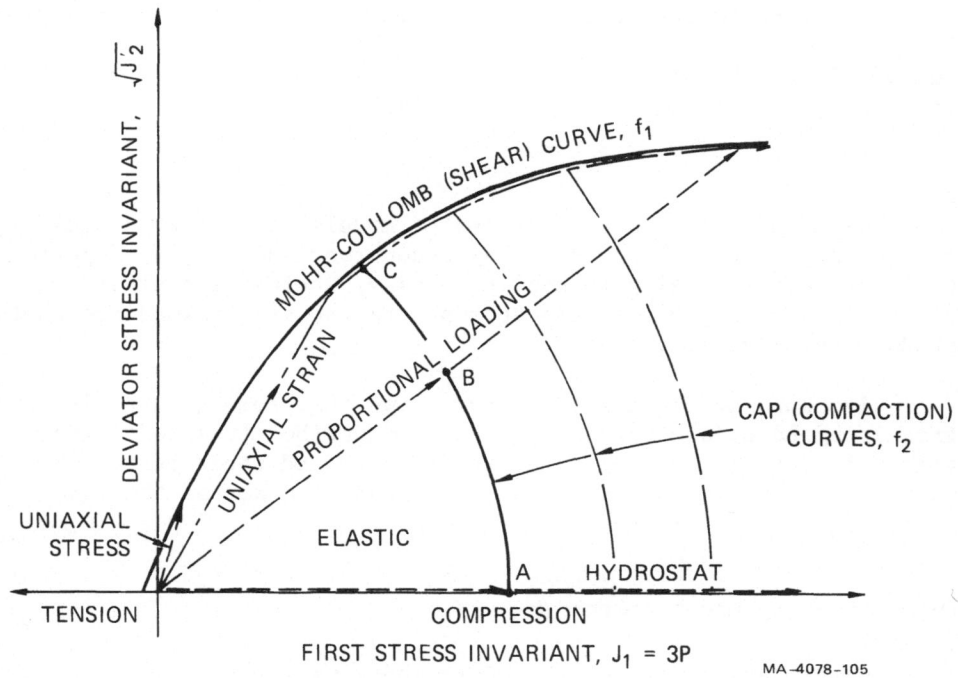

FIGURE 2   MOHR-COULOMB AND CAP YIELD CURVES WITH FOUR LOADING
PATHS THAT INTERSECT THE YIELD CURVES

Modern developments in cap models have been provided by many
researchers: a representative model has been given by Sandler et
al. [23]. Their model has a Mohr-Coulomb curve for shear strength
as a function of pressure, plus a movable cap curve that is
tangent at one point to the Mohr-Coulomb curve. This cap model is
used with a normality flow law in some regions, but not in all, to
avoid nonphysical results. The main drawbacks of the cap model
are the difficulty in fitting it to data and the basically
isotropic nature of the model.

For simulating projectile penetrations into reinforced
concrete, Gupta and Seaman [12] generated an isotropic cap model
with a Mohr-Coulomb curve and a movable cap. The main advantages
of this model (CAP1) are that all the parameters have clear
physical interpretations. The physical meaning of the parameters
provides benefits:

(1)   When the user wishes to extend the model to treat a
      similar material, the user knows what parameters to
      alter and what the direction and magnitude of the
      alteration mean.

(2)   The model can be readily fitted to standard data,
      because the parameters pertain to curves from such data.

The model has the disadvantage of isotropy, so that fracture can
be treated only in a very simplified fashion.

D.   Other Models

Bazant has introduced a series of models to represent various
aspects of concrete behavior. The endochronic model [24] appeared
to be suited especially to cyclic loading, whereas the plastic-
fracturing model [25] is more appropriate for single loadings that
produce tensile damage.

Wittmann and Zaitsev [26] have shown that some of the microfracturing behavior of concrete can be described by a model that
treats in detail the aggregate, the cement, and their bond. They
allowed cracks to propagate along the bonds and across the cement
matrix in response to the local stresses.

DESCRIPTION OF MPCAP PROPERTIES

In the development of the MPCap model we are treating
separately the shearing and volumetric components of the stress.
Below we present first the shear stress solution procedure, and
then the volumetric solution. This separation is possible because
we regard the intact, solid material as isotropic: it is the
presence of shearing and tensile damage with a nonuniform spatial
orientation that causes the apparent anisotropy.

A.   Shear Stress Formulation

The shear strength and shear stress formulations are based on
the multiple-plane concept. A set of orientations are chosen
initially for the shear strength calculations. Then all shear
stress operations are made on these planes: plastic flow, rate-
dependence, shear cracking, tensile cracking. A possible set of
such planes is shown in Figure 3. All these orientations can
occur at every point in the material; they are not distributed in
location as in the figure. We first present the choice of planes
for use in the model, and then describe the stress-strain calculations that occur on each plane.

Table 1 lists some of the possible sets of orientations that
could be used in the model. Each of these sets of orientations is
complete and isotropic: no one quadrant around a point is
favored. For the MPCap model we have selected the third set with
seven planes in the two-dimensional case. These orientations are
shown in Figure 4. The full nine planes of the three-dimensional

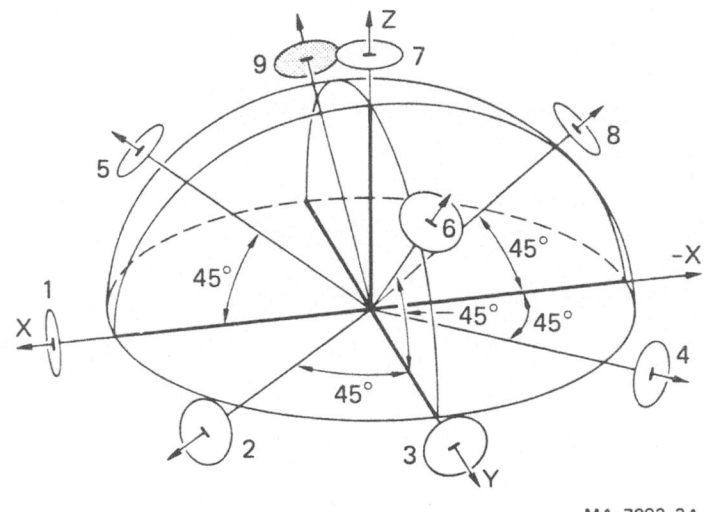

MA-7893-3A

FIGURE 3   RELATIVE LOCATIONS OF THE COORDINATE
DIRECTIONS AND INITIAL ORIENTATION
OF THE SHEARING PLANES

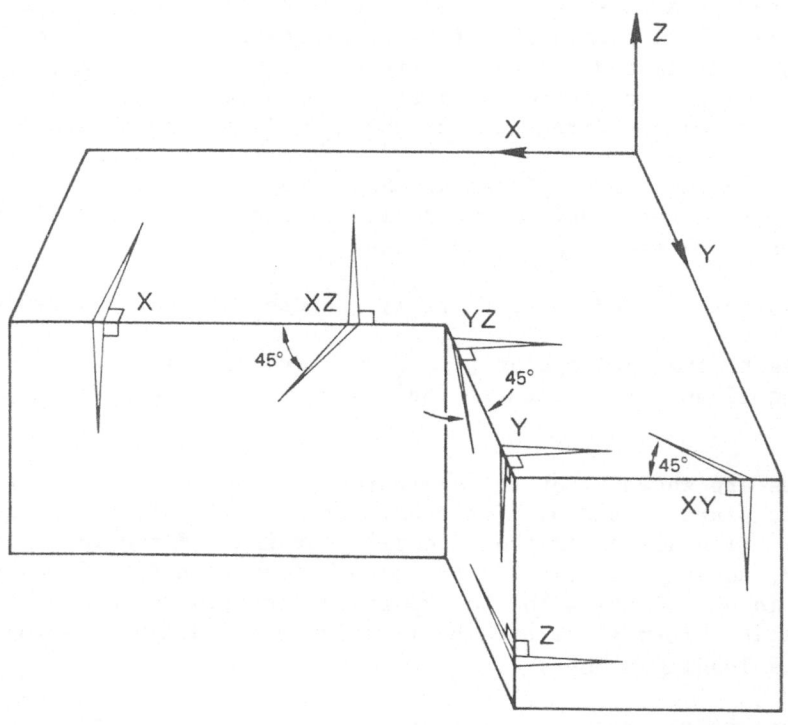

MA-7893-171

FIGURE 4   MULTIPLE SHEAR PLANES CONSIDERED IN THE MPCAP MODEL

Table 1

ORIENTATION OPTIONS FOR SHEAR BAND MODELING[a]

No. of Planes 2-D	3-D	Coordinate Axes	45° Planes[b]	Equiangular Planes[d]
3	3	X, Y, Z	None	None
5	7	X, Y, Z	None	(X,Y,Z), (-X,Y,Z) [(X,-Y,Z), (-X,-Y,Z)][c]
7	9	X, Y, Z	XY, YZ, XZ, (-X,Y), [(-Y,Z), (-X,Z)][c]	None
9	13	X, Y, Z	XY, YZ, XZ, (-X,Y), [(-Y,Z) (-X,Z)][c]	(X,Y,Z), (-X,Y,Z), [(X,-Y,Z), (-X,-Y,Z)][c]

[a]The 2-D state is taken as either plane strain or axisymmetry. Here there are no shears in the YZ and XZ directions. This symmetry makes certain that the 3-D planes are all identical: for example, YZ and (-Y,Z). Therefore, for 2-D problems, these two planes can be treated as a single plane. Orientations are designated by the directions of the normals to the damage planes.

[b]Normals to these band planes are at 45 degrees between the coordinate directions. For example, -X,Y means the normal is at 45 degrees between the -X and Y axes.

[c]Orientations listed in brackets are needed only for 3-D symmetry.

[d]Normals to the equiangular planes are directed along lines that have equal angles to each of the listed coordinate axes.

problems are shown. For two dimensions, planes 8 and 9 are identical to planes 5 and 6. For a calculation, the planes are given initial orientations in the external coordinate directions. However, during the calculations each orientation is allowed to rotate in accordance with the imposed deformations: hence, the orientations themselves and the relative orientations change during a loading process.

The shear stress calculations treat elastic, plastic, and rate-dependent processes. First, a standard incremental elastic calculation of the stress tensor is made, using the stress tensor from the previous time increment and the strain increment tensor.

Then the shear and normal stresses are computed on each orientation. The following paragraphs treat mainly this shear stress calculation on the planes: later the combination of the tensor computation and the computation by planes is described.

For the plastic and rate-dependent behavior on each orientation, a work-hardening Coulomb model is used to describe the shear strength. This strength, τ_{iy} on the ith plane is a sum of two functions—a cohesion that varies with shear strain, and a normal stress effect that acts in the Coulomb fashion, as shown below.

$$\tau_{iy} = Y_s \, (\varepsilon_i^p) + S_{in} \tan \phi_i \tag{1}$$

Here Y_s is the usual shear yield strength and work-hardening function, except that it pertains only to the ith plane, and work-hardening is a function only of the plastic strain on that plane. S_{in} is the normal stress on the ith plane. ϕ_i is the angle of friction, which may be a function of the plastic strain and of the shearing and fracture damage. Figure 5 shows some possible yield functions and their variations with plastic strain and normal stress. This proposed strength variation preserves the standard form, but generalizes it to allow matching more complex behavior. Because of the two functional dependencies in Eq. (1), it is evident that the strength will be different on each plane and that it will develop based on the history of straining on the plane.

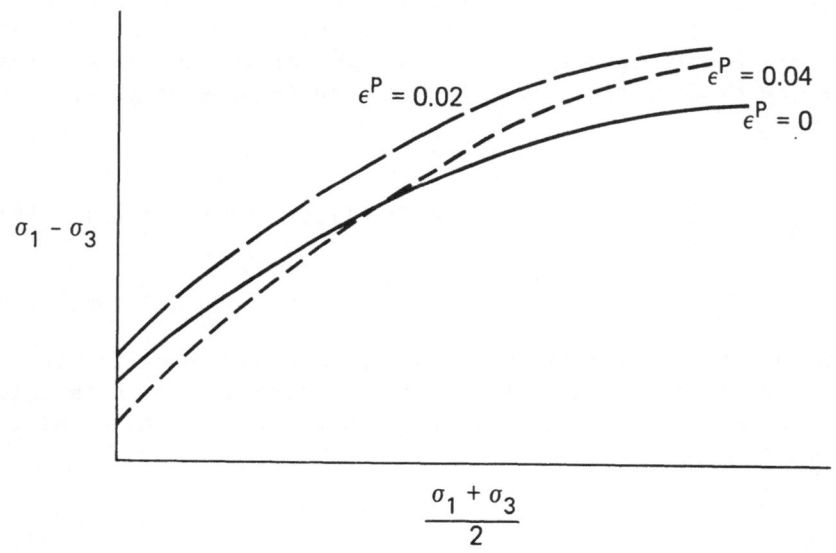

JA-314583-35

FIGURE 5   CHANGES IN A MOHR-COULOMB CURVE AS A FUNCTION OF PLASTIC STRAIN

Damage on a plane is evidenced by a reduced deviatoric strength and no tensile strength. Yet under compression, the plane still can maintain a shearing load.

With a prescription of the shear strength on each plane, we can now account for plastic flow and rate-dependence in computing the shear stress. These features are combined into a rate-dependent model of the linear type proposed by Malvern [27].

$$\frac{d\sigma_i'}{dt} = 2G \frac{d\varepsilon_i'}{dt} - \frac{\sigma_i' - \tau_{iy}}{T} \tag{2}$$

where σ_i' is the vector sum of the shear stress on the ith orientation plane, G is the shear modulus, ε_i' is the tensor shear strain acting along the ith plane, T is the time constant, and τ_{iy} is the current shear yield strength on the ith plane. This one-dimensional expression for shear stress will be solved initially and then later we will account for the two-dimensional nature of the shear stress on the plane. The yield strength is assumed to work-harden linearly during the time step, Δt, as follows:

$$\tau_{iy} = \tau_{iyo} + \frac{\Delta\tau}{\Delta t} t \tag{3}$$

where τ_{iyo} is the shear yield strength at the beginning of the time interval, $\Delta\tau$ is the incremental increase in yield strength, and t is the time running from 0 to Δt during the time interval. To perform the integration of Eq. (2) to obtain the shear stress at the end of the interval, multiply each term by $\exp(t/T)$. The solution is

$$\sigma_i' = \sigma_{io}' X + (2GT\dot{\varepsilon}_i' + \tau_{iyo})(1 - X) + \Delta\tau_{iy}\left[1 - \frac{T}{\Delta t}(1 - X)\right] \tag{4}$$

where $\dot{\varepsilon}_i'$ is the applied deviator strain rate, and $X = \exp(\Delta t/T)$.

Having the stress from Eq. (4), we can now determine the plastic strain that occurs in the time interval. For this calculation we separate the applied strain into elastic and plastic components:

$$\Delta\varepsilon' = \Delta\varepsilon'^E + \Delta\varepsilon'^P \tag{5}$$

and use the elastic stress-strain relation:

$$\sigma_i' = \sigma_{io}' + 2G\Delta\epsilon_i'^E \tag{6}$$

By subtracting the two equations for σ_i', (4) and (6), and using the expression for the plastic strain implied in Eq. (5), we can obtain

$$(\sigma_{io}' - 2GT\dot{\epsilon}_i' - \tau_{iyo})(1 - X) - \Delta\tau_{iy}\left[1 - \frac{T}{\Delta t}(1 - X)\right]$$

$$+ 2G\left(\Delta\epsilon_i' - \Delta\epsilon_i'^P\right) = 0 \tag{7}$$

Then we can solve for the plastic strain:

$$\Delta\epsilon_i'^P = \frac{\left(\sigma_{io}' - \tau_{iyo}\right)(1 - X) + 2G\dot{\epsilon}_i'\left[\Delta t - T(1 - X)\right]}{2G + M_w\left[1 - \frac{T}{\Delta t}(1 - X)\right]} \tag{8}$$

where $M_w = \Delta\tau/\Delta\epsilon_i'^P$ is a work-hardening modulus. Thus, the plastic strain absorbed on each plane can be computed.

Now that we have completed the one-dimensional calculation, we can explore the actual two-dimensional shear stress state on the ith plane. Figure 6 is a sketch of the plane with two shear stresses in orthogonal coordinates. The stress σ_{io}' existed at the beginning of the time interval and the stress $\sigma_i'^E$ was computed elastically for the end of the interval. Some portion of the difference between these two stresses represents an elastic change and another portion represents a plastic change. Only the plastic change will be affected by the plasticity and rate-dependent processes. (Note that on the plane we are considering a normality or radial return rule for the plastic flow.) Hence we wish to perform our stress-relaxation calculation along the direction of the radial vector to $\sigma_i'^E$. Thus we can use Eq. (2) and its solution Eq. (8), once we correctly interpret its terms in the two-dimensional context. The first term in Eq. (8) on the right-hand side is replaced by

$$2G\frac{d\epsilon_i'}{dt} = \frac{\sigma_i'^E - \sigma_{io}'}{\Delta t} \tag{9}$$

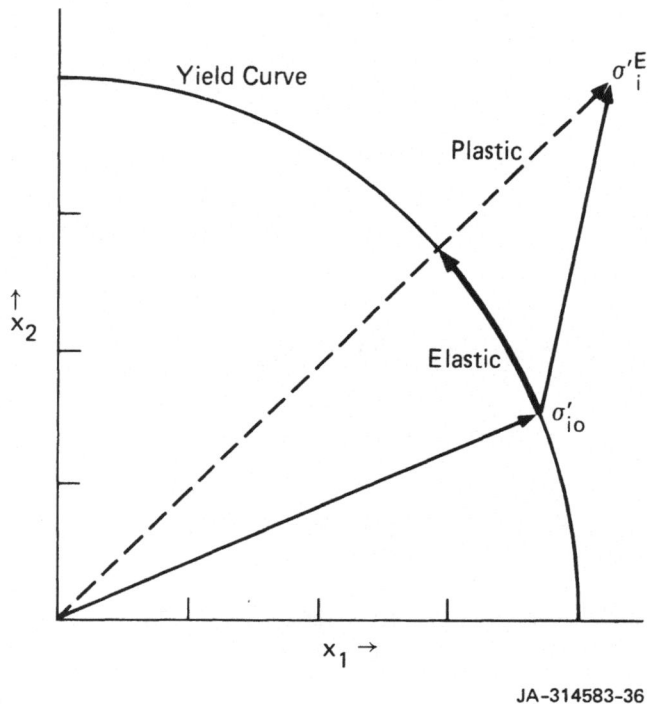

JA-314583-36

FIGURE 6 ONE OF THE ORIENTATION PLANES SHOWING LOADING
FROM AN INITIAL STATE TO A FINAL STATE BY A
COMBINATION OF ELASTIC AND PLASTIC STRAINS

Now we can interpret the plastic strain computed in Eq. (8) as the resultant plastic change in the direction of $\sigma_i^{\prime E}$. The individual components of plastic strain can be computed because the direction and magnitude of the resultant are now known.

B.   Multiple-Plane Relaxation Solution Procedure

In the foregoing derivation a one-dimensional solution was developed for computing the shear stress and plastic strain on each plane. Now it is necessary to account for all the active planes (those with stresses above the yield strength) simultaneously to determine the stress and plastic strain tensors.

The stress solution procedure must fit into a standard finite difference or finite element computer program. Therefore, the stress will be computed at each cell or element at many time intervals. For each time interval, our solution procedure must take the preceding stress tensor and the strain increment tensor, and compute a new stress tensor appropriate to the current time. For this solution procedure we have in mind the following sequence of operations:

(1) Compute the new stress tensor from the imposed strain increment tensor, treating the strains as entirely elastic. Include here the pressure computed from the hydrostatic calculations described later.

(2) Transform the stresses to determine the normal and shear stresses (two components) acting on each of the designated planes.

(3) Compute the plastic shear strains on each of the planes.

(4) Transform the plastic shear strains from each plane to the tensor orientation and sum to form the terms of the tensor.

(5) Recompute the stress tensor, using for the elastic strain increment tensor the imposed strain minus the plastic strain tensor.

In this sequence steps (1), (2), (4), and (5) are standard operations; step (3) is more complex, and hence is the topic of this section.

Before presenting a method for obtaining the stress solution on each plane, we will describe some other attempts to solve this problem and some of our requirements for an acceptable scheme. We do not believe that an exact solution method is available in general, although there are methods that can handle special cases. Taylor [28] showed that five planes are needed in general to absorb an arbitrary strain. For fewer planes, no solution is available; for more planes, there is no unique solution. Peirce et al. [29] have solved a special case for two slip planes in a two-dimensional geometry. We are attempting to determine the stress state with a large (and indefinite) number of planes without a fixed orientational relation between the planes and with different strengths on each plane. We are aiming for an engineering solution, that is, not an exact solution, but a physically realistic approximation.

To clarify the purpose for the solution procedure, let us explore several multiple plane cases and try to estimate what the physically correct result should be in each case. Our solution procedure must be able to handle all of these with reasonable accuracy. Consider the case in which there are two orthogonal planes with equal shear stresses and yield strengths on each plane. Then if one plane were allowed to yield in the one-dimensional fashion of the equations in the preceding section, that plane would absorb all the plastic strain and reduce the shear stresses on both planes to the yield level. Probably a more reasonable solution would be obtained if each plane received just one-half of the plastic strain.

Now consider a second case with two planes, but here one plane has an elastically computed shear stress twice as high as the shear on the other plane. Under a gradually applied (quasi-static) loading, the first plane might absorb all the plastic strain. This plastic strain would relax both shear stresses, and probably the shear stress on the second plane would actually never reach yield. If the loading were applied rapidly enough, the shear on the second plane would also exceed yield and so some plastic strain should also accumulate on that plane. Thus, we would expect that the plastic strain should be more uniformly distributed over the planes for higher rate loadings.

A relaxation solution procedure is proposed for determining the shear stresses on all planes simultaneously. First, we will treat the case of rate-independent plastic flow, and later modify this analysis to handle the rate-dependent plastic processes in the preceding section. The relaxation process begins after step (2) above, in which the elastically computed stresses have been determined on each plane. Then we move from plane to plane and use Eq. (8) (or its rate-independent equivalent) to determine the plastic strain that would allow complete relaxation on each plane (the plastic strain requested by each plane). This requested plastic strain is then multiplied by a fraction f. The value of f is usually in the range of 1/2 to 1, depending on the number of active planes and on the precision we are intending. With these fractions of the plastic strain, the calculation proceeds on through steps (4) and (5), and then back to step (2). Now at step (3) again, the shear stresses on all planes are compared with the yield strength on that plane. If yield has not been exceeded on any plane, the solution is complete. However, if the shear stress exceeds yield on one or more planes, the preceding steps are repeated until all the excess shear stress has been relaxed.

The precision of the preceding relaxation solution is controlled by three factors:

(1) The imposed strain increment size. If the strains from the calling program are too large, these strains can be divided into subincrements and imposed in a series of subcycles. The acceptable strain size can be related to the acceptable precision on the shear stress.

(2) The fraction f of the requested plastic strain absorbed on each cycle. Large values near 1.0 will produce more rapid convergence, but also more over-relaxation (excessive plastic strain). We have

used $f = 1/\sqrt{n}$, where n is the number of active
planes.

(3)   The convergence criterion, based either on the
       shear stress above yield, or the increment of
       plastic strain generated during the current
       iteration.

These three factors should be coordinated to represent about the
same level of precision for an efficient calculation.

The preceding solution method can be readily modified to
account for rate-dependent effects. During the first time that
step (3) is taken, the plastic strain is computed from Eq. (8), and
the shear stress is computed from Eqs. (4) or (6). This partially
relaxed shear stress is used as the stress threshold in the subsequent relaxation calculations. Thus the shear stress is made to
gradually approach this relaxed value during the subsequent
iterations.

The foregoing solution procedure may be lengthy and require
considerable computing time in some cases. The accuracy obtained
as well as the computing time will strongly depend on the settings
of the precision factors. Because the number of subcycles and the
number of active planes at any time naturally adjust to handle the
imposed strain, the computing time is dependent on the amount of
strain. For practical problems there are usually only a few cells
or elements that are undergoing large strains. For such a case the
increase in the computing time over that for a simpler plastic
procedure is not very important.

The proposed relaxation solution procedure has a number of
beneficial features. Although no mathematical proof has been
attempted, the solution obtained should converge to a unique set of
values for decreasing strain increment sizes. The accuracy of the
method can be adjusted simply by modifying the strain increment
size and the other precision factors. Any number of active planes
can be accommodated in the method, and the planes can be at any
orientation with respect to each other. Thus, an initially
orthogonal system need not remain orthogonal to be solvable.
Additional, and highly nonlinear processes, can be added without
requiring modifications in the solution method. Such processes are
thermal softening, tensile or spall opening, and cross-hardening
(work hardening of one plane caused by shearing strain on another
plane).

C.   Cap Yield Curve and Solution Procedure

The shear-enhanced compaction often observed in geologic
materials and concrete is provided by a cap work-hardening yield

surface plus a hydrostatic pressure-volume relation. We begin the
discussion of these model features by first outlining the standard
cap model approach and the underlying experimental basis for these
features. Then the actual processes in the MPCap model are
described.

The usual cap curves and hydrostat are shown in Figures 2 and
7. The shear strength is controlled primarily by a Mohr-Coulomb
curve and the pressure is controlled by a cap curve. Because of
the curvature of both lines, the pressure and shear are interdependent.

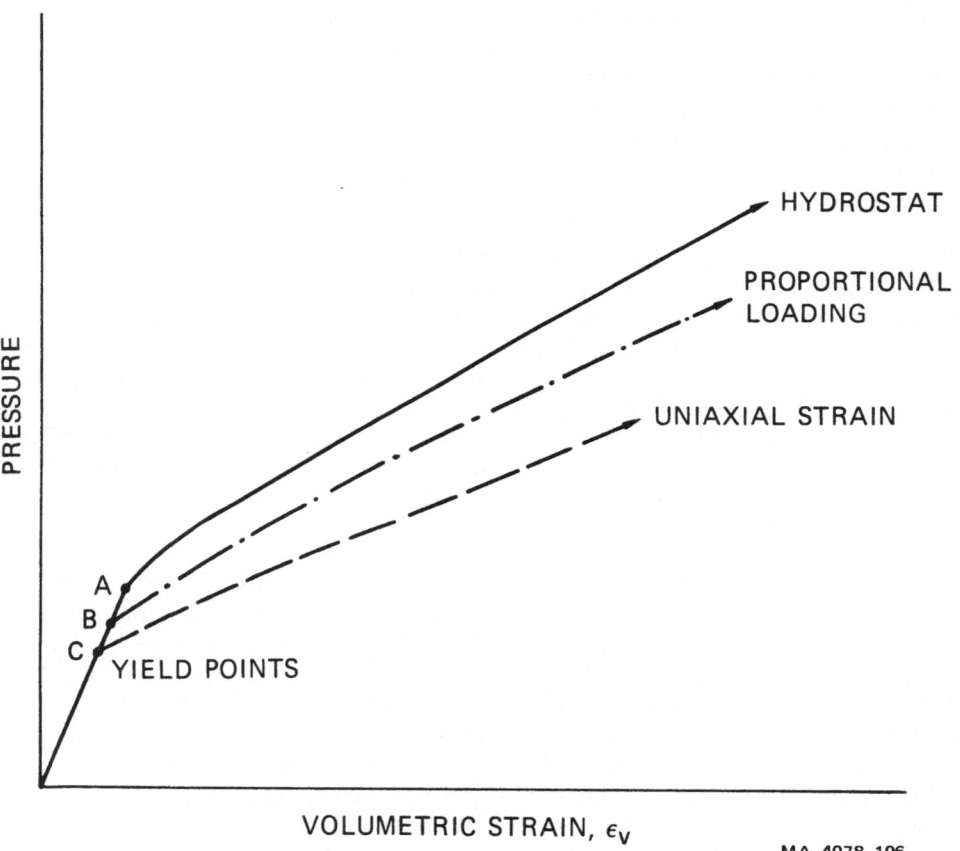

FIGURE 7   TYPICAL PRESSURE-VOLUME CURVES FOR LOADING PATHS
WITH DIFFERENT AMOUNTS OF DEVIATORIC STRESS

The pressure calculations are made in the following way. For
virgin loading the stress point begins at the origin in Figure 7
and is moved elastically until it meets the hydrostatic compaction
curve. For a ductile porous material this compaction curve corresponds to a microscopic yielding process around pores. The small

particles (or material with voids) have a highly nonuniform stress state with an average hydrostatic state P. When the compaction curve is reached, the particles (or material around the voids) begin to yield and fill in the voids. This plastic compaction process leads to the inelastic behavior. As continued loading occurs, the pressure usually rises, representing an increasing resistance to the compaction. If unloading occurs from a point beyond the initial yield, then the stress point follows an essentially elastic process as shown. These elastic slopes in Figure 7 gradually increase with the consolidation of the material. For some fairly high pressure, the voids may all be eliminated and the stress point is then on the equation-of-state surface for the solid.

This hydrostat model is defined only for compressive loading; the behavior in tension is assumed to be elastic up to tensile failure. However, the tensile failure is governed by the shear plane treatment, and hence no special tensile behavior is required for the pressure.

Shear-enhanced compaction is provided by the curvature of the cap curve in the $\bar{\sigma}$ - P plane. For example, Figure 7 shows three pressure-volume paths appropriate to points A, B, and C in Figure 2. The pure hydrostat exhibits the highest pressure, whereas the A curve with a large shear stress shows the lowest pressure. This shear-enhancement of the compaction process in real material probably arises because shear strains tend to collapse the material around the voids more readily than a purely hydrostatic strain can.

The MPCap model provides for shear-enhanced compaction with the following features:

- Hydrostat and associated processes for bulk elastic loading, plastic compaction, elastic unloading, and consolidation. The hydrostatic curve is given by parabolic segments for ease in fitting arbitrarily complex laboratory data.

- Shear enhancement by a cap work-hardening process. An ellipse was selected for the shape of the cap.

The pressure is determined from the hydrostat and the cap curve. First, the hydrostatic pressure is determined from the density, considering both the elastic processes and the compaction curve. The equivalent stress, $\bar{\sigma}$, is computed from the deviator stress tensor determined from the solution of the shear stresses on all the shear planes. Then the pressure is computed from the current cap ellipse. The expression for this ellipse is:

$$P = \sqrt{P_H^2 - \bar{\sigma}^2}$$

where P is the pressure at the current state point, and P_H is the pressure on the zero shear stress hydrostat. With this solution procedure we are assuming that the equivalent stress from the multiple planes is correct, and that we only need to adjust the pressure to account for this stress. Thus, we do not solve for $\bar{\sigma}$ and P simultaneously, but in sequence.

The cap solution procedure outlined above does not contain the plastic flow rules based on normality as most cap models do; however, a similar response is provided by the MPCap method. For example, let us assume that the stress point is at A in Figure 8 on a cap curve. If pure shear strain is applied, the elastically computed stress state is moved to a point such as B. Stress relaxation on the planes due to yielding will reduce this stress to a level such as at C. However, this stress is still on the same cap curve because no additional volume strain occurred; therefore, the pressure has reduced to point C. This motion of the stress point from points A to B and then to C is very similar to the behavior that would occur for an imposed normality condition. A similar quasi-normal behavior occurs when a combination of hydrostatic and shearing strains is applied. Then the stress would move

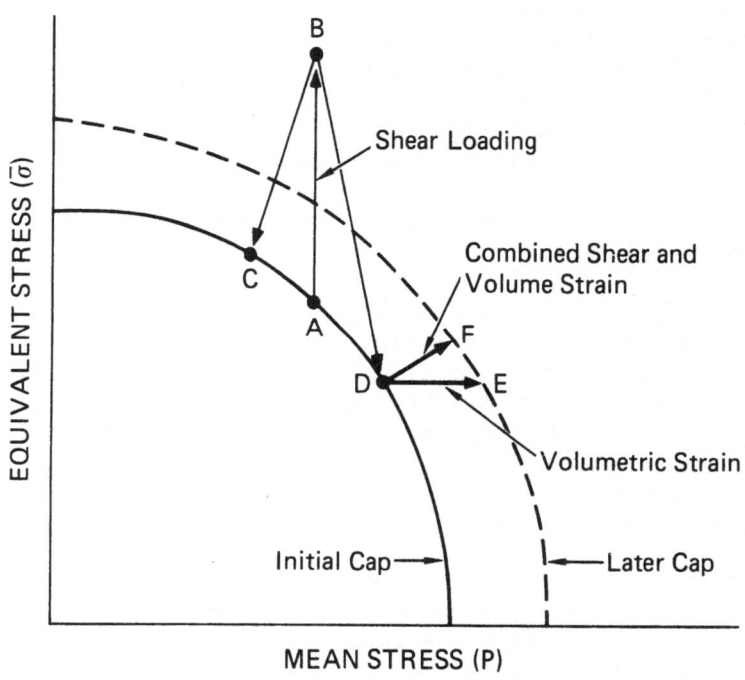

JA-314583-44

FIGURE 8 STRESS PATHS GENERATED IN THE MP CAP
MODEL FOR SEVERAL STRAIN LOADINGS

from a point such as D on the first cap curve to a point F on the second cap curve defined by the new volume strain. From these considerations it appears that the MPCap model response is physically reasonable, and much like that of the usual cap plasticity model.

D. Inclusion of Microstructural Features

The present MPCap model was formulated to provide a framework from which to build a microstructural model for concrete. We wish to include three types of micro features: crack-like, void-like, and a transition type between cracks and voids. The following steps are planned for inclusion of these features.

Crack-like features include tensile and shear cracks and shear bands. Distributions of sizes of cracks and bands are already provided on the planes (because they were part of the SHEAR4 basic model). The next step is to formulate nucleation and growth processes for these microcracks. It is expected that nucleation is a function of the stress and strain states and their histories at each point in the material. In some cases the cracks will already be present for the initial loading. Growth or enlargement of the area of the crack is also expected to be a function of the stress and strain states in the material. Under tensile loading it is expected that opening will occur. This opening may correspond to the elastic relation of Sneddon and Lowengrub [30] or at high rates to the viscous relation of He and Hutchinson [31]. Under shear stresses there will be a shear displacement of the crack faces and this displacement can also be computed by elastic or viscous relations. The provision of opening and shear displacements are already present. Nucleation and growth processes are present, but they are pertinent to submicrosecond loading times, and not to the longer loading times normally experienced in concrete.

The void-like features are microvoids with a spherical shape. Such voids can be used in determining the hydrostat, and thus these features primarily influence the pressure component of stress. The first step will be to include a void model such as that of Bhatt, Carroll and Schatz [32]. In their model they considered a micro-element consisting of a block of Coulomb friction material containing a spherical void. They applied external pressures (or volume changes) and computed the resulting elastic and plastic stress and strain distributions. The detailed simulation led to a macro pressure-volume relation that was very similar to that observed in porous rocks. We expect to include a similar computation into the MPCap model to compute the hydrostatic pressure.

The third micro feature to be included is a block with an ellipsoidal void. The ellipsoidal void provides for the interaction between the shear stresses and the pressure. This interaction occurs in two ways:

(1) Shear stresses distort the void and can lead to volumetric changes.

(2) Shear stress concentrations around the void lead to premature yielding near the void, and then to overall yielding.

The second of these effects has been treated by Gurson [33] in his model for ductile fracture by void growth in metals. The first has not been satisfactorily handled for plastic flow, although He and Hutchinson [31] have treated the case for a viscous material. Although inclusion of this third micro feature will require a major effort, it is essential for representing shear-enhanced compaction in the model.

Considering the foregoing list of features to be added, we clearly foresee a period of continuing development for the model.

DISCUSSION

The MPCap model is in an early stage of development, yet it already includes many features that represent real concrete behavior. The two most important tasks remaining are delineation of the necessary micro features, and then demonstration of its ability to match concrete data.

One immediate advantage of the model is a means to display and organize three-dimensional information about concrete. By providing discrete planes on which to project shearing behavior, the model allows us to work with much more manageable concepts than the usual three-dimensional ones. We are able to address flow and fracture on specific material planes, rather than restricting attention to invariants which obscure the anisotropic behavior.

ACKNOWLEDGEMENTS

The SHEAR4 model used here was developed under a contract with Dr. George Mayer of the Army Research Office. The CAP model development was sponsored by Dr. George Sliter of the Electric Power Research Institute.

REFERENCES

1. Drucker, D. C. and W. Prager. Soil Mechanics and Plastic Analysis or Limit Design. Quarterly of Applied Mathematics 10 (1952) 157-165.

2. Chen, W. F. Plasticity in Reinforced Concrete (McGraw-Hill, New York, 1981).

3. Chen, W. F. Plasticity in Reinforced Concrete. Proceedings of a Workshop on Constitutive Relations for Concrete, New Mexico Engineering Research Institute, University of New Mexico, Albuquerque, N.M., April 28, 29, 1981.

4. Lindberg, H. E. and L. E. Schwer. Choice and Construction of Yield Functions for Underground Cavity Analysis. Poulter Laboratory Technical Report (SRI International, Menlo Park, California 94025).

5. Schleicher, F. The Energy Limit of Elasticity (in German), Zeitschrift fuer Angewandte Mathematik und Mechanik 5, 6 (1925) 478-479.

6. Seaman, L. One-Dimensional Stress Wave Propagation in Soils. Final Report DASA 1757 by SRI International for Defense Atomic Support Agency (Washington, D.C. 20301, February 1966).

7. Shockey. D. A, D. R. Curran, L. Seaman, J. T. Rosenberg, and C. F. Petersen. Failure of Rock under High Tensile Loads. Int. J. Rock Mech. Sci. Geomech. Abstr. 11 (1974) 303-317.

8. Linde, R. K., L. Seaman, and D. N. Schmidt. Shock Response of Porous Copper, Iron, Tungsten and Polyurethane. J. of Appl. Phys. 43, 8 (August 1972) 3367.

9. Bazant, Z. P. Mathematical Modeling of Concrete and Its Experimental Basis. Proceedings of a Workshop on Constitutive Relations for Concrete (New Mexico Engineering Research Institute, University of New Mexico, Albuquerque, N. M., April 28, 29, 1981).

10. Seaman, L., D. R. Curran, and D. A. Shockey. Computational Models for Ductile and Brittle Fracture. J. Appl. Phys. 47 (1976) 4814-4826.

11. Murri, W. J., C. Young, D. A. Shockey, R. E. Tokheim, and D. R. Curran. Determination of Dynamic Fracture Parameters for Oil Shale. SRI International Final Report to Sandia Laboratories (Alburquerque, New Mexico, 1977).

12. Gupta, Y. M. and L. Seaman. Local Response of Reinforced Concrete to Missile Impact. SRI International Final Report No. EPRI NP-1217 (Electric Power Research Institute, Palo Alto, California, October 1979).

13. Curran, D. R., D. A. Shockey, and L. Seaman. Dynamic Fracture Criteria for a Polycarbonate. J. Appl. Phys. 44 (1973) 4025.

14. Batdorf, S. B. and B. Budiansky. A Mathematical Theory of Plasticity Based on the Concept of Slip. National Advisory Committee for Aeronautics Technical Note No. 1871 (Washington, April 1949).

15. Como, Mario and Salvatore D'Agostino. Strain Hardening Plasticity with Bauschinger Effect. Meccanica 4, 2 (1969) 146-158.

16. Como, Mario and Antonio Grimaldi. Analytical Formulation of the Theoretical Subsequent Yield Surfaces of Metals with Strain-Hardening and Bauschinger Effect in an Ideal Tension-Torsion Test. Meccanica 4, 4 (1969) 286-297.

17. Naghdi, P. M., F. Essenburg, and W. Koff. An Experimental Study of Initial and Subsequent Yield Surfaces in Plasticity. J. Appl. Mech. 25 (1958) 201-209.

18. Bazant, Z. P. and B. H. Oh. Microplane Model for Fracture Analysis of Concrete Structures. Proceedings of the Symposium on the Interaction of Non-Nuclear Munitions with Structures (U.S. Air Force Academy, Colorado, May 10-13, 1983).

19. Seaman, L., D. R. Curran, and W. J. Murri. A Continuum Model for Dynamic Tensile Microfracture and Fragmentation. J. Appl. Mech. ASME, to appear.

20. Seaman, L., D. R. Curran, and D. A. Shockey. Scaling of Shear Band Fracture Processes. Proc. of the 29th Sagamore Conference on Material Behavior Under High Stress and Ultrahigh Loading Rates, eds. J. Mescall and V. Weiss (Plenum Press, New York, 1983).

21. Seaman, L. and J. L. Dein. Representing Shear Band Damage at High Strain Rates, IUTAM Symposium on Nonlinear Deformation Waves (Tallinn, Estonia, August 22-28, 1982).

22. Drucker, D. C., R. E. Gibson, and D. J. Henkel. Soil Mechanics and Work-hardening Theories of Plasticity. Trans of the Amer. Soc. of Civil Eng. 121 (1956) 338-346.

505

23. Sandler, I., F. L. DiMaggio, and G. Y. Baladi. Generalized Cap Model for Geological Materials. J. Geotech. Div., ASCE 102, GT7 (July 1976) 683-699.

24. Bazant, Z. P. and P. D. Bhat. Endochronic Theory of Inelasticity and Failure of Concrete, J. Eng. Mech. Div., ASCE 102 (1976) 701-722.

25. Bazant, Z. P. and Sang-Sik Kim. Plastic-Fracturing Theory for Concrete. J. Eng. Mech. Div., ASCE 105 (1979) 407-428.

26. Wittman F. H. and Yu. V. Zaitsev. Crack Propagation and Fracture of Composite Materials such as Concrete, 5th Int. Conf. on Fracture, ed. D. Francois, 5, 2261-2274 (Cannes, France, Mar. 29-Apr. 3, 1981).

27. Malvern, L. E. The Propagation of Longitudinal Waves of Plastic Deformation in a Bar of Material Exhibiting a Strain-Rate Effect. J. Appl. Mech. 18 (1951) 203-208.

28. Taylor, G. I. Plastic Strain in Metals. J. Inst. Metals 62 (1938) 307.

29. Peirce, D., R. J. Asaro, and A. Needleman. Material Rate Dependence and Localized Deformation in Crystalline Solids. Acta Metall. 31 (1983) 1951-1976.

30. Sneddon, I. N. and M. Lowengrub. Crack Problems in the Classical Theory of Elasticity (John Wiley and Sons, Inc., New York, 1969).

31. He, M. Y. and J. W. Hutchinson. The Penny-Shaped Crack and Plane Strain Crack in an Infinite Body of Power-Law Material. J. Appl. Mech., Trans. of the ASME 48 (Dec. 1981) 830-840.

32. Bhatt, J. J., M. M. Carroll, and J. F. Schatz. A Spherical Model Calculation for Volumetric Response of Porous Rocks. Paper No. 75-APMW-49, J. Appl. Mech. Trans. of the ASME (March 1975).

33. Gurson, A. L. Continuum Theory of Ductile Rupture by Void Nucleation and Growth: Part I - Yield Criteria and Flow Rules for Porous Ductile Media. J. Eng. Materials and Tech., Trans. of the ASME (January 1977) 2-15.

APPLICATION OF FRACTURE MECHANICS
TO CEMENTITIOUS COMPOSITES
NATO-ARW - September 4-7, 1984
Northwestern University, U.S.A.
S. P. Shah, Editor

APPLICATION OF CONTINUOUS DAMAGE MECHANICS TO STRAIN AND FRACTURE
BEHAVIOR OF CONCRETE

Jacky MAZARS - Jean LEMAITRE

Laboratoire de Mécanique et Technologie
ENSET-Université Paris 6-CNRS
61, avenue du Président Wilson
94230 CACHAN - FRANCE

ABSTRACT

   Elasticity and damage by microcracking constitute the essential phases of the mechanical behavior of concrete.

   We show here the interest to describe these phenomena by
continuous damage mechanics. The formulation proposed uses the
effective stress concept, the damage is assumed to be isotropic
and as a consequence represented by a scalar variable D.

   The importance of strain extensions $(\varepsilon > 0)$ in the degradation of the material appears in the formulation of the damage
threshold surface. To take into account the sensitivity of concrete to gradient and volume effects we propose a probabilistic
definition of the initial damage threshold.

   When the local rupture of the material is complete, we use
linear fracture mechanics together with a definition of an equivalent crack to the actual damaged zone.

   Experimental results show the interest of these approaches.

1 - INTRODUCTION

Generally speaking, the behavior of a material is directly linked to its microstructural state and its evolution under solicitations.

For concrete, the heterogeneity of its microstructure associated with the great porosity of the binding material and with the presence of granulates, is an essential factor of the phenomenological aspect of the behavior.

To be physically realistic, a model must be compatible with the local mechanical phenomena. We propose here to discuss the interest to use the continuous damage mechanics to describe the behavior and the fracture of concrete.

2 - A FEW ASPECTS OF THE BEHAVIOR OF CONCRETE

2.1 - Modes of damage and behavior

From experimental observations, from micromechanical models proposed by several authors [1], [2], [3] ..., and from a general study in [4] schematically we can describe

- a state of initial degradation (defects of compactness, microcracks in the paste created by dilation and shrinkage);
- a propagation of the microcracks around the biggest grains under load;
- a dependence of the microporous structure of the cement paste to the hydrostatic pressions.

The main characteristic deriving from that phenomenological considerations are:

- damage is the principal aspect of the behavior of the material;
- it only appears beyond a certain threshold of solicitation;
- it influences the macroscopic mechanical characteristic of the material;
- damage modes differ according to the type of solicitation.

Type A: The solicitations applied allow extensions ($\varepsilon > 0$) in one main direction at least; the formation of microcracks in mode I is then possible and the behavior of the material shows an instability more or less important as important are the tension or the compression.

Type B: The solicitations applied allow no extension, the strong
hydrostatic pressure associated leads to the local initiation of
microcracks in mode II and III, the friction between the lips
yielding a ductile macroscopic behavior.

Type C: The solicitation is an hydrostatic pressure, the essential phenomenon is then the collapse of the microporous structure
which leads to a consolidation of the material.

2.2 - Damage and cracking

The cracking of concrete results from the creation, propagation and coalescence of microcracks.

The solicitations leading to that kind of phenomenon is of
type A previously defined.

Many authors have been interested in the study of the propagation of cracks, amoung the latest studies we can note Francois
et al [5], Hillerborg et al [6], Mazars et al [7]. They have
shown that:

- damage appears far ahead of the front of the macrocrack.
- The part of the macrocrack which appears on the surface is
not significant in regard to the zone of complete separation of
the material inside the thickness of the structure.

We thus propose a scheme showing all those phenomena, as well
as the theoretical approaches we will use to model them.

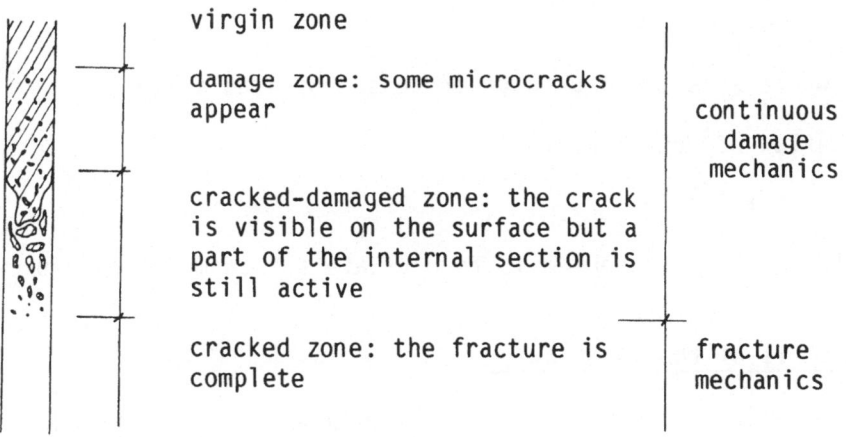

virgin zone

damage zone: some microcracks
appear

cracked-damaged zone: the crack
is visible on the surface but a
part of the internal section is
still active

cracked zone: the fracture is
complete

continuous
    damage
    mechanics

fracture
mechanics

3 - CONTINUOUS DAMAGE MODEL FOR CONCRETE

3.1 - Effective stress

Kachanov [8] was the first to use the concept of effective stress to describe the uniaxial rupture by creep of metallic materials. That theory has since been developed especially by Lemaitre [9], [10] and used in other cases (static rupture, fatigue ...).

That concept introduces a damage variable D. The relation between the usual stress σ and the effective stress $\hat{\sigma}$ can be written:

- $\hat{\sigma}$ = IM(D): σ in the general case of anisotropic damage for which the damage variable is a tensor D (IM is a fourth order tensor)

- $\hat{\sigma} = \frac{\sigma}{1-D}$ is the case of isotropic damage for which the damage variable is a scalar.

$$D = 0 \qquad \text{for the virgin material}$$
$$0 < D < D_c \qquad \text{for the damaged material}$$
$$D = D_c \qquad \text{for the cracked material}$$

Thus the behavior of an isotropic linear elastic damaged material is represented by:

- the law of elasticity coupled with damage

$$\hat{\sigma} = \frac{\sigma}{1-D} = \Lambda : \varepsilon$$

where Λ is the fourth order tensor of elasticity

- a law of damage evolution

$$\overset{\circ}{D} = F (\varepsilon, \overset{\circ}{\varepsilon})$$

The difficult problem is to evaluate and to model the function F.

3.2 - Basis of the formulation

The model proposed has been designed upon the following considerations:

- the behavior is essentially elastic-damageable ($\varepsilon = \varepsilon^e$);
- the damage appears beyond a threshold and it is isotropic;
- the damage mode corresponds to the type A, hence it is

linked to the existence of extensions.

3.3 - Constitutive equations

Only the major aspects developed in [4] are described here

3.3.1 - Thermodynamical aspects

* Classification of the variables for the volume element

Observable variables	internal variable	associated variables
ε_{ij} strain tensor		σ_{ij} stress tensor
T temperature assumed to be constant	D (damage)	S entropy
		Y damage strain energy release rate

* Specific free energy

$$\psi = \frac{1}{2\rho} \Lambda_{ijkl} \, \varepsilon_{ij} \, \varepsilon_{kl} \, (1-D)$$

ρ is the density

* Equations of state

$$\sigma_{ij} = \rho \, \frac{\partial \psi}{\partial \varepsilon_{ij}} = \Lambda_{ijkl} \, \varepsilon_{kl} \, (1-D) \tag{1}$$

$$Y = \rho \, \frac{\partial \psi}{\partial D} = - \frac{1}{2} \Lambda_{ijkl} \, \varepsilon_{ij} \, \varepsilon_{kl} = - \bar{Y} \quad (\bar{Y} > 0)$$

* Clausius-Duhem inequality

$$\bar{Y} \, \dot{D} > 0 \quad (\dot{D} = \frac{dD}{dt})$$

or $\Lambda_{ijkl} \, \varepsilon_{ij} \, \varepsilon_{kl} \, \dot{D} > 0 \tag{2}$

let us use the partition of stresses introduced by Bazant [11] (see figure below):

$$d\sigma_{ij} = d\sigma^e_{ij} - d\sigma^d_{ij}$$

we can deduce from (1)

$$d\sigma^e_{ij} = \Lambda_{ijkl} \, d\varepsilon_{kl} \, (1-D)$$

and

$$d\sigma_{ij}^d = \Lambda_{ijkl} \, \epsilon_{kl} \, dD$$

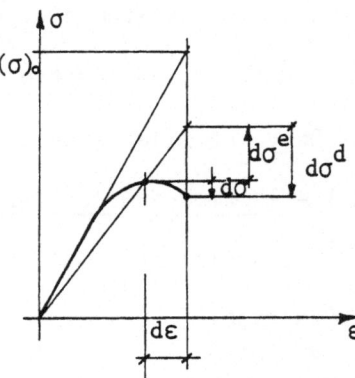

The partition of stresses

Besides, let us write $\Lambda_{ijkl} \, \epsilon_{kl} = (\sigma_{ij})$.

then
$$\overset{od}{\sigma}_{ij} = (\sigma_{ij})_0 \, \overset{o}{D} \tag{3}$$

and (2) leads to
$$\overset{od}{\sigma}_{ij} \, \epsilon_{ij} > 0 \tag{4}$$

3.3.2 - <u>Damage surface threshold</u>. Due to the importance of the extensions in the damage mode chosen we propose in [12] the concept of equivalent strain:

$$\tilde{\epsilon} = \sqrt{\sum_i < \epsilon_i >^2_+}$$

ϵ_i is a principal strain

$<\epsilon_i>_+ = 0$ if $\epsilon_i < 0$

$<\epsilon_i>_+ = \epsilon_i$ if $\epsilon_i > 0$

The damage threshold for a given state D can be thus defined by:

$$f(\tilde{\epsilon}, D) = \tilde{\epsilon} - K(D) = 0$$

If $D = 0$, then $K(0) = \epsilon_D$. (limit of elasticity or damage of the material).

3.3.3 - <u>Law of evolution</u>. The conditions to observe are:

- in case of evolution

$f = 0 \Rightarrow \tilde{\epsilon} = K(D)$
$\overset{o}{f} = 0 \Rightarrow \overset{o}{\tilde{\epsilon}} > 0$
$\Rightarrow \overset{od}{\sigma}_{ij} \neq 0$

- in case of non evolution

$f < 0$ or

$f = 0$ and $\overset{o}{f} < 0 \Rightarrow \overset{o}{\tilde{\epsilon}} < 0$
$\Rightarrow \overset{od}{\sigma}_{ij} = 0$

The expression for the damage strain rate must be:

$$\overset{o}{\sigma}{}^d_{ij} = g_{ij} \, \langle \overset{o}{\widetilde{\varepsilon}} \rangle_+ \tag{5}$$

So we choose

$$g_{ij} = (\sigma_{ij})_o \, F(\widetilde{\varepsilon}) \tag{6}$$

From (3), (5) and (6) we derive

$$\overset{o}{D} = F(\widetilde{\varepsilon}) \, \langle \overset{o}{\widetilde{\varepsilon}} \rangle_+$$

In order to respect (2) $F(\widetilde{\varepsilon})$ must be a positive continuous function of $\widetilde{\varepsilon}$.

The behavior is then completely defined by:

$$d\sigma_{ij} = d\sigma^e_{ij} - d\sigma^d_{ij}$$

with

$$d\sigma^e_{ij} = \Lambda_{ijkl} \, (1-D) \, d\varepsilon_{kl}$$

and

$$d\sigma^d_{ij} = \quad 0 \text{ if } f < 0 \text{ or } f = 0 \text{ and } \overset{o}{f} < 0$$

$$(\sigma_{ij})_o \, dD \text{ if } f = 0 \text{ and } \overset{o}{f} = 0$$

$$\overset{o}{D} = F(\widetilde{\varepsilon}) \, \langle \overset{o}{\widetilde{\varepsilon}} \rangle_+$$

4 - DAMAGE AND FRACTURE

4.1 - Thermodynamic approach of fracture

* classification of the variable for the whole body

observable variables	internal variable	associated variable
q (displacement)		Q load
T Temperature sup-	A (crack area)	S entropy
posed constant		G strain energy release rate

* free energy for the whole body

514

$$\psi = U - TS \quad \text{with } U = \frac{1}{2} Qq$$
with the rigidity

$K = Q/q$, we have:

$$U = \frac{1}{2} K(A) q^2$$

* equations of state

$$Q = \frac{\partial \psi}{\partial q}$$

$$G = \frac{\partial \psi}{\partial \Lambda} = \frac{1}{2} q^2 \frac{\partial K}{\partial A} = - \overline{G} \qquad (\overline{G} > 0)$$

* Clausius-Duhem inequality

$$\overline{G} \, \overset{\circ}{A} > 0$$

* evolution laws

In the special case of brittle rupture,

$$\overline{G} < G_c \quad \Rightarrow \quad \overset{\circ}{A} = 0$$

$$\overline{G} = G_c \quad \Rightarrow \quad \overset{\circ}{A} > 0$$

4.2 - Damage and fracture under a constant load

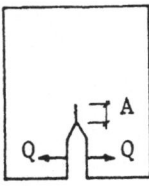

Let us consider a structure with a crack of surface A and let us suppose that everywhere else, the material is virgin.

We study two cases:

4.2.1 - The material is elastic-damageable

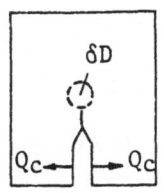

Given δD, a local evolution of damage under a constant load Q_c. The total dissipation of the resulting energy is:

$$D(\delta D) = \int_V \overline{Y} \, \delta D \, dv$$

4.2.2 - The material is perfectly brittle

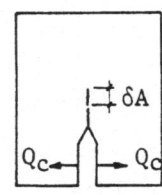

Given δA, the evolution of the crack under the load Q_c. The dissipation energy is:

$$D\,(\delta A) = G_c\,\delta A$$

4.2.3 - Energetic equivalence of both phenomena

$$D\,(\delta D) = D\,(\delta A) \Rightarrow \delta A = \frac{\int_v \tilde{Y}\,\delta D\,dv}{G_c}$$

δA is the increase of the crack "equivalent" to the damage increase δD.

As shown in [4] in the case of a linear elastic material, the damageable structure and the equivalent fracturable structure have the same global behavior. Then for a given state we can deduce:

$$K(A) = K(D)$$

5 - PROBABILISTIC ASPECT OF DAMAGE

Many experimental studies have shown the sensitivity to the effects of gradient and volume. That phenomenon is due to the presence of local defects (microvoids, initial microcracks...) and can be described by considering a random distribution of local strength in terms of equivalent strain $\tilde{\varepsilon}_D$.

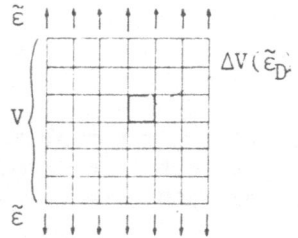
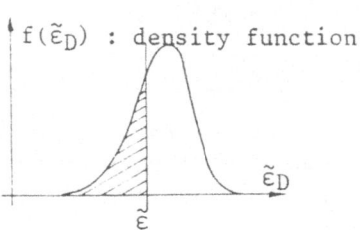

Random distribution of local strength

If we choose a Weibull distribution, the damage probability of a volume ΔV under the effect of a uniform solicitation $\tilde{\varepsilon}$ can be written:

$$Pd\,(\tilde{\varepsilon},\,\Delta V) = k\,\tilde{\varepsilon}^m\,\Delta V \qquad k \text{ and } m \text{ being material constants.}$$

For a given volume V submitted to any solicitation $\tilde{\varepsilon}$, we show in [13]:

$$Pd\ (\tilde{\varepsilon},\ V) = 1 - \exp\ (-k \int_V \tilde{\varepsilon}^m\ dv)$$

* Statistical estimation of the initial damage threshold

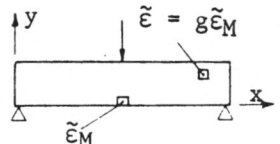

Given $\tilde{\varepsilon}_M$ the maximum local solicitation inside a structure.

In the case of a linear behavior and a radial loading, we can express the solicitation $\tilde{\varepsilon}$ at any point by:

$$\tilde{\varepsilon}(x,\ y,\ z) = g(x,\ y,\ z)\ \tilde{\varepsilon}_M$$

Suppose that the damage threshold is reached in the most loaded element, we show that it can be determined by:

$$\varepsilon_{Do} = [\frac{W_o}{\int_v g^m\ dv}]^{1/m}$$

m and Wo are characteristics parameters of the materials.

The standard deviation of ε_{Do} is:

$$s = \varepsilon_{Do}\ \sqrt{\frac{\Gamma\ (1 + 2/m)}{\Gamma^2\ (\ + 1/m)} - 1}$$

Γ is the mathematical gamma function

6 - SOME APPLICATIONS

6.1 - Damage law of evolution

It can be determined from the behavior of the material in tension identified with the results given by a test in bending (more stable and less sensitive to defects than other tests) we propose:

$$\dot{B} = [\frac{\varepsilon_{Do}\ (1 - A)}{\tilde{\varepsilon}^2} + \frac{A\ B}{\exp\ [B(\tilde{\varepsilon} - \varepsilon_{Do})]}]\ <\dot{\tilde{\varepsilon}}>_+$$

A and B are characteristics parameters of the material.

6.2 - Ability of the model to describe the behavior

We intend to illustrate that part on a notched test plate of C.T. type (see figure below). We compare some results given by experiments, and by F.E.M. calculation. The parameters A and B were identified as shown above, but in order to eliminate the effects of discrepancy ε_{Do} was adjusted to have the same maximum load as the test shown.

6.2.1 - Local behavior

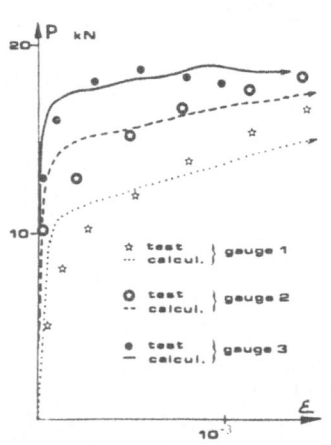

experimental crack

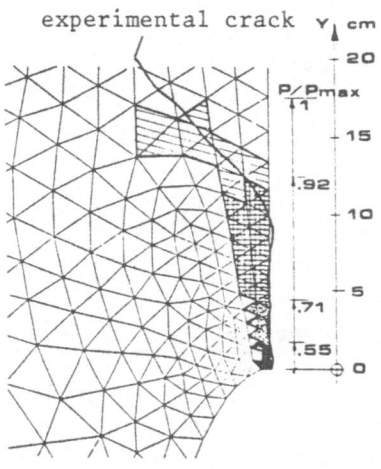

Strains evolutions ahead the initial notch.

Evolution of the damaged zone (calculation) and path of the crack.

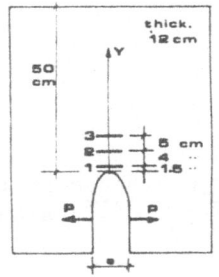

Local behavior of a CT specimen

6.2.2 - Global behavior

We use here the combination of Damage mechanics (to the initiation of the crack) and Fracture mechanics (after). In this last case the damaged zone created at the top of the curve corresponds to an "equivalent crack" A_o. Then the propagation was obtained by considering G_c as a constant. We have shown in [7] that it was correct for this kind of test.

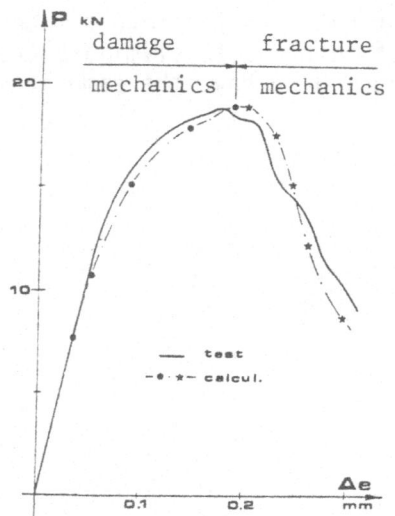

Global behavior of a CT
specimen

6.3 - Previsonal calculation

Here we show the interest of the model as regard to the
prediction of the global behavior in comparison to the discrepancy
to be expected from the test results. The parameters A, B, W_o and
m have been determined by means of preliminary test in tension and
bending. Thus no adjustment has been performed on these results.
Those concern three plates CT made with the same concrete.

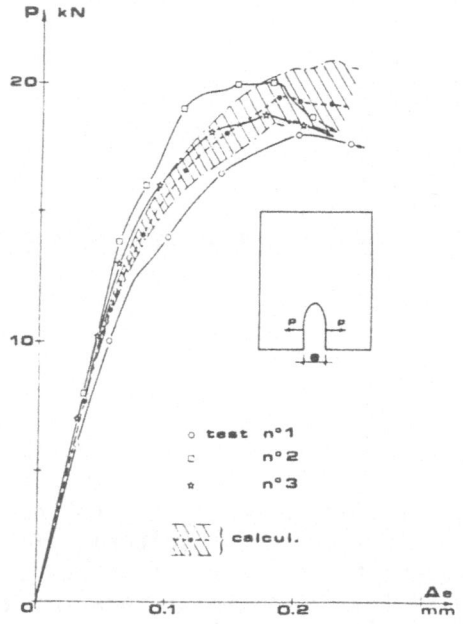

Performance of the
previsonal calculation

The dashed zone on the figure represents the influence of the standard deviation on the threshold ε_{D_0}. That result gives us information about the discrepancy of results.

7 - CONCLUSION

The results presented show the interest to use continuous damage mechanics to describe the degradation of concrete to rupture and its consequences on the behavior of the structures. It allows in particular to predict the initiation of cracks. Their propagation is then described by the use of linear fracture mechanics, from the concept of equivalent crack to the damaged zone.

REFERENCES

1. Lino, M. A model for a microcracked material. Revue Francaise de Géotechnique N. 11 (1980) pp. 29-41. Un modèle de matériau microfissuré.

2. Modeer, A. A fracture mechanics approach to failure analyses of concrete materials. Report Univeristy of Lund (1979).

3. Buyukozturk, O., A. H. Nilson, and F. O. Slate. Stress strain response and fracture of concrete model in biaxial loading. ACI Journal (August 1971) pp. 590-599.

4. Mazars, J. Application of damage mechanics to the non linear behavior of concrete. Thèse d'Edat, Université Paris 6 (1984) Application de la mécanique de l'endommagement au comportement non linéaire et à la rupture du béton de structure.

5. Francois, D., S. Chhuy, M. E. Benkirane and J. Baron. Crack propagation in prestressed concrete; interaction with reinforcement. Advances in Fractue Mechanics (Fracture 81) vol. 4. Pergamon Press (1982) pp. 1507-1514.

6. Hillerborg, A., M. Modeer and P. Petersson. Analysis of crack formation and crack growth in concrete by means of fracture mechanics and finite elements. Cement and concrete research. vol. 6 (1976) pp. 773-782.

7. Mazars, J. and D. Legendre. Damage and fracuture mechanics for concrete - a combined approach. ICF 6, New Delhi (1984).

520

8. Kachanov, L.M. Time of the rupture process under creep condi-
    tions. Isv. Akad. Nauk. SSR. Ofd. Tekh. Nauk, n° 8 (1958) pp.
    26-31.

9. Lemaitre, J. and J. L. Chaboche. Phenomenological aspect of
    rupture by damage. Journal de mécanique appliquée. vol. 2,
    n°3, (1978). Aspects phénoménologique de la ruptue par
    endommagement.

10. Lemaitre, J. Damage modelling for prediction of plastic or
    creep fatigue failure in structures. 5th conference on
    structural mechanics in reactor technology (SMIRT 5) (1979).

11. Bazant, Z. P. Work inequalities for plastic fracturing
    materials. Int. J. Solids Structures. vol. 16 (1980) pp.
    873-901.

12. Mazars, J. Mechanical damage and fracture of concrete
    structures. Advances in Fracture Mechanics (Fracture 81) vol.
    4. Pergamon Press (1982) pp. 1499-1506.

13. Mazars, J. Probabilistic aspects of mechanical damage in
    concrete structures. Fracture Mechanics Technolgy Applied to
    Material Evaluation and Structure Design. Martinus Nijhoff
    Publishers (1983) pp. 657-666.

APPLICATION OF FRACTURE MECHANICS
TO CEMENTITIOUS COMPOSITES
NATO-ARW - September 4-7, 1984
Northwestern Univeristy, U.S.A.
S. P. Shah, Editor

ON A CONTINUUM MODELLING OF DAMAGE

Mark Kachanov

Department of Mechanical Engineering
Tufts University
Medford, MA, 02155
USA

ABSTRACT

A brief review of the papers presented in the Session "Damage
and Continuum Modelling" is given. It is followed by several remarks of general character on continuum modelling of damage.

KEYWORDS

Damage Mechanics, Fracture Mechanics, Concrete Fracture, Microcrack, Micromechanics.

A brief review of the papers presented in the Session "Damage
and Continuum Modelling" followed by several remarks of the author
on continuum modelling of damage is given below.

It has become clear that the conventional theory of plasticity describing deformation of metals beyond the elastic limit is
not an adequate mechanical model for the cementitious materials.
Development of more appropriate constitutive models for inelastic
behavior and fracture of such materials constitutes the topic of
the Session reviewed. Papers given in the Session represent various approaches to the problem. In the paper of L. Seaman, J. Gran
andD.Curran, the macroscopic behavior of a material is modelled by
a computer simulation scheme suited for incorporation of the micromechanical information. In the papers of J. Mazars and J. Lemaitre
and of D. Krajcinovic, damage in brittle-elastic materials is considered in the framework of irreversible thermodynamics. In the
first paper, the simplest case of isotropic damage is analyzed
whereas the second one addresses a more complex case of anisotropic

microcrack-related damage.

In the paper of L. Seaman, J. Gran and D. Curran, a computational model for behavior of concrete-like materials is proposed. It is suggested to construct the macroscopic stress-strain laws on the basis of a computer simulation scheme, namely, by summing the contributions from individual micro-sources of inelasticity. These sources include shear-related slips, microcracking and compaction of pores; incorporation of the latter into the model is viewed by the authors as an important refinement. The computational technique is based on a multiplane concept, i.e. a certain set of orientations is initially chosen and inelastic stress-strain relations are formulated for each of the planes belonging to this set. Material parameters contained in these relations are allowed to be different for different planes.

For a shear-related plastic flow on a given plane, a work-hardening Coulomb type relation is used for prediction of the shear strength (yield limit). The work hardening function and the angle of friction can vary with the plane orientation. For loading exceeding the yield limit on a given plane a constitutive law as a linear rate dependent plastic flow is accepted and a step-by-step integration of this law (based on finite time increments) is suggested. To account for a multitude of planes of plastic flow, an approximate computer simulation scheme for distribution of plastic strain among the slip planes (called "relaxation procedure") is proposed.

In addition to plastic flow, the compaction of pores is incorporated into the model. The compaction is treated as enhanced by two factors: 1) shear stresses and 2) hydrostatic pressures exceeding certain elastic limit. To account for the shear-enhanced compaction, a cap work-hardening yield surface is used (introduced originally for soils by D. Drucker et al. [1]). For a hydrostatic stress-enhanced compaction, a non-linear constitutive pressure-volume curve, with a yield point, is used. As a future development, the authors plan to add other micro-sources to the computer-simulation scheme including cracks, shear bands and voids of various configurations.

The model provides a convenient means to display and incorporate the micromechanical information on inelastic behavior of concrete, with the ability to absorb a multitude of constitutive laws for various possible micromechanisms. In the reviewer's opinion, this makes the model potentially very useful, provided more quantitative knowledge of the actual micromechanics is available. Without such knowledge, the model represents simply a curve-fitting tool with a large number of constitutive parameters playing the role of adjustable coefficients.

In the paper of J. Mazars and J. Lemaitre the problem of concrete damage by microcracking is considered in detail for the case of isotropic damage when microcracks have no preferential orientations (a structure of constitutive relations for the general case of anisotropic damage is also briefly discussed). The formulation is applicable to the loadings when tensile microcracking is a dominant mode of deterioration of microstructure; these loadings are characterized by the condition that at least one of the principal strains ε_i is positive. The structure of constitutive relations for an elastic-damageable material is discussed in the framework of irreversible thermodynamics, with the damage parameter D (scalar, in the case of isotropic damage) being an internal variable. The essential and the most difficult part of modelling is concretization of the general thermodynamic relations; at this point, certain physical hypotheses on the damage evolution laws are to be introduced. To characterize the damage threshold, the authors propose the concept of "equivalent strain" $\tilde{\varepsilon}$ which is positive only if one of the principal strains is tensile and is zero otherwise. The threshold is reached when $\tilde{\varepsilon}$ reaches certain critical value; the latter depend on the level of damage D. The law of damage evolution is proposed in the form relating the rate of change, $\dot{D}$, to $\tilde{\varepsilon}$ and $\dot{\tilde{\varepsilon}}$. The specific form of this law contains three parameters (A, B, ε_{D_0}) to be chosen by fitting the experimental data in tension. The model has been applied to a notched test plate with a crack growing from the notch for which the results are reasonably good. The advantages of the scheme appear to be its efficiency, relative simplicity, low number of constants and a possibility to determine the latter from the experimental data on tension. This simplicity is a consequence of using a scalar damage parameter. If the damage field has preferential orientations, such a scalar description becomes inadequate (at least for non-proportional loadings). The general thermodynamical scheme (involving tensorial description of damage) suggested for this case in the reviewed article may not be as simple and unambiguous in applications.

The authors also introduce the energy release rate associated with the damage growth, analogous to the energy release rate associated with crack propagation. By equating these two rates the concept of an "equivalent crack" is introduced. This concept, in the reviewer's opinion, may not be very useful, since evolution of the damage zone involves more "degrees of freedom" (like "thickening") than propagation of a crack and the phenomenology of its growth may therefore be different from that of a crack. It can also be remarked that applicability of the condition $G = G_c$ (suggested for the "equivalent crack") to the crack propagation in concrete-like materials is not quite clear (see, for example, extensive experimental data of D. Barker et al. [2]).

The paper of D. Krajcinovic 1) addresses the important problem of anisotropic damage due to microcracks. A general framework of irreversible thermodynamics with internal variables is used. The internal variables include sets of parameters {ω} and {γ} characterizing microcracks and elementary plastic slips, respectively. "Perfectly brittle" and "perfectly ductile" solids are characterized as corresponding to γ = 0 and to ω = 0. A "perfectly brittle" material is considered in the reviewed paper; damage in such material is constituted by a field of microcracks. It is proposed to describe a microcrack field contained in an elementary (representative) volume by approximately representing it as π/Δθ sets of parallel microcracks, with orientation increments Δθ . Each family of parallel microcracks is characterized, in a certain integral sense, by one axial vector. Namely, a vector representing such a family has a direction normal to the cracks and its length is defined by the following averaging procedure. A plane S parallel to the microcracks and corresponding to a certain cross-section in the material is specified. The contribution of a given crack to the length of the vector representing the family is proportional to the area of a certain projection of this crack onto S. This projection on S is constructed in a special way: it is described by a cone formed by straight lines emanating from the crack's edge and inclined 45° to the crack's plane. (This angle is called "The angle of diffusion of stresses" and its choice is not discussed). Thus, the contribution of a given crack decreases linearly with distance from S and vanishes completely at the distance equal to the crack's radius. The underlying idea is that weakening of a cross-section (its reduced ability to transmit stresses) is related to its distance from the microcracks. This appears to be a sound physical idea but the proposed solution based on the concepts of "cone of influence" and "angle of diffusion of stresses" raises some questions.

First, the actual "shielding" zone (crack's "shadow") where stresses are reduced by the presence of the crack is substantially different from the proposed cone (see formulas for a full stress field in a solid with a penny shaped crack subjected to uniform remote tension perpendicular to the crack [5]). For example, at the apex of the cone where, according to the proposed model, the tensile stress perpendicular to the crack assumes its full remote field value, the actual value of this stress, for mode I remote loading, constitutes only 50% of the remote value (and only 35% if a two dimensional configuration is considered). Second, if the mode of loading is changed (from tension to shear) the zone of

---

1. The present review is based, also, on [3] and on the latest and more detailed presentation of the model [4] which was kindly made available to the reviewer.

"shielding" may become a zone of "amplification" where stresses are
raised by the presence of the crack. In this case, the concepts of
"cone of influence" and "angle of diffusion of stresses" seem to
become unclear. Finally, these concepts are not clear if the cones
of influence of different cracks overlap.

With the selection of the set of vectors $\{\underset{\sim}{\omega}^{(j)}\}$ as internal
variables, it is suggested to derive the constitutive equations
from the thermodynamic potential - Helmholtz free energy. Thus,
the problem of constitutive equations is formulated as concretization of the potential. Such a concretization is proposed in the
form involving combined invariants of strain and vectors $\{\underset{\sim}{\omega}^{(j)}\}$.
It is given by formula (16) of [3] and, in a slightly different
form, by formula (19) of [4]. According to the latter, the poten-
tial-strain energy is given by the expression:

$$\rho\psi = \frac{1}{2}(\lambda + 2\mu)\,\varepsilon_{kk}\varepsilon_{\ell\ell} + \mu(\varepsilon_{kk}\varepsilon_{\ell\ell} + \varepsilon_{k\ell}\varepsilon_{k\ell}) +$$

$$(\omega_p^{(\alpha)}\omega_p^{(\beta)})^{-1/2}\,[C_1\,\varepsilon_{k\ell}\varepsilon_{mm}\,\omega_k^{(\alpha)}\omega_\ell^{(\beta)} + C_2\,\varepsilon_{k\ell}\,\varepsilon_{\ell m}\,\omega_k^{(\alpha)}\,\omega_m^{(\beta)}]$$

This formula is derived from the following two requirements:
(1) the potential should be invariant with respect to all orthogonal transformations. This follows from isotropy of material in the
absence of damage and is analogous to the corresponding requirement
of the theory of isotropic elasticity; since both strain $\underset{\sim}{\varepsilon}$ and
vectors $\{\underset{\sim}{\omega}^{(j)}\}$ are arguments of the potential the latter will depend
not only on the invariants of ε but on the full set of invariants
of both $\underset{\sim}{\varepsilon}$ and $\{\underset{\sim}{\omega}^{(j)}\}$ including their combined ones.
(2) Written as a sum of all such invariants taken with some (unde-
termined) coefficients the potential is further concretized by
imposing two conditions: (a) it is quadratic in strains (for the
stress-strain relations to be linear if the damage field is constant) and (b) it is linear in damage.

The above written expression for the potential constitutes
the proposed concretization of the general thermodynamical relations. The following remarks representing the opinion of the
reviewer can be made in connection with this expression.

(I) The number of undetermined coefficients in the model
(playing the role of adjustable parameters) appears to be substantially underestimated. Indeed, there seems to be no reason to
assume that the undetermined coefficients C_1 and C_2 of the combined
invariants have to be the same for the invariants with different
α, β, i.e. for the combined invariants involving different vectors
$\underset{\sim}{\omega}^{(j)}$. Then, the bracket will contain $n(n+1)$, rather than two,
coefficients $C_1^{(\alpha\beta)}$, $C_2^{(\alpha\beta)}$ where n is the number of families of
parallel microcracks into which the original microcrack field is
broken. On the other hand, differentiation of the potential with

respect to strains yields stresses. The total number of elastic constants contained in thus obtained stress-strain laws is restricted by the symmetry conditions for the tensor of elastic constants and cannot exceed 21. Since the elastic constants will be combinations of vectors $\underset{\sim}{\omega}^{(J)}$ and coefficients $C_k^{(\alpha\beta)}$, the latter, in view of the restricting relations, will have to be re-adjusted every time the damage field is changed.

Thus, the actual number of undetermined coefficients contained in the model may be quite large. This appears to be a natural consequence of using general thermodynamic relations with a large number of parameters of state-vectors $\underset{\sim}{\omega}^{(J)}$.

(II) In the case when damage ω is <u>isotropic</u>, linearity of the potential in damage is a direct consequence of the original L. Kachanov's concept of the actual stress elevated due to a cross-sectional area loss: relation $\sigma_{actual} = \sigma_{nominal}/(1-\omega)$ contains ω linearly. If damage is <u>anisotropic</u>, the assumption of linearity of the potential in the damage parameter(s) seems to be a generalization of the actual stress concept. In this case, however, validity of the linearity assumption appears to depend on how the damage parameter(s) used in the model reflects the area losses in different directions. The relation between the damage parameters $\{\omega^{(J)}\}$ and the area losses used in the model involves the additional concept of "cone of influence of a crack" and "angle of diffusion of stresses" (discussed above). Further examination of the linearity assumption seems therefore desirable.

(III) Assuming that the hypothesis of linearity is accepted, note that the constructed expression for the potential does not appear to be really linear with respect to $\underset{\sim}{\omega}^{(J)}$'s; rather, it contains ratios of terms quadratic in $\underset{\sim}{\omega}^{(J)}$'s to square roots of terms quadratic in $\underset{\sim}{\omega}^{(J)}$'s. The difficulty in forming combined invariants of $\underset{\sim}{\varepsilon}$ and $\{\underset{\sim}{\omega}^{(J)}\}$ seems to be a consequence of vectorial, rather than tensorial representation of damage. For instance, if a second order tensorial parameter $\underset{\sim}{\alpha}$ were used, the combined invariants of $\underset{\sim}{\varepsilon}$ and $\underset{\sim}{\alpha}$ linear in $\underset{\sim}{\alpha}$ and quadratic in $\underset{\sim}{\varepsilon}$ could have been easily constructed as $\varepsilon_{ij}\,\varepsilon_{jk}\,\alpha_{ik}$, $\alpha_{kk}\,\varepsilon_{ij}\,\varepsilon_{ij}$, $\varepsilon_{ii}\,\varepsilon_{jk}\,\alpha_{jk}$ and $\alpha_{kk}\,(\varepsilon_{ii})^2$.

(IV) Since differentiation of the potential with respect to strains yields stresses, the expression for $\rho\psi$ written above incorporates the effect of damage on the elastic properties. This, as stated in [4], establishes a link between the proposed model and literature on effective elastic properties of solids with many cracks. In the reviewer's opinion, there are some difficulties in establishing such a link; one of them is the necessity to re-adjust the coefficients $C_k^{(\alpha\beta)}$ when the damage field is changed (discussed above). It can also be remarked that, although a substantial progress has been made recently [10,11] determination

of the effective elastic response of a cracked solid remains a
difficult problem in the general case of anisotropy and non-small
crack concentration.

Note, in conclusion, that, although the framework of irreversible thermodynamics may provide a general structure of constitutive
equations, one cannot obtain sufficient concretization of these
equations on the basis of thermodynamics alone. Such attempts
result in the introduction of undeterminate constants playing,
essentially, the role of adjustable coefficients; their number
depends on the complexity of the damage parameter used in the model.
In the simplest case of one scalar parameter (isotropic damage) this
number is relatively low (two, in the original L. Kachanov's model
and three, in the model of J. Mazars and J. Lemaitre). This allows
one to determine the coefficients from simple experiments and thus
makes the modelling practical. If, however, a more complex case
of anisotropic damage is considered, the number of coefficients
increases dramatically. It seems that further concretization of
the constitutive equations calls for investigations of the actual
micromechanics of damage, including the problem of microcrack
interactions.

The following remarks on continuum modelling of damage represent the opinion of the reviewer.

Damage parameters used in constitutive equations can be
treated either phenomenologically or on the basis of micromechanical considerations. In the latter case the damage parameter
appears as a result of certain procedure of statistical averaging
over a representative volume. The choice of such procedure appears
to be of importance.

The simplest averaging procedure is taking a volume average
without regard to the defect positions inside the representative
volume. If, for instance, the defects considered are microvoids,
then such averaging results in a volumetric density of pores
(porosity). For a medium with cracks, averaging of this kind was
introduced by A. Vakulenko and M. Kachanov (1971) and, in a more
clear form, by M. Kachanov (1980); the result of averaging is a
symmetric second order tensor ("crack density tensor") defined as
follows. A single crack with surface S is described by a tensori-
al δ-function concentrated on S: $\underset{\sim}{n}\,\underset{\sim}{n}\,\delta$ (S) where $\underset{\sim}{n}$ is unit normal
to S and $\underset{\sim}{n}\,\underset{\sim}{n}$ denotes a dyad (tensor product); note that it is
independent of choice of sense of $\underset{\sim}{n}$. If the crack surface is flat
then $\underset{\sim}{n}\,\underset{\sim}{n}$ is constant along S, otherwise it is variable along S.
Geometry of an array of N cracks is fully described by a sum
$\sum\limits_{i=1}^{N}\underset{\sim}{n}_i\,\underset{\sim}{n}_i\,\delta(S_i)$ of δ -functions concentrated on the crack surfaces.

Crack density tensor $\underset{\sim}{\alpha}$ is introduced as a volume average of this sum: 2)

$$\underset{\sim}{\alpha} = \left\langle \sum_i n_i n_i \ \delta(S_i) \right\rangle_V = \frac{1}{V} \sum_i \int_{S_i} n_i n_i \ dS = \binom{\text{for flat}}{\text{cracks}} \frac{1}{V} \sum_i S_i n_i n_i \quad (1)$$

Where summation may be substituted by integration over orientations if the crack distribution can be approximated by a continuous one.

Thus, $\underset{\sim}{\alpha}$ describes geometry of a crack field averaged over a representative volume V. This description takes into account the preferential orientations of the cracks.

The following properties of $\underset{\sim}{\alpha}$ can be mentioned:

(a) the linear invariant of $\underset{\sim}{\alpha}$ is $\rho = (\sum S_i)/V$ - total area of cracks per unit volume characterizing concentration of cracks.

(b) for one system of parallel cracks, $\underset{\sim}{\alpha} = \rho \underset{\sim}{n}\underset{\sim}{n}$.

(c) in the general case, being a sum of symmetric dyads, $\underset{\sim}{\alpha}$ is symmetric and can, therefore, be represented in its principal axes as $\underset{\sim}{\alpha} = \alpha_1 \underset{\sim}{e}_1 \underset{\sim}{e}_1 + \alpha_2 \underset{\sim}{e}_2 \underset{\sim}{e}_2 + \alpha_3 \underset{\sim}{e}_3 \underset{\sim}{e}_3$. It means that any field of cracks within the elementary volume V can be represented by three mutually orthogonal systems of parallel cracks. For the isotropic ("chaotic) crack array, $\underset{\sim}{\alpha} = \frac{1}{3} \rho \underset{\sim}{I}$ where $\underset{\sim}{I}$ is a unit tensor so that the scalar ρ provides an adequate description of the array.

Note that this tensor was used by A. Dragon [12] in the studies of brittle fracturing of rocks and by S. Murakami and M. Ohno [13] in a theory of anisotropic creep damage (where it was called "damage tensor" and denoted by $\underset{\sim}{\Omega}$).

As has been remarked in [7], the crack density tensor $\underset{\sim}{\alpha}$ can be modified depending on the physical property considered (effective elastic properties, fluid filtration, conductivity). Indeed, according to (1), the contribution of a given crack into $\underset{\sim}{\alpha}$ is proportional to S_i . If the effective elastic properties of a medium with cracks are considered, then the relative "weight" of the i-th crack is generally proportional to $S_i^{3/2}$ (at least for dilute

---

2. According to properties of a δ-function concentrated on a surface S, integral of any function $f(x) \ \delta(S)$ over a volume V containing S is reduced to an integral over S:

$$\int_V f(x) \ \delta(S) \ dv = \int_S f(x) \ dS$$

concentrations). In this case, it is appropriate to take $\underset{\sim}{\alpha}$ as a volume average of the δ-field $\sum n_i n_i \sqrt{S_i} \, \delta(S_i)$, i.e.

$\underset{\sim}{\alpha} = \frac{1}{V} \sum S_i^{3/2} \, \underset{\sim}{n}_i \underset{\sim}{n}_i$. On the other hand, if $\underset{\sim}{\alpha}$ were used for characterization of damage and the contribution of the i-th crack into the reduction of the effective cross-sectional area were assumed to be proportional to S_i , then (1) gives the appropriate definition of $\underset{\sim}{\alpha}$.

The simplicity of description of a crack field - by a symmetric second order tensor $\underset{\sim}{\alpha}$ - is achieved at the expense of certain information not reflected in $\underset{\sim}{\alpha}$. In particular, although $\underset{\sim}{\alpha}$ describes, in the average sense, orientations and sizes of individual cracks, it is insensitive to their <u>locations</u> within the representative volume V. How appropriate is it to use $\underset{\sim}{\alpha}$ as a parameter of state in mechanical models? It depends on the physical property considered, and, in particular, on the sensitivity of this property to mutual locations of microcracks.

Consider, for example, the <u>effective elastic properties</u> of a medium with cracks. Assume, for simplicity, that the cracks are far apart so that the approximation of non-interacting cracks is applicable and each crack can be treated as embedded into the remotely applied stress field. Consider two crack configurations shown (Fig. 1), each containing only two cracks. The crack density tensor $\underset{\sim}{\alpha}$ will be the same for both of them. The effective response of the specimen is determined by the average relative displacement of its upper and lower edges. If the cracks do not interact, this displacement is a sum of the displacements due to each of the cracks considered as single ones and subjected to the remotely applied stress. Therefore, the effective elastic properties will

Figure 1. Two crack configurations with the same crack density tensor $\underset{\sim}{\alpha}$.

be the same for both configurations. From the point of view of
damage, however, these configurations may be different (depending
on the rheology of the material) since average stress in the
dangerous cross-section is higher in the second configuration.

This simple example illustrates that the effective elastic
moduli (representing the volume average of the material's response)
are, generally, less sensitive to "details" of the crack array
geometry than the fracture-related properties. Therefore, although
the use of simple volume average-tensor $\underset{\sim}{\alpha}$ as a parameter of state
may provide some insight into the effective elastic properties [7],
it can be expected to be less successful for the description of
damage. It seems that more sophisticated averaging procedures reflecting the statistics of microcracks and their interactions within
the elementary volume V are more appropriate for characterization
of local damage.

It appears that the further progress in continuum approach to
damage calls for the studies in micromechanics. In particular, the
problem of microcrack interactions should be mentioned. Its solution may lead not only to a proper characterization of a local
damage, but, also, to calculation of the energy release rates associated with propagation of the fracture zone. It could also enhance
the computer simulation schemes (of the type developed by L. Seaman
et al.) by incorporating the actual micromechanics into such schemes.
A general method of analysis of the microcrack interactions was
outlined in [8] (see also [9] for further results).

Note that development of the continuum description of damage
can, generally, be broken into three problems: (a) introduction of
a damage parameter based on an appropriate statistical averaging
over a representative volume; (b) formulation of the evolution
equations for such parameter; (c) formulation of a criterion of
local failure (local thermodynamical instability). These problems
have been discussed by A. Chudnovsky [14, 15].

Acknowledgment. The author thanks the Director of the Workshop
and the Air Force Office of Scientific Research for partial support.

References

1. Drucker, D.C., Gibson, R.E. and Henkel, D.J., Soil
Mechanics and Work-Hardening Theories of Plasticity. Transactions of the American Society of Civil Engineers, vol. 121
(1956), 338-346.
2. Barker, D.B., Hawkins, N.M., Fure-Lin Jeang Kyu Zong Cho,
and Kobayashi, A.S. Concrete Fractures in a CLWL Specimen.
To appear in Engineering Mechanics Division, ASCE.
3. Krajcinovic, D. Constitutive Equations for Damaging Materials. Journal of Applied Mechanics, vol. 50 (1983), 355-360.

4. Krajcinovic, D. Continuum Damage Mechanics Revisited: Basic Concepts and Definitions. Submitted to the Journal of Applied Mechanics.

5. Sneddon, I.N. and Lowengrub, M. Crack Problems in the Classical Theory of Elasticity, J. Wiley & Sons, 1969.

6. Vakulenko, A.A. and Kachanov, M.L. Continuum Theory of Medium with Cracks. Mekhanika Tverdogo Tela, no. 4 (1971), 159-166 (in Russian).

7. Kachanov, M.L. Continuum Model of Medium with Cracks. Journal of Engineering Mechanics Division, ASCE, vol. 106 (1980), 1039-1051.

8. Chudnovsky, A. and Kachanov, M. Interaction of a Crack with a Field of Microcracks. Letters in Applied Engineering Sciences, vol. 21, no. 8 (1983), 1009-1018.

9. Chudnovsky, A., Dolgopolsky, A. and Kachanov, M. Interaction of a Crack with a Microcrack Array. NASA Technical Report (1984). Also, submitted to the International Journal of Solids and Structures.

10. Kanaun, S.K. Elastic Medium with Random Fields of Inhomogeneities. Chapter 7 in the book by I. Kunin. Elastic Media with Microstructure, vol. 2, Springer-Verlag (1983).

11. Budiansky, B. and O'Connell, R.J. Elastic Moduli of a Cracked Solid. International Journal of Solids and Structures, vol. 12 (1976), 81-97.

12. Dragon, A. On Phenomenological Description of Rock-Like Materials with Account for Kinetics of Brittle Fracture. Archiwum Mechaniki Stosowanej, Warsaw, Poland, vol. 28, no. 1 (1976), 13-30.

13. Murakami, S. and Ohno, N. A Continuum Theory of Creep and Creep Damage. In "Creep in Structures", Springer, Berlin (1981), 422-444.

14. Chudnovsky, A.I. A Theory of Long-Time Strength in Fatigue and Creep. Workshop on a Continuum Mechanics Approach to Damage and Life Prediction, Carrollton, Kentucky (1980), 79-87.

15. Chudnovsky, A.I. Statistics and Thermodynamics of Fracture. To appear in Letters in Applied Engineering Sciences.

SECTION VI

STRAIN-RATE AND DYNAMIC EFFECTS ON
CRACK PROPAGATION

**APPLICATION OF FRACTURE MECHANICS
TO CEMENTITIOUS COMPOSITES**
NATO-ARW - September 4-7, 1984
Northwestern University, U.S.A.
S. P. Shah, Editor

DYNAMIC EFFECT IN CONCRETE MATERIALS

Robert L. Sierakowski

Department of Civil Engineering, The Ohio State Univesity
470 Hitchcock, 2070 Neil Ave., Columbus, OH 43210

INTRODUCTION

The subject of predicting and evaluating the influence of dynamic effect in materials appears to be a subject of increasing
awareness and importance to the scientific community. In Civil Engineering Technology this problem is considered as a most important
and timely topic particularly as it relates to the establishment of
appropriate design methodology associated with masonry and concrete
material structures of both the reinforced and non-reinforced
types.

Dynamic effects are produced by introducing rapidly applied
loads of short duration into structrual elements. Therefore, by
impact or impulsive loading is implied the interactions occurring
between impacting bodies and structures, and the corresponding
forces acting at the contacting surfaces of these structures. The
loading of structures subjected to these type of events are thus
generally classified as Impact/Impulsive Loads.

Of principal interest to Engineers is the mechanical behavior
of the structural materials, as exposed to these rapidly applied
loads. A quantification of these events can generally be cate-
gorized either by the time repsonse of the imposed load relative to
the natural period of the structure, or alternatively by an established strain rate. Summary tables identifying these characteristics have been noted in Tables 1 and 2.

LOAD CLASSIFICATION

Load Classification	τ/T*	Type of Load
Quasi-Static	>4	Conventional testing of concrete
Quasi-Impact	~ 1	Transient loading on structures
Impulsive and Impact	<0.25	Kinetic Energy, blast loads
Shock Loads	$<10^{-6}$	High energy explosives

* τ/T is the ratio of load duration (τ) to characteristic response time (T).

Table 1.

STRAIN RATE REGIMES

Load Classification	Strain Rate	Material Characterization
Quasi-Static	$10^{-4} - 10^{-1}$/sec	Constitutive Eq.
Intermediate	$10^{0} - 10^{1}$/sec	Constitutive Eq.
High	$10^{2} - 10^{4}$/sec	Constitutive Eq.
Shock Waves	10^{5}/sec	Eq. of State

Table 2.

One of the more significant characteristics of impact or impulsive type loads in the regime where material constitutive laws are applicable is the complicated material responses observed including also modes of deformation and fracture mechanisms. Generally., these responses can be broadly classified as near field and far field effects. The near field response can be characterized by the geometry, rigidity, and mass of the target media as well as the intensity, mass, and velocity of the defined impact/impulsive load. The far field response is controlled primarily by the stress wave propagation, bond strength between aggregate/reinforcement, material constitutive equations and failure criteria of the constituent materials. In addition, the velocity/intensity of the loading influences the strain rate which has an effect on the stress wave propagation and the corresponding failure mechanisms occurring in the structure.

A key feature associated with dynamic loadings is thus the interactive nature of the events which occur. This can graphically be represented in Figure 1. Figure 1 displays the interactive role of the important events necessary to understand and model the behavior of materials subjected to impact/impulsive loads. That is, in order to determine dynamic properties necessary as modeling parameters for dynamic events one must have knowledge of Wave Propagation Effects as well as knowledge of the Material Constitutive Equations. However, to understand wave propagation events the very same dynamic properties which are being sought must be known a priori as input to the Constitutive Equations which are to be determined. Thus input from all three events must be synthesized in order to understand and quantify dynamic effects. Each of these topics are selectively highlighted in the accompanying paragraphs in order to review the current ongoing work in these areas as well as for potentially recognizing where further work and deficiencies in our existing data base is necessary.

DYNAMIC PROPERTIES

While it is important to recognize that a number of dynamic properties could be examined the following remarks will be limited to the influence of material strain rate as it effects dynamic properties.

To develop an understanding of strain rate effects in materials recourse is generally made to observations using appropriately designed experimental devices. Early rate effect studies on the behavior in concrete were investigated using controlled rate of loading tests in what we now call conventional testing machines. Only impact type tests, however, achieve rates comparable to those in the impact/impulsive loading regime. Table 2 displays a classification schedule for identifying the various and important strain rate regimes.

At the high end of the quasi-static load spectrum and for measuring dynamic events, a dynamometer placed in tandem with the specimen in testing machines provides an indicator of the impact force. At strain rates above approximately 10^1 sec^{-1} wave propagation effects in specimen and dynamometer complicate the stress determination and must be taken into account. This has led to the

538

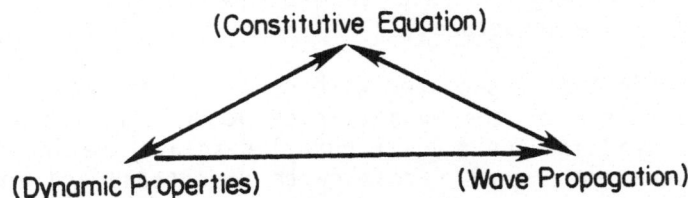

Figure 1.

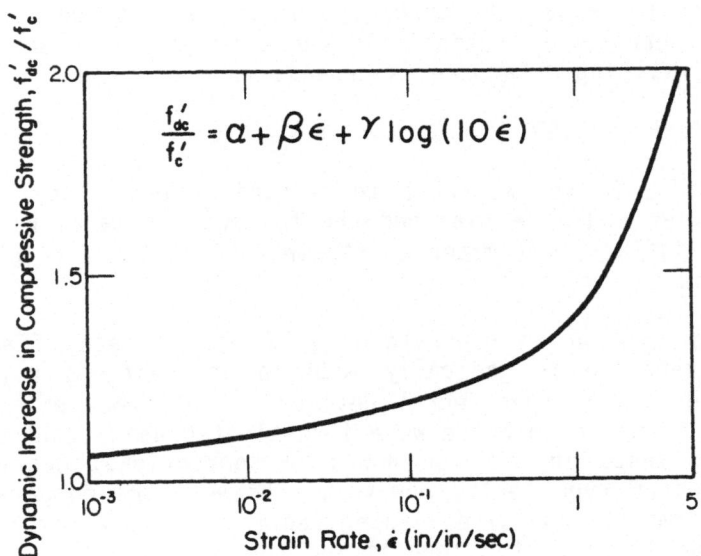

Figure 2.

development of specific testing devices for various high rate regimes. For intermediate strain rates, drop tests using instrumented striker weights have been used to obtain strain rate data. An important element involved with analysis of data using this type of test equipment is the requirement for accountability of inertia effects. For higher strain rates, a useful experimental apparatus is the split Hopkinson pressure bar system which has been used extensively for dynamic tests of metals. The long transmitter bar used with this system serves as a dynamometer measuring the force, and hence the average stress over the cross section, at the specimen face in contact with the pressure bar. For very high strain rates, that is above 10^3 sec^{-1}, flyer plate impact tests have been used as an experimental procedure. Gas guns are generally used as the propelling devices for the flyer plates with stresses generated ranging up to the vicinity of 10^3 kbars. Each of these test procedures is discussed dependently.

Drop Tests

Some of the earliest work on rate effects in concrete were reviewed in an ASTM symposium dedicated to the effects of speed in Mechanical Testing of concrete (24). At stress loading rates from 1 to 1000 psi per second, using conventional testing machines, the compressive strength was reported to be a logarithimic function of the loading rate, with recorded strength increases up to 100% of the strength reported at so-called standard rates, that is, 20 to 50 psi per sec. By using cushioned impact tests Watstein (1953) obtained strengths 185% of the standard (41). Watstein also reported rate effects on the measured secant modulus of elasticity, energy absorption value and strain to failure. These cushioned impact tests used a 140-lb hammer striking a steel capped specimen set in plaster of paris atop a dynamometer mounted on a 3200-lb anvil. Green (15) and Hughes and Gregory (20) used a ballistic pendulum with a hammer having a one-inch diameter face, which impacted one face of a 4-inch cube specimen mounted against an anvil. The anvil was also suspended as a pendulum to measure retained kinetic energy. Atchley and Furr (1) used a drop tester with drop heights up to 20 ft and found dynamic strengths up to more than 1.6 times the static strengths.

Seabold (32) studied rate effects related to the unconfined compressive strength in concrete up to strain rates of about 5 sec^{-1} and introduced empirical formula containing both a linear term and a logarithmic term in the strain rate. Figure 2.

Most recently Hughes and Gregory (20) introduced a drop tester with a 48.2-lb hammer falling through a height of up to more than 6 feet. The specimens tested were mainly 4-inch cubes, although some 4 x 4 x 8 inch prisms were also studied. Each of the specimens were fitted with metal plates on both the top and bottom and mounted atop a steel load column fitted with strain gages to record the transmitted force. The load column used was not of uniform diameter, and was also short in geometric length to avoid reflected waves interacting during the recording. By using simple bar-wave theory they were able to estimate the transient stresses occurring and concluded that impact strengths averaged about 1.92 times the static compressive strength of the material. This number is comparable to the dynamic strength increases reported by Watstein in his drop test experiments. It was found by Hughes and Gregory that in some of their strong concretes values greater than 37,000 psi were found. Hughes and Watson (21) used a similar test technique and observed that the increase in compressive strength in dynamic tests over the static strength was greater for low-strength concrete than for the corresponding high-strength concrete.

Pressure Bar Tests

Dynamic elastic properties of concrete bar specimens have also been studied using wave propagation tests by Goldsmith (11, 12, 13, 14). Analyzing one dimensional wave propagation induced by impacting a long bar with a steel sphere the wave speeds, fracture strains, and energy required to fracture geological and cementious materials have been studied. The dynamic tensile strength and strain energy required to fail concrete specimens have been measured in similar longitudinal impacts using various scaled projectiles by Birkimer and Lindemann (6) and by Griner (17) and Sierakowski (33, 34). The latter tests (33, 34) have applied the split Hopkinson bar to unconfined 3/4-inch-diameater specimens of fine-grained concrete and established the feasibility of the method, but no other tests in the literature have given data for concrete tested in this manner, and very little data is available for the range from 10^1 to 10^3 sec^{-1}. Results obtained from these tests showed that dynamic compressive strengths around 26,400 psi could be reached using 3/4-inch-diameter specimens. This was found to be 1.93 times the static compressive strength of 13,690 psi. The specimens used were controlled both in size and aggregate, and had been aged for four years in laboratory storage in a secure environment. They were cut from some of the same bars previously tested by Sierakowski (33) who had found dynamic strengths around 9000 psi as compared to a static strength of 7900 psi for 28-day concrete. The static strengths of the four-year aged concrete were also measured on 3/4-inch-diameter specimens comparable to the impact loaded specimens.

Bhargava and Rehnstrom (4) have recently used a large sized split Hopkinson pressure bar to test concrete specimens. For their experiments, they used a 4-inch-diameter, 8-inch-long specimen and a 10-inch-long striker bar. This bar configuration produced an impact pulse length of the order of 100 microseconds, which was approximately twice the transit time of an elastic loading pulse through the specimen and thus complicated interpretation of the test data. They did find however, dynamic strenghts of 1.46 to 1.67 times the static strength in plain concrete as well as for fiber-reinforced and polymer modified concrete. Their so-called failure strengths were identified as the maximum amplitude of short-pulse loads that could be transmittted through the specimen and were associated with initial cracking rather than with complete crushing of the cylinders tested. The vertical bar apparatus used was 108 ft high, and the 10 inch long striker bar could be dropped through heights of up to 32.8 ft to impact the 6 ft long incident bar.

Among the very few other applications of split Hopkinson bar technology to concrete is that made by Kormeling (22), for the case of dynamic tensile tests. In these tests, the reported dynamic tensile strengths were more than twice the static value at strain rates of approximately 0.75 sec^{-1}. It should also be mentioned that Suaris and Shah (35, 36) have recently surveyed mechanical properties of materials subject to impact and rate effects in fiber-reinforced concrete. Noteworthy is the fact that these investigators as well as others have found significant increases in the impact tensile strength in fiber-reinforced concrete over that of plain concrete (19).

It should also be mentioned that Drop and Pressure bar tests have generally been performd in a loading environment for which confined pressure loading has been absent. In many situations it would appear relevant to consider localized impact loadings of concrete cylinders with various degrees of lateral confinement. Some limited data has been obtained by Takeda (37) on triaxial testing of concrete cylinders loaded by compressive and tensile axial loads for axial straining rates of 2×10^{-5} sec^{-1}, 2×10^{-2} sec^{-1} and 2 sec^{-1}. These loads were applied after statically applying a confining pressure up to 1.5 times the uniaxial compressive strength had been imposed. Their results when plotted in the form of octahedral shear stress versus octahedral normal stress, normalized by dividing by the uniaxial compressive strength at the particular rate of straining, showed no significant differences among the relations at the three rates of straining.

Flyer Plate Tests

At the higher stress levels associated with strain rates of the order of 10^3 sec^{-1} and greater, Read and Maiden (31) have surveyed the state of the art. They were unsuccessful in modelling the observed dynamic behavior and concluded that additional experiments were needed. Higher stress levels than those examined by the previous authors were studied most extensively from experimental wave propagation data generated by Gregson (16). The latter's test procedure utilized a gas gun to provide flyer plate impacts against thin concrete plates in order to determine the uniaxial strain response of the material at pressures from 40,000 to 8,000,000 psi. Below 360,000 psi a multi-wave system was confirmed with a dispersive elastic precursor observed.

WAVE PROPAGATION

In order to investigate dynamic effects in concrete materials over impulse/impulsive loading ranges where constitutive equations modelling retains importance studies associated with wave propagation events are useful. Among the more important parameters for study are pulse speed, attenuation, gemometric/material dispersion effects, as well as fracture characterization.

One of the earliest experimental efforts examining the effects of pulse propagation in concrete was reported on by Landon and Quinney (23) Fig. 3. In this experiment the effects of a pressure pulse generated by detonation of an eight ounce primer of gun cotton placed in contact with one end of a concrete bar 36 inches in diameter was discussed. Spalled pieces of concrete generated by the pressure pulse were used to reconstruct a pressure versus distance curve. The early research led to establishing a data base on fracture response of brittle materials.

Comprehensive studies concerning wave propagation in certain classes of rock and concrete were performed later by Goldsmith (11, 12, 13, 14). At a presentation to the IUTAM Symposium on Stress Waves in Anelastic Solids in 1963 test results on the rock diorite were reported (11). The rock material was cut into laboratory test specimens from geological outcrops into the shape of cylindrical bars 0.75 in diameter and 22 inches long. Surface strain gages were mounted on test specimens and longitudinal waves were introduced by spherical and flattened projectiles using an air gun apparatus. Impact velocity was held constant at approximately 330 in/sec for the test program. Pulse transmission and attenuation were monitored and wave speeds calculated. Results of these tests indicated that the pulse propagates without dispersion and with an attenuation proportional to some power of the ratio of the distance traversed to bar diameter for initial wave passage.

Subsequently, Goldsmith (12) presented wave propagation data on a number of geologic materials including diorite, a coarse grained leucogranite, a fine grained spessartite, and a fine grained basalt. The test procedures used in these studies were similar to that as used initially in (11) for diorite. Results from these tests indicated that in leucogranite wave attenuation

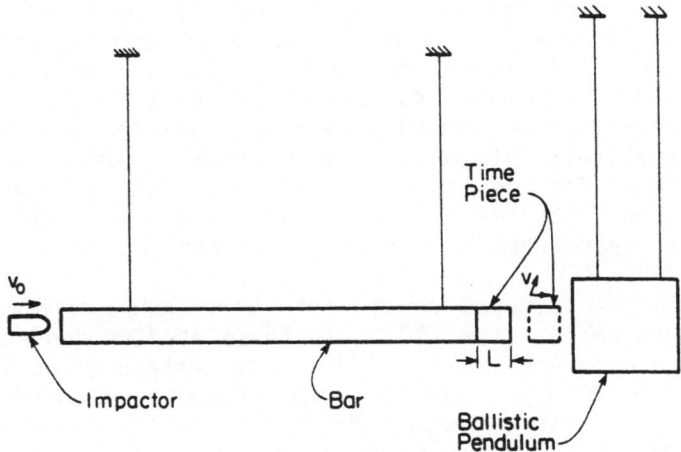

Pressure Bar Test Arrangement

Figure 3.

was significant. Spessartite was found to be highly elastic, and exhibited virtually no attenuation or dispersion during initial pulse passage, and was not affected by repeated passage of stress waves. Basalt showed slight attenuation and no dispersion. The rock types were ranked according to their elastic properties as follows: spessartite (most elastic), basalt, diorite, and leuco-granite (least elastic). The factors which most influenced the dynamic behavior of the igneous rock types were found to be grain size, grain bond strength, and the effect of repeated shocks.

Following these studies on rock specimens, Goldsmith and co-workers (14) conducted a similar series of tests on laboratory pre-pared specimens of concrete. The specimens were again 0.75 inches in diameter and 24 inches in length. Pulse propagation data was obtained from strain gages mounted to the surface of the con-crete. A spherical impactor was used with impact velocities of 1600 and 3300 inches per second. Stress amplitudes of approxi-mately 5000 psi were calculated (based upon the static modulus). The pulse propagated with essentially little discernible dispersion and relatively little or no attenuation, similar to the results of that found for the igneous rock basalt. The wave propagation speed was found to be similar to that of the coarse grained igneous rocks and was approximately 144,000 inches per second.

Further studies on concrete bars were performed by Goldsmith and reported in 1968 (13). In these studies conventional concrete was again investigated with focus on certain effects such as 1) the influence of aggregates on gage response imbedded directly below the gage, 2) strain amplitude versus increasing impact velocity, 3) the use of a protective metal cap on the impact end of the bar versus an uncapped bar, and 4) prestressing using steel rods. Concrete composed of a lightweight aggregate was also studied and compared with the dynamic properties of conventional concrete. For the concrete composed of lightweight aggregate the resulting data indicated that the wave speed was 123,000 inches per second and the ratio of the dynamic Young's modulus to the static value was 1.2. For conventional concrete the wave speed noted was found to be 144,000 inches per second. For these materials pulse attenuation was found to increase with increase in the strain level. It was further concluded that for some rocks a threshold strain level was reached where upon any further increase in the impact generated by the striker resulted only in greater comminution of the impact end. Placement of a metallic cap on the impact end reduced this effect and resulted in a 100 percent increase in the strain level reached within the bar. Also, significant differences in data were noted for strain gages placed on pieces of aggregate as distinct from results obtained with gages located on the cement proper.

More recent studies conducted by Griner (18) and Sierakowksi

(33) examined wave propagation data at velocities in the vicinity of the threshold level for producing first tensile fracture in concrete bar specimens 3/4" and 1½" diameter by 36" long. The effects of impactor length on wave shape, attenuation, and pulse speed for several controlled cement aggregate specimens have been considered. Also evaluated was the influence of end protective caps and epoxy rebonding as effecting the pulse passage along the bars. Some of the effects noted in these studies were that compressive pulse speeds were changed by up to 10% in shifting from medium to coarse grain aggregates and that cure time was similarly influenced. In addition, compressive pulse speeds were found to increase while tensile pulse speeds decreased.

CONSTITUTIVE EQUATIONS

Appropriate material modeling as delineated by constitutive equations plays an important role over the strain rate range extending from 10^{-4} sec^{-1} up to 10^4 sec^{-1}. Beyond this latter range, that is in the shock wave regime, the material can be considered as a fluid medium for response purposes with strength considered as a principal defining parameter. In the following paragraphs a review of selective areas in the constitutive equation modelling of concrete are described under the topical headings of Linear/Non-Linear Elastic Models, Plastic Models, Axiomatic Models, Rheological Models, and Empirical Models.

Linear/Non-Linear Elastic Models

The observed nonlinear behavior of certain reinforced concrete structural elements, for which the stress state can be basically idealized as of the compression-tension type, is mainly due to tensile cracking of the concrete and the resulting reduction in concrete stiffness. To model this observed behavior, a number of researchers have developed and employed linear elastic fracture models as for example Ngo and Scordelis (25) and Phillips and Zienkiewicz (28). In these models, the stress-strain relations for uncracked concrete have been developed based upon a linear theory of elasticity and defined completely by Young's modulus E, Poisson's ratio ν, or bulk modulus $K = E/3(1-2\nu)$ and the corresponding shear modulus $G = E/2(1 + \nu)$. Expressions involving the volumetric strain, σ_{kk}, mean normal stress $\rho = \sigma_{kk}/3$, deviatoric stress $S_{ij} = \sigma_{ij} - \sigma_{kk}\sigma_{jj}/3$, and shear deformation defined by $\ell_{ij} = \sigma_{ij} - \sigma_{kk}\delta_{ij}/3$ have been developed. The same type of linear elastic constitutive relations has been used for cracked concrete; however, the instantaneous tangent modulus has been modified in such a manner that the stresses in the direction perpendicular to the crack are zero. This model does not take into account the sudden stress release due to propagation of the crack which may induce further cracking. Attempts have been made to modify this model by incorporating the sudden release of fracture stresses in a global manner or by including provision for additional crack generation by the opening or closing of existing cracks. These criteria have been introduced by Chen and Suzuki (9) and by Chen and Ting (10).

Other constitutive equation relations have been developed in terms of the stress rate and strain rate. That is,

$$\dot{\rho} = K_t \dot{e} kk$$

$$\dot{S}_{ij} = 2G_t \dot{e}_{ij}$$

In these equations, $K_t = K_t(G_{ii})$ represents the tangent bulk modulus which is a function of the first stress invariant while $G_t = G_t(S_{ij}S_{ij}/2)$ is the shear modulus which is a function of the second stress invariant. Using the current values of the tangent moduli, similar modifications have been used for the inclusion of the cracks in the model as in the linear elastic fracture model.

Inelastic/Plastic Models

Such models take into account the limited plastic flow observed in the stress-strain behavior of concrete before crushing. The simplest model constructed is based upon the perfectly plastic model. The resulting constitutive relation is divided into three parts, the first of these regimes being before plastic flow occurs, the second during plastic flow, and the third after fracture. Included in such relations are an inherent yield and fracture criteria (Drucker/Prager criterion, and Coulomb criterion). For such models, the incremental plastic flow law which expresses the normality of the plastic deformation rate vector to the yield surface can be written as,

$$d\varepsilon^p_{ij} = \lambda \frac{\partial g}{\partial \sigma_{ij}}$$

where $d\varepsilon^p_{ij}$ is the incremental plastic strain tensor, σ_{ij} is the stress tensor, g is the yield function, and λ is a positive proportionality factor. This model has been argued as not meeting certain requirements of continuum mechanics when applied to brittle concrete materials, as discussed for example by Chen and Ting (10).

To modify the perfectly plastic model and remove some of these deficiencies, an elastic strain hardening plastic model has been introduced by Chen and Chen (8). Using the normality condition of the incremental theory of plasticity, it is possible to obtain the following flow rule:

$$d\sigma_{ij}^{p} = \frac{1}{h} \frac{\frac{\partial g}{\partial \sigma_{ij}}}{\sqrt{\frac{\partial g}{\partial \sigma_{mn}} \frac{\partial g}{\partial \sigma_{mn}}}} dg$$

Here the function $f(\varepsilon_p) = g(\sigma_{ij})$, $h = df/d\varepsilon_p$, and $\varepsilon_p = d\varepsilon_{ij}^{p} d\varepsilon_{ij}^{p}$. Matrix constitutive equations based on this model have been presented by Chen and Chen (8).

To describe the dynamic behavior of concrete, as well as the non-linear effects occurring in elastic-viscoplastic fracture models analytical methods have been developed by Nilsson (26). In this model a rate hardening parameter, H_r, is introduced in order to take into account the effects of strain rate on the ductile yield and brittle fracture surfaces.

$$H_r = f_{cv} [c_1 + c_2 \ln \dot{\varepsilon}_{ef} + c_3 (\ln \dot{\varepsilon}_{ef})^2]$$

Where $\dot{\varepsilon}_{ef}$ is the effective strain rate expressed in terms of the octahedral normal and shear strains, $\dot{\varepsilon}_0$ and $\dot{\gamma}_0$, respectively.

$$\dot{\varepsilon}_{ef} = w_1 \dot{\varepsilon}_0^2 + w_2 \dot{\gamma}_0^2$$

Here the quantities w_1 and w_2 are weight coefficients introduced to take into account different microfailure mechanisms. The parameters c_1, c_2, and c_3 are determined by fitting of the data from dynamic compression tests of concrete.

Using a modification of Perzyna's theory of elastic-viscoplasticity Perzyna (27), Bicanic and Zienkiewicz (5) have developed a rate and history dependent model for plain concrete. The material behavior in this model is described using two surfaces in principal stress space, these being, the discontinuity surface defining the departure from elasticity, and the strength limit surface which monitors the initiation of the softening regime.

Axiomatic Models

In the classical incremental flow theory of plasticity, a yield criterion is postulated and a hardening rule is specified for the subsequent yield surfaces. Any constitutive law based upon such elastoplastic behavior is considered a discontinuous model consisting of separate stages for loading, unloading, and re-loading. Experimental evidence indicates that separation of the elastic deformation from plastic deformation is a mathematical ide-alization of the complicated actual phenomena. In order to circum-vent some of the difficulties associated with the classical the-ories of plasticity, that is, postulation of a yield surface and its motion in stress space, hardening rules, and unloading cri-teria, several investigators have attempted to present continuous models. Among these axiomatic models, the endochronic theory pre-sented by Valanis (39) has found considerable applications.

In its basic format endochronic theory can be considered as a generalization of the classical theory of viscoplasticity which is both history dependent and strain rate dependent. Using irrevers-ible thermodynamics, Valanis (39) introduced the notion of a pseudo-time scale, the so-called intrinsic time, and presented a constitutive relation in integral/differential form and in terms of an intrinsic time. Such a constitutive model can describe dif-ferent characteristics of material behavior including strain hard-ening, unloading, reloading, cross hardening, and continued cyclic straining.

Bazant (2,3) extended and applied the endochronic theory to different types of materials including sand, rock, plain concrete, and reinforced concrete.

The simple endochronic theory results in an unloading response which is not elastic at the beginning of unloading. Consequently, infinitesimal hysteresis loops in the first quadrant of the stress-strain space are not closed which is in contrast to experimental evidence available on the behavior of metals. In a recent paper, Valanis and Lee (40) presented a modified endochronic model based on a new intrinsic time scale which is a measure of length in the plastic strain space. This model leads to the closure of these in-finitesimal hysteresis loops.

Empirical Models (Equivalent Uniaxial Model)

The majority of current analysis techniques use an equivalent uniaxial constitutive model for analyzing two-dimensional problems such as beams, plates, and thin shells. In this model, the biaxial stress-strain behavior of concrete is treated by a single equivalent stress-strain relationship. Many expressions have been proposed in the literature. The following expression of a model of this type can be cited, as for example, the model presented by Popovics (29) and Buyukozturk and Tseng (7) where,

$$\frac{\sigma}{\sigma_p} = \frac{n}{n-1 + (\frac{\varepsilon_e}{\varepsilon_p})^n} \frac{\varepsilon_e}{\varepsilon_p} \tag{1}$$

Here σ is the principal stress, σ_p is the maximum principal stress, ε_p is the corresponding equivalent strain, ε_e is the equivalent strain, and n is the shape factor which is given by

$$n = \frac{1}{1 - \frac{\sigma_p}{E_0 \varepsilon_p}} \tag{2}$$

The quantity E_0 is the initial slope of the σ-ε_e curve. By differentiating Eq. (1), the slope at a point with equivalent strain ε_e is found to be

$$E = \frac{[1 - (\frac{\varepsilon_e}{\varepsilon_p})^n] \, n(n-1) \frac{\sigma_p}{\varepsilon_p}}{\left[n-1 + (\frac{\varepsilon_e}{\varepsilon_p})^n\right]^2}$$

Due to significant influence of hydrostatic pressure on the behavior of concrete under triaxial state, this model cannot be accurately used in three dimensional analyses.

Rheological Models

These models have been developed based upon studies of the propagation of one dimensional stress pulses generated in impact tests of long bar specimens. Several models can be cited in the literature. The solid friction constitutive model developed by Goldsmith et al (12) and Pozzo (30) seems to describe the dissipation mechanism in concrete better than other models. The stress-strain relation for a solid friction model can be represented in following form:

$$\sigma = E \, \varepsilon + \frac{\dot{\varepsilon}}{\beta\omega}$$

Here ω is the natural frequency and β a damping constant.

Concluding Remarks and Recommendations

The preceeding sections have been used to identify some of the complexities associated with impact/impulsive loadings as well as to selectively review technological issues. This review albeit cursory in scope can be used as a basis for identifying some of the research issues necessary for developing a more complete understanding of the events occurring. Some remarks and recommendations in relation to these issues are indicated in the accompanying paragraphs.

Strain Rate Sensitivity

It is generally accepted that brittle materials such as concrete and mortar are strain rate sensitive with a definitive response defined for different loading modes. This is illustrated in Fig. 4 which highlights schematically the differences in rate sensitivity effects for concrete in terms of tension, compression, and flexural response modes. (Mortar behaves similarly). Omitted from this diagram is data for the dynamic shear mode which is currently unavailable.

Some recent compression test work by Tang (38) for mortar indicates a linear dependence for the ultimate strength as affected by strain rate as shown in Fig. 5 and can be considered as a representative behavior pattern. At a strain rate of 10^3 sec^{-1} the dynamic factor would be 1.92 and lead to a very useful and simplistic correction factor in constitutive equation models which in turn could be used in existing computer algorithms.

Also important are results related to dynamic overload factors. These have been summarized in Fig. 6 which shows the influence of strain rate sensitivity as related to the dynamic load factors couched in terms of the ratios $\sigma_{dyn}/\sigma_{static}$ and based upon maximum/ultimate stress values.

It can be readily observed that a wide range of values have been reported on in the literature these being obtained using various test techniques over a wide range of strain rates. In addition to the range of values a number of gaps in the data base exist. It should also be mentioned that much of this data has been acquired without regard to extensive consideration of such important material parameters as aggregate sizes, cement mix preparations, material aging, bond strength of aggregates or other reinforcements, confining strengths and the effect of repeated shocks. Development of an understanding of these issues is necessary but appears limited by the number of investigators

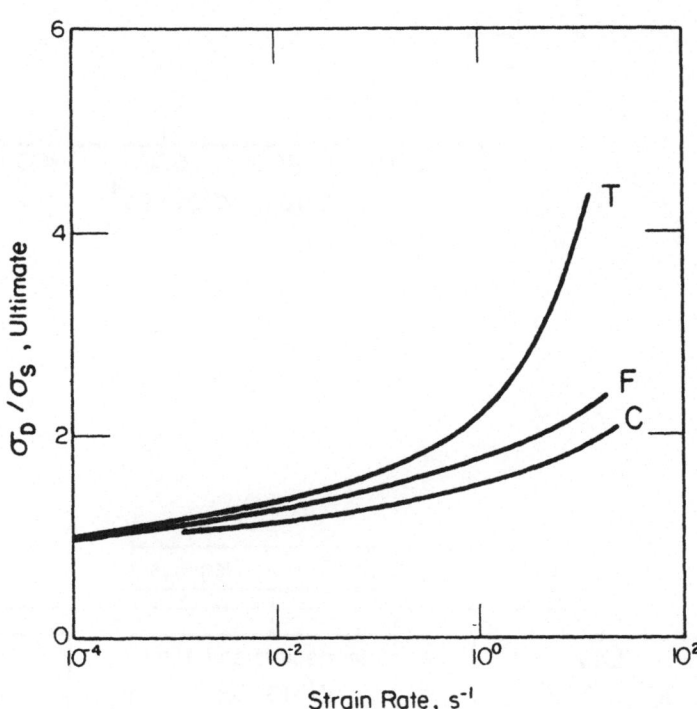

Figure 4.

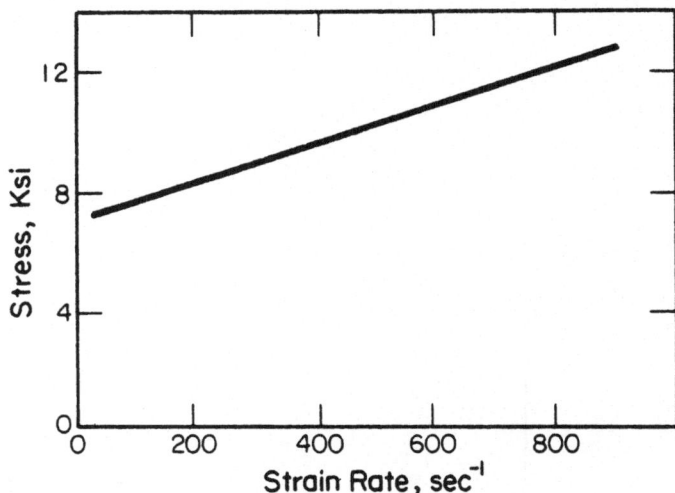

Figure 5.

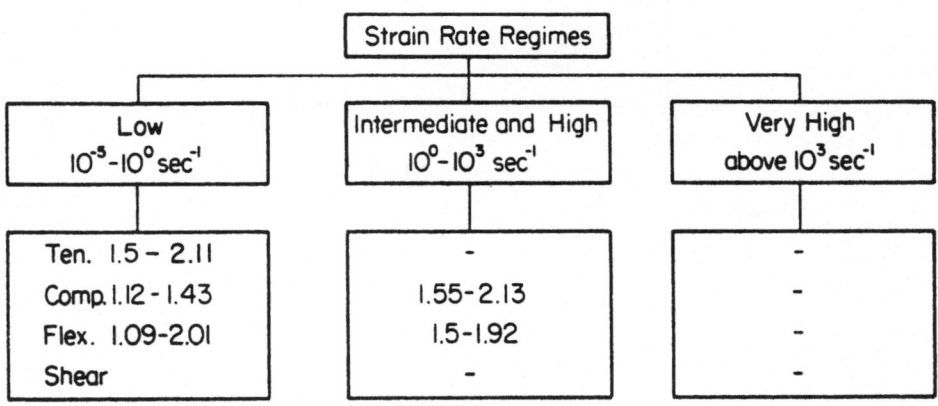

Figure 6.

studying these research areas.

Wave Propagation

Studies on wave propagation as previously noted are important for a number of reasons. If we omit the manner in which the pulse is introduced into the material proper, the principles of wave mechanics can be used as a means for developing predictive methodology for determining particle displacement, wave reflections and transmissions, for dynamic loaded concrete elements. In the case of impact/impulsive loading of materials the effect of localized loads can produce extremely high local stresses resulting in front face spall, comminution and fracturing as well as such far field effects as scabbing.

The propagation of stress waves represent a key feature in producing the effects which result in and control the above failure/fracture mechanisms as well as in developing experimental procedures for determining dynamic properties and constitutive equations.

Some of the important parameters that need to be determined from wave studies for concrete, using analytical or experimental procedures, include data on tension/compressional wave velocitites, effects of microvoids and microcracks on wave progression, effect of aggregate size and reinforcement on pulse attenuation, geometric and material pulse dispersion, damping, effect of material aging, size effects, and scaling parameters.

A number of these factors have been investigated in the papers cited in the section on wave propagation; however, only a selective data base exists. For example, if we examine the compressive wave speeds in concrete bars of fixed diameter, variable aggregate, and cure cycle only the following wave speed data has been observed (Table 3). These results clearly demonstrate the necessity for a more exhaustive evaluation of developing the methodologies for acquiring the necessary data base for the parameters cited.

Table 3

	Cure Time	Compressive Wave Speed
Medium Aggreg.	7 day	140,000 in/sec
	28 day	____
Coarse Aggreg.	7 day	167,000 in/sec
	28 day	____

Constitutive Equation Modelling

It would be ideal if in developing models for constitutive equations subjected to impulse/impulsive loads that a universally applicable methodology could be developed to synthesize all loading modes.

For the present, when considering the development of constitutive equations for use in dynamic loading of concrete, a number of important issues are in need of further study. Principal among these is the testing of a number of the models currently available beyond the one dimensional stress state to multiaxial stress states. The incorporation of previous strain history, that is, the number of prior impacts/impulsive loadings should be known in order to expand predictive modelling capability. In addition, the recognized rate sensitivity as it relates to different loading modes should be incorporated in any model development along with a representation of the Bauschinger effect to account for differences in loading and reversed loading. Also important are the size of the components being tested. That is, the response of concrete and similar structural materials is dependent on the size of the geometric element as well as the distribution of defects such as developed microcracks and inherent microvoids within the materials.

Additional research thrusts are also necessary to develop a more complete understanding of empirical and semi-empirical constitutive laws. Considerable physical evidence for modelling purposes may be gleaned by examining the microstructural response among the tools and methodologies available for establishing physical characterization of material response at the microstructural level are the TEM and SEM, X-ray radiography, and Neutron scattering techniques. Data obtained using such test techniques can be effectively used for both establishing a better physical understanding of the important process variables, such as microvoids and microcracks, as well as determining how these physical results may be interpreted and extended for use in phenomenological models. These data can then be used to realistically assess parameters as used in empirical and semi-empirical models.

It is hoped that the above remarks may prove a stimulus to increasing overall awareness of the research issues and needs in an important technological area.

REFERENCES

1. Atchley, B. L., and Furr, H. L., Strength and Energy Absorption Capabilities of Plain Concrete Under Dynamic and Static Loading, ACI Journal November (1967) 745-756.
2. Bazant, Z. P., Endochronic Inelasticity and Incremental Plasticity, International Journal of Solids and Structures 16 no. 9, Sept. (1978), 151-165.
3. Bazant, Z. P., A New Approach to Inelasticity and Failure of Concrete, Sand, and Rock: Endochronic Theory Proceedings of the 11th Annual Meeting, Society of Engineering Science, G. J. Dvorak, ed., (Duke University, Durham, NC, 1876), pp. 158-159.
4. Bhargave, J., and Rehnstrom, A., Dynamic Strength of Polymer Modified and Fiber-Reinforced Concretes, Cement and Concrete Research 7 (1977), 199-207.
5. Bicanic, N. and Zienkiewicz, O. C., Constitutive Model for Concrete Under Dynamic Loading, Earthquake Engineering and Structural Dynamics 1 (1983), 689-710.
6. Birkimer, D. L. and Lindemann, R., Dynamic Tensile Strength of Concrete Materials, ACI, Journal, Proc. 68 (1971), 47-49 and Supplement No. 68-8 (1971).
7. Buyukozturk, O. and Tseng, T. M., Concrete in Biaxial Cyclic Compression Journal of Structural Engineering, ASCE, 110 no. 3, Mar. (1984), 461-476.
8. Chen, A. C. T. and Chen, W. F., Constitutive Equations and Punch-Indentation of Concrete, Journal of Engineering Mechanics Division, ASCE, 101 no. EM6, Dec. (1975), 889-906.
9. Chen, W. F. and Suzuhi, H., Constitutive Models for Concrete, presented at the October 16-20, 1978, ASCE Annual Convention and Exposition and Continuing Education Program, Chicago, Illinois.
10. Chen, W. F. and Ting, E. C., Constitutive Models for Concrete Structures, Journal of Engineering Mechanics Division, ASCE, 106 no. EM 1, Feb. (1980), 1-19.
11. Goldsmith, W. and Austin, C. F., Some Dynamic Characteristics of Rocks, Stress Waves in Anelastic Solids, (Springer, Berlin, Germany, 1964), p. 277.
12. Goldsmith, W., Austin, C. F. , Wang, C. C. and Finnegan, S., Stress Waves in Igneous Rocks, Journal of Geophysical Research 71, no. 8 (1966), 2055.
13. Goldsmith, W., Kenner, V. H. and Ricketts, E. E. , Dynamic Loading of Several Concrete-Like Mixtures, Proc. ASCE, J. Structural Division, 94, ST7 (1968), 1803-1827.
14. Goldsmith, W., Polivka, M., and Yang, T., Dynamic Behavior of Concrete Experimental Mechanics, Feb. (1966), 65-79.
15. Green, H., Impact Strength of Concrete, Institution of Civil Engineers, Proc. 28 (1964), 361-396.
16. Gregson, V. G., Jr., A Shock Wave Study of Fondu-Fyre WA-1 and a Concrete, (General Motors Materials and Structures Laboratory, Report MSL, 1971).

17. Griner, G. R., Dynamic Properties of Concrete Master's Thesis, University of Florida (1974). Also Griner, G. R., Sierakowski, R. L., and Ross, C. A., Dynamic Properties of Concrete Under Impact Loading Bulletin No. 45, (The Shock and Vibration Information Center, NRL, Washington, D. C., June 1975).

18. Griner, G. R., Sierakowski, R. L. and Ross, C. A., Dynamic Properties of Concrete Under Impact Loading, Shock and Vibration Bulletin Part 5, (1975), 131-142.

19. Hoff, G. C., Selected Bibliography on Fiber-Reinforced Cement and Concrete, Supplement No. 2, Miscellaneous paper C-76-6, (U. S. Army Waterways Experiment Station, Vicksburg, Miss, 1979).

20. Hughes, B. P. and Gregory, R., Concrete Subjected to High Rates of Loading in Compression, Magazine of Concrete Research 24, (1972), 25-36.

21. Hughes, B. P. and Watson, A. J., Compressive Strength and Ultimate Strain of Concrete Under Impact Loading, Magazine of Concrete Research 30, (1978), 189-199.

22. Kormeling, J. A., Zielinski, A. J., and Reinhardt, H. W., Experiments on Concrete Under Single and Repeated Impact Loading, Report No. 5-80-3, (Delft University of Technology, Stevin Laboratory, May 1980).

23. Landon, J. W. and Quinney, H., Experiments with the Hopkinson Pressure Bar., Proc. Roy. Soc. A., 103, (1923), 622.

24. McHenry, D. and Shideler, J. J., Review of Data on Effect of Speed in Mechanical Testing of Concrete, ASTM STP 185, (1956), 72-82.

25. Ngo, D. and Scordelis, A. C., Finite Element Analysis of Reinforced Concrete Structures, ACI Journal 64 no. 3, Mar. (1967).

26. Nilsson, L., Impact Loading of Concrete Structures, Publication No. 79, Department of Structural Mechanics, Chalmers University of Technology, (1979).

27. Perzyna, P., Fundamental Problems in Viscoplasticity, Advances in Applied Mechanics 9, (1966), 263-377.

28. Phillips, D. V. and Zienkiewicz, O. C., Finite Element Nonlinear Analysis of Concrete Structures, Proceedings, Institute of Civil Engineers, 61 part 2, Mar. (1976), 59-88.

29. Popovics, S. A., Numerical Approach to the Complete Stress-Strain Curve of Concrete Cement and Concrete Research 3 (1973), 583-599.

30. Pozzo, E., Rheological Model of Concrete in the Dynamic Field, Mecanica, June (1970) 143-158.

31. Read, H. E. and Maiden, C. J., The Dynamic Behavior of Concrete, (Topical Report 3 SR-707, Systems, Science and Software, La Jolla, CA, August, 1971).

32. Seabold, R. H., Dynamic Shear Strength of Reinforced Concrete Beams - Part III, Tech. Report R-695, Naval Civil Engineering Lab., Port Hueneme, CA, Sept. (1970).

33. Sierakowski, R. L., Malvern, L. E., Collins, J. A., Milton, K. E. and Ross, C. A., Penetrator Impact Studies of Soil/Concrete, Final Report, U. S. AFOSR Grant No. 77-3209 and AFAL TR-78-9, University of Florida, Gainesville, FL, November 30 (1977), 109-110.

34. Sierakowski, R. L., Malvern, L. E. and Doddington, H., Hopkinson Bar Tests of Three-Fourths Inch Diameter Concrete Specimens, Unpublished (1981).

35. Suaris, W. and Shah, S. P., Properties of Concrete Subjected to Impact, ASCE Jounal of Structural Engineering, v. 109, July (1983), pp 1727-1741.

36. Suaris, W., and Shah, S. P., Impact Test Methods for Fibre Reinforced Concrete, ACI Special Publication, 1984.

37. Takeda, J., Tachikawa, J., and Fujimoto, K., Mechanical Behavior of Concrete Under Higher Rate of Loading than in the Static Test, Proceedings of the Symposium on the Mechanical Behavior of Materials, Kyoto, August, 2124, Vol. II (1974), 479-486.

38. Tang, T., Malvern, L. E., Jenkins, D. A., Dynamic Compressive Testing of Concrete and Mortar, Proceedings Fifth Engineering Mechanics Division, ASCE, Laramie, WY, August 1-3, 1984, pp. 663-666.

39. Valanis, K. C., A Theory of Viscoplasticity Without a Yield Surface, Part I - General Theory, and Part II - Application to Mechanical Behavior of Metals, Archives of Mechanics 23, no. 4, (1971), 517-551.

40. Valanis, K. C. and Lee, C. F., Some Recent Development of the Endochronic Theory with Applications, Nuclear Engineering and Design 69, (1982) 327-344.

41. Watstein, D., Effect of Straining Rate on the Compressive Strength and Elastic Properties of Concrete, ACI Proc. 47, no. 52, April (1953), 729-744.

APPLICATION OF FRACTURE MECHANICS
TO CEMENTITIOUS COMPOSITES
NATO-ARW - September 4-7, 1984
Northwestern Univeristy, U.S.A.
S. P. Shah, Editor

TENSILE FRACTURE OF CONCRETE AT HIGH RATES OF LOADING

Hans W. Reinhardt

Stevin Laboratory
Department of Civil Engineering
Delft University of Technology
Delft, The Netherlands

Introduction

Strictly speaking almost all structures are subjected to loads
which vary with time. Road and rail traffic, cranes, pedestrians
cause loads fluctuating in time; goods, furnature, partition walls
are removed and replaced which also causes varying loads. However,
all these loads change rather slowly which allows to treat them
quasi-statically, at most with a coefficient that counts for possible dynamic effects. Material properties are taken from static testing. On the other hand, there are loading cases lasting only a very
short period of time which may be intended or may occur only acci-
dentally. Pile driving or certain machinery can cause short loading
pulses. Extraordinary short time loading can occur by blast waves
due to explosions, by collisions of ships, cars, trains or air
planes, by tornade born missiles, falling objects in industrial
buildings or on offshore structures, by wave attack on coastal
structures or by earthquake. The question arises whether the short
loading duration may affect strength and deformation behaviour of
the materials involved. For demolition of concrete structures sometimes explosives are used in order to bring a structure down or to
comminute structural parts for the separation of the reinforcing
steel and possible reuse of the concrete. In this case, the energy
demand depends on the properties of the materials which may be a
function of the loading rate.
The pulse duration of these impact and impuls loads covers a range
between one millisecond and about ten seconds [1]. Depending on the
mass of the striking bodies, the material properties, the geometry
and the velocity during impact or depending on rise time of blast

waves or the ignition rate of the explosive, the strain rate may vary between 10^{-4} s^{-1} and 10^{+4} s^{-1}. During earthquakes strain rates up 10 s^{-1} may occur. In the elastic range of concrete behaviour these strain rates can be translated into stress rates depending on the elastic modulus. Roughly speaking, stress rates between 3 N/mm²s and 400.10^6 N/mm²s can occur, but in practical engineering an upper value of 10^6 N/mm²s may be appropriate.

While the structural engineer is mostly interested in the relation between mechanical properties and strain or stress rate, from a materials science point of view, the cause of the strain rate effect should be examined. It is the aim of this contribution to reveal some mechanisms which provoke the strain rate effect. Attention will be paid to the dynamic behaviour of cracks, to the damaging of materials in impact loading and to energy consumption under high rates of loading. Furthermore, experimental results will be presented and compared to the theoretical prediction.

Modelling

Structure of concrete

Concrete is a composite material consisting of aggregate particles embedded in a matrix of hardened cement paste and fine inert material. The paste contains pores of various sizes. The paste contains pores of various sizes. The interface between aggregate and paste may be weaker than the bulk paste, it may even be cracked due to differential thermal movement and shrinkage during hydration and drying. Concrete can be modelled on different levels [2]: on the macro-level it is regarded as a homogeneous isotropic material, in the meso-level pores, inclusions, and cracks have to be considered, and finally, on micro-level cement paste is composed of colloidal particles, polysilicates, water in various types of appearance and voids.

To study the strain rate effect on concrete all three levels should be treated, of course, each of them with an appropriate approach. On the macro-level linear and/or non-linear fracture mechanics may apply and statistical mechanics as well; on the meso-level attention should be focussed on fracture energy of matrix, aggregate particles, and bond between these constituents; the behaviour on the micro-level may be described implicitly by the rate theory although the different particles are not considered explicitly. The following will show some theoretical approaches to model the rate dependent properties of concrete. The relevant properties are strength, stress-strain curve, crack propagation, energy consumption and fragmentations.

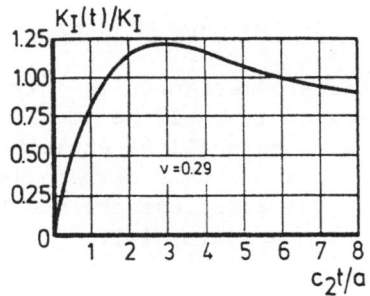

Fig. 1   Variation of dynamic to static stress intensity factor with
time for normal impact loading (ν = Poisson ratio) [3]

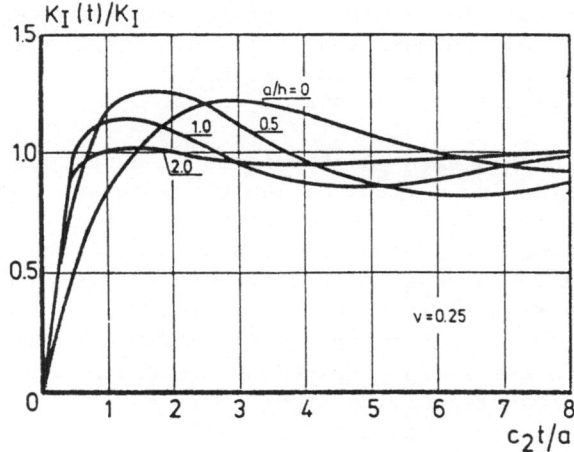

Fig. 2   Dynamic to static stress intensity factor versus time for
strip subjected to normal displacement [3]

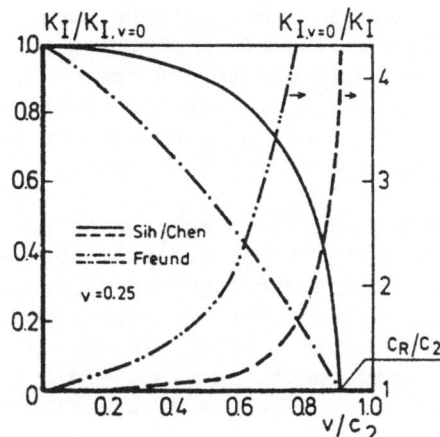

Fig. 3   Semi-infinite crack propagating in a finite strip [4]

Concrete as a homogeneous material (macro-level)

Linear fracture mechanics

Although linear fracture mechanics is only a tool for very large concrete cross-section it shall be considered as a reference case. Some basic loading cases shall be explained. The first two examples [3] pertain to cracks with a certain length 2a which are stressed by a sudden load or displacement, the third treats a crack propagating with a constant velocity [4].

A finite crack of length 2a in an infinite sheet is loaded at its faces by an impact load σ_0 (heaviside unit step function) at time t = 0. The time-dependent stress intensity factor K_I (t) divided by the static factor K_I is shown in figure 1 as function of normalised time $c_2 t/a$ with c_2 the propagation velocity of a shear wave. It can be seen that the dynamic stress intensity factor rises steeply reaching a maximum at $c_2 t/a = 3$ and then decreases in a damped oscillation to the static value.

A sudden displacement of the faces of a strip with finite height 2h containing a crack of length 2a causes a dynamic stress in-tensity factor which is shown in fig. 2 in normalised form as a function of time $c_2 t/a$. The same features are revealed as in fig. 1, i.e. an early increase of the dynamic stress intensity factor and an oscillating approach to the static value. The amplitude of the oscillation depends on the ratio a/h. The larger this ratio, the thinner is the strip compared to the crack length. This means that stress waves travelling from the faces of the strip to the crack and vice versa interact with each other and lead to a rather stable situation. In a very thick strip this takes more time and the variation of the stress intensity is larger.
It should be pointed out that these two examples are based on a constant crack length during the event. Assuming a constant fracture toughness K_{IC} for the concrete impact loading would mean that the critical situation of $K_I = K_{IC}$ would be reached at lower stresses than in the static case. On the other hand, it should also be noted that the stress σ_0 is kept constant during the impact loading. This is a special loading case, but not typical for impact on structures where σ_0 is a function of time.

In the following example a semi-infinite crack is considered in a finite strip. At t = 0 the faces of the strip are moved by an instantaneous displacement of magnitude σ_0 which is kept constant. The crack propagation velocity v is also kept constant. Fig. 3 gives the normalised dynamic stress intensity factor versus crack velocity v/c_2. It shows unity at v = 0 and a steep decay with larger crack velocities. When the crack velocity reaches the Rayleigh wave speed c_R the dynamic stress intensity factor becomes zero. According to Freund [49] the dynamic stress intensity factor varies almost linearly with the crack speed which is kept constant, from the static value at zero speed to zero at the Rayleigh wave speed for arbitrary loads which are in equilibrium.

This theoretical result may be interpreted by considering the stress waves which cause a stress intensity at the crack tip. At low crack velocities the information from the stressed body and from the points which are unloaded due to crack extension, is transferred to the crack tip rather quickly compared to the crack velocity. At higher crack velocities the crack faces may not move (open up) fast enough to provide the strains in the neighbourhood of the crack tip which are necessary for a high stress intensity factor. Since the Rayleigh wave is responsible for the energy transport along the crack surfaces the stress intensity drops to zero at c_R.

Since $K_I \sim \delta_o$ the result means that a running crack causes less stress intensity at the same displacement than a stable crack. Or expressed in another way, if fracture is defined by $K_I = K_{Ic}$ this would mean that the loading capacity is larger the greater the (constant) crack velocity is. The dashed lines in fig. 3 give an impression of such an interpretation. It cannot be more than a rough indication of a strength increase with higher rate of loading, the more because test results about the crack propagation velocity in concrete are scarce.

The theory of constant stress can be extended to arbitrary stress loading by an appropriate use of the stress-time relation. For the special loading case of a constant strain rate $\dot{\varepsilon}_o$ which results in a constant stress rate $\dot{\sigma}_o$ in an elastic material, Kipp and Grady [10] derived the relation for the stress intensity factor for a penny-shaped crack

$$K_I(t) = \alpha \frac{2}{\sqrt{\pi}} \dot{\sigma}_o \sqrt{c_s} \frac{2}{3} t^{3/2}$$

where α is a geometric coefficient equal to 1.12 for the penny-shaped crack, c the shear wave velocity and t the loading time. If K_{Ic} is regarded as a fracture criterion a relation between strain rate $\dot{\varepsilon}_o$ and fracture stress can be established

$$\sigma_c = \left\{ \frac{9\pi \ EK_{Ic}^2}{16 \ \alpha^2 \ c_s} \right\}^{1/3} \dot{\varepsilon}_o^{1/3}$$

which holds for high strain rates and/or sufficiently large cracks. By use of the known relations between stress intensity factor and stress $K_{Ic} \sim \sigma_c \sqrt{\pi a}$, it can be shown that a relation exists

$$a \sim \{ \frac{c_s K_{Ic}}{E \dot{\varepsilon}_o} \}^{2/3}$$

which links crack length a and strain rate $\dot{\varepsilon}_o$. For average values of concrete

$$a \sim 0.1 \times \dot{\varepsilon}^{-2/3}$$

i.e. strain rate of 1 s^{-1} requires a crack length of 0.1 m, strain rate of 100 s^{-1} a one of 5 mm.

Non-linear fracture mechanics

Linear fracture mechanics do not take any plastic deformation or micro-cracking in the region of the crack tip into account. Whereas linear elastic material description may apply to glass and some high strength metals it is certainly not applicable to concrete and mortar. It has been demonstrated in theoretical and numerical analysis [5, 6] that a micro cracking zone exists around the tip of a visible crack in concrete. This has also been confirmed by deformation controlled experiments [7, 8]. This micro-cracking zone (or fracturing zone, process zone, softening zone) depends mostly on the descending branch of the stress-strain curve of concrete under tensile loading. The smoother the decay of this curve and the smaller the deformation where no stress can be transferred anymore, the smaller this zone will be. Besides concrete mix, maximum aggregate size, humidity and temperature, the strain rate influences the shape of the stress-deformation curve. Thus, also the softening zone size should depend on the loading or straining rate of concrete. Before this concept is worked out some ideas shall be recalled which have been developed for plastic material. Although concrete does not behave as a plastic material some aspects can be transferred to it.
In a paper by Broberg [9] five aspects are discerned accounting for the influence on the behaviour of a running crack. Comparing two crack tips, one travelling with speed v_1 and the other with v_2 a point with distance r from the tip experiences the same stress time history if the stress intensity factors are in the ratio of $K_2/K_1 = (v_2/v_1)^{1/2}$. This follows from the fracture mechanics description of stress distribution $\sigma = K/(2\pi r)^{1/2}.f(\Theta)$ (Θ = angle coordinate) and from time place relation $r_2/r_1 = v_2/v_1$. If r_1 is a point of the process zone boundary r_2 is a homologous point for crack speed v_2. That means that the size of the process zone depends on the crack speed. This similarity argument is depicted in fig. 4.

The process zone of concrete is characterized by numerous small cracks which develop in the matrix and between matrix and

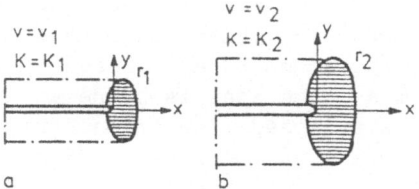

Fig. 4   Damage regions at crack velocity v_1 (a) and v_2 (b) [9]

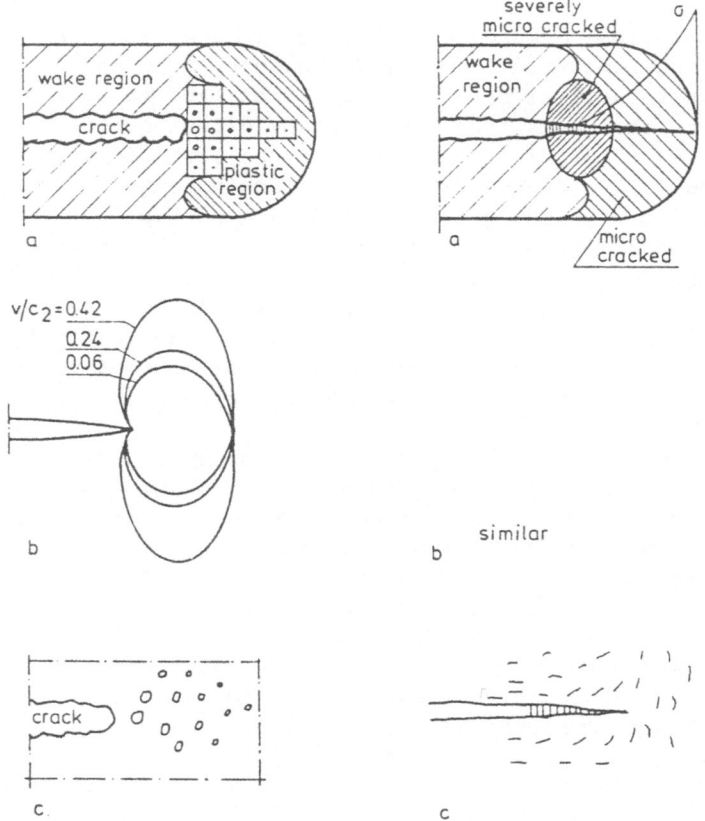

Fig. 5   Phenomena in plastic (left) and microfracturing materials
          (right)
          a) process region at fast growth
          b) curves of constant principal stress
          c) holes at fast crack growth.

aggregate. In plastic material these are imperfections and voids
(holes) which are opened by cleavage and coalescense to a main
crack. If the crack tip velocity is not small compared to the wave
speed in the plastic region more holes can open up because unloading of the region around a hole propagates also with wave speed.
Therefore the unloading reaches other stressed places with a delay
only, which means that a large zone is prone to crack extension.
Another aspect is the wave speed itself. In terms of concrete, the
influence of micro-cracking on the wave speed should be considered.
As a first estimate, wave speed depends on the square root of the
elastic modulus for equivoluminal waves. Micro-cracking means a decrease of stiffness and hence a decrese in wave speed. Energy transport to the crack tip shall be impeded. The micro-cracking zone
will then be stretched more uniformly and the dominant role of the
main crack diminishes.

On theoretical grounds, the stress field around the crack tip
could be established for the plastic region. It turned out that at
higher crack velocities the higher stresses tend to spread more perpendicular to the crack front than in the direction of it. This is
a consequence of the different speed of the irrotational and equivoluminal waves. This effect causes the cracking zone to widen up.
This means also that the stresses ahead of the tip favor microcracks to open up perpendicular to the main crack. Broberg calls
this a change in micro-separation morphology. Fig. 5 shows on the
left hand side a few pictures by Broberg which apply to plastic material and on the right hand side a possible translation to concrete. The comparison is only qualitative, of course, but may help
to develop a theory for concrete with all relevant features.

As a result it may be expected that the energy dissipation
during fracture is strongly dependent on crack velocity, i.e. at
high rates of loading and therefore higher crack velocities the
strength should be higher and the deformation as well since more
volume is micro-cracked. The question arises whether this approach
is only valid for one discrete crack or whether it may also hold
for the softening zone around micro-cracks itself. In that case an
even higher strength and deformation capacity should be expected.
In regard to a quantitative treatment it should be pointed out that
the input is not complete: wave speed as a function of softening
should be known, the stress field in the micro-cracked zone can
only can analysed tentatively and the orientation of micro-cracks
as a function of crack velocity should be studied. Without this information the axis of the graph of fig. 6 can be divided only as an
indication.

It should be kept in mind that these considerations are done
for a constant crack velocity. Constant stress rate or constant
rate of deformation of a specimen does not necessarily mean constant crack speed, it depends on loading configuration and specimen

geometry. At low stress rate cracks can propagate fast as soon as a critical stage is reached [45, 46]. On the other hand high stress rate can accelerate a crack to propagate with high speed.

Energy criteria

Birkimer [11] considers the strain energy W during the rise time t_r of the stress which is given by

$$W = A \int_o^{t_r} \sigma dx$$

with A the cross-sectional area of a specimen. Using the relation dx = vdt, where v is the particle velocity which can be expressed by v = c.ε with c the speed of the irrotational wave, a relation can be obtained between the fracture stress σ_c and a constant strain rate $\dot{\varepsilon}_o$:

$$\sigma_c = \{ \frac{3E^2 W_f}{A_c} \}^{1/3} . \dot{\varepsilon}_o^{1/3}$$

This relation is only valid if the <u>fracture energy</u> W_F is a material constant which does not depend on strain rate. This formula resembles the one derived by Kipp and Grady since both show the dependence of fracture stress on the cubic root of strain rate.

Explosive loading is common practice in mining. This explains why rate effects have been studied extensively in that field. However, the aim of the study may be different. One of the questions is the size of fragments which occur during blasting. Grady [12] has put forward a theory which regards the <u>kinetic energy</u> as the value that determines fracture process and fragment size. Starting from the kinetic energy of an expanding sphere which was pre-stressed, he is able to link the fragment size d and strain rate $\dot{\varepsilon}$ by

$$d \sim \dot{\varepsilon}^{-2/3}$$

i.e. the higher the strain rate the smaller the fragment diameter. For a review of more models on fragmentation see Grady [13].

Ehrlacher [48] reviewed various references and emphasized also the importance of kinetic energy at high rates of loading. If the strain energy release rate G_{stat} applies for static loading in case of a crack propagating with speed v the portion

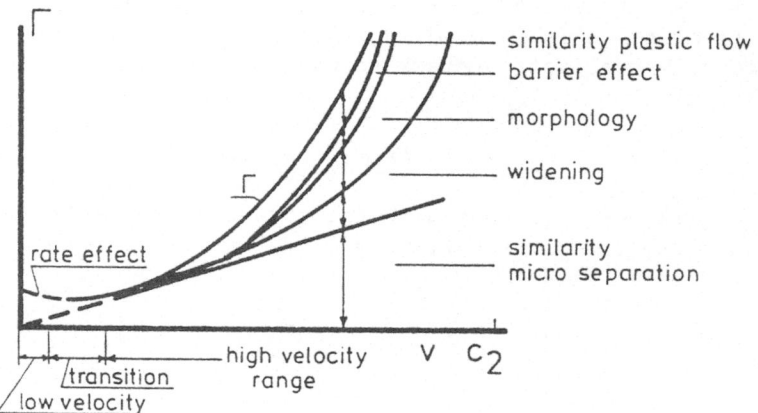

Fig. 6   Energy dissipation per unit of crack growth [9]

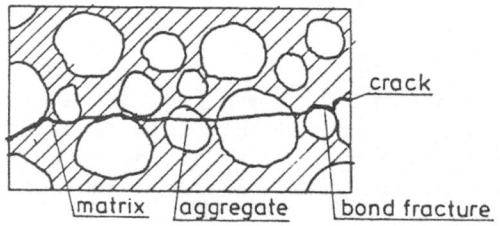

Fig. 7   Crack path in concrete

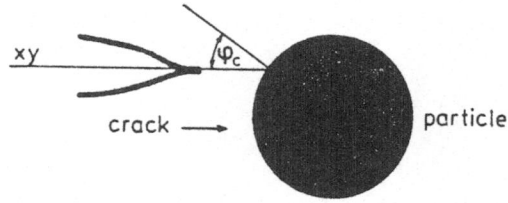

Fig. 8   Crack approaching a particle

$$G_I = G_{stat} \ (1 - v/c_R)$$

is dissipated through crack tip propagation whereas the reminder

$$G_{II} = G \ stat \cdot v/c_R$$

is transferred into kinetic energy. As soon as the crack propagates with the Rayleigh wave speed c_R all energy would be transformed into kinetic energy.

Concrete as multiphase material (meso-level)

On the meso-level concrete is regarded as composite material of aggregate, hardened cement paste and the interface between these two constituents. In terms of fracture one should know the contribution of these three components to the fracture stress. Since the aggregate material is usually much tougher than the cement paste the crack path has an important influence on strength, i.e. the more aggregate particles are fractured the larger the fracture stress shall be. This concept is the starting point of a fracture model worked out by Zielinski [14].
He models concrete by spheres embedded in a matrix. During fracture new surfaces are created which consist partly of aggregate, partly of cement paste, and partly of bond zones between aggregate and paste, fig. 7. The chance that an aggregate particle fractures depends on the aggregate content, the fracture plane in regard to a particle position and the stress rate. It is obvious that a higher aggregate content increases the chance that a particle is intersected by a crack plane. Whether an embedded particle fractures or is torn off from the matrix depends on the crack propagation velocity and the angle of the crack plane in respect to the normal of a sphere, fig. 8. It is argued that inertia forces impede a crack from deviating from a straight line more with increasing velocity. It is postulated that a relation between a critical angle ϕ_c and the stress rate σ exists, fig. 9. $\phi_c \sim 0$ means that fracture occurs by debonding in any case and ϕ_c 90° means that fracture of aggregate takes place in any case. The first is expected at sustained loading whereas the last pertains to very fast crack growth. With these assumptions the crack path can be predicted as function of loading rate.

The second step is to attribute an appropriate specific surface energy γ to each constituent. Values taken from other researchers [15] indicate that bond is the weakest zone, whereas aggregate is superior to cement paste. Summing up the energies of the corresponding areas the total energy is derived as a function of

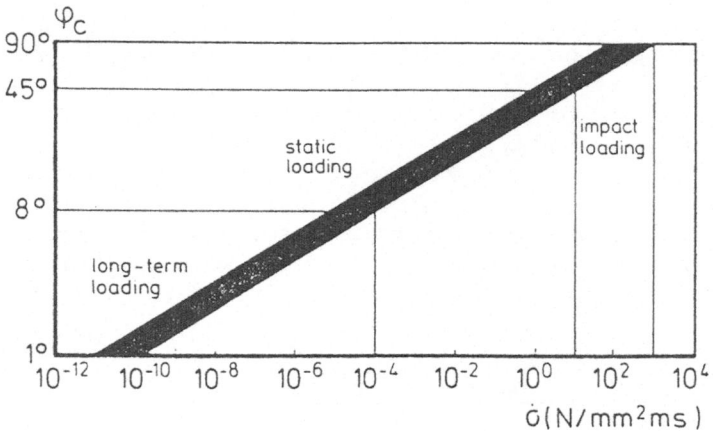

Fig. 9   Critical angle ϕ_c versus stress rate $\dot\sigma$ [14]

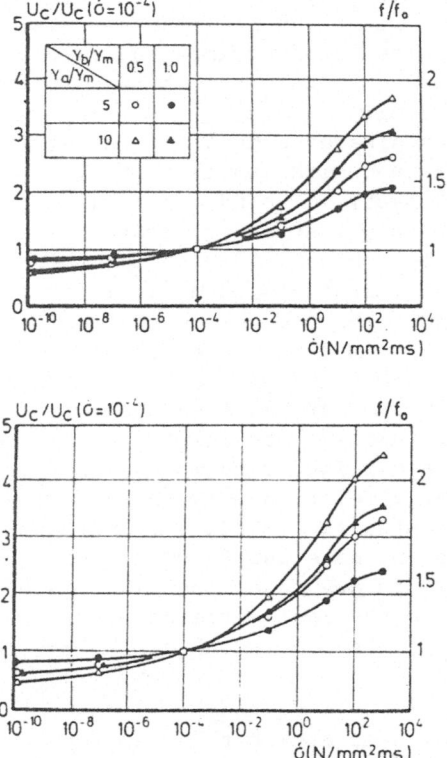

Fig. 10   Relative fracture energies [14]
          a) aggregate volume ratio 0.5
          b) aggregate volume ratio 0.75

loading rate. As can be seen from fig. 10 the energy increases with loading rate; the increase is the greatest for a large ratio γ_a/γ_b and a low γ_b/γ_m (a = aggregate, m = matrix, b = bond), it is the smallest for a low value γ_a/γ_m and γ_b/γ_m equal to unity. That means that a low quality matrix with poor bond properties is most sensitive for rate effects whereas a high quality matrix with good bond is least sensitive. The lower part of fig. 9 shows the same dependancies for an aggregate volume ratio 0.75 instead of 0.50. It is clear that more aggregate increase the rate sensitivity.

Assuming similar stress-strain behaviour in static and impact loading it may be stated that tensile strength and fracture energy are linked by $f \sim \sqrt{U_c}$. As an approximation fig. 10 may be interpreted in terms of strength increase the ratio of it is given on the right hand vertical axis. By the aid of this model a stress rate of $10^6 N/mm^2 s$ would lead to a strength increase of 40 to 110% in res-pect to static loading.

A model by Eibl and Gödde [19] belongs also to the meso-level. The authors consider a unit element of a two phase material consisting of an aggregate disc surrounded by matrix material to form a stip. Since aggregate is stiffer than matrix, waves propagate faster in aggregate. This phenomenon causes differential movement of the two phases under impact loading which leads to a stress distribution which is different from static loading. Together with inertia forces the strength is considerably increased and the fracture tends to propagate through aggregate with higher loading rates, fig. 11. This model is not yet finalized but may serve as a guide-line for further study.

Concrete on micro-level

Hardened cement paste consists of colloidal elements, silicates and absorbed water layers if considered on a micro-level. The paste is attached to the aggregate particles by adhesion. The behaviour and properties of all constituents should be taken into account to make a real physical model on this level. Besides the hard skeleton of hydration products which possesses primary and secondary bonds between the gel particles also absorbed and capillary water interacts with the solid material.

This task seems rather impossible. On the other hand, one could schematize the real structure by composing it of a group of elements with fictitious size. To account for voids and defects in the micro-structure each element contains a circular crack. The elements are connected as a series system. This concept has been developed by Mihashi and Wittmann [16], fig. 12. It is assumed that the micro-crack length depends on the elastic modulus and the surfaces energy of the material and has a given statistical distribution.

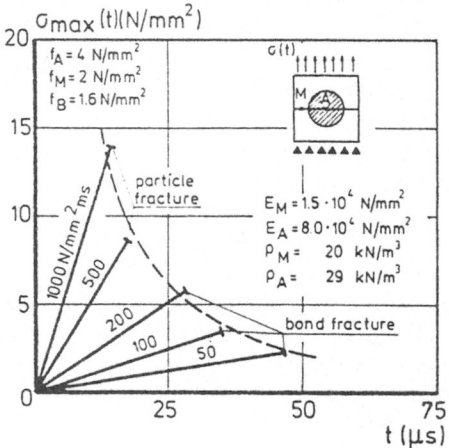

Fig. 11   Tensile strength vs. loading time [19]

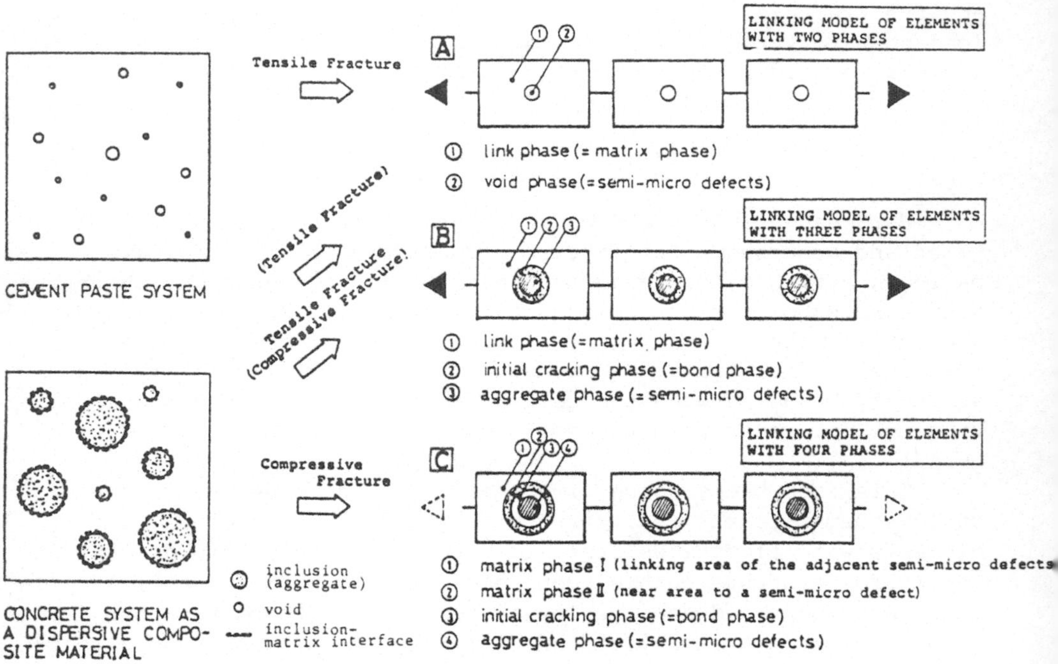

CEMENT PASTE SYSTEM

CONCRETE SYSTEM AS
A DISPERSIVE COMPO-
SITE MATERIAL

Tensile Fracture

(Tensile Fracture)

Tensile Fracture
(Compressive Fracture)

Compressive
Fracture

⊚ inclusion
    (aggregate)
○ void
--- inclusion-
    matrix interface

Ⓐ
① ②
LINKING MODEL OF ELEMENTS
WITH TWO PHASES

① link phase (= matrix phase)
② void phase (=semi-micro defects)

Ⓑ
① ② ③
LINKING MODEL OF ELEMENTS
WITH THREE PHASES

① link phase (=matrix phase)
② initial cracking phase (=bond phase)
③ aggregate phase (= semi-micro defects)

Ⓒ
① ② ③ ④
LINKING MODEL OF ELEMENTS
WITH FOUR PHASES

① matrix phase I (linking area of the adjacent semi-micro defects)
② matrix phase II (near area to a semi-micro defect)
③ initial cracking phase (=bond phase)
④ aggregate phase (=semi-micro defects)

Fig. 12   Linking models of elements [16]

The material defects and the characteristic properties of each
element have the same statistical distribution in the whole region.
Whereas stress redistribution is possible under compressive loading
this is not the case in tensile loading.

The mathematical treatment starts with the theory of rate process being a function of the activation energy of the stressed media. The rate of the crack nucleation is assumed to be proportional
to the number of molecules in the vicinity of the tip of the pre-
existing cracks. With these assumptions the probability of failure
in a certain time interval is computed. The influence of loading
rate on strength of concrete turns out to be

$$f \, / \, f_o = (\dot\sigma/\dot\sigma_o)^\lambda$$

where f denote strength, $\dot\sigma$ the rate of loading, subscript o static
reference and λ a material parameter which depends on concrete composition, age, temperature and humidity.

Körmeling [17] uses also the rate theory for the prediction
of concrete tensile failure under high stress rates. In order to
point on the different approach of Mihashi/Wittmann it is necessary
to recall the fundamentals of rate theory. Herein, it is assumed
that a state of a molecule is characterized by an activation energy
level. In order to change its position energy has to be supplied,
for instance by thermal activation. If mechanical stress is applied
the energy is increased and the chance that a molecule will move in
the directionof the stresses is greater than inthe other direction.
In terms of fracture, a forward activation leads to bond breaking,
a backward activation to bond mending [18]. Körmeling tries to
consider both effects which enables the application of this theory
to fibre reinforced concrete.

Extrapolating from other material the use of this theory seems
promising. Experimental results and the prediction of the theory
are given in fig. 13 for the case of constant stress. Two features
can be seen: first that the lines for various temperatures converge
at very short loading times (10^{-12} s) and, second, that this loading time is the same for all materials. The strength at that time
is the highest which can be achieved [15].

From measured microfracture data, relations are obtained by
Shockey et al [22] for the nucleation and growth rate of microcracks. Actually this approach may not belong to this chapter because it is an empirical treatment without intended physical background. On the other hand, it resembles some features of the theory
above which might link micro-cracks to unit elements. The nucleation rate is given by

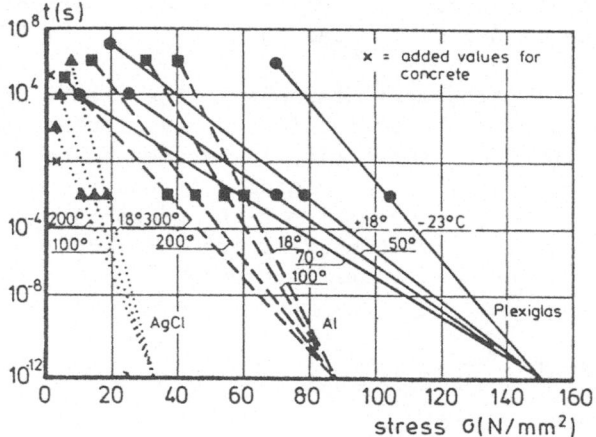

Fig. 13 Stress temperature life time relations [25, 17]

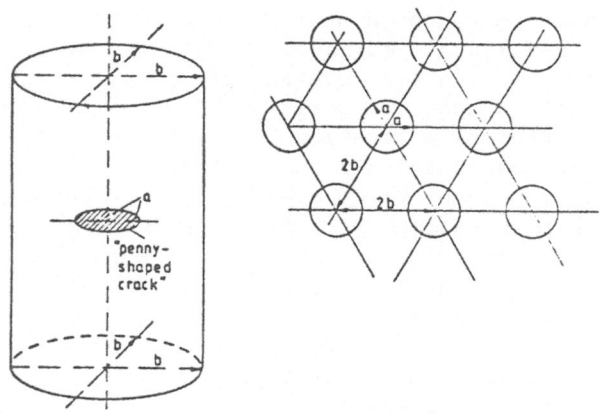

Fig. 14 Geometry of cracked concrete representative elements [50]

$$\dot{N} = \dot{N}_o \exp \frac{\sigma - \sigma_{no}}{\sigma_1}$$

and the growth rate of circular cracks with radius R is described by

$$\dot{R} = R \frac{\sigma - \sigma_{go}}{4\eta}$$

where $\dot{N}_o$ is a threshold value, σ the stress and the rest are material properties. Thus N grows exponentially with stress whereas the crack expansion rate is a linear function of stress.

Grady and Kipp [47] set up a damage theory using a two parameter Weibull distribution of flaws as a starting point

$$n = k\varepsilon^m$$

with n the number of flaws which can be activated at or below a tensile strain level of ε. The constants k and m are regarded as material parameters. The damage D is expressed as

$$D = N.V$$

where N is the number of idealized penny-shaped flaws per unit volume and $V = \frac{4}{3}\pi r^3$ is the spherical region surrounding the flaw of radius r. This volume approximates the stress-relieved volume around the crack. The value of D is zero for undamaged material while D = 1 corresponds to complete fracture. The elastic modulus of the material is given by

$$E = E_o (1 - D)$$

where E_o is the elastic modulus for intact material. By analysing the rate of flaw activation and integrating over loading time an expression is received for the damage parameter D as a function of time. The general espression is worked out for a constant strain rate $\dot{\varepsilon}_o$

$$\varepsilon (t) = \dot{\varepsilon}_o t$$

which leads to the strain rate dependent fracture stress

$$f = E_o (m + 3)(m + 4)^{-\frac{m+4}{m+3}} \alpha^{-\frac{1}{m+3}} \dot{\varepsilon}^{\frac{3}{m+3}}$$

and the fracture time

$$t_f = (m + 4)^{-\frac{1}{m + 3}} \alpha^{-\frac{1}{m + 3}} \dot{\varepsilon}_o^{-\frac{m}{m + 3}}$$

where

$$\alpha = \frac{8\pi c_g^3 k}{(m + 1)(m + 2)(m + 3)}$$

c_g can be considered as a constant fracture growth velocity after activation. k and m are material parameters again. Since E_o, m and α are constants for a certain material

$$f \sim \dot{\varepsilon}^{\frac{3}{m + 3}}$$

the power of $\dot{\varepsilon}$ being 1/3 for m = 6 which yields the same relation as has been received from linear fracture mechanics. Above relation which has been verified for several rocks, did not obey exactly the cube root law but is a good approximation (for instance: oil shale with k = 1.7 x 10^{27}/m^3, c = 8, c_g = 1.3 km/s). In addition to experiments by Birkimer [26] no verification for concrete could be found.

A continuous damage theory is developed by Suaris and Shah [24, 52] based on the Helmholtz free energy function. The authors derive a damage evolution equation which considers the inertia associated with the micro-crack growth. The micro-cracks are represented in vectorial forms which allows to distinguish between cracks due to compression, bending or uniaxial tension. The coefficients in the equations are determined with the aid of experimental results. It is shown that the influence of loading rate is different at different loading conditions, it is largest in tensile and least in compressive loading.
A model which takes also account of the porous structure of concrete and uses linear fracture machanics parameters is worked out by Weerheijm [50]. This model is a continuation of the early work by Mott [51] and considers the specific features of the structure of concrete. Concrete is modelled as a material containing penny-shaped cracks at a constant distance 2b diameter 2a (fig. 14). It is argued that the energy balance consists of three terms, i.e. the fracture energy G_{Ic}, the elastic energy release G_I and the kinetic energy E_{kin}:

$$E_{kin} - \int_{a_c}^{a} (G_I - G_{Ic})\, da = 0$$

The last two terms have to be integrated during crack propagation from a critical starting crack length a_c till the extended crack length a. G_{IC} is a material property and is taken constant. G_I is a function of geometry, stress state and Young's modulus.
In order to determine E_{kin} the material adjacent to the crack should be considered which moves when the crack propagates. From this condition a relation is obtained between elastic properties, crack length, crack propagation velocity and loading rate. By inputting this relation into the energy balance equation a integral integration emerges which can be solved numerically if starting values are available. These are deduced from the assumption that cracks start to propagate when the stress is equal to 60% of the static tensile strength. Furthermore the ratio $a/b = n^{1/3}$ when n is the porosity of the concrete.

Final remarks on modelling

Regarding material on a macro-level there are some empirical formulae which link strength and strain rate, for instance for flow stress of plastic material

$$\frac{\sigma}{\sigma_o} = \{\frac{\dot{\varepsilon}}{\varepsilon_o}\}^n$$

where o denotes static testing and n a coefficient [20]. There are others (see review in [21]) which sometimes fit experimental data better. However, these formulae do not reflect the physical background.

Linear fracture mechanics together with the wave equation provides information on the behaviour of a crack of constant length which is subjected to a step load and to an increasing load with constant stress or strain rate. Linear fracture mechanics predict a relation of the shape $f \sim \dot{\varepsilon}\ 1/3$ which means a rather strong influence of strain rate on strength. A similar relation is derived by energy consideration assuming a constant value for fracture energy.

Non-linear fracture mechanics is a suitable tool in order to show the relevant points involved in energy dissipation during high rates of loading. There is a crack velocity range where rate effects on micro-cracking (ductility in plastic material) are important. This is the range with rather low velocity which can be important in regard to civil application. That would mean that stress-deformation curves in tensile testing should be available as function of strain rate by means of which the extension and shape of the softening zone could be calculated. The high velocity range suffers from lack of data as concrete is concerned; therefore the

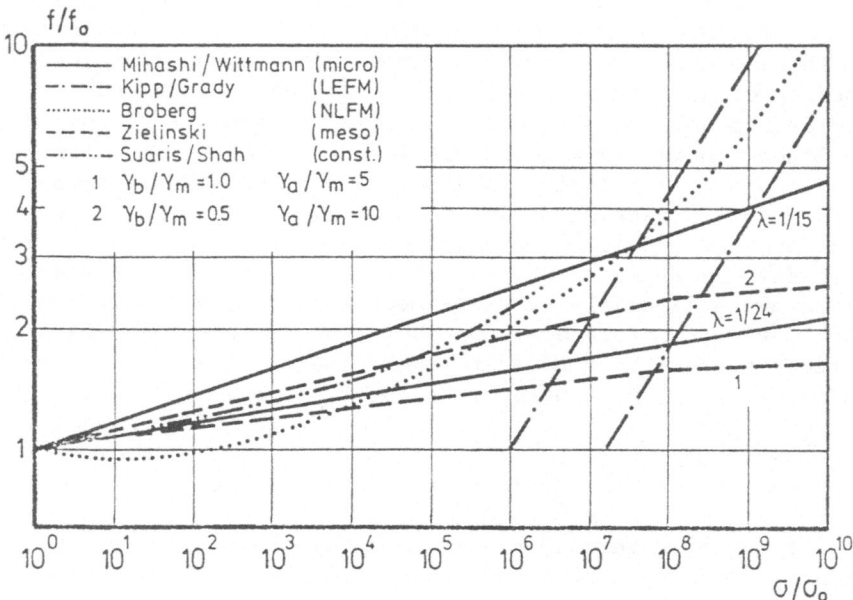

Fig. 15   Comparison of theoretical predictions for stress rate
effect on tensile strength of concrete

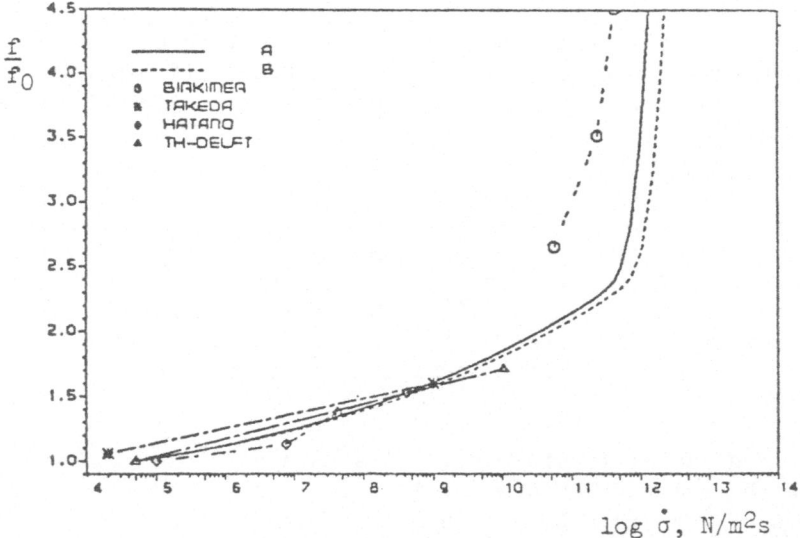

Fig. 16   Prediction of tensile strength as function of loading
rate [50].

Line A: f_{stat} = 5 N/mm^2, $\dot{\sigma}_{stat}$ = 0.05 N/mm^{2}s, n = 6.4%.

$$K_{IC} = 0.3 \times 10^{12} \ N/m^{3/2}.$$

Line B: same as A, except n = 13%

non-linear fracture mechanics approach can only be qualitative at
the moment.

   In contrary to that, the consideration on the meso-level can
produce quantitative relations between high and low stress rate
loading. The same is true for models based on the micro-level of
concrete which use the rate theory for predicting rate effects. For
some of them, limits of applicability are given, for others not.

   Fig. 15 exhibits a few theoretical predictions of the tensile
strength of concrete according to five models. All of them predict
an increase in strength with increasing stress rate. In a log-log
plot the Mihashi-Wittman theory reproduces as straight lines the
slope of it being a function of λ which depends on material and environment. With lower λ the loading rate effect is smaller. Also
the theory based on linear fracture mechanics (Kipp/Grady) leads to
straight lines, however, with a much steeper inclination. According
to the author the theory applies only to high stress rates, the
range depending on crack size. Examples of rock are in the range as
shown in the plot. The exact place of the line depends on flaw
size. The theory which considers fracture energy on a meso-level
(Zielinski) predicts about the same result as Wittman/Mihashi. The
place of the lines depends on the ratio between surface energy of
aggregate, matrix and bond respectively. Two examples are given,
both with an aggregate volume ratio p_a = 0.75. The portion until
σ/σ_o = 10^7 follows the expectations whereas the break and the decay
after that stress rate are questionable. Suaris and Shah predict a
steady increase of strength with loading rate. About the same result is achieved by qualitative considerations on the grounds of
non-linear fracture mechanics (Broberg) with the difference that in
a low rate region the strength even may be lower than in the static
case. This may be attributed to rate effects on the process zone
which are not yet compensated by inertia effects as discussed in
the preceding chapter. At very high rates of loading this theory approaches the prediction by linear elastic fracture mechanics
(Grady/Kipp).
An example of the theoretical prediction by Weerheijm [50] is given
in fig. 16, which shows the strength increase as function of stress
rate. In the first seven orders of magnitude the strength increase
goes up to about 2.2 times the static strength. Thereafter a steep
increase occurs which is even stronger than the one predicted by
Kipp/Grady. This steep increase is due to the fact that most of the
energy is converted into kinetic energy rather than fracture energy,
a feature which is in agreement with the prediction by Ehrlacher
[48].

   The following paragraph will provide some experimental data on
various mechanical properties of concrete as function of stress or
strain rate. Since there is a quite recent review of data by Suaris
and Shah [23] the treatment will be more exemplified than exhaustive.

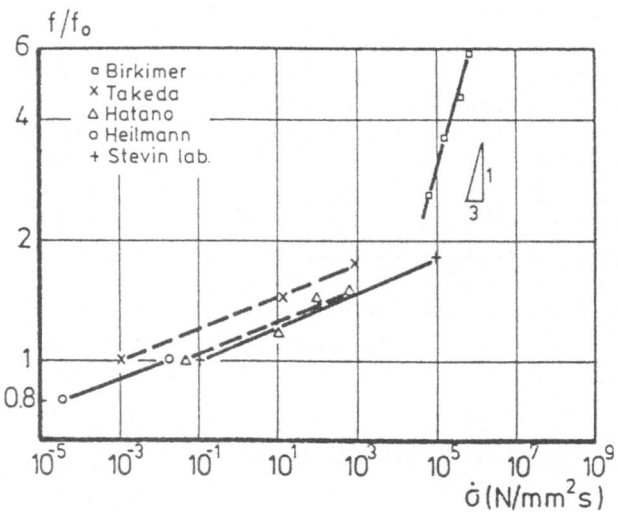

Fig. 17 Tensile strength vs. stress ratio, f_o = static strength

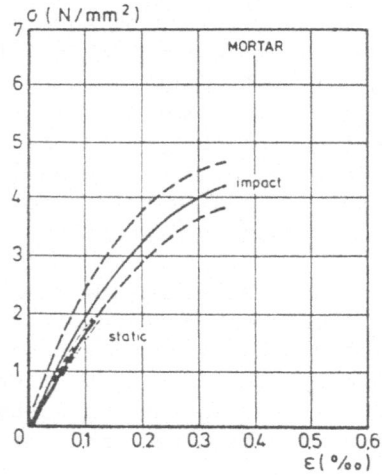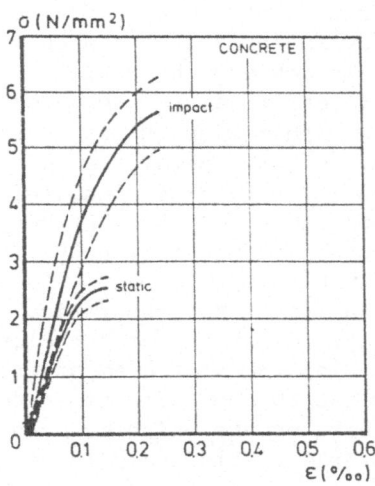

Fig. 18 Stress strain curves in static and impact tensile testing [36]

Experimental results

General remarks

There are at least two ways to set up experiments, one with the intention to establish the values of the basic parameters of a physical model, another to determine technological properties, as tensile strength, compressive strength and others, and to deduce the values of the model parameters. Most investigations of the past belong to the second category with the limitation that evaluation of results often stopped at empirical relations. Even today only little information is attainable in regard to rate effects on basic parameters. Therefore most experimental results will be interpreted in a more qualitative way than in a quantitative one.

The subchapters will follow the same pattern: first data from experiments are presented, then, influences which result from mix, moisture temperature will be discussed. It will be tried to point on different models explaining the observed phenomena.

Tensile strength and strain

The uniaxial tensile strength of concrete has been tested by several researchers [26 till 30]. The results are presented in normalized form in fig. 17 as function of stress rate. All investigations show an increase in strength. Up to $\dot{\sigma} = 10^5$ N/mm^{2}s the increase is about twofold compared to static testing. Beyond this point tensile strength seems to follow the prediction of the theory put forward by Grady [13] and Birkimer [11] which gives a $f \sim \dot{\varepsilon}^{1/3}$ relation. The lower stress rate region is best represented by the theory of Mihashi/Wittmann [16] with a power $\lambda = 1/23$. It also fits quite well into the predicted range of Zielinski [14].

The authors indicate that concrete is more sensitive for stress rate effects with a high water cement ratio, a smaller maximum aggregate size and higher cement content [31]. To explain this behaviour the non-linear fracture mechanics may be used. Recalling fig. 5 it is obvious that the widening of the micro-cracking zone and the change in morphology cause the largest increase in fracture energy dissipation at high rates of loading. Both phenomena depend on the amount of plastic deformation.
In terms of concrete the post-peak portion of the stress-strain curve is an appropriate measure for micro-cracking. If that portion exhibits a steep decay, micro-cracking is extended over a larger region at relative low stresses nearby the crack tip than in case of less decay. Static results [32, 33] indicate that the post-peak tensile stresses decay more with lower water cement ratio. A steeper decay means then less sensitivity to creation of a micro-cracking zone. A slow decay (lower concrete quality, higher water cement

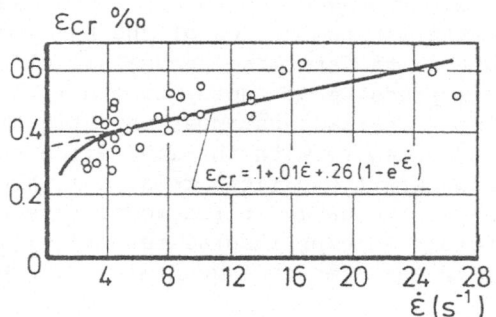

Fig. 19 Critical strain vs. strain rate in uniaxial tensile testing
[26]

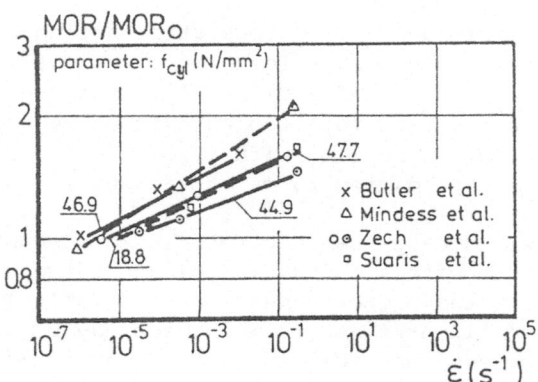

Fig. 20 Relative modulus of rupture vs. strain-rate [37]

ratio) of the stress-strain curve confines the micro-cracking region at static loading [34] but gives rise to extensive change in morphology and widening of the cracking zone due to high stress rates. Higher cement content (at same water cement ratio) causes a similar effect as higher water cement ratio which makes concrete more ductile in uniaxial tension. Larger particle size leads to more cracking at a relative low stress level [35] which makes concrete less sensitive to rate effects.

Another phenomena which has been found in concrete testing is multiple cracking. Whereas in static tensile loading only one distinct crack occurs during impact testing many cracks start propagating (see Broberg). The chance that more than one crack goes through the specimen grows with increasing stress rate [36]. The creation of more new surfaces demands more energy and means a higher strength and a larger strain at maximum stress. This effect can have a considerable influence on the result at high stress rates. Fig. 18 shows some examples of stress-strain curves for static ($\dot{\sigma}$ = 0.1 N/mm^{2}s) and impact loading ($\dot{\sigma}$ = 30 000 N/mm^{2}s). The equipment did not allow to measure the complete stress-strain curve in a deformation controlled manner.

An increase of strain at maximum stress is also found by Birkimer [26], fig. 19. The best fit to the data was the formula shown in the diagram which is obviously valid only for high strain rates. Compared to static tests where $\dot{\varepsilon}_{cr}$ is roughly .1 o/oo the critical strain at $\dot{\varepsilon}$ = 4 s^{-1} is abour four times and at $\dot{\varepsilon}$ = 24 s^{-1} about six times the value found in static testing. From the attainable data it is not possible to distinguish between the cracking zone and the rest of the material. Therefore it is hard to say whether it should be attributed to a very local phenomenon or to an equally distributed one.

Flexural strength

Three point and four point bending tests have been performed in order to establish the influence of rate of loading on the modulus of rupture. Fig. 20 shows some results as received and compiled by Suaris and Shah [37 till 40]. The vertical axis is the relative modulus of rupture, the horizontal axis gives the strain rate of the outmost fibre in s^{-1}. As in uniaxial tensile loading the strength increases with increasing strain rate. There is some scatter or maybe systematic difference between four point and three point bending. The upper two lines belonging to four point bending are a little more sensitive to strain rate than the remaining curves from three point bending. Also, the absolute concrete quality seems to have a certain effect. Suaris/Shah have noted that, the higher the concrete compressive strength is, the lower the relative increase in the flexural strength with increasing strain rate.

These authors also conclude from tests in various environment of curing that wet concrete is more sensitive to strain rate than dry concrete. Nevertheless, this influence is rather slight and has not been noticed in uniaxial testing [31].

The strain rate influence in bending seems to be almost the same as in uniaxial tensile testing. This can be concluded from comparing fig. 17 and fig. 20 assuming a constant Young's modulus which is of course only an approximation. The four point bending tests may be a little more sensitive to strain rate than axial loading. Suaris and Shah [24] attribute this difference to the vectorial behaviour of micro-cracks which are oriented differently for the various types of loading.

The experimental data fit within the band of prediction by various theories, see fig. 12, within the range of low to moderate strain rate. The data do not reach the region of fast fracture where much higher strength should be expected.

Since reinforcing of concrete and mortar by fibres is a widely used technique in order to improve the flexural strength testing has been carried out on a variety of products, such as concrete containing steel fibres of different shape, aspect ratio, quality, or reinforced with organic fibres, glass fibres or other anorganic fibres. All these reinforced concretes exhibited a strain rate sensitivity of the flexural strength which is comparable to or higher than that of plain concrete, of course, with higher absolute strength values which depend on the amount and type of fibres. For more details see for instance references [37, 41] which contain several more references on impact behaviour of fibre reinforced concrete.

Discussion

There are several models on the three levels of consideration of concrete which enable to predict the influence of stress or strain rate on the tensile properties of concrete. All of them have certain capabilities which make them suitable for different rates of stress. In the low and moderate range all theories predict results which are close together, in the high range fracture mechanics models are obviously superior to others. As has been noted already the use of models needs quantitative input values in order to receive quantitative answers. Up till now only little information is available for a quantitative prediction. Some models can be tuned by a few experimental results in order to predict the rate influence over a large range of stress rates, but it is not possible to predict the influence of the constituents on the rate sensitivity. There is one exception, that is the model by Zielinski which considers concrete on the meso-level taking aggregate, matrix and bond into account separately. Although some assumptions are not rigorous the model can predict the rate influence in a way which is

in agreement with experimental findings. Table 1 shows the
influence of some technological parameters on fracture energy and
tensile strength at static and dynamic loading in the way as they
follow from the model.

Table 1. Predicted influence of some parameters on fracture energy
and tensile strength [42]

parameter	fracture energy		tensile strength	
	static	dyn.	static	dyn.
aggregate toughness	+	++	+	++
aggregate content	+	++	+	++
max. particle size of aggregate	-	-	--	--
aggregate angularity	+	0	+	+
water cement ratio	--	-	--	-0
moisture	+	0	+	0
static preloading < 0.6 f_o	0	0	-0	0

Legend: ++ pronounced increase
+ slight increase
0 no
- slight decrease
-- pronounced decrease

Summary and conclusion

According to Wittmann[2], concrete can be analysed on three
levels: a macro, a meso and a micro level which differ from each
other by the degree of heterogeneity. It is tried to attribute a
physical model to each level which is able to predict the stress or
strain rate influence on the tensile fracture of concrete. On the
macro-level linear fracture mechanics, non-linear fracture mechanics and global energy criteria may be applicable. On the meso-
level the constituents of concrete like aggregate, matrix and bond
are considered. On these grounds models are developed which take account of the specific surface energies of the constituents when a
crack occurs. Furthermore inertia effects play an important role.
On the micro-level the hydration products should be distinguished.
There are theories based on the rate theory which may be attributed
to that level. On the other hand, a precise correlation between activation centres and hydration products cannot be given. Nevertheless, these theories are ranked in that group. All theories predict
similar relations between strength and stress rate in the low and

moderate stress rate range. In the high stress rate (high crack velocities) range linear and non-linear fracture mechanics are most capable. It seems that kinetic energy plays an important role, more than crack propagation velocity can play since measured crack velocities in concrete were not larger than two tenths of the shear wave velocity.

Experimental results are compiled for uniaxial tensile and flexural testing in an exemplified way. The models are checked against the experiments leadint to some conclusion in regard to the applicability of the models and the influence of various parameters on the stress rate sensitivity of concrete. It became clear that the fracture mechanics approach seems valid at high stress rates whereas the treatment on the meso-level gives most insight into the behaviour of concrete at moderate rates. As the influences on the stress rate sensitivity is concerned a low water cement ratio and/or coarse aggregate reduces the sensitivity whereas higher cement content and moisture increase the sensitivity. Non-linear fracture mechanics are used to try to explain this behaviour.

References

[1] Struck, W., Voggenreiter, W. Examples of impact and impulsive loading in the field of civil engineering. RILEM Materials and Structures 8 (1975), no. 44, pp 81-87.

[2] Wittmann, F.H. Creep and shrinkage mechanisms. In: Creep and shrinkage in concrete structures, ed. by Z.P. Bazant and F.H. Wittmann, John Wiley & Sons, Chichester, 1982, pp 129-162.

[3] Chen, E.P., Sih, G.C. Transient response of cracks to impact loads. In: Mechanics of fracture 4, Elastodynamic crack problems, ed. G.C. Sih, Noordhoff, Leyden, 1977, pp 1-58.

[4] Sih, G.C., Chen, E.P. Cracks moving at constant velocity and acceleration. In: Mechanics of fracture 4, Elastodynamic crack problems, ed. G.C. Sih, Noordhoff, Leyden, 1977, pp 59-117.

[5] Hillerborg, A. et al. Analysis of crack formation and crack growth in concrete by means of fracture mechanics and finite elements. Cement and Concrete Res. 6 (1976), pp 773-782.

[6] Bazant, Z.P., Oh, B.H. Concrete fracture via stress-strain relations. Report no. 81-10/665c, The Techn. Inst., Northwestern University, Evanston, October 1981.

[7]   Entov, V.M., Yagust, V.I. Experimental investigation of laws
      governing quasi-static development of macrocracks in con-
      crete. Mech. of Solids 10 (1975), no. 4, pp 87-95.

[8]   Reinhardt, H.W. Crack softening zone in plain concrete under
      static loading. To appear in Cement and Concrete Res.

[9]   Broberg, K.B. On the behaviour of the process region at a
      fast running crack tip. IUTAM Symp. High velocity deformation
      of solids, ed. by K. Kawata and J. Shioiri, Springer, Berlin,
      1978, pp 182-194.

[10] Kipp, M.E. Grady, D.E., Chen, E.P. Strain-rate dependent
      fracture initiation. Int. J. Fracture 16 (1980), pp 471-478.

[11] Birkimer, D.L. A possible fracture criterion for the dynamic
      tensile strength of rock. Proc. 12th Symp. on Rock Mechanics,
      ed. by G.B. Clark, Am. Inst. of Mining, 1971, pp 573-590.

[12] Grady, D.E. Local inertial effects in dynamic fragmentation
      J. Appl. Physics 53 (182), pp 322-325.

[13] Grady, D.E. The mechanics of fracture under high-rate stress
      loading. Preprints William-Prager-Symp. on Mechanics of Geo-
      materials: Rocks, concretes, soils, ed. by Z.P. Bazant, The
      Techn. Inst., Northwestern University, Evanston, 1983, pp 149-
      188.

[14] Zielinski, A.J. Fracture of concrete and mortar under uni-
      axial impact tensile loading. Delft University Press, Delft,
      1982.

[15] Hillemeier, B., Hilsdorf, H.K. Fracture mechanics studied on
      concrete compounds. Cement and Concrete Res. 7 (1977), no. 5,
      pp 523-536.

[16] Mihashi, H., Wittmann, F.H. Stochastic approach to study the
      influence of rate of loading on strength of concrete. HERON
      25 (1980), no. 3.

[17] Körmeling, H.A. The tensile behaviour of concrete at high
      strain rates. Part II: Deformation kinetics. Stevin report
      5-83-12, Delft University of Technology, Delft, 1983.

[18] Krausz, A.S., Eyring, H. Deformation kinetics. J. Wiley &
      Sons, New York, 1975.

[19] Eibl, J., Gödde, P. Investigation of the tensile strength
      of concrete under high loading rates (in German).
      Universität Karlsruhe, September, 1983.

588

[20] Alder, J.F., Phillips, K.A. The effect of strain rate and temperature on the resistance of aluminium, copper and steel to compression. J. Inst. Metals 83 (1954), pp 80-85.

[21] Johnson, W. Impact strength of materials. Edward Arnold, London, 1972.

[22] Shockey, D.A., Seaman, L., Curran, D.R. Computation of crack propagation and arrest by simulating microfracturing at the crack tip. ASTM STP 627 East fracture and crack arrest, ed. by G.T. Hahn and M.F. Kanninen, Philadelphia, 1977, pp 274-285.

[23] Suaris, W., Shah, S.P. Mechanical properties of materials subject to impact. Introd. Report Intern. Symp. Concrete structures under impact and impulsive loading. Berlin, 1982, pp 33-62.

[24] Suaris, W., Shah, S.P. A constitutive model for concrete under dynamic loading. SMIRT 7th Conf., Chicago, August 1983.

[25] Zhurkov, S.N. Kinetic concept of the strength of solids. Proc. 1st Intern. Conf. on Fracture, Sendai, 1965, Vol. 2, pp 1167-1184.

[26] Birkimer, D.L. Lindeman, R. Dynamic tensile strength of concrete material. ACI J., Jan. (1971), title no. 68-8, pp 47-49.

[27] Takeda, J., Tachikawa, H. Deformation and fracture of concrete subjected to dynamic load. Proc. Intern. Conf. on Mech. Behaviour of Materials, Kyoto, 1971, Vol. IV, pp 267-277.

[28] Hatano cited by Kvirikadze, O.P. Determination of the ultimate strength and modulus of deformation of concrete at different rates of loading. Int. Symp. Testing in situ of concrete structures. Budapest, 1977, pp 109-117.

[29] Heilmann, H.G., Hilsdorf, H., Finsterwalder, K. Strength and strain of concrete under tensile stress (in German). DAFStb, Bulletin 203, Berlin, 1977.

[30] Reinhardt, H.W. Concrete under impact loading - Tensile strength and bond. HERON 27 (1982), no. 3.

[31] Körmeling, H.A., Zielinski, A.J., Reinhardt, H.W. Experiments on concrete under single and repeated uniaxial impact tensile loading. Stevin Laboratory Report 5-80-3, Delft University of Technology, 2nd print, Aug. 1981.

[32] Evans, R.H., Marathe,M.S. Microcracking and stress-strain
      curve for concrete in tension. RILEM Materials and Struc-
      tures 1 (1968), no. 1, pp 61-64.

[33] Hughes, B.P., Chapman, G.P. The complete stress-strain curve
      for concrete in direct tension. Bulletin RILEM no. 30, March
      1966, pp 95-97.

[34] Reinhardt, H.W. Plain concrete in uniaxial post-peak cyclic
      tensile and tensile-compressive loading. Preprint W. Prager
      Symp. Mechanics of Geomaterials: Rocks, concretes, soils. Ed.
      by Z.P. Bazant, Northwestern University, Evanston, 1983, pp
      639-642.

[35] Petersson, P.E. Fracture energy of concrete: Practical per-
      formance and experimental results. Cement and Concrete Res.
      10 (1980), pp 91-101.

[36] Zielinski, A.J., Reinhardt, H.W. Impact stress-strain
      behaviour of concrete in tension. Proc. RILEM Symp. Concrete
      structures under impact and impulsive loading. Berlin, 1982,
      pp 112-124.

[37] Suaris, W., Shah, S.P. Properties of concrete subjected to
      impact. J. of Structural Eng., ASCE, 109 (1983), no. 7, pp
      1727-1741.

[38] Butler, J.E., Keating, J. Preliminary data derived using a
      flexural cyclic loading machine to test plain and fibrous
      concrete. Materials and structures 74 (1981), no. 79, pp 25-
      33.

[39] Mind ess, S., Nadeau, J.S. Effect of loading rate on the
      flexural strength of cement and mortar. Bulletin Am. Ceramic
      Soc. 56 (1977), no. 44, pp 429-438.

[40] Zech, B., Wittmann, F.H. Variability and mean value of
      strength as a function of load. J. ACI 97 (1980), no. 5, pp
      358-362.

[41] Naaman, A.E., Gopalaratnam, V.S. Impact properties of steel
      fibre reinforced concrete in bending. Intern. J. of Cement
      Composites and Lightweight Concrete 5 (1983), no. 4, pp 225-
      233.

[42] Zielinski, A.J. Behaviour of concrete at high rates of ten-
      sile loading. Stevin Laboratory, Delft University of Techno-
      logy, Report no. 5-83-5, Delft, 1983.

[43] Hahn, G.T., Kanninen, M.F. ed. Fast fracture and crack
      arrest. ASTM, STP 627, Philadelphia, 1977.

[44]   Sih, G.C. ed. Dynamic crack propagation. Noordhoff Intern.
       Publ., Leyden, 1972.

[45]   Achenbach, J.D., Brock, L.M. On quasistatic and dynamic
       fracture. In: Dynamic crack propagation, ed. G.C. Sih, Noord-
       hoff Intern. Publ., Leyden,1972.

[46]   Vardar, O. Finnie, I. The prediction of fracture in brittle
       solids subjected to very short duration tensile stresses.
       Intern. J. of Fracture 13 (1977), no. 2, pp 115-131.

[47]   Grady, D.E., Kipp, M.E. Continuum modelling of explosive
       fracture in oil shale. Int. J. Rock Mech. Mining Sci &
       Gemech. Abstracts 17 (1980), pp 147-157.

[48]   Ehrlacher, A. Behaviour of solids with a system of cracks.
       Preprints W. Prager Symp. Mechanics of Geomaterials: Rocks,
       concretes, soils, ed. Z.P. Bazant, Northwestern University,
       Evanston, Illinois, 1983, pp 554-566.

[49]   Freund, L.B. Crack propagation in an elastic solid subjected
       to general loading - I. Constant rate of extension. J. Mech.
       Phys. Solids 20 (1972), pp 129-140.

[50]   Weerheijm, J. Crack model for concrete under dynamic tensile
       load (in Dutch), Report TNO-PML 1984-15, 's-Gravenhage, 1984.

[51]   Mott, N.F. Fracture of metals: Theoretical Consideration
       Engineering 165 (1948), pp 16-18.

[52]   Shah, S.P. Constitutive relations of concrete subjected to a
       varying strain rate. Symp. Proc. The interaction of non-
       nuclear munitious with structures. U.S. Air Force Academy,
       Colorado, May 10-13, 1983, pp 81-84.

SECTION VII

STRESS-CORROSION, TIME AND TEMPERATURE
EFFECTS ON FRACTURE

**APPLICATION OF FRACTURE MECHANICS
TO CEMENTITIOUS COMPOSITES**
NATO-ARW - September 4-7, 1984
Northwestern University, U.S.A.
S. P. Shah, Editor

INFLUENCE OF TIME ON CRACK FORMATION AND FAILURE OF CONCRETE

Folker H. WITTMANN

Swiss Federal Institute of Technology, Lausanne
Laboratory for Building Materials

ABSTRACT

First of all experimental results are summarized and discussed
all of which document the influence of time on crack formation and
failure of concrete. The observed behaviour can be explained qualitatively on the basis of a phenomenological approach. In this case
the time-dependent degradation of the composite structure of concrete
is considered. Refined methods are necessary for a quantitative
prediction of the influence of time on crack formation and failure
of concrete. It is shown that crack theory provides us with a powerful tool to study crack propagation under sustained load. If the
elastic modulus is replaced by a time-dependent operator crack growth
can be linked with creep of the material. In this way life-time of
concrete under sustained load can be predicted realistically. In
addition a stochastic model based on rate theory is outlined. This
model enables us in principle to predict failure under arbitrary
loading conditions including high rate and impact. Variability of
strength can be determined on the basis of structural parameters.
Mechanisms of crack growth are finally discussed. It is concluded
that crack growth is a rate process similar to creep. So far. there
is no direct evidence of stress corrosiom. Degradation mechanisms
of porous materials such as crystallisation pressure, however, can
be fortified by an external load.

1. INTRODUCTION

Usually strength of concrete is determined by applying a monotonically increasing load. Loading time until failure is often chosen
to be two to five minutes. We will call a value obtained in this way
standard strength or reference strength of concrete. If a much
smaller rate of loading is chosen or if a sustained load is applied
failure will occur at lower values of load. To describe the decrease
of maximum load under sustained load the term static fatigue is used.
Under cyclic loading conditions dynamic fatigue (Wöhler-line) is
observed. Finally under high rate of loading and under impact loading
conditions the observed strength increases considerably.

In all the above mentioned cases the decisive parameter is time
to failure. After having briefly summarized experimental findings
theoretical concepts to describe influence of time on crack formation
and failure will be discussed. Then the question will be raised if
there is stress corrosion involved when concrete fails under sustained load.

2. EXPERIMENTAL FINDINGS

2.1. Influence of time on strength
and ultimate deformation of concrete

If identical concrete specimens are tested in compression with
varying rate of loading a result similar to the one shown in Fig. 1
will be observed (1). It can be seen that the ultimate stress decreases while the failure strain increases as the rate of loading becomes
slower. It has to be mentioned that the exact behaviour depends
among other influences on moisture content and age of concrete.

More often the behaviour of concrete under high sustained load
has been studied. It is well known that time-dependent deformation
under these conditions can be subdivided into three stages : (I)
primary creep or transitional creep where creep rate decreases
steadily, (II) secondary creep or steady state creep with a constant
creep rate and (III) ternary creep with rapidly increasing rate of
creep until failure. In Fig. 2 some characteristic results are shown.
At higher loads the duration of all three stages decreases. If we use
similar results and plot the life-time of a specimen as function of
the related applied stress σ/σ_0 where σ_0 is the ultimate stress under
standard loading conditions we obtain the well-known static fatigue
line as shown in Fig. 3. It has been observed that there is good correlation between rate of steady creep (stage II) and life-time under
high sustained load.

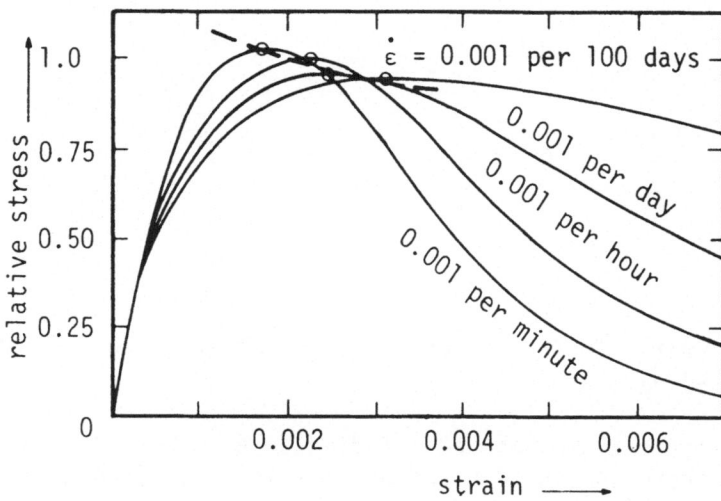

Fig. 1 : Stress strain diagrams of identical concrete specimens
obtained under different rate of loading after Ref. 1.
As shown by the dashed line the failure strain increases
as the rate of loading decreases.

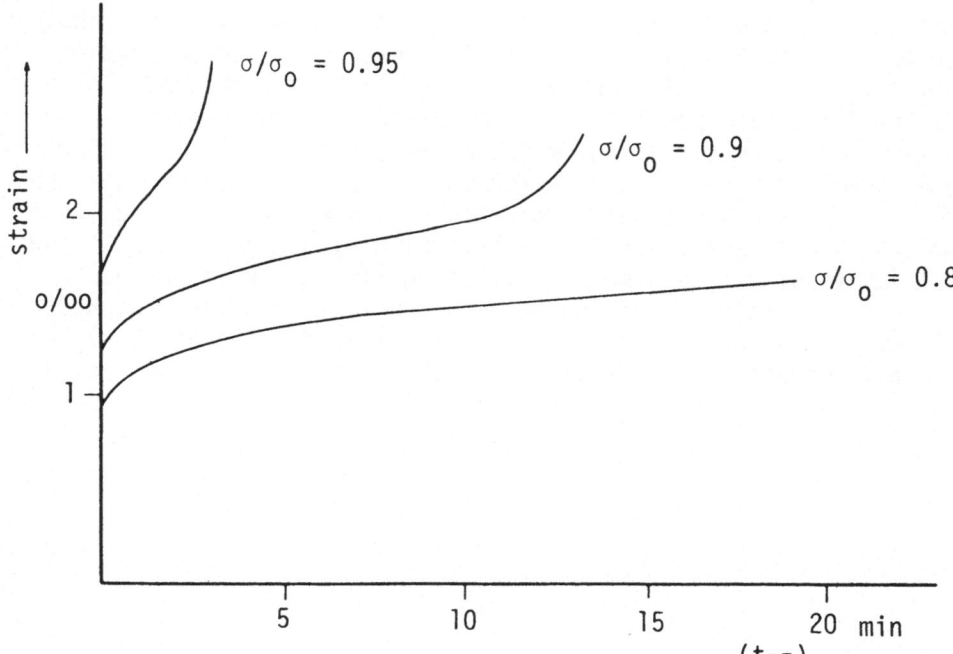

Fig. 2 : Time-dependent strain of concrete under high sustained
load after Ref. 2.

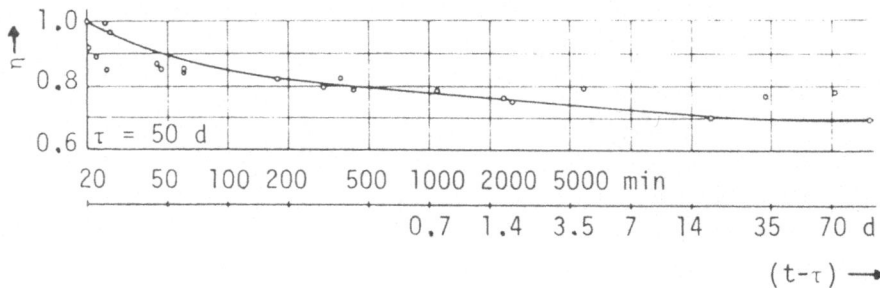

: Static fatigue of concrete loaded at an age of about
50 days after Ref. 3.

Dynamic fatigue has been studied by a number of authors. For a
given concrete composition and under well controlled climatic conditions Wöhler-lines can be determined. A typical example is shown
in Fig. 4. It has been shown that Miner's rule can be applied to
obtain an approximate estimation of service life of a cyclically
loaded concrete structure (4). On the basis of several Wöhler-lines
the Smith-diagram for concrete can be plotted. A typical example
for hardened cement paste is shown in Fig. 5.

As observed on many other materials strength of concrete
increases at high rate of loading and under impact loading conditions. In Fig. 6 a typical result is shown. Under these loading
conditions the elastic modulus is also slightly increased. These
modified material properties are of particular interest in the
context of seismic loading, missile impact, and explosions. For
some time there has been disagreement on the question whether the
ultimate strain increases or decreases with rate of loading. By
now it is well established that beginning with high values at low
rate of loading (see also Fig. 1) the ultimate strain decreases
first as the rate of loading increases but after having reached a
minimum value further increase of rate of loading leads to higher
ultimate strains. This relation is schematically shown in Fig. 7.

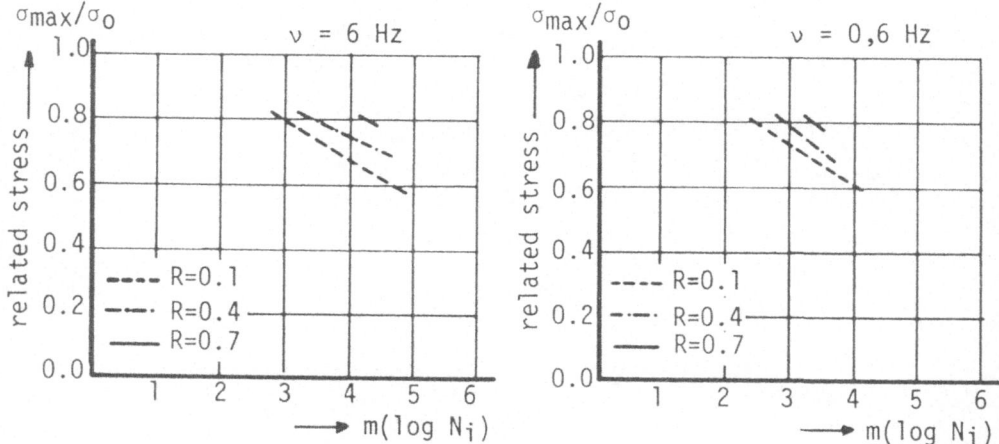

Fig. 4 : Typical Wöhler-lines for concrete after Ref. 4 and Ref. 5. The parameter R stands for $\sigma_{min}/\sigma_{max}$. It is obvious that frequency of the applied stress has a big influence on dynamic fatigue strength.

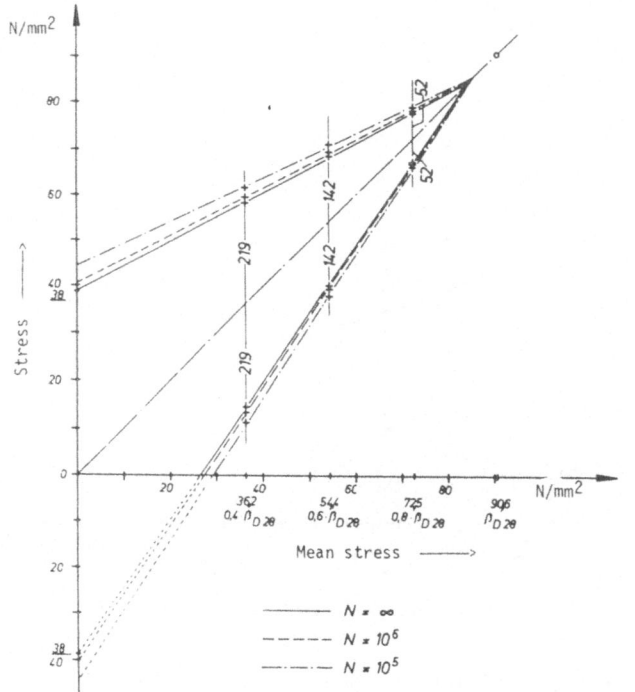

Fig. 5 :

Smith diagram for hardened cement paste with a water-cement ratio of 0.3 after Ref. 6.

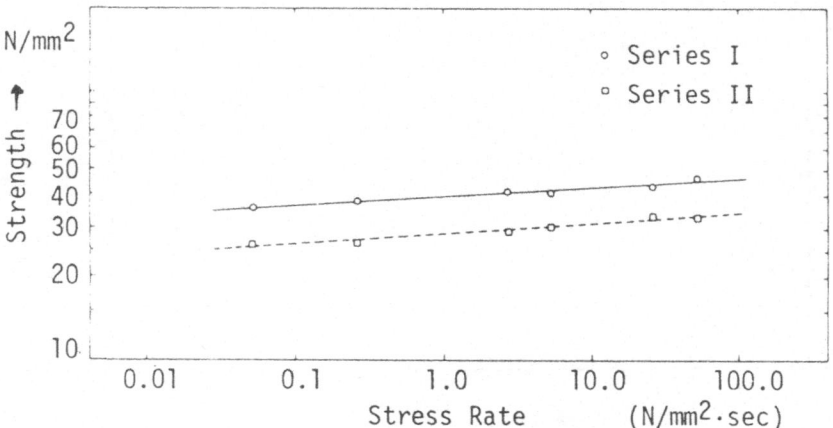

Fig. 6 : Increase of strength as function of stress rate for two different types of concrete after Ref. 14.

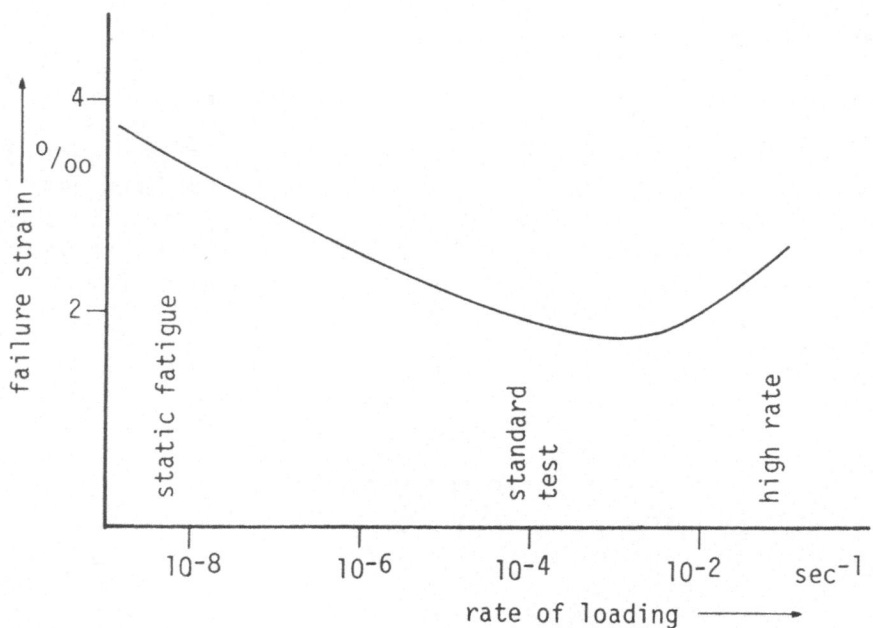

Fig. 7 : Failure strain as function of rate of loading.

2.2. Phenomenological explanation

The complex structure of concrete does not facilitate an explanation of material properties in terms of mechanisms. Therefore early investigations concentrated on phenomenological explanations. In Fig. 8 the stress strain diagram of concrete until ultimate load is given. In addition to the longitudinal strain ε_ℓ transversal strain ε_t and volumetric strain ε_v ($\varepsilon_v = \varepsilon_\ell - 2\,\varepsilon_t$) are shown. Based on these relations different stages of a concrete specimen under load can be distinguished :

Stage I : At very low stress the composite material reacts like a homogeneous and isotropic solid. No new cracks are formed. This stage is limited by the turning point of the ε_v-relation. It has to be noted that the limits between different stages are not clearly defined and they should rather be looked upon as being more or less broad transitional bands.

Stage II : As the load overcomes a critical value corresponding to the turning point of the ε_v-relations existing cracks will propagate progressively and new cracks will be formed. This crack formation has been observed directly and it is also expressed by an increase of the Poisson ratio ν. Additional evidence is provided by measurements of the pulse velocity. In stage II cracks may propagate but they will be stopped from further development by different crack arresting mechanisms. It is a situation of progressive stable and subcritical crack growth. Stage II is limited by the point where the volumetric strain becomes a maximum.

Stage III : At even higher loads further crack growth and crack coalescence lead finally to a total degradation of the composite structure. At this stage crack arresting mechanisms have only a slightly retarding effect but they cannot avoid further degradation any more. The dashed area in Fig. 8 indicates the degree of structural degradation.

It is obvious that the influence of time on crack formation and failure of concrete is most important at stage II and in particular in stage III. If we accept that the degradation of the composite structure under a given load depends both on the intensity of the applied load and on time we can explain the above mentioned experimental results qualitatively on this phenomenological level (7).

For a reliable prediction of material properties and for a systematic development of new and better composite materials, however, this approach is not sufficient. Therefore we need a more detailed description of the processes involved. Some concepts which may help us to reach this goal will be described in the following sections.

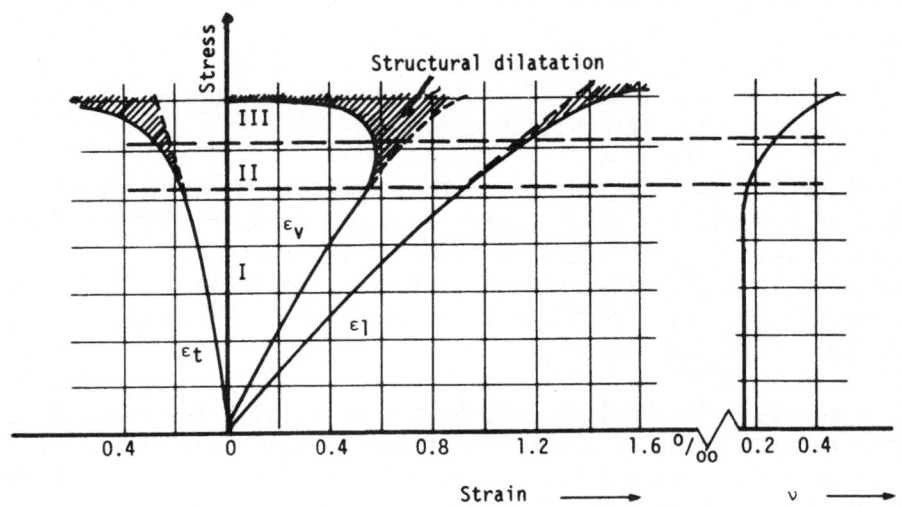

Fig. 8 : Longitudinal strain ε_ℓ, transversal strain ε_t, and volumetric strain ε_v as function of the applied stress. Three stages of structural degradation are indicated. On the right side Poisson ratio is shown as function of applied stress. The apparent Poisson ratio increases as structural degradation begins.

3. THEORETICAL CONCEPTS

3.1. Crack theory

Crack theory has proved to be a very powerful tool to study failure processes in porous and in composite materials (8) and (9). Let us consider first one isolated cylindrical pore with a radius r. If a compressive load is applied above a critical value two symetrical cracks will be created as shown in Fig. 9. Now we introduce a related crack length λ which is the real crack length ℓ divided by the pore radius :

$$\lambda = \frac{\ell}{r} \tag{1}$$

Then the related crack length can be given as function of the applied load in an implicit way

$$q = \frac{\pi E \gamma}{2r} \sqrt{\frac{(1+\lambda)^7}{(1+\lambda)^2-1}} \tag{2}$$

It can be seen that in this case we have stable and steady crack growth as the load increases. Therefore equ. (2) cannot be used as failure criterion. In a real porous material, however, above a critical crack length nearest cracks will first coalesce and finally unstable crack growth takes place when a critical average crack length is reached.

In a composite material cracks can be arrested by strong inclusions. Further increase of load is necessary before these cracks can further propagate. This crack arresting mechanism is schematically shown in Fig. 10.

So far crack growth has been discussed as function of load and the influence of time has not yet been taken into consideration. If we want to study the influence of creep of the material in the vicinity of the crack tips we have to replace the elastic modulus E in equation (2) by a creep operator $\tilde{E}$:

$$\frac{1}{\tilde{E}} = \frac{1}{E(\tau)} \{1 + E(\tau) \cdot C(t,\tau)\} \tag{3}$$

In this equation τ means time of application of load (age of concrete) and $C(t,\tau)$ stands for the specific creep function. After introducing several simplifying assumptions the following equation for static fatigue is obtained (7) and (8) :

$$\frac{\sigma}{\sigma_0} = \frac{\beta(t)}{\beta(\tau)} \; m(t,\tau) \sqrt{\frac{E(\tau)}{E(t)}} \; \frac{1}{1+\psi(t,\tau)} \tag{4}$$

$\beta(t)/\beta(\tau)$ represents strength increase of concrete due to progressing hydration while under load, $\psi(t,\tau)$ is defined in the following way :

$$\psi(t,\tau) = C(t,\tau) \cdot E(\tau) \tag{5}$$

The term $m(t,\tau)$ represents the strengthening effect of an applied load. Creep of the material in the vicinity of crack tips can lead to crack propagation and thus can weaken a loaded specimen but at the same time creep can reduce high stress concentrations which causes an increase of the maximum supportable stress. These two opposing effects

depend on the composition, the age and the moisture content of concrete. For young concrete $m(t,\tau)$ can have values of up to 1. 3.
$m(t,\tau)$ is much bigger if a tensile load is applied (10).

In Fig. 11 static fatigue as measured on specimens of hardened
cement paste loaded at an age between 1 day and 7 days is shown (2).
Solid lines have been obtained by using equ. (4). It can be seen that
the calculated relation for τ = 1d and τ = 3d increases after having
passed through a minimum. This effect is caused by the strength gain
due to further hydration. For older specimens hydration has slowed
down and therefore strength decrease dominates all over the period
investigated. It can be noted that a young concrete specimen under
high sustained load either fails very quickly or it does not fail
any more.

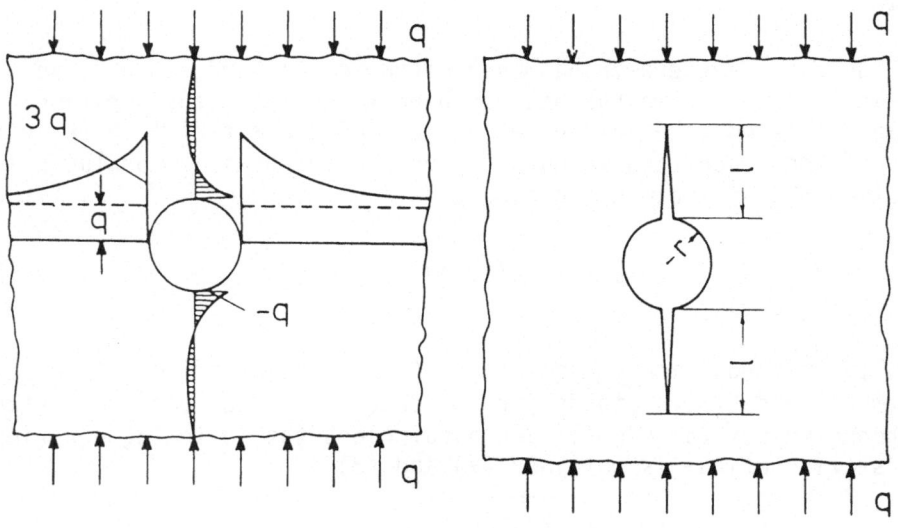

<u>Fig. 9</u> : Stress distribution around a cylindrical pore and stable
crack formation.

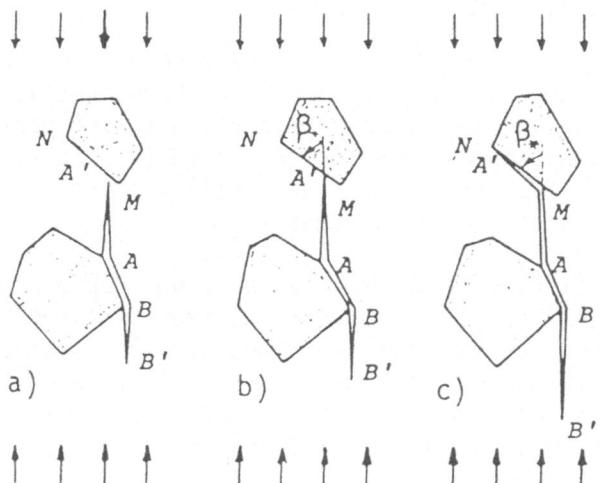

Fig. 10 : An interfacial crack grows into the surrounding matrix (a)
and is stopped by a neighbouring inclusion (b). Further
increase of load is necessary before the arrested crack
can further propagate (c).

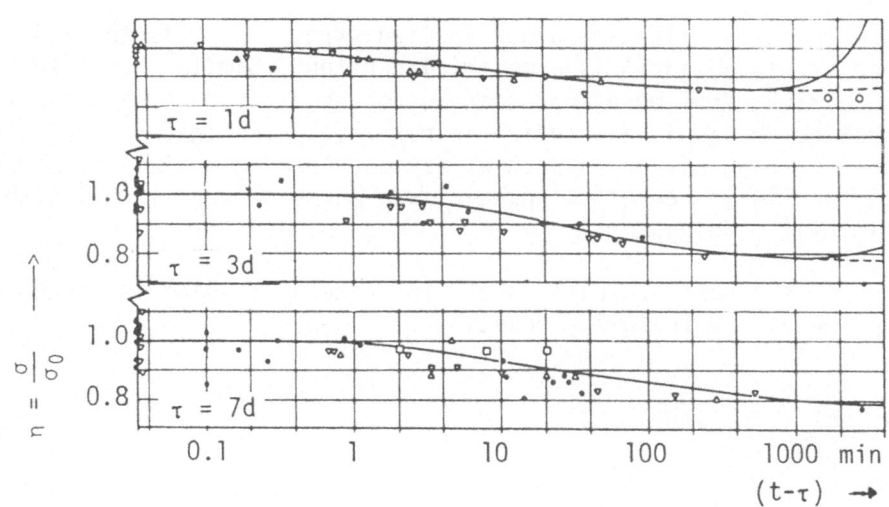

Fig. 11 : Static fatigue of hardened cement paste loaded at three
different ages. The solid line has been calculated by using
equation (4).

3.2. Rate theory and stochastic approach

We have seen that by means of crack theory crack propagation in a composite structure can be simulated. Random structures representing concrete with a given mix proportion taking particle geometry into consideration can also be generated (9) and (11). In this way degradation of concrete structure until failure can be studied in principle as a stochastic process.

A totally different approach has been developed by Mihashi and Izumi (12) and (13). Three models to represent the heterogeneous structure of concrete have been proposed. In the simplest case local stress concentration is simulated by a Weibull-type chain model. This model can be applied to describe failure of a brittle material. The more complex models take crack arresting mechanisms into consideration.

If we introduce a realistic probability density function for local stresses, rate theory can be applied to describe crack propagation. In this way time-dependent crack growth is introduced as a thermally activated stochastic process. Relative lifetime t/t_0 of a brittle material is found to be :

$$\frac{t}{t_0} = \left(\frac{\sigma_0}{\sigma}\right)^\beta \tag{6}$$

where t_0 is the lifetime under applied stress σ_0, t is the lifetime under stress σ, and β is a material constant. Similar formulae have been derived from more comprehensive materials models with crack arresting. Mihashi has studied in particular the influence of different loading histories such as fatigue loading on failure strength (13) and (14). Theoretical predictions agree generally well with experimental findings.

The strength increase under high rate of loading is found to be given by the following equation :

$$\frac{\sigma}{\sigma_0} = \left(\frac{\dot{\sigma}}{\dot{\sigma}_0}\right)^{\frac{1}{1+\beta}} \tag{7}$$

For a realistic reliability assessment variability of material strength is of major importance (15). The stochastic approach of Mihashi and Izumi (12) predicts that variability increases as the strength increases at high rate of loading but that the coefficient of variation remains constant.

Some experimental results obtained on concrete bars under three point bending condition are shown in Fig. 12.

The dashed line and the solid line are calculated by using equ. (7). A least square fit lead to values for β of 20 and 25 respectively. Within the range of accuracy the coefficient of variation is found to be constant.

In the case of crack mechanics we have studied growth of discrete cracks as function of time. With Mihashi's model the overall behaviour of a large number of cracks is investigated statistically. It is obvious that these two apparently totally different approaches describe in fact identical material properties and probably they can even be combined in the futur.

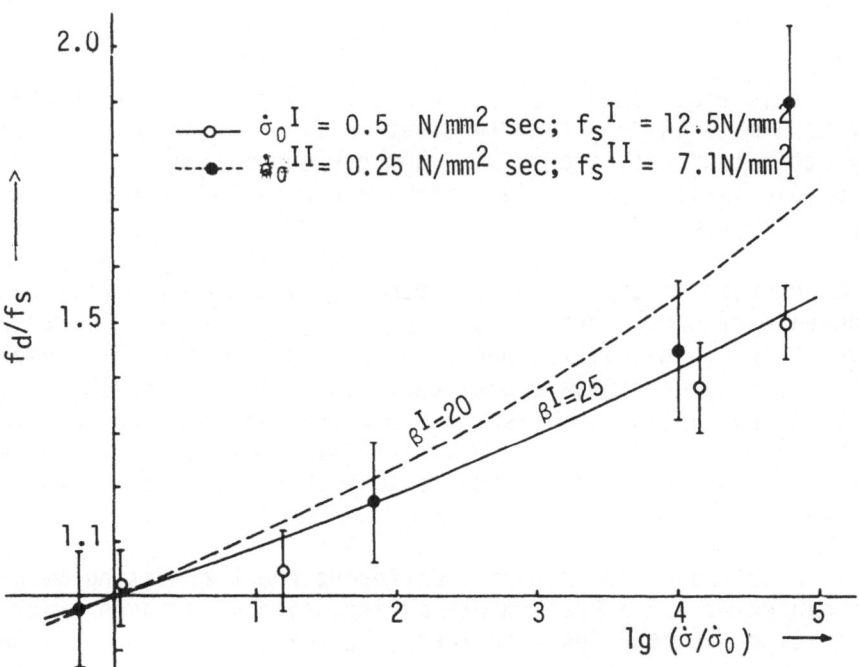

Fig. 12 : Strength increase of concrete bars under three point bending condition after Ref. 16.

4. CRACK GROWTH MECHANISMS

In the vicinity of a crack tip high stress concentrations exist. If we use 2c for the crack length and ρ for the radius of curvature and if a stress σ is applied the maximum local stress is given by the following expression :

$$\sigma_m = \sigma \left(1 + 2 \sqrt{\frac{c}{\rho}}\right) \qquad (8)$$

In Fig. 13 a section through a crack tip is shown. High local stress near the crack tip leads to a finite probability that bond strength is overcome at a stress well below the critical stress. Successive bond failure can be considered to be a major mechanism of subcritical crack growth. The probability of bond failure can be estimated on the basis of rate theory :

$$P = const \quad \exp \left\{- \frac{Q - V\sigma_m}{RT}\right\} \qquad (9)$$

In this equation Q stands for the activation energy of the bond and V is a material constant (activation volume). In this way thermally activated crack growth of a homogeneous material can be described. Similarity of equ. (9) with the corresponding expression for creep is obvious.

Microcreep processes do not necessarely and exclusively cause an increase of crack length 2c, they can also increase the radius of curvature ρ. This latter mentioned process decreases the high local stress σ_m and thus reduces the probability for further crack growth. A substantial increase of ρ can even increase the maximum load bearing capacity of a porous material. In equ. (4) the parameter m (t, τ) had been introduced to take this phenomenon into consideration.

In a heterogeneous material different crack arresting mechanisms can exist. If a crack meets an inclusion, if a sharp crack runs into a large pore for instance further crack growth can be either neglegible or at least slowed down considerably. In Fig. 14 crack velocity in a composite material is schematically shown as function of stress intensity factor (17). First the crack velocity increases monotonically with increasing K. As the crack meets an obstacle further crack growth is hindered and the growth rate drops sharply. This process is repeated until a crack has reached a critical length above which crack arresting mechanisms do not work any more.

While one crack has stopped to grow other cracks may further propagate. In a porous and composite material failure is never caused

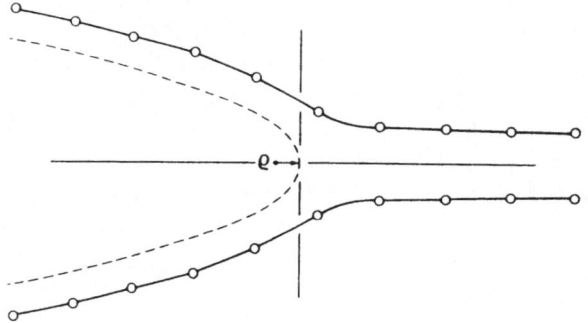

Fig. 13 : Section through a crack tip.

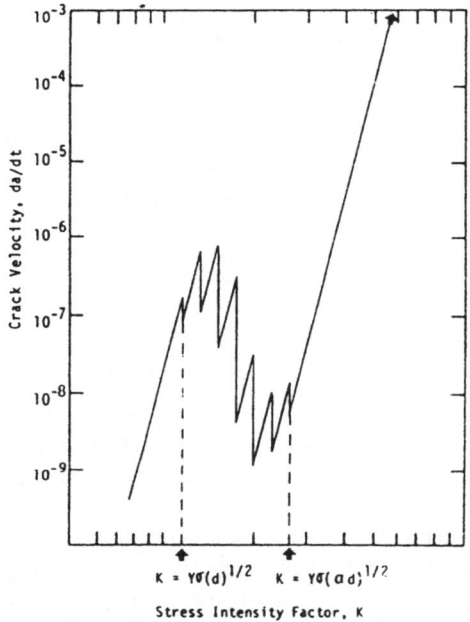

Fig. 14 : Crack velocity in a composite material as function of
stress intensity factor after Ref. 17.

by crack growth of one critical macroscopic crack. Ahead of a macroscopic crack a fracture process zone is developed by microcrack
formation and large plastic deformation. The extent of the fracture
process zone and its significance for the failure process depend on
the stress distribution ahead of the crack and it is not a material
constant. Crack coalescence in the fracture process zone is thus a
second major mechanism of crack growth.

Charles (18) proposed a mechanism of dissolution to explain water corrosion in soda-lime glass. It is postulated that an applied stress induces a preferential direction of growth of an existing flaw to bring about delayed failure. This stress corrosion mechanism is supposed to be a three step reaction. At the beginning at the terminal end which associates Na+ ion to the network hydrolysis reaction takes place according to Charles (18) :

$$-[Si—O—[Na]] + H_2O \rightarrow -[Si—OH] + Na^. + OH^- \tag{10}$$

The free hydroxyl ion formed can induce the second step in glass dissolution :

$$-[Si—O—Si]— + OH^- \rightarrow -[Si—OH] + -[SiO^-] \tag{11}$$

In this step the strong Si-O-Si bond is broken. As a result a silanol end and an end structure capable of dissociating another water molecule are formed. As a consequence new hydroxyl ions can be produced :

$$-[Si—O^-] + H_2O \rightarrow [HO—Si]— + OH^- \tag{12}$$

As these reactions are enhanced in an expanded glass network stress induced preferential crack growth can be explained.

Michalske and Freiman (19) have suggested another three step mechanism of water induced stress corrosion in glass. According to this model an adsorbed water molecule donates an electron to the silicon and a proton to the oxygen in the streched linkage unit Si-O-Si. This transfer produces two new bonds, one between the original silicon and the oxygen from the water, the other between the original oxygen and a hydrogen. The newly created fracture surface in this case is saturated with hydroxyl groups. In Fig. 15 these three steps of water induced stress corrosion are schematically shown in a crack tip.

It has been speculated that hydroxyl ion attack is also a decisive mechanism in static fatigue of hardened cement paste and concrete (21), (22) and (23). More recently Tait (24) measured crack

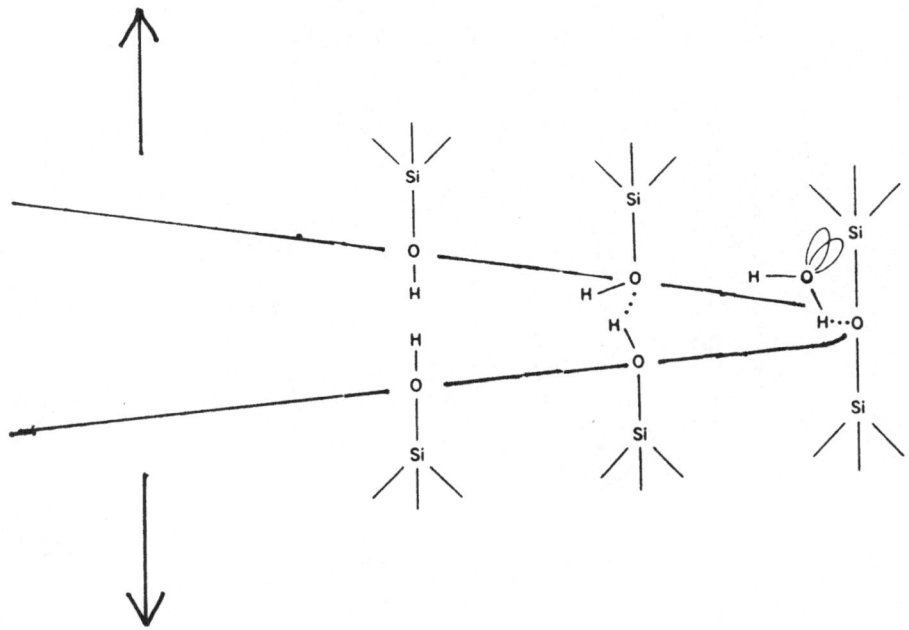

Fig. 15 : The three stages of water induced stress corrosion after
Ref. (19) schematically shown in a crack tip (see also
Ref. 20).

velocity of hardened cement paste in dry and in wet specimens. Results are summarized in Fig. 16.

From earlier results of Krokosky (21) and (22) and from the data taken from Tait (24) and shown in Fig. 16 it is obvious that moisture content has a significant influence on crack growth velocity. This is in good agreement with experiments to determine static fatigue of concrete. Strength decrease of dry specimens under sustained load is small or not existent. Until now it cannot be decided, however, whether the increased crack growth rate at high humidity content is really caused by chemical stress corrosion.

Creep of hardened cement paste can be represented by a power law :

$$\varepsilon_c = a\, t^n \tag{13}$$

In Fig. 17 the parameter a of equ. (13) is plotted as function of RH. It can be seen that creep increases considerably at higher RH. It is known that at high relative humidity disjoining pressure of water separates a large number of particles in the xerogel (26)

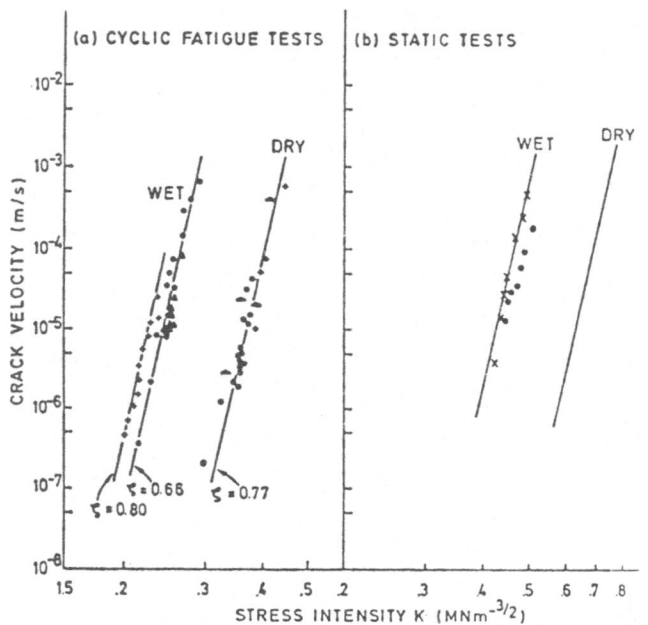

<u>Fig. 16</u> : Crack velocity as function of stress intensity as deter-
mined in dynamic and static fatigue test on wet and dry
specimens of cement mortar after Ref. 24.

and as a consequence the structure is weakened. The density of bonds
decreases while the average distance of points of contact increases
at high RH. Possibly this fact alone can explain the influence of
moisture content on static fatigue and on creep. Further research is
needed to clarify the role of chemical stress corrosion in context
of subcritical crack growth.

Two types of chemical attack can be distinguished in hardened
cement paste : dissolution and expansive action of newly formed
crystals. The weakening action of both mechanisms is randomly oriented in an unloaded specimen, but a preferential direction can be imposed to these deteriorating processes by an external load. Undou-
tedly long term static fatigue will be influenced by crystallization
pressure. In the xerogel of hardened cement paste there are enough
water soluble components that it remains questionable of dissolution
of strong Si-O-Si bonds has any significance in static fatigue.

If time-dependent alterations of the microstruture of concrete
can be neglected the phenomenon of delayed fracture can be represented in three different ways. In Fig. 18 the v-K plot (a), the
life-time under high sustained load (b), and the increase of

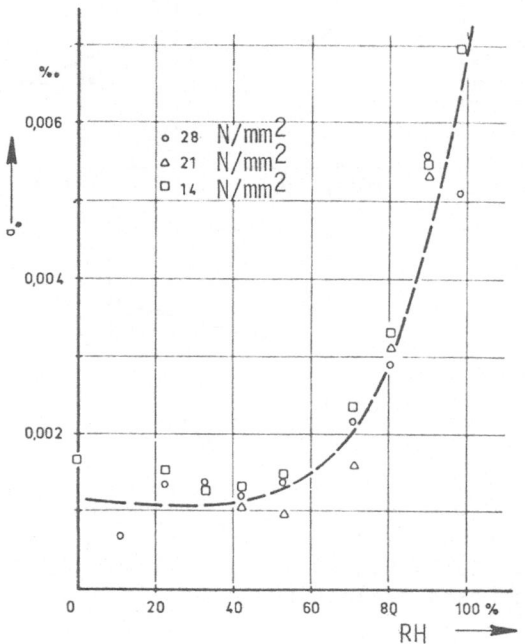

Fig. 17 : Creep of hardened cement paste at time t = 1h (see equ. (13)) as function of RH after Ref. 25.

strength at high rate of loading (c) are schematically shown. Mindess, Nadeau, and Barrett (27) have shown that these three relations are interdependent and that the slope in a log-log-scale representation can be described by a unique value of n.

If we use data from Mihashi and Wittmann (14) shown in Fig. 6 we obtain a value of n = 24. From results of Tait (24) shown in Fig. 16 we deduce a value of n = 27. Finally if we use experimental data of life-time under high sustained load (3) we find a value of n = 21. An average value of n = 24 is thus found. It seems that all three relations of Fig. 18 can be described approximately by one single value of n. In Ref. (28) a formula is given to calculate life-time under high sustained load if n is known.

Further research is necessary to verify the general validity of this concept. In particular the influence of concrete composition and influence of moisture content on n has to be studied in detail.

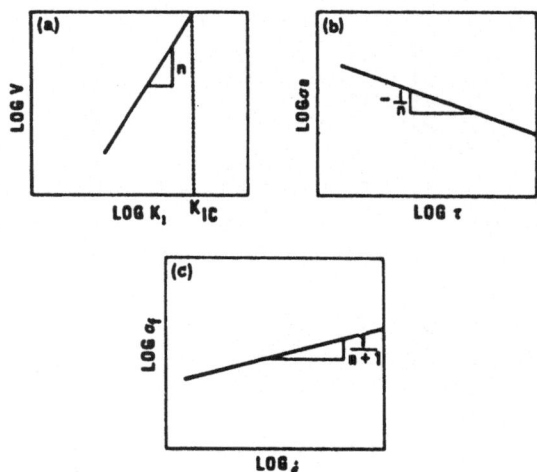

Fig. 18: Relation between the v-K-plot (a), the life-time under high sustained load (b), and the strength increase at high rate of loading (c).

5. CONCLUSIONS

A phenomenological approach to describe failure of concrete is able to explain qualitatively the influence of time on the degradation of the composite structure. Crack theory as developed by Zaitsev (8) and a stochastic theory proposed by Mihashi (13) form a solid basis for a quantitative treatment of this problem. It is possible to predict concrete strength under various types of loading such as static or dynamic fatigue or impact realistically. The presented theoretical concepts do not take stress corrosion into consideration. Further research and special investigations are needed to clarify the role of chemical stress corrosion in the context of time-dependent crack growth, if there is any.

REFERENCES

1. Rasch, C., Spannungs-Dehnungs-Linien des Betons und Spannungs-
    verteilung in der Biegedruckzone bei konstanter Verformungsge-
    schwindigkeit, Deutscher Ausschuss für Stahlbeton, Schriften-
    reihe, Heft 154 (1962).

2. Wittmann, F.H., and Y.B. Zaitsev, Verformung und Bruchvorgang
    poröser Baustoffe bei kurzeitiger Belastung und Dauerlast,
    Deutscher Ausschuss für Stahlbeton, Schriftenreihe, Heft 232
    (1974).

3. Rüsch, H., Sell, R., Rasch, C., Grasser, E., Hummel, A.,
    Wesche, K., and Flatten, H., Festigkeit und Verformung von
    unbewehrtem Beton unter konstanter Dauerlast, Deutscher Ausschuss
    für Stahlbeton, Schriftenreihe, Heft 198 (1968).

4. Siemes, A.J.M., vermoeiing van beton, deel 1 : drukspanningen,
    CUR-VB, rapport 112 (1983).

5. Siemes, A.J.M., Fatigue of Plain Concrete in Uniaxial Compression,
    Proc. IABSE Colloquium Fatigue of Steel and Concrete Structures,
    Lausanne (1982).

6. Holzapfel, F., Das Verhalten von Zementstein unter dynamischer
    Beanspruchung, Dissertation, RWTH, Aachen (1970).

7. Ziegeldorf, S., Phenomenological Aspects of the Fracture of
    Concrete, in Fracture Mechanics of Concrete, ed. F.H. Wittmann,
    Elsevier Science Publishers, Amsterdam (1983).

8. Zaitsev, Y.B., Crack Propagation in a Composite Material, in
    Fracture Mechanics of Concrete, ed. F.H. Wittmann, Elsevier
    Science Publishers, Amsterdam (1983).

9. Zaitsev, Y.B., and Wittmann, F.H., Simulation of crack propaga-
    tion and failure of concrete, Mat. and Struct., 14 (1981),
    357-365.

10. Wittmann, F.H., and Zaitsev, Y.B., Behaviour of hardened cement
    paste and concrete under high sustained load, Proc. Int. Conf.
    on Mechanical Behaviour of Materials, Kyoto, Japan, Vol. IV
    (1971), 84-95.

11. Wittmann, F.H., Roelfstra, P.E., and Sadouki H., Simulation and
    analysis of composite structures, (to be published in Mat. Sci.
    Eng. 1984).

12. Mihashi, H., and Izumi, M., A stochastic theory for concrete fracture, Cem. Concr. Res. 7 (1977), 411-422.

13. Mihashi, H., A stochastic theory for fracture of concrete, in Fracture Mechanics of Concrete, ed. F.H. Wittmann, Elsevier Science Publishers, Amsterdam (1983).

14. Mihashi, H., and Wittmann, F.H., Stochastic approach to study the influence of rate of loading on strength of concrete, Heron (The Netherlands) 25 (1980) No 3.

15. Zech, B., and Wittmann, F.H., Probabilistic approach to describe the behaviour of materials, Nucl. Eng. Design 48 (1978), 575-584.

16. Zech, B., and Wittmann, F.H., Variability and mean value of strength of concrete as function of load, J. Amer. Concr. Inst. 77 (1980) 358-362.

17. Okada, T., and Sines, G., Crack coalescence and microscopic crack growth in the delayed fracture of alumina, J. Amer. Cer. Sce. 66 (1983) 719-725.

18. Charles, R.J., Static fatigue of glass I, J. Appl. Phys. 29 (1958) 1549-1553.

    Charles, R.J., Static fatigue of glass II, J. Appl. Phys. 29 (1958) 1554-1560.

19. Michalske, T.A., and Freiman, S.W., A molecular interpretation of stress corrosion in silica, Nature 295 (1982) 511-512.

20. Lawn, B.R., Physics of fracture, J. Amer. Cer. Soc. 66 (1983) 83-91.

21. Krokosky, E.M., Static fatigue in hydrated Portland cement, Mat. and Struct. 6 (1973) 447-452.

22. Barrick II, J.E., and Krokosky, E.M., The effects of temperature and relative humidity on static strength of hydrated Portland cement, Journal of Testing and Evaluation 4 (1976) 61-73.

23. Shah, S.P., and Chandra, S., Fracture of concrete subjected to cyclic and sustained loading, J. Amer. Concr. Inst. 67 (1970) 816-825.

24. Tait, R.B., Fatigue and fracture of cement mortar, PhD dissertation, University of Cape Town (1984).

25. Wittmann, F.H., Bestimmung physikalischer Eigenschaften des Zementsteins, Deutscher Ausschuss für Stahlbeton Schriftenreihe Heft 232 (1974).

26. Wittmann, F.H., Grundlagen eines Modells zur Beschreibung charakteristischer Eigenschaften des Betons, Deutscher Ausschuss für Stahlbeton Schriftenreihe Heft 290 (1977).

27. Mindess, S., Nadeau, J. S., and Barrett, J. D., Effect of constant deformation rate on the strength perpendicular to the grain of Douglas-fir, Wood Science 8 (1976) 262-266.

28. Nadeau, J. S., Fracture Mechanics: An overview, Proc. First International Conference on Wood Fracture, August 1978, Banff, Alberta, Canada, pp. 175-186.

APPLICATION OF FRACTURE MECHANICS
TO CEMENTITIOUS COMPOSITES
NATO-ARW - September 4-7, 1984
Northwestern University, U.S.A.
S. P. Shah, Editor

RATE OF LOADING EFFECTS ON THE FRACTURE OF CEMENTITIOUS MATERIALS

Sidney Mindess

Department of Civil Engineering
The University of British Columbia
Vancouver, British Columbia V6T 1W5, Canada

1.   INTRODUCTION

The effects of the rate of loading on the strength and
fracture of cementitious materials have been studied extensively
at least since the work of Abrams (1) in 1917. Other significant
early contributions in this area were made by Jones and Richart
(2), Watstein (3), Wright (4) and Evans (5). Most of the early
studies (before 1955) have been reviewed by McHenry and Shideler
(6). These studies, and the many that have since been carried
out, are in general qualitative agreement, in that almost all of
them show that the apparent strengths of cement, mortar and
concrete increase as the rates of loading are increased. Most
commonly, researchers have found that their data can be
represented adequately by a linear relationship between the
strength and the logarithm of the stress rate (or strain rate),
i.e.

$$\sigma = A + B \log R \qquad (1)$$

where σ = strength (in tension, compression, or flexure)
         R = stress rate, (or strain rate, or rate of cross-head
               deflection of the testing machine)
         A,B = constants

However, the data are not particularly consistent quantitatively
from study to study. There is a wide variability in the values of
the constants A and B, and some data cannot be represented by
Equation 1. While the reasons for this variability cannot be
stated with certainty, they are probably related to:

(i) the type of material tested, and

(ii) the test methods themselves, which have varied widely from study to study.

In addition, Rusch (7) showed that for the limiting case of constant compressive stress, concrete could sustain indefinitely loads of only about 75-80% of the strength at the usual rates of loading. He also found that the modulus of elasticity, E, increased as the rate of loading increased; this had been reported previously by others (2,3,5).

The early studies were, nevertheless, consistent in one respect: none of them provided any real explanation for the phenomena that they reported. Perhaps the earliest explanation for these rate of loading effects was provided by Newman (8), and it is worthwhile quoting it extensively:

> "At standard or medium rates of loading, failure is initiated when the small flaws and microcracks attain an unstable length and propagate rapidly until they interconnect and the structure becomes completely disrupted.... As the rate of loading is increased, however, the disruptive effect of the microcracks is delayed, and the specimen is momentarily able to support a higher load. The creep deformations, which have insufficient time to develop at medium and fast rates of loading, become the most important factor at slow rates. At slow and medium rates of loading, stress concentrations can cause microcracks to propagate slowly, even at stresses below 50-60% of the ultimate. The slow crack propagation process is the cause of the small degree of non-linearity of the stress-strain curve below 50% of the ultimate and the failure of concrete under long-term loading at stresses only 75-80% of the short-term ultimate strength."

This is, indeed, the simple explanation. The rate of loading phenomena which have been observed must be interpreted by using the principles of fracture mechanics. The purpose of the present work is to use fracture mechanics to provide a more detailed explanation of rate of loading effects on the fracture of cementitious materials.

2.   FRACTURE MECHANICS APPROACH TO RATE OF LOADING EFFECTS

The fracture mechanics approach to rate of loading effects involves a combination of the classical Griffith theory with an empirical relationship describing subcritical crack growth. For perfectly brittle materials, fracture is governed by the Griffith equation,

$$\sigma_c = \left(\frac{2E\gamma}{\pi a}\right)^{1/2} \tag{2}$$

which may also be written in the form

$$\sigma_c = \left(\frac{EG_c}{\pi a}\right)^{1/2} \tag{3}$$

where σ_c is the fracture strength, E is the modulus of elasticity, "a" is the crack length, γ is the fracture surface energy, and $G_c (= 2\gamma)$ is the critical strain energy release rate. We may than define an intrinsic property of a brittle material, the fracture toughness, K_c, as

$$K_c = \sqrt{EG_c} \tag{4}$$

Substituting this relationship back into Equation (3), we can write

$$\sigma_c = \frac{K_c}{(\pi a)^{1/2}} \tag{5}$$

which states that fracture will occur when the crack length "a" reaches some critical value.

During subcritical crack growth in brittle materials, the crack velocity can be described by the empirical relationship

$$V = AK_I^N \tag{6}$$

where $V = \frac{da}{dt}$ = rate of crack extension, and A and N are constants. Subcritical crack growth (the slow growth of cracks that are too small to cause failure under the prevailing stress) leads to a dependence of the failure stress on the loading rate, since in specimens loaded slowly, more time is available for slow crack growth than in specimens loaded rapidly. Thus, the rate of loading effect in cementitious materials must be due, at least in part, to the growth of a crack as governed by Equation (6) until it reaches the critical value as defined by Equation (5).

As has been shown by Nadeau, Bennett and Fuller (9), the dependence of strength on the rate of loading can be described by the logarithmic expression

$$\ln\sigma_c = \frac{1}{(N+1)} \ln B\dot{\sigma} + \frac{1}{(N+1)} \ln\left(\sigma_i^{N-2} - \sigma_f^{N-2}\right) \tag{7}$$

where $\dot{\sigma}$ is the stress rate, the subscripts i and f refer to the initial condition (before loading) and the final condition (at fracture), respectively, and

$$B \equiv 2K_c^{2-N}(N+1)/AY^2(N-2) \qquad (8)$$

where Y is a geometric parameter which depends on the specimen dimensions and the method of loading. Equation 7 implies that a plot of $\ln\sigma_c$ versus $\ln\dot{\sigma}$ would have a slope of $\frac{1}{N+1}$ at low values of $\dot{\sigma}$, and would reach a constant value of $\ln\sigma_c$ (slope = 0) at high values of $\dot{\sigma}$. That is, at very high loading rates, the strength would be largely independent of loading rate because there would be virtually no time for subcritical crack growth to occur.

Unfortunately, when trying to apply the above theoretical considerations to cementitious materials, complications appear. First, while hardened cement paste may be considered to be essentially a brittle material, concrete itself is not particularly brittle, and its stress-strain (σ-ε) curve is far from linear. Second, and perhaps even more important, cementitious materials exhibit creep under load, the magnitude of the creep increasing with the applied stress. Thus, in regions where the local stress is concentrated by cracks or other flaws in the material, creep deformations may be considerable. While the creep may serve to relax the stress locally, the creep strain may also contribute to failure if indeed the failure of cement and concrete are governed by a maximum strain criterion.

3. REVIEW OF THE LITERATURE ON RATE OF LOADING EFFECTS

Based on work by Evans (10) and Wiederhorn (11), it can be shown that there are three independent methods for determining the constant B in Equation 7, as indicated in Figure 1:

i)   from constant load tests, by a logarithmic plot of the applied stress, σ_a, versus the time to failure, τ: the slope of this plot is $-1/N$;

ii) by direct observations of controlled crack growth, where A and N are the intercept and slope, respectively, of a plot of log V vs. log K_I; and

iii) from constant rate of loading tests to failure, in which the first two terms of Equation 7 are plotted, giving both the slope N and, from the intercept, B.

Over the years, all three types of tests have been carried out on cementitious materials by various investigators.

3.1 Constant Load Tests

The results of the many sustained load investigations that

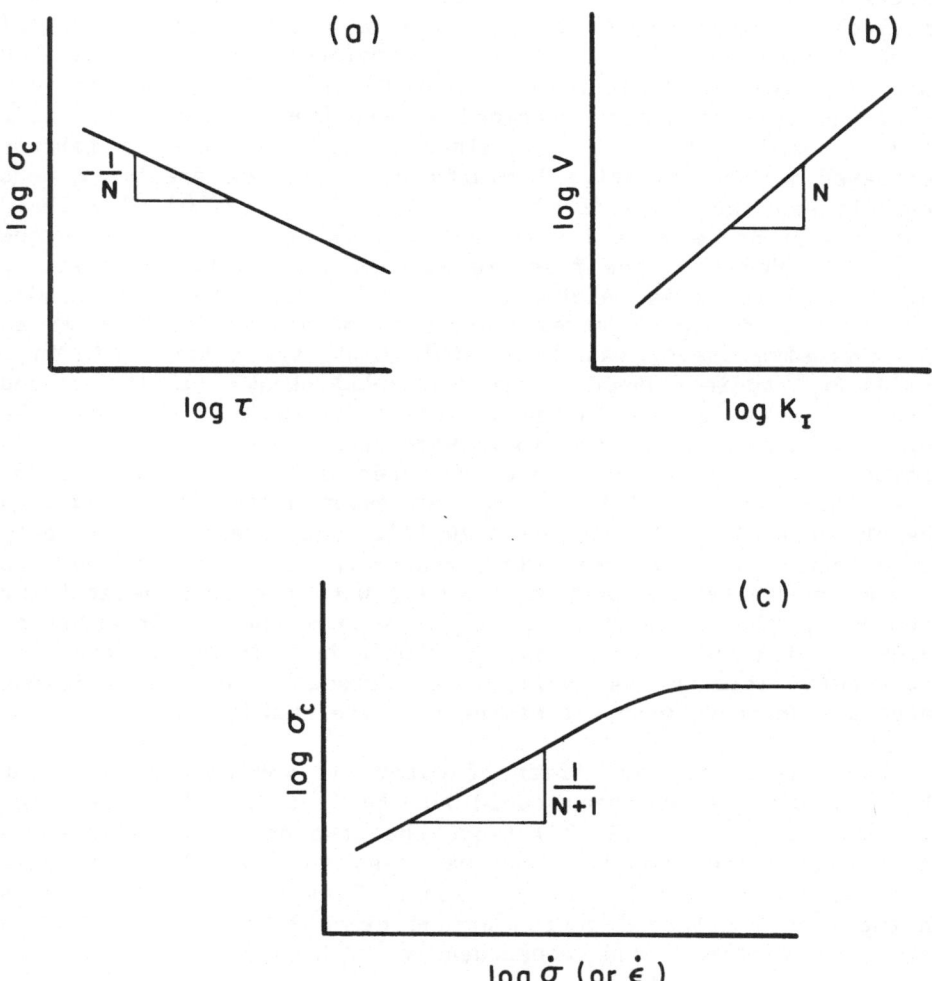

Figure 1.   Three ways of determining the constant B(Equation 7):
(a) Constant load tests; σ_a is applied stress, τ is
time to failure.   (b) Direct observation of controlled
crack growth.   (c) Constant rate of loading tests; σ_c
is failure stress; $\dot{\sigma}$ is stress rate, $\dot{\epsilon}$ is strain rate.

have been reported (12–24) are fairly consistent.   Most of the
studies show that failure under sustained stress will occur at
sustained stress levels of 70–90% of the short term strength, in
tension, flexure, and compression.   Sustained loads slightly below
those required for static fatigue failure to occur in compression
will actually strengthen concrete slightly (12,13), presumably due
to some additional bond formation under pressure.

The precise mechanisms of crack growth in cementitious materials are not very well understood. There is considerable evidence, however, that the presence of water plays a critical role. It has been observed that, in completely <u>dry</u> specimens, the rate of crack development diminishes to zero (14), and for such specimens failure under sustained stress does not occur (13,15). It was noted that, under sustained loads, the time to failure decreased as the relative humidity at which the specimens were equilibrated increased (15-19). Cook and Haque (20) also found that the presence of moisture assisted crack growth, and Domone (21) concluded that the fracture process was moisture sensitive. The reasons for this behaviour are not clear. Shah and Chandra (13) suggested that a stress-corrosion mechanism, with water as the corrosive agent, was responsible. On the other hand, in a series of studies, Husak, Barrick, and Krokosky (15-19) argued that while there was indeed a stress-corrosion mechanism, the corrosive agent was the OH^- ion. Since their test results (17,19) showed that the time to failure increased with increasing temperature, they argued that the stress corrosion mechanism depended on the presence of $Ca(OH)_2$ (whose solubility decreases as the temperature increases). In their view, static fatigue is due to OH^- ion attack on highly stressed Si-O bonds, with the rate determining step being the production of OH^- ions from the dissociation of $Ca(OH)_2$. [In this connection, it should be pointed out that for short-term loading as well, the strengths of cementitious materials decrease with increasing relative humidity.]

In addition to the effect of water, it was also pointed out that the effects of creep could not be ignored. For instance, Wittmann and Zaitsev (22,23) found that the decrease in strength under a sustained tensile load was less than would be expected from their theoretical considerations. They concluded that creep in the highly stressed region around crack tips brought about a stress relaxation, and consequently a higher load would be required to cause failure.

Although most of the studies cited above did not present their experimental data in the form of Figure 1(a), the results of Al-Kubaisy and Young (24) gave a value for the constant N (Equation 6) of 33.6.

3.2 Controlled Crack Growth Tests

Relatively few controlled crack growth studies have been carried out on cementitious materials; most have used double torsion specimens (25-29), though some tests on large, specially designed double cantilever beams have also been reported (30-31). The data, generally presented in the form of a log V vs log K_I plot, can be defined by the slope (N) and intercept (A) of the linear portion of this plot (Figure 1(b)). Most of the data available are summarized in Table 1. These studies (with the exception of ref. 26) show roughly comparable results: N values of the order of 30 for both cement and mortar, in the range of crack velocities from about 10^{-6} to 10^{-2} m/s. It was also found that neither polymer impregnation (27) nor fibre reinforcement (29) had any significant effect on N, though a fair degree of scatter is inherent in these tests.

Studies carried out on very large (2.8 x 1.1 x 0.3 m) DCB specimens (30) showed an extensive damage zone ahead of the crack tip, of the order of 0.2 - 0.5 m. Subsequent tests on somewhat larger prestressed DCB specimens (31) showed a microcracked region of about 0.2 m ahead of the crack; this region stayed about the same size as the crack propagated.

3.3 Constant Rate of Loading Tests to Failure

By far the most common technique of determining the effect of loading rate on the strength and fracture characteristics of cementitious materials has been by tests at a constant rate of loading till failure. Some of these test data are summarized in Table 2. It may be seen that, while there was considerable variability as between investigations, the flexure and tension tests

Table 1. Values of N (slope of the log V vs. log K_I plot) obtained from double torsion tests.

Reference	Specimen Size (mm) l x d x w	Material	N
25	229 x 13 x 76	hcp* (saturated)	36
26	229 x 13 x 76	hcp (saturated) hcp (dry)	75 64
27	105 x 6 x 57	mortar polymer impregnated mortar }	~30
28	813 x 38 x 152	mortar	20-30
29	1219 x 51 x 406	hcp (saturated) fibre reinforced concrete	36.5 15-80

*hcp = hydrated cement paste

Table 2.   Values of N from constant rate of loading tests.

Reference	Material	Type of Test	N
2	7-day concrete	Compression	29.9
	"	"	49.3
	"	"	40.1
	28-day concrete	"	30.0
	"	"	28.6
	"	"	33.0
3	"strong" concrete	Compression	26.8
	"weak" concrete	"	25.5
32	Concrete	Compression	43.4
33	Mortar	Compression	25-27
	Lightweight concrete	"	59
	Normal concrete	"	26
4	7-day concrete	Flexure	28.2
	28-day concrete	"	23.0
33	Mortar	Flexure	18-22
	Lightweight concrete	"	25.6
34	7-day concrete	Flexure	21.0
35	Cement paste	Flexure	17.7
	Mortar	"	14.9
36	Mortar	Flexure	20-25
32	Concrete	Tension	21.9

tended to yield lower values of N than did the compression tests, perhaps because of the more complex strain fields and larger amounts of cracking and damage in the compression specimens (which tend not to fail by the propagation of a single main crack, as is the case for the other two types of loading).

3.4 Impact Tests

At the opposite pole from the sustained load tests, where the loads might be applied for days or weeks before failure occurs, lie the impact tests, in which the entire loading history of the specimens is measured in milliseconds.   Much of the available impact data on concrete has been reviewed by Suaris and Shah (37). Some values of N obtained from constant rate of loading studies using various impact tests are summarized in Table 3.

As may be seen from Table 3, the values of N obtained from impact tests are essentially the same as those obtained from constant rate of loading tests at much lower stress rates. This would suggest that even at these very high stress rates, the fracture processes are much the same. However, it has been suggested that at very high rates of loading, there is a decrease in the amount of microcracking (40). On the other hand, Zielinski (41) has argued that the higher impact strengths are caused by simultaneous extensive microcracking in the whole volume of the stressed specimen, as indicated by the much larger impact strains compared to the static strains. This difference in interpretation may be due to the very different tests: an instrumented Charpy apparatus (40) versus the Hopkinson split bar (41). It was suggested (42) that the crack arresting action of tough aggregate particles was of great importance for concrete fracture. Zielinski (43) has pointed out that since the time in which the fracture occurs is extremely short, there will be no time for stress relaxation in the crack region due to creep or other processes.

Table 3.   Values of N from impact tests, analyzed by the constant rate of loading technique.

Reference	Material	Type of Test	N
38	Concrete, 16 mm aggregate	High velocity projectiles striking ballistically suspended specimens	20
	Concrete, 24 mm aggregate		40
39	Fibre reinforced concrete	Instrumented drop-weight apparatus	10-20
40	Mortar and Concrete	Instrumented Charpy apparatus	25-35
41	Mortar	Hopkinson split bar	21
42	Concrete	Hopkinson split bar	20-30
43	Mortar and Concrete	Instrumented drop-weight apparatus	20-30

Suaris and Shah (44) have suggested, however, that the model described above [Equation 7] to explain the rate of loading effect is not an accurate description of the behaviour of concrete. Equation 7 is based on a constant value of N. However, Suaris and Shah (44) have shown that the value of N for mortar decreases with increasing strain rate, from about 47 at low strain rates to about 16 at high strain rates. They suggest instead a continuum damage theory to explain rate of loading effects.

4.   Effect of Loading Rate on Fracture Toughness

While many loading rate studies have been carried out as described above, relatively little work has been done on measuring the effects of loading rate on the fracture of underlined notched cementitious materials. The few studies that have been carried out show that K_c also appears to increase with increased loading rates (45-47). Assuming that the inverse slope of a plot of K_c versus loading rate is also N+1 (Figure 1(c)), these results are shown in Table 4. These data are based on very few specimens, and so they are not very reliable; nonetheless, the values of N (with the possible exception of Ref. 46) are in the same range of values reported for the other types of tests described above.

Recently, more extensive tests of this type were carried out by the author. Beams of hydrated cement paste, notched to half the beam depth, were tested to failure in 3-point bending on a 406.4 mm span, as shown schematically in Figure 2, using a testing machine with strain rate control. Six different rates of crosshead motion were used, ranging from 5.54×10^{-7} m/s to 3.81×10^{-5} m/s, with twenty specimens tested at each. From the maximum load data, K_c was calculated using the method of Brown and Srawley (48). In addition, using a linear finite element analysis of the notched test specimens, it was possible to estimate not only the

Table 4.   Values of N from plots of log K_c versus log $\dot{\sigma}$

Reference	Material	Test	N
45	Mortar	Circumferentially notched round bar in bending	~20
46	Cement	Notched beam in bending	~55
47	Soil Cement	Notched beam in bending	~20

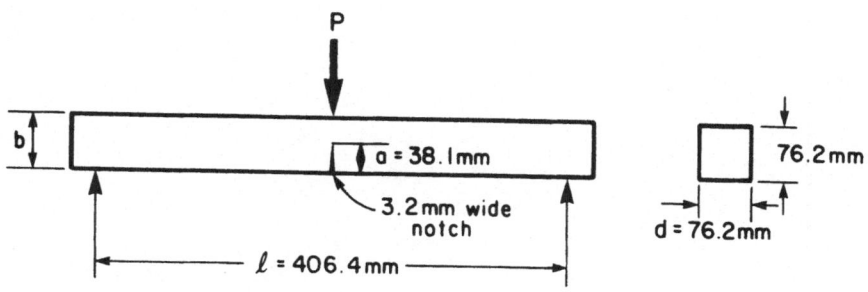

$$K_c = \frac{3P\ell\sqrt{\pi a}}{2b^2d} \left[1.10 - 1.67\left(\frac{a}{b}\right) + 8.04\left(\frac{a}{b}\right)^2 - 13.97\left(\frac{a}{b}\right)^3 + 14.47\left(\frac{a}{b}\right)^4 \right]$$

Figure 2. Determination of K_c in 3-point bending.

failure stress and strain at the tip of the notch, but also the strain rates and stress rates at the tip of the notch, corresponding to the different cross-head deflection rates. These results are summarized in Table 5. It should be noted that, due to the nature of these calculations, the coefficient of variation of the maximum stress at the crack tip is the same as that of the maximum load; the coefficient of variation of the maximum strain at the crack tip is the same as that of the time to failure.

The cumulative probability distributions of K_c values for these specimens are shown in Figure 3. While the data for three of the intermediate loading rates (2.03×10^{-6} m/s, 3.81×10^{-6} m/s and 1.27×10^{-5} m/s) are very similar, there is a definite increase in K_c between the loading rates of 5.54×10^{-7} m/s and 2.03×10^{-5} m/s. However, the highest rate of loading (3.81×10^{-5} m/s) resulted in the lowest value of K_c.

The mean values of K_c are plotted against the rate of cross-head deflection ($\dot\delta$) and the computed strain rate at the notch tip ($\dot\varepsilon$) in Figure 4. Considering only the five lowest loading rates, this data can be represented by the equation

$$\log K_c = -0.00033 + 0.0267 \log \dot\delta \qquad (9)$$

The correlation coefficient, based on all 100 data points is, however, rather low: 0.34. This is due to the fairly high degree

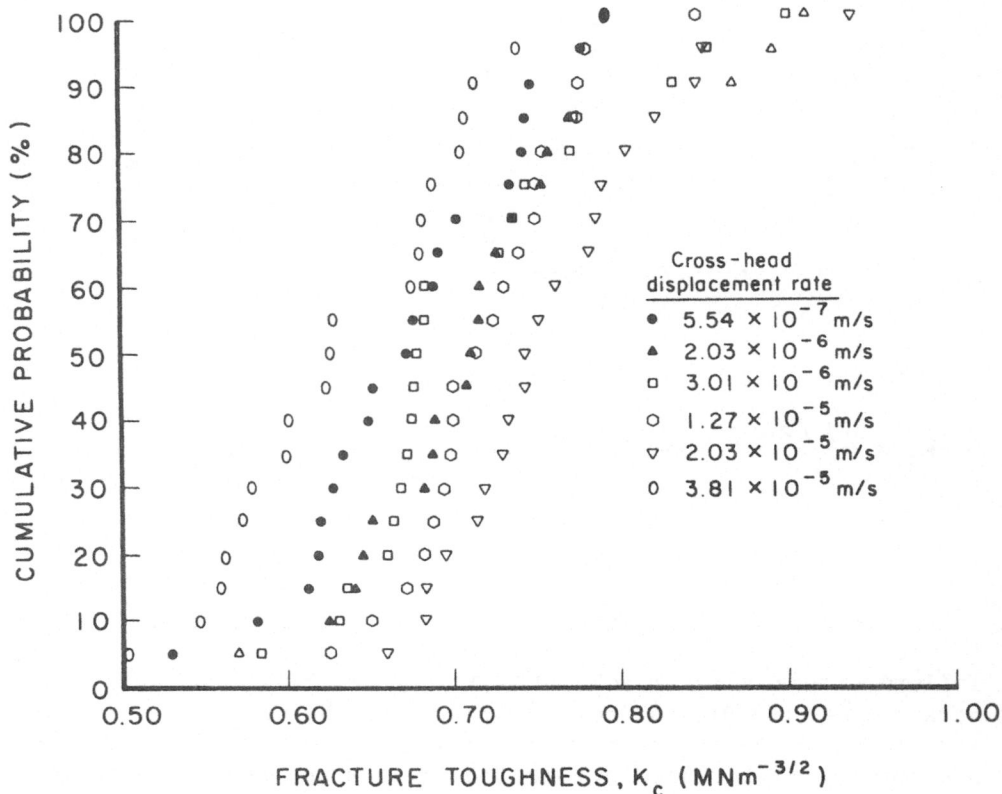

Figure 3. Cumulative probability distributions of K_c for hardened cement paste at different rates of loading.

of scatter for each set of twenty data points.

The value of N calculated from the inverse slope of the line in Figure 4 is about 36. This is within the range of data shown in Table 4, and is very close to the values reported in the literature as described earlier: ref. 25, N ≅ 36; ref. 24, N ≅ 34; ref. 27, N ≅ 30; and ref. 29, N ≅ 37. This close agreement supports the view that classical fracture mechanics can be used to describe the fracture of hardened cement paste. Moreover, it indicates that subcritical crack growth is indeed responsible for the rate of loading effect.

On the other hand, the data of Table 5 show that the failure strains at the crack tip for all of the loading rates were very

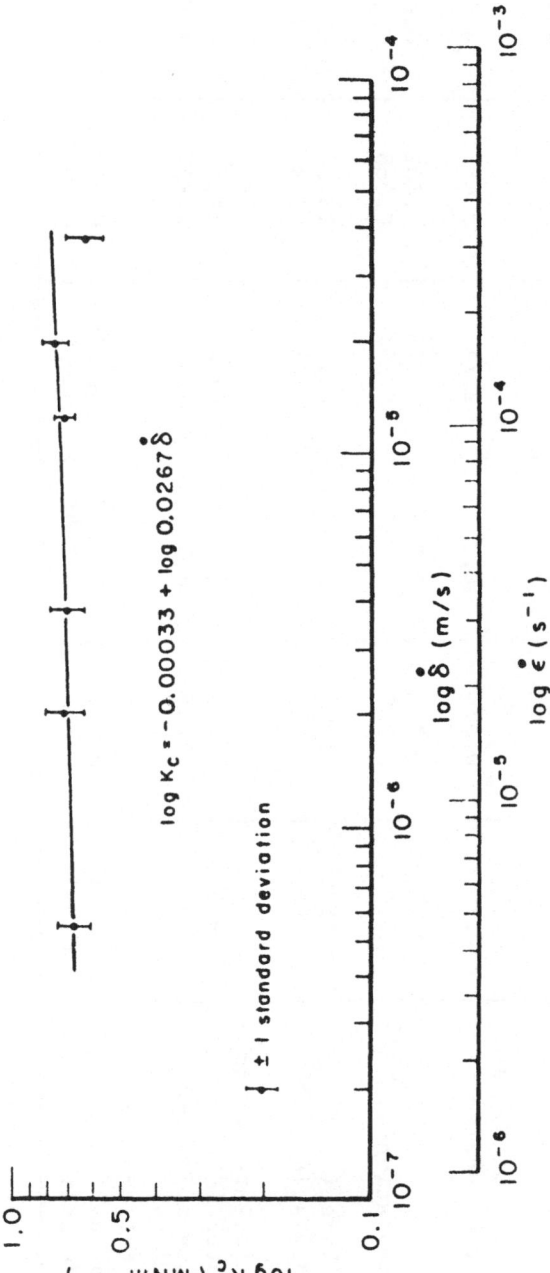

Figure 4. Logarithmic plot of K_c versus rate of cross-head deflection ($\dot{\delta}$) and calculated strain rate ($\dot{\epsilon}$) for hardened cement paste.

Table 5.   Rate of loading data for notched cement paste beams.

cross-head displacement rate (m/s)	5.54×10^{-7}	2.03×10^{-6}	3.81×10^{-6}	1.27×10^{-5}	2.03×10^{-5}	3.81×10^{-5}
calculated strain rate at root of notch (s^{-1})	4.65×10^{-6}	1.72×10^{-5}	3.27×10^{-4}	1.09×10^{-4}	1.74×10^{-4}	3.28×10^{-4}
calculated stress rate at root of notch (MPa/s)	0.10	0.37	0.76	2.74	3.46	6.40
K_c ($MNm^{-3/2}$) standard deviation ($MNm^{-3/2}$)	0.675 0.068	0.724 0.088	0.713 0.080	0.723 0.051	0.762 0.068	0.639 0.075
time to failure (s) standard deviation (s)	387 86	116 25	55 12	15.6 3.3	13.2 3.3	5.8 2.0
calculated failure stress at root of notch (MPa) standard deviation (MPa)	39.4 4.48	42.7 6.05	41.9 6.54	42.8 4.30	45.7 4.82	37.1 5.13
calculated failure strain at root of notch standard deviation	0.0018 0.0004	0.0020 0.0004	0.0018 0.0004	0.0017 0.0004	0.0023 0.0006	0.0019 0.0007

nearly the same, approximately 0.002 (0.2%). Thus, it may also be possible to interpret these results by a maximum limiting strain theory.

The reason for the low value of the K_c obtained at the highest loading rate is unclear. It may be that this loading rate is high enough so that no further increase in K_c can take place because there is not enough time for subcritical crack growth to occur, as indicated by Equation 7. Dwivedi and Pratt (49) have obtained rather similar data, also showing a decrease in K_c at the highest strain rates, with values of N in the range 20-40. However, they have suggested that this may be related to relaxation processes occuring within the material. Further study is required to elucidate this point.

5. CONCLUDING REMARKS

From the results described above, it may be seen that the physical fracture processes in cementitious materials do not appear to vary strongly with changes in the rate of loading. The principal difference seems to be that at higher loading rates, there is somewhat less internal microcracking (44) and there is a tendency for more coarse aggregate particles to be fractured (50). However, the work of fracture is not particularly rate sensitive (51), and there is very little difference between static and dynamic values of the strain energy release rates (52).

The increase in strength and in K_c with increases in loading rate is clearly associated with the phenomenon of subcritical crack growth, as was discussed earlier. As has been pointed out (44), tensile strengths appear to be more sensitive to rate of loading effects than do compressive strengths, and this must be associated with the different crack patterns that develop in these two cases. Unfortunately, however, the slow crack growth model described by Equation 7 does not appear to be a very good representation of the fracture processes in cement and concrete. This model assumes a constant value of N. However, as has been reported by Suaris and Shah (37), K_c is dependent on the crack velocity, increasing as the crack velocity increases, while N appears to decrease considerably at very high rates of loading. Moreover, Alford (52) has reported that crack velocities vary considerably along the crack path, though in general they tend to increase as the crack progresses. This variability in crack velocity is related to the heterogeneous nature of concrete, and to the extensive branch cracking that is known to occur (53,54). Therefore, while we now have a reasonably good empirical knowledge of stress rate effects in cementitious materials, a satisfactory theoretical model has not yet been formulated.

6.   ACKNOWLEDGEMENTS

This work was supported by a grant from the Natural Sciences and Engineering Research Council of Canada.

REFERENCES

1.   Abrams, D.A. Effect of Rate of Application of Load on the Compressive Strength of Concrete. Proceedings, American Society for Testing and Materials 17, Part II (1917) 364–377.

2.   Jones, P.G. and F.E. Richart. The Effect of Testing Speed on Strength and Elastic Properties of Concrete. Proceedings, American Society for Testing and Materials 36, Part II (1936) 380–392.

3.   Watstein, D.   Effect of Straining Rate on the Compressive Strength and Elastic Properties of Concrete. Journal of the American Concrete Institute 49 (1953) 729–744.

4.   Wright, P.J.F.   The Effect of the Method of Test on the Flexural Strength of Concrete. Magazine of Concrete Research 4 (1952) 67–76.

5.   Evans, R.H.   Effect of Rate of Loading on Some Mechanical Properties of Concrete, in W.H. Walton, ed., Proceedings of a Conference on the Mechanical Properties of Non-Metallic Brittle Materials, London, 1958 (London: Butterworths Scientific Publications, 1958), pp. 175–192.

6.   McHenry, D. and J.J. Shideler. Review of Data on Effect of Speed in Mechanical Testing of Concrete, in Symposium on Speed of Testing of Non-Metallic Materials, ASTM Special Technical Publication No. 185 (Philadelphia: American Society for Testing and Materials, 1956), pp. 72–82.

7.   Rusch, H.   Researches Toward a General Flexural Theory for Structural Concrete.   Journal of the American Concrete Institute 57 (1960) 1–28.

8.   Newman, K.   Concrete Control Tests as Measures of the Properties of Concrete.   Proceedings of a Symposium on Concrete Quality, (London: Cement and Concrete Association, 1964), pp. 120–138.

9. Nadeau, J.S., R. Bennett and E.R. Fuller, Jr. An Explanation of the Rate-of-Loading and Duration-of-Load Effects in Wood in Terms of Fracture Mechanics. Journal of Materials Science 17 (1982) 2831-2840.

10. Evans, A.G. Slow Crack Growth in Brittle Materials Under Dynamic Conditions. International Journal of Fracture Mechanics 10 (1974) 251-259.

11. Wiederhorn, S.M. Subcritical Crack Growth in Ceramics, in R.C. Bradt, D.P.H. Hasselman and F.F. Lange, eds., Fracture of Ceramics, II (New York, N.Y.: Plenum Press, 1974), pp. 613-646.

12. Coutinho, A. de Sousa. Note sur la Rupture de Béton Maintenu a une Contrainte Constante. Matériaux et Constructions 2 (1969) 49-57.

13. Shah, S.P. and S. Chandra. Fracture of Concrete Subjected to Cyclic and Sustained Loading. Journal of the American Concrete Institute 67 (1970) 816-825.

14. Ruetz, W. The Two Different Physical Mechanisms of Creep in Concrete, in A.E. Brooks and K. Newman, eds., The Structure of Concrete (London: Cement and Concrete Association, 1968), pp. 190-201.

15. Husak, A.D. and E.M. Krokosky. Static Fatigue of Hydrated Cement Concrete. Journal of the American Concrete Institute 68 (1971) 263-271.

16. Husak, A.D. Static Fatigue of Portland Cement Concrete. Ph.D. Thesis, Carnegie-Mellon University, Pittsburgh, 1969.

17. Barrick, J.E. The Effects of Temperature and Relative Humidity on Static Fatigue of Hydrated Portland Cement. Ph.D. Thesis, Carnegie-Mellon University, Pittsburgh, 1972.

18. Krokosky, E.M. Static Fatigue in Hydrated Portland Cement. Matériaux et Constructions 6 (1973) 447-452.

19. Barrick, J.E. and E.M. Krokosky. The Effects of Temperature and Relative Humidity on Static Strength of Hydrated Portland Cement. Journal of Testing and Evaluation 4 (1976) 61-73.

20. Cook, D.J. and M.N. Haque. The Tensile Creep and Fracture of Dessicated Concrete and Mortar on Water Sorption. Matériaux et Constructions 7 (1974) 191-196.

21. Domone, P.L. Uniaxial Tensile Creep and Failure of Concrete. Magazine of Concrete Research 26 (1974) 144-152.

22. Wittmann, F.H. and Ju. V. Zaitsev. Behaviour of Hardened Cement Paste and Concrete Under High Sustained Load, in Mechanical Behaviour of Materials, Vol. IV (Japan: The Society of Materials Science, 1972), pp. 84-95.

23. Zaitsev, Ju. V. and F.H. Wittmann. Fracture of Porous Viscoelastic Materials Under Multiaxial State of Stress. Cement and Concrete Research 3 (1973) 389-395.

24. Al-Kubaisy, M.A. and A.G. Young. Failure of Concrete Under Sustained Tension. Magazine of Concrete Research 27 (1975) 171-178.

25. Mindess, S., J.S. Nadeau and J.M. Hay. Effects of Different Curing Conditions on Slow Crack Growth in Cement Paste. Cement and Concrete Research 4 (1974) 953-965.

26. Nadeau, J.S., S. Mindess and J.M. Hay. Slow Crack Growth in Cement Paste. Journal of the American Ceramic Society 57 (1974) 51-54.

27. Evans, A.G., J.R. Clifton and E. Anderson. The Fracture Mechanics of Mortars. Cement and Concrete Research 6 (1976) 535-548.

28. Wecharatana, M. and S.P. Shah. Double Torsion Tests for Studying Slow Crack Growth of Portland Cement Mortar. Cement and Concrete Research 10 (1980) 833-844.

29. Yam, A.S.-T. and S. Mindess. The Effects of Fibre Reinforcement on Crack Propagation in Concrete. International Journal of Cement Composites and Lightweight Concrete 4 (1982) 83-93.

30. Sok, C., J. Baron and D. Francois. Fracture Mechanics Applied to Concrete (in French). Cement and Concrete Research 9 (1979) 641-648.

31. Chhuy, S., M.E. Benkirane, J. Baron and D. Francois. Crack Propagation in Prestressed Concrete. Interaction with Reinforcement, in Advances in Fracture Research, Vol. 4 (Pergamon Press, 1981), pp. 1507-1514.

32. Takeda, J. and H. Tachikawa. Deformation and Fracture of Concrete Subjected to Dynamic Load, in Mechanical Behaviour of Materials, Vol. IV (Japan: The Society of Materials Science, 1972), pp. 267-277.

33. Mihashi, H. and F.H. Wittmann. Stochastic Approach to Study the Influence of Rate of Loading on Strength of Concrete. HERON (The Netherlands) 25 (1980).

34. McNeely, D.J. and S.D. Lash. Tensile Strength of Concrete. Journal of the American Concrete Institute 60 (1963) 751-760.

35. Mindess, S. and J.S. Nadeau. Effect of Loading Rate on the Flexural Strength of Cement and Mortar. American Ceramic Society Bulletin 56 (1977) 429-430.

36. Zech, B. and F.H. Wittmann. Variability and Mean Value of Strength of Concrete as Function of Load. Journal of the American Concrete Institute 77 (1980) 358-362.

37. Suaris, W. and S.P. Shah. Mechanical Properties of Materials Subjected to Impact, in RILEM-CEB-IABSE Symposium on Concrete Structures Under Impact and Impulsive Loading (Berlin, 1982), pp. 33-62.

38. Birkimer, D.L. and R. Lindeman. Dynamic Tensile Strength of Concrete Materials. Journal of the American Concrete Institute 68 (1971) 47-49.

39. Naaman, A.E. and V.S. Gopalaratnam. Impact Properties of Steel Fibre Reinforced Concrete in Bending. International Journal of Cement Composites and Lightweight Concrete 5 (1983) 225-233.

40. Gopalaratnam, V.S., S.P. Shah and R. John. A Modified Instrumented Charpy Test for Cement Based Composites. Experimental Mechanics (in press).

41. Zielinski, A.J. Experiments on Mortar Under Single and Repeated Uniaxial Impact Tensile Loading. Report 5-81-3, Department of Civil Engineering, Delft University of Technology, The Netherlands, 1981.

42. Kormeling, H.A., A.J. Zielinski and H.W. Reinhardt. Experiments on Concrete Under Single and Repeated Uniaxial Impact Tensile Loading. Report 5-80-3, Department of Civil Engineering, Delft University of Technology, The Netherlands, 1980.

43. Zielinski, A.J. Model for Tensile Fracture of Concrete at High Rates of Loading. Cement and Concrete Research 14 (1984) 215-224.

44. Suaris, W. and S.P. Shah. Properties of Concrete Subjected to Impact. Journal of Structural Engineering 109 (1983) 1727-1741.

45. Barr, B. and T. Bear. A Simple Test of Fracture Toughness. Concrete 10 (1976) 25-27.

46. Higgins, D.D. and J.E. Bailey. Fracture Measurements on Cement Paste. Journal of Materials Science 11 (1976) 1995-2003.

47. George, K.P. Theory of Brittle Fracture Applied to Soil Cement. Journal of the Soil Mechanics and Foundation Division, ASCE 96 (1970) 991-1010.

48. Brown, W.F. and J.E. Srawley. Plane Strain Crack Toughness Testing of High-Strength Metallic Materials, in American Society for Testing and Materials STP No. 410 (Philadelphia: ASTM, 1966), 129 pp.

49. Dwivedi, V.S. and P.L. Pratt. Private Communication, 1983.

50. Zielinski, A.J. Fracture of Concrete and Mortar Under Uniaxial Impact Tensile Loading. Ph.D. Thesis, Delft University of Technology, 1982.

51. Hibbert, A.P. and D.J. Hannant. Impact Resistance of Fibre Concrete. TRRL Supplementary Report 654, Transport and Road Research Laboratory, Crawthorne, Berkshire, 1981.

52. Alford, N.McN. Dynamic Considerations of Fracture in Mortars. Materials Science and Engineering 56 (1982) 279-287.

53. Mindess, S. and S. Diamond. A Preliminary SEM Study of Crack Propagation in Mortar. Cement and Concrete Research 10 (1980) 509-519.

54. Mindess, S. and S. Diamond. The Cracking and Fracture of Mortar. Materiaux et Constructions 15 (1982) 107-113.

SECTION VIII

IMPLICATIONS FOR CONCRETE STRUCTURES

**APPLICATION OF FRACTURE MECHANICS
TO CEMENTITIOUS COMPOSITES**
NATO-ARW - September 4-7, 1984
Northwestern Univeristy, U.S.A.
S. P. Shah, Editor

THE ROLE FOR FRACTURE MECHANICS IN
CONVENTIONAL REINFORCED CONCRETE DESIGN

Neil M. Hawkins

Chairman
Department of Civil Engineering
University of Washington
Seattle, WA 98195, USA

ABSTRACT

Fracture mechanics role within the framework of the ACI Building
Code will be significant if fracture mechanics provides answers to
public safety issues, and relatively insignificant if fracture
mechanics only provides a different theoretical viewpoint for provisions which already provide an adequate margin of public safety.
Public safety issues for which fracture mechanics can provide answers
include size effects, appropriate finite element analyses procedures,
new materials, minimum reinforcement ratios, maximum reinforcement
spacings, and time dependent effects. Issues for which fracture
mechanics can provide a consistent theoretical basis include the
diagonal tension cracking capacity for shear, torsion, and combined
shear and torsion loadings.

INTRODUCTION

In the U.S.A. conventional reinforced concrete design can be best
described as design in accordance with ACI Code 318 "Building Code
Requirements for Reinforced Concrete" (1); the ACI Code provisions
also serve as the prime basis for the development of design provisions
for bridges (2); nuclear containment structures (3); liquid retaining
structures (4);etc. This paper describes the role possible for fracture mechanics' concepts within the framework of ACI Code 318.

First, it must be recognized that the ACI Code 318 is a consensus
code, and that its governing body is loath to incorporate new provisions into that Code unless it can be demonstrated that existing

provisions are non-conservative or non-definitive. Through an examination of the regulatory history for prestressed concrete, the writer has shown (5) that the rapidity with which new knowledge on a particular subject is incorporated into the ACI Code is more a function of whether or not provisions already exist on that subject than a function of the validity of that new knowledge. If no relevant specifications exist and a given piece of research is perceived as providing definitive answers to public safety concerns such as strength and ductility, then that new information is readily accepted by the regulatory authorities. However, once a specification exists, human inertia and the weight of prior practice delay the incorporation of that new knowledge particularly if that knowledge is perceived as associated primarily with serviceability rather than public safety concerns. The future role for fracture mechanics in conventional design will depend largely on whether regulatory authorities see a need for fracture mechanics in relation to public safety, or whether they see fracture mechanics as simply a vehicle for providing a different viewpoint on provisions which already incorporate an adequate margin for public safety.

FRACTURE MECHANICS AND PUBLIC SAFETY ISSUES

Size Effects

As Bazant (6) has noted, failure criteria in terms of stresses or strains require artificial manipulations for applicability to materials with strain-softening characteristics involving a decline in stress with increasing strain. Classical cases, for example, are the development of the equivalent rectangular stress block to handle problems of concrete in combined bending and compression, and the traditional use of a bending tensile strength for concrete about 50 percent greater than concrete's direct tensile strength. The necessity for such manipulations suggests that the resultant failure criteria may not be conservative for all practical situations, and may be size dependent. A fracture mechanics' approach can allow for strain softening and automatically incorporate size dependency. If the strength or ductility of reinforced concrete members, controlled by the properties of the concrete as opposed to the reinforcement, are size dependent then public safety issues are involved and there is a definite role for fracture mechanics in conventional concrete design.

For shear, for example, there is clear evidence that the strength of members without web reinforcement decreases with increasing depth (7). Test results obtained by Kani are shown in Fig. 1. All test beams had the same concrete strength, steel percentage, width, and shear span length to depth ratio (a/d). For an a/d ratio of 4 the shear strength decreased 26% from 200 psi at an overall depth of 12 inches to 148 psi at an overall depth of 48 inches. What is not clear is how far that trend continues beyond the limits shown and the

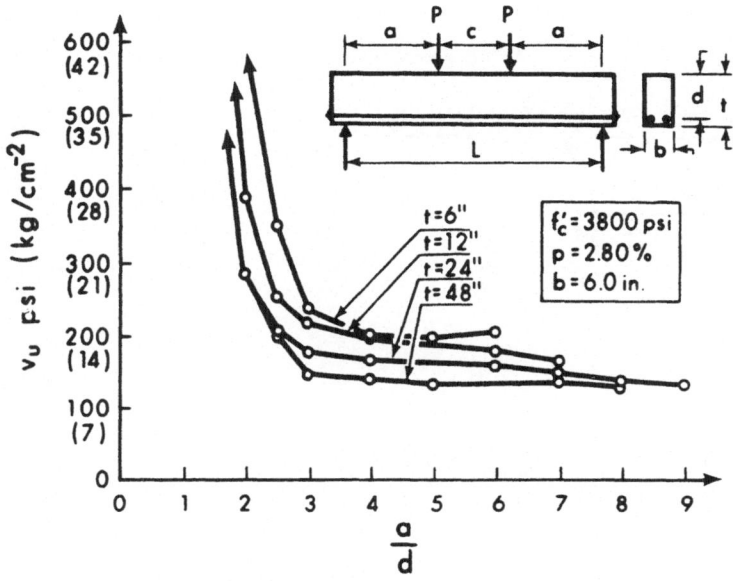

FIG. 1   SHEAR STRESS AT FAILURE V_u VERSUS a/d FOR
BEAMS OF VARIOUS DEPTHS (7)

significance of that trend for the shear strength of slabs, one-way
joists, or prestressed hollow-core planks which usually do not contain
shear reinforcement and have depths less than 12 inches, or assessments of the shear strength of beams with depths greater than
12 inches but containing minimum shear reinforcement as required by
Sec. 11.5.5.1 of ACI 318-83. If fracture mechanics' concepts are
applicable to reinforced concrete then there is immediate reason to
doubt the conservatism of such clauses as Sec. 11.5.5.2 of ACI 318-83
which permits waiver of minimum shear reinforcement requirements if
it can be shown "by test that required nominal flexural and shear
strengths can be developed when shear reinforcement is omitted," and
Sec. 19.1.9 which permits analysis for shells and folded plates based
upon the measured deformations and/or strains of a model. Indeed,
the acceptance of fracture mechanics' concepts for concrete would
immediately challenge the validity of model testing for concrete
structures except for the establishment of flexural behavior.

Bazant (6) has illustrated the significance of the application of
fracture mechanics principles, either linear or non-linear, to
concrete using Fig. 2. When standard strength criteria are used,
failure occurs when the nominal stress, σ_N, characterizing the stress
state at a critical point reaches the limiting strength for the
material. σ_C. For any loading condition, flexure, shear or torsion,
the nominal stress state can be described by σ_N is equal to C_N P/bh
where C_N is a coefficient independent of size and characterizing the

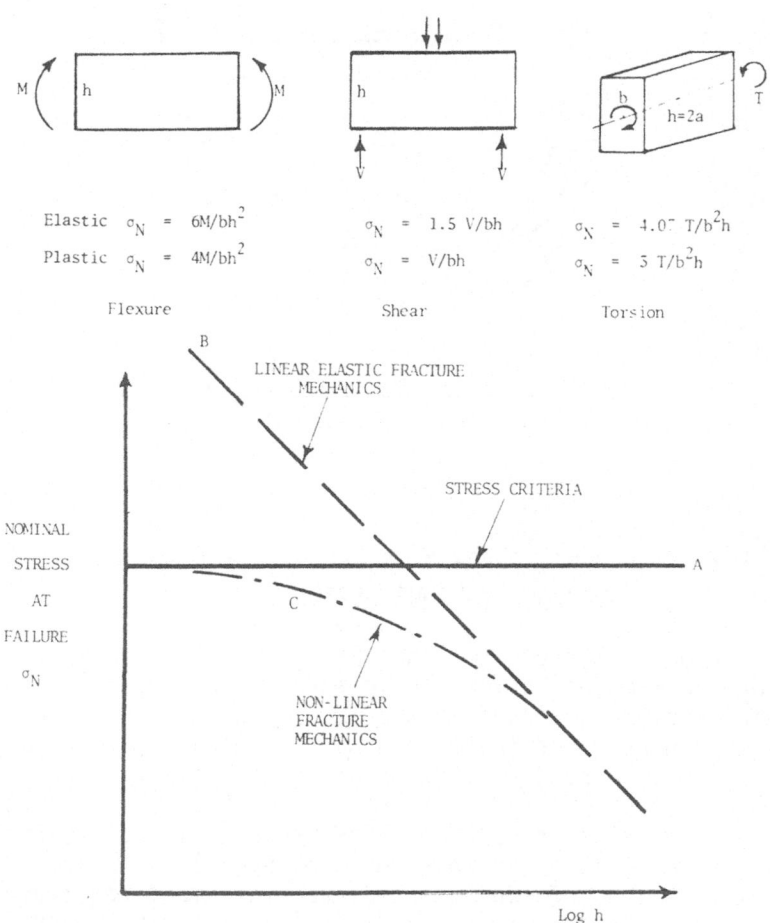

$$\text{Elastic} \quad \sigma_N = 6M/bh^2$$
$$\text{Plastic} \quad \sigma_N = 4M/bh^2$$

$$\sigma_N = 1.5 \, V/bh$$
$$\sigma_N = V/bh$$

$$\sigma_N = 4.0^- \, T/b^2h$$
$$\sigma_N = 5 \, T/b^2h$$

Flexure   Shear   Torsion

FIG. 2   SIZE EFFECT FOR DIFFERING FAILURE CRITERIA (6)

structural shape and type of loading (shear, torsion or flexure);
P is the loading parameter (M, V or T); h is the characteristic
dimension; and b is the thickness.

When the log of the nominal stress at failure, σ_N, is plotted
against some size parameter such as log h, then the failure state
according to any strength criteria, elastic, plastic or viscoplastic,
is the streight line A. Different strength criteria can only change
the level of the line A not its slope. For linear elastic fracture
mechanics σ_N varies inversely as $\sqrt{h}$ and therefore the plot of log σ_N

versus log h is the straight line B with a slope of -1/2. Researchers are far from agreement on what is the appropriate fracture model for concrete. They are, however, in agreement with the concepts indicated by curve C. At one extreme linear elastic fracture mechanics' concepts are applicable when the size of the specimen becomes large enough, and at the other extreme strength concepts are applicable when the size of the specimen becomes small enough. The curvature of line C, and therefore its precise location, is also known to be related to the heterogeneity of the concrete, the characteristic size of the aggregate, and the aggregate's volume fraction (8). Thus, different locations for curve C are to be expected for different materials. The task confronting fracture mechanics is determination of the factors dictating the characteristics of curve C.

The validity of a non-linear fracture mechanics' approach, such as curve C Fig. 2, has been demonstrated by Bazant (6) by examination of the three-point bending results reported by Walsh (9). For Bazant's non-linear approach the total potential energy release caused by fracture is taken as a function of the crack length and a blunt crack band width determined by the maximum aggregate size. Failure is assumed to occur when the potential energy release rate equals the energy consumed per unit crack-band extension. For the center cracked plain concrete panel of width b shown in Fig. 3, the crack length is 2a and the crack-band width is nd_a.

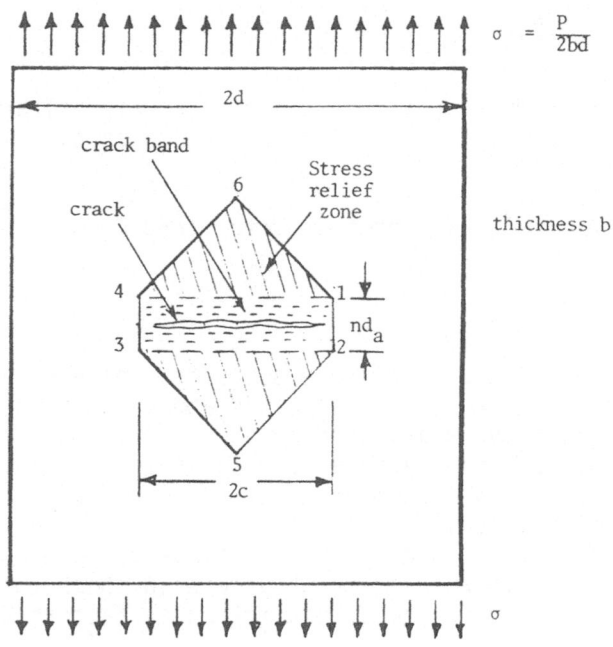

FIG. 3   CRACK BAND PROPAGATION (6)

The crack extension causes a stress relief area shown as hatched in Fig. 3. The energy W, released from the panel with cracking, is given approximately by the formula:

$$W = 2a^2 b \sigma^2/2E_c + 2nd_a ab \sigma^2/2E_c \tag{1}$$

If the energy consumed per unit advance of the crack band is called the fracture energy G_f, then the criterion for failure becomes:

$$\frac{\partial W}{\partial a} = G_f b \tag{2}$$

Bazant has shown that Walsh's results satisfy neither a strength nor a non-linear fracture approach but do give reasonable agreement with his prediction.

The writer (10) has also reported a non-linear approach similar to Bazant with fracture again assumed to occur when the strain energy release rate exceeds the incremental increase in surface energy. For the center cracked concrete cylinder of diameter 2R shown in Fig. 4, the crack length is 2c. Extension of the crack causes a stress relaxation zone which is spherical in shape and the criterion for rapid crack propagation becomes:

$$p^2 = \frac{2G_c c}{\partial F} \tag{3}$$

where G_c is the fracture toughness of the material and ∂F is the flexibility of the cylinder for transverse sections passing through poles A and B in Fig. 4.

The critical value of c is determined from the requirement that:

$$\frac{d}{dc} (p^2) = 0 \tag{4}$$

The writer used a fracture toughness value established from direct tensile data. That value was about 20% greater than the G_f value subsequently established by Bazant from more recent observations of the nature of the fracture process zone. For the writer's approach, the stable crack was assumed to extend in a direction perpendicular to the maximum principal tensile stress. Where there were tensile stress gradients in two perpendicular directions as in a modulus of rupture beam, the relative rate of crack growth in those two directions was taken as the ratio of the stress intensity factors for those two directions.

The writer found good agreement between strength predictions made using his hypothesis and the results of modulus of rupture tests

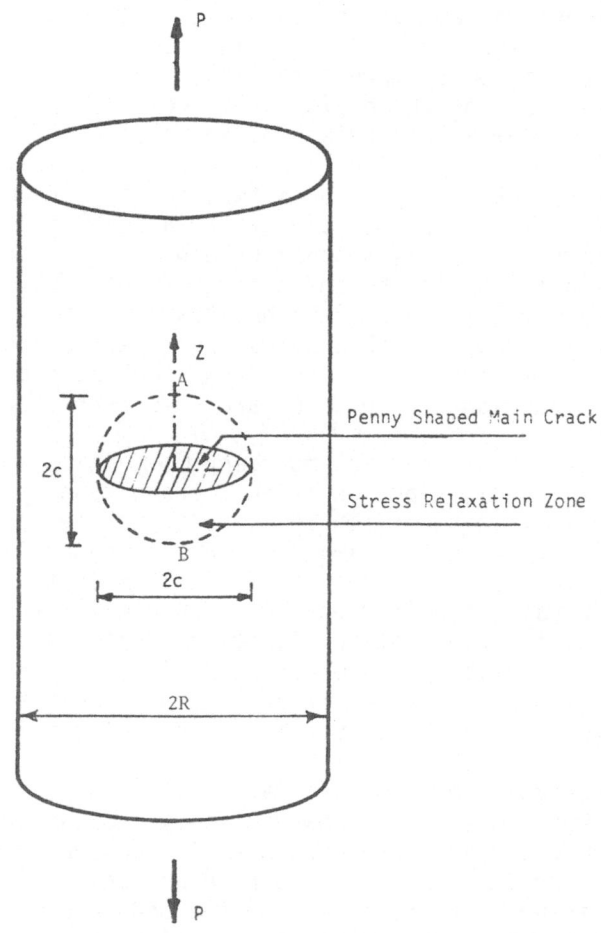

FIG. 4   STRESS RELAXATION WITH CRACKING (10)

reported by Hsu (11). The predicted extreme fiber stresses for failure of 10 in. x 10 in. cross-section beams were 22% less than those of 6 in. x 6 in. cross-section beams. This result again indicates clearly the significant size effect predicted by a fracture mechanics' approach.

Finite Element Analyses and New Materials

Current conventional reinforced concrete codes are best described as low-tech codes. They provide little incentive to the designer for the use of improved materials or sophisticated methods of analysis. In the former case the code often becomes inapplicable and in the latter case the code frequently penalizes the designer. The conventional code is intended only for use with "ordinary" materials and members of "normal" proportions. It is not intended that ACI Code 318 be used for "special" materials such as ultra high strength concretes, polymer impregnated concretes, or expansive cement concretes, or for the design of "special" structures such as nuclear containment or liquid retaining vessels.

It is inherently inconsistent to simultaneously apply finite element analyses and failure criteria developed from ACI Code 318 provisions. There is a spurious dependence of results on mesh size. For example, should the shear strength of a member be associated with the development of the limiting shear stress in one element, or with the development of such a stress averaged across 50% or more of the depth of the member? Further, how should the designer handle the high stresses developed at notches and re-entrant corners?

With a smeared crack approach (12), it is possible to determine the stress redistribution effects brought on by cracking and provide a more logical union of finite element analyses and ACI Code failure criteria. However, ultimately what is needed is a discrete crack model, and a fracture mechanics' failure criterion that can predict ACI Code failure criterion. Only then will it be possible to provide a consistent union between finite element analyses and ACI Code criteria. Until such a model is available, the increasing use of finite element analyses without the development of associated guide-lines for assessing failure is an obvious public safety issue.

Increasingly "special" materials are being used in "ordinary" structures. Most mixes now routinely include admixtures, and while most common admixtures, when used in accordance with manufacturer's instructions, are known not to have any deleterious effects on public safety issues, at the same time the Code provides no means for recog-nizing any positive public safety effects of such admixtures or special materials. Admixtures and special materials in fairly common use today include super-plasticizers, silica fume, fly ash, light-weight aggregate concrete, no-fines concrete, polymer impregnated concrete, fiber reinforced concrete, expansive cement concretes, and ultra-high strength concretes. Obviously the development of separate building code provisions for each of those special materials is not likely, although it should be noted that Japan and the Soviet Union have developed separate codes for expansive cement concretes because of the better cracking behavior and consequently the different reinforcement details that can be used with such materials.

ACI Code 318-83 contains many inconsistencies that relate to "special" material issues. The Code imposes widely varying cut-offs on the maximum concrete strengths that can be used with different provisions. Section 10.2.7.3 covering assumptions in the equivalent rectangular stress block for compressive loading has been developed specifically to be applicable to concrete strengths in excess of

8,000 psi. However, many of the benefits that might be associated with the use of such high strengths in columns are negated by Sec. 10.13 dealing with transmission of column loads through floor systems. Equation (11-37) dealing with punching shear of slabs and footings is restricted to concrete strengths not greater than 5,000 psi and designers using the shear friction provisions of Sec. 11.7 must refer to the Commentary to the Code if full utilization is needed of the potential capacity of a concrete with a strength greater than 4,000 psi. The horizontal shear strength provisions for composite concrete members of Sec. 17.5 restrict the shear stress at intentionally roughened surfaces lacking minimum ties to 80 psi regardless of the concrete strength.

Probably one of the greatest inconsistencies in ACI Code 318-83 concerns lightweight concrete. The use of splitting tensile tests to establish relative tensile strength values for lightweight concrete is specified in Sec. 9.5.2.3 where a formula is given for the cracking moment; in Sec. 11.2 where blanket procedures are specified for shear and torsion calculations; and in Sec. 12.2.2.3 where procedures are specified for multiplication factors on the basic development length. There are, however, other places in the Code where $\sqrt{f_c'}$ values are specified, obviously relating to the concrete's tensile strength, without accompanying corrections for the use of lightweight concrete. Examples are in the flexural reinforcement distribution requirements for beams and one-way slabs in Sec. 10.6; the permissible stress limitations for flexural prestressed concrete members in Sec. 18.4; and the membrane reinforcement requirements for shells in Sec. 19.4.10. While the use of splitting test data to establish relative tensile strengths, f_{ct}, for lightweight concrete is apparently encouraged, ACI Code 318 also specifies that where such data are not available for a given lightweight concrete, then $\sqrt{f_c'}$ values should be multiplied by 0.75 for "all-lightweight" and 0.85 for "sand-lightweight." For the shear and torsion provisions the continued use of splitting test data is a concession to vested lightweight concrete interests. Committee 426 (13) has recommended use of the 0.75 and 0.85 factors only since they give considerably better agreement with available test data for shear and torsion in lightweight concrete than the use of $\sqrt{f_c'}$ values multiplied by $f_{ct}/6.7$.

In reality, ACI Code 318 assumes the following relationships between various tensile strength values for concrete made with normal weight aggregates: direct tensile strength 5.7 $\sqrt{f_c'}$; tensile splitting strength 6.7 $\sqrt{f_c'}$; and modulus of rupture 7.5 $\sqrt{f_c'}$. Obviously, the true inter-relation between those values is a function of the tensile capacity-crack width relationship for the concrete or the strain softening portion of the pseudo stress-strain curve for that concrete in tension. The foregoing values are those associated with the limestone aggregate commonly used in the Midwest of the U.S.A. There is ample evidence that other values are appropriate for different aggregate types. For example, routine testing at the University of

Washington over more than a decade has established the following values for the hard, well-rounded glacial aggregate available in Seattle: direct tensile strength 5.5 $\sqrt{f'_C}$; tensile splitting strength 6.3 $\sqrt{f'_C}$; and modulus of rupture 7.1 $\sqrt{f'_C}$.

For the crushed basalt aggregate commonly used in the Sydney area of Australia, the corresponding values are: tensile splitting strength 7.1 $\sqrt{f'_C}$; and modulus of rupture 9 $\sqrt{f'_C}$.

The wide differences between Seattle and Sydney tensile strength values for normal aggregate concretes and the inconsistencies between tensile strengths and the observed shear and torsion values for lightweight concretes suggests strongly that standardized procedures need to be developed that permit the designer to adjust the tensile strength assumptions of the Code in accordance with the properties of the material being specified for a given structure. Fracture mechanics' techniques involving K_I and K_{II} tests offer the best hope for developing "rational" standardized procedures.

Minimum Reinforcement Ratios

ACI Code 318 contains a multiplicity of provisions specifying minimum reinforcement ratios and maximum spacings of reinforcement. Table 1 contains a listing of some of those provisions. In four instances, provisions are given which permit waiver of those requirements if the adequacy of the resulting design can be established by test and analyses. However, because of the specific and prescriptive nature of the minimum ratio and maximum spacing limitations, it is often extremely difficult to persuade regulatory authorities to permit use of those waivers and almost impossible to invoke such waivers, if a construction or detailing error has resulted in those limitations being violated in an existing structure. With predictions of the growth, both in length and width of a discrete crack, and the spacing of cracks, a rational means would exist for assessing reinforcement ratio effects and the implications of different spacings for the reinforcement.

Bazant and Oh (14) have demonstrated the realism of that assertion. By using the energy criterion of fracture mechanics, as well as the strength criterion, they found that realistic values for the spacing and width of cracks in a parallel crack system could be obtained. Their energy criterion indicated that crack spacing was a function of the axial strain in a bar, the bar spacing, bar diameter, and the concrete's fracture energy and modulus of elasticity. Both energy and strength criteria predict a minimum strain in the bar necessary to produce any cracks while the energy criterion and bond slip conditions also yield a lower bound on the possible spacing of continuous cracks. Bazant and Oh found that crack spacing was proportional to the cube root of the effective tension area of concrete concentric with the bar. That finding was consistent with a finding by Gergeley and Lutz

TABLE 1

Section	Provision
7.6.5	Maximum spacing of primary flexural reinforcement in slabs and walls.
*7.10.3	Waiver on lateral reinforcement requirements for compression members where established as feasible by tests and analyses.
7.12	Shrinkage and temperature reinforcement, minimum amounts and maximum spacing of non-prestressed. Minimum compressive stress and maximum spacing of prestressed.
8.10.5.2	Maximum spacing of transverse reinforcement in T-beam construction.
10.5	Minimum amount of reinforcement for flexural member.
10.6	Distribution of flexural reinforcement in beams and one-way slabs.
11.5.5.1	General requirements on minimum shear reinforcement for flexural members.
*11.5.5.2	Waiver of minimum shear reinforcement requirements where established as feasible by tests.
11.5.5.3	Minimum shear reinforcement amount for non-prestressed.
11.5.5.4	Minimum shear reinforcement amount for prestressed members.
11.5.5.5	Minimum stirrup reinforcement amount for combined shear and torsion.
11.6.8.2	Spacing limitations on torsion reinforcement.
11.8.8	Limitations on minimum amount and maximum spacing of transverse shear reinforcement in deep flexural members.
11.8.9	Limitations on minimum amount and maximum spacing of longitudinal shear reinforcement in deep flexural members.
11.9.4	Tie reinforcement distribution requirements for brackets and corbels.
11.10.9.2	Minimum transverse shear reinforcement ratio for walls.
11.10.9.3	Maximum transverse shear reinforcement spacing for walls.
11.10.9.4	Minimum longitudinal shear reinforcement ratio for walls.
11.10.9.5	Maximum longitudinal shear reinforcement spacing for walls.
13.4.2	Maximum spacing of reinforcement in two-way slabs.
*14.2.7	Waiver on quantity of reinforcement and thickness limitations for walls when shown to be adequate by strength and stability considerations.
14.3	Minimum vertical and horizontal reinforcement requirements for amount, spacing, and layout in walls.
14.3.7	Minimum additional reinforcement requirements around window and door openings in walls.
17.6	Minimum tie area and maximum spacing requirements for horizontal shear strength in composite concrete flexural members.
18.8.3	Minimum amount of prestressed and non-prestressed reinforcement in prestressed members.
18.9	Minimum bonded reinforcement ratios for one-way and two-way prestressed members and limitations on spacing.
*18.11.2	Limitations on minimum reinforcement for prestressed compression members.
18.12.4	Spacing limitations and minimum average prestress requirements for two-way prestressed concrete slab systems.
19.4.3	Minimum shell reinforcement ratios, maximum spacing and special requirements.

* Waiver provisions

established only after extensive statistical curve-fitting of numerous test data (15). Intermediate partial depth cracks between full tension area depth cracks were also shown to get shorter and shorter, as the crack spacings became denser. That behavior was consistent with the behavior observed by Broms (16) in bending tests on beams. The ultimate local bond force obtained as the optima for fitting results to observed crack data was 95 percent of the concrete compressive strength per unit length of a deformed bar. Again, that result is consistent with the writer's findings from local bond tests on bonded over-short lengths deep within well reinforced concrete blocks (17). Finally Bazant and Oh showed how to use their approach to predict the formation and spacing of skew cracks in a biaxially stressed and biaxially reinforced plate.

Time Dependent Effects

Fracture testing of concrete is very much in its infancy. Most tests to date have involved short-term loadings and normal temperatures. Yet from the few tests that have been made with loading history, loading rate and temperature as variables, it is obvious the fracture mechanics' test results can offer interesting new insight into loading history and time effects. Fracture mechanics may well provide a means for better characterizing those effects and for answering public safety issues related to such questions as the safe operating period, safe crack length, and remaining life for a structure where it is known that there is less than a desirable amount of reinforcement crossing a given crack plane. Through fracture mechanics' techniques, it should be possible to answer such questions as how fast a structure will fail when loaded above its safe operating load, or what size of crack it is reasonable to permit in a prestressed hollow core plank when it leaves the factory.

For reinforced concrete in bending, traditional thinking relates the increase in deformations with time to the level of applied stress in compression. Fracture testing shows that reduction in the stiffness contribution of the tension zone concrete with time is important for both time-dependent effects and cyclic load effects. Some exploratory time-dependent studies have been made at the University of Washington using CLWL-DCB fracture specimens (18). It has been observed that most of the change in the wedge load occurs in the first 5 to 10 minutes after a new position is reached. However, crack opening displacements increase almost continuously for several hours after the new position is reached. Shown in Figs. 5a and 5b are the load-compliance and crack length-time relationships obtained in test No. 7 in that investigation. The replica crack length, which has been shown to equal the total crack length including the fracture process zone (19) and the effective crack length (the elastic equivalent crack length including any non-linear crack-tip region effects), increase with time at a constant ratio. There is reason to suspect that increase continues until a steady energy state is achieved in the

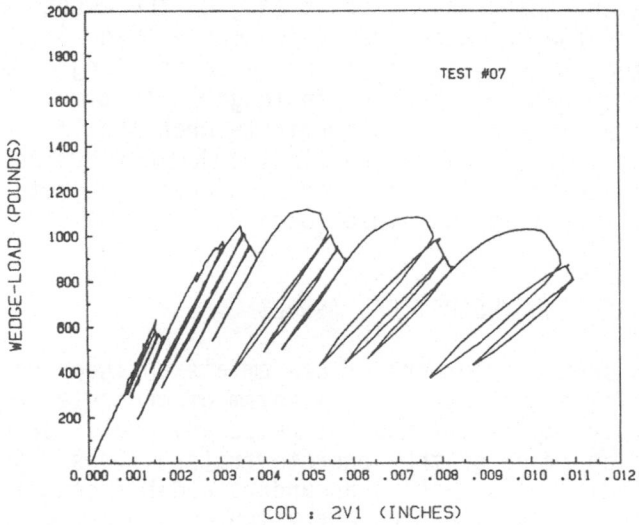

FIG. 5a   LOAD-DISPLACEMENT COMPLIANCE FOR SPECIMEN 7 (18)

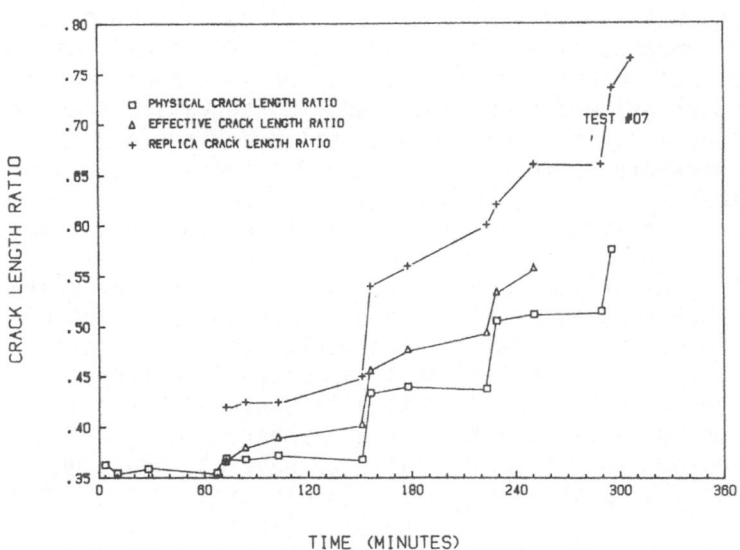

FIG. 5b   CRACK LENGTH-TIME RELATIONSHIP FOR SPECIMEN 7 (18)

specimen. However, the physical crack length, the length established
from the slope of the unloading-reloading curve (Fig. 5a), and presumably the effective elastic traction-free crack length excluding any
crack-tip effects, remains constant in length with time. Physical
crack length extensions appear to terminate when the loading stops.
These observations help explain the finding that deflection and crack
width increases can be significant for reinforced concrete beams for
cyclic loading as well as sustained loading.

FRACTURE MECHANICS AND THEORETICAL ISSUES

The more theoretical members of academia's engineering community
frequently criticize the blatant empiricism of concrete building
codes. While that empiricism may not raise any public safety issues,
it does lead to an image of the concrete fraternity as a group of
simple minded individuals lacking an understandable central philosophy
and concerned primarily with the generation of complex rules and cook
book solutions. Fracture mechanics, in combination with rational
models for post cracking response, offers possibilities for dramatically reducing the empiricism of the Code. Rational models for the
design of the shear and torsion reinforcement of prestressed and non-
prestressed beams, based on compression field theory, have been
developed by Collins and Mitchell (20). Their model is based on truss
equilibrium conditions. It assumes a section cracked due to diagonal
tension and utilizes directly material stress-strain properties and
geometric compatability conditions. However, because of the assumption of a section cracked in diagonal tension, their model can lead to
overly conservative designs and overly conservative predictions of
service load behavior when the diagonal cracking load is close to the
ultimate strength of the member. Those situations could be corrected
through the use of fracture mechanics' approaches to predict cracking,
service load behavior, and minimum reinforcement requirements.

In 1977 the writer reported the use of the energy criteria of
fracture mechanics to predict diagonal tension cracking of reinforced
concrete beams (10). In associated work, Wyss (21) used the same
approach to predict the torsional capacity of rectangular reinforced
concrete beams and to demonstrate a theoretical basis for the circular
interaction curve between torsion and shear at diagonal cracking that
has been established experimentally by several researchers. It is to
be expected that with more precise information on the nature of the
fracture process zone, it will be possible to obtain good agreement on
both diagonal cracking strength and consequent serviceability issues
for a wide variety of shear and torsion loading conditions. Greater
confidence can then be attached to the use of discrete crack models in
finite element analyses. In this section, results of the writer's and
Wyss' predictions are described and some of the more obvious
deficiencies in those approaches noted.

The energy model utilized by the writer was shown in Fig. 4. While that model, and the similar model of Bazant, yield reasonable diagonal cracking results, that model should be recognized as being fundamentally inconsistent with the Dugdale-Barenblatt model for the fracture process zone determined recently by the writer and his colleagues as a result of tests on CLWL-DCB specimens (19). When the model of Fig. 4 is used, a limiting value is assumed for G_c. However, in CLWL-DCB tests, and in tests reported by many other researchers, no limiting value of the crack resistance K_R has been achieved for the wide range of specimen dimensions tested. The exception is Walsh's results (9). In the writer's tests, no limiting value of K_R was achieved for crack extensions as large as 8 inches. Thus, while the writer's approach and that of Bazant may point the way for future rigorous analyses of diagonal cracking, it is to be expected that the K_R value for such analyses will be a function of crack extension, rather than a unique value as assumed in the approach discussed here.

Diagonal Tension Cracking in Shear

For predictions of diagonal tension cracking in reinforced concrete beams of normal proportions using a limiting value of G_c, the writer (10) has found that the boundary conditions represented by the physical width and depth of the beam, and the intensity of flexural loading relative to the co-existing shear loading, necessitate separate treatments for beams with narrow as opposed to wide webs and for sections cracked in flexure as opposed to those uncracked in flexure. The need for separate treatments for those four cases has also been established experimentally (7).

(a) Section Uncracked in Flexure: It is assumed that the maximum principal stress criterion determines the position at which the crack initiates and the direction in which it grows. Thus, for a rectangular section, at a point of contraflexure, the crack forms at mid-depth at a 45 degree angle. It is further assumed that crack continues to grow in essentially the initial principal stress direction. That assumption is also consistent with findings of the writer and his colleagues from tests on CLWL-DCB specimens (23) subject simultaneously to K_I and K_{II} loadings.

The growth of the principal tension stress crack reduces the shear flexibility of the section. In other words, no aggregate interlock is assumed across the diagonal tension crack. Again that assumption is consistent with findings of the writer and his colleagues (23). Thus, the growth of the diagonal tension crack creates a zone of stress relief as shown in Fig. 6. An accurate analysis, recognizing the differing stress intensities for crack growth in the horizontal and vertical directions would require use of the ellipsoid indicated by broken lines in Fig. 6. However, simplistically, the effects of crack growth can be recognized by assuming a spherical or cylindrical stress-relief zone.

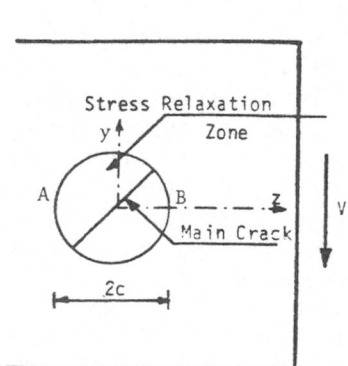 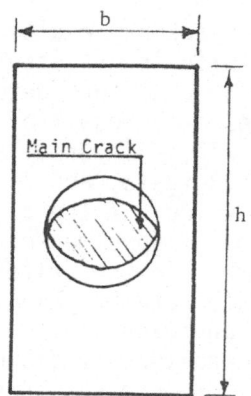

FIG. 6   CRACK AND STRESS RELIEF GROWTH FOR SHEAR LOADING (10)

Use of the energy criteria and the critical G_c value described previously gives (10) for thick web beams (b > h/2) where the stress relief zone does not extend to the side surfaces and is therefore spherical:

$$c_{crit} = 0.222 \sqrt{bh} \tag{6}$$

and

$$V_c = 8.0 \, (bh)^{0.75} \sqrt{f'_c} \tag{7}$$

and for thin web beams (b < h/4) where the stress relief zone quickly extends to the side surfaces and can therefore be approximated as cylindrical:

$$c_{crit} = h/6 \tag{8}$$

and

$$V_c = 8.9 \sqrt{b} \sqrt{f'_c} \tag{9}$$

where c_{crit} is the principal tension crack length at which diagonal tension cracking occurs and V_c is the diagonal tension cracking shear.

If the effective depth, d, of a member is taken as the overall depth; h, less 2.0 inches; and b as h/2 (the ratio commonly used in laboratory tests for shear), then V_c/bd calculated from Eqs. (7) and (9) varies as follows:

$$\begin{array}{ccc} & \dfrac{V_c/bd}{} & \\ d = 10 \text{ in.} & d = 28 \text{ in.} \end{array}$$

Eq. (7) $3.3 \sqrt{f_c'}$ $1.86 \sqrt{f_c'}$

Eq. (9) $6.7 \sqrt{f_c'}$ $2.3 \sqrt{f_c'}$ for $b = h/4$

In general, $V_c/bd \sqrt{f_c'}$ values decrease as the b/d ratio decreases for Eq. (7) and increase as the b/d ratio increases for Eq.(9) but these effects are small. Hence, it is not unreasonable that Committee 426 (7) suggests the effect of b/d should be ignored. Committee 426 (13) has also recommended, based on a review of test data, that when M/Vd is greater than two and the reinforcement ratio greater than 1.3%, V_c/bd should be taken as $2.3 \sqrt{f_c'}$. That finding is consistent with use of the energy criterion of fracture mechanics for the member sizes likely to be critical in shear in practice.

(b) Section Cracked in Flexure: Recent research has shown that the reduction in shear capacity that occurs for a section cracked in flexure is not due to a weakening effect for shear caused by the flexural crack. Taylor (24), for example, carefully designed an investigation in which he measured independently for a flexurally crack beam the contributions to the total shear strength of the compression zone, the dowel force in the reinforcement, and the aggregate interlock forces. A typical result is shown in Fig. 7. He found good agreement, as indicated in Fig. 7, between the sum of those three components and the total applied shear. Thus, it is apparent that flexural cracking causes a redistribution of shear but not a weakening of the shear strength. The writer has proposed (10) that weakening is due to an interaction of shear and flexure effects. The energy criterion of fracture mechanics is ideally suited to handling those interaction effects because the changes in shear and bending flexibility are additive.

In the writer's approach, the shear flexibility is assumed to be the same as that for a beam uncracked in flexure. Thus, consistent with Taylor's findings no change in shear flexibility is assumed due to flexural cracking. The bending flexibility, F_B, is taken as:

$$F_B = \frac{2}{E} \int_0^C \frac{dz}{\left(I_0 - \frac{\pi r^4}{2}\right)} \tag{10}$$

where I_0 is the moment of inertia for a section cracked in bending but uncracked in diagonal tension and $r^2 = c^2 - Z^2$ where c and Z are as shown on Fig. 6. Thus, F_B also changes with the growth of the diagonal crack, and the likely location for that crack is immediately above the flexural crack where I_0 is a minimum. The resulting energy equilibrium equation is:

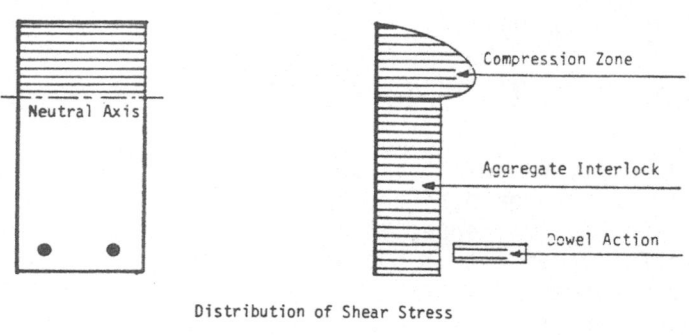

Distribution of Shear Stress

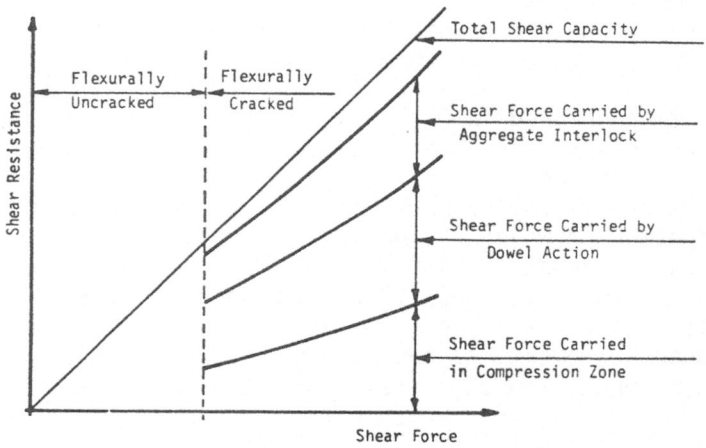

Shear Resistance Components

FIG. 7   SHEAR CONTRIBUTIONS FOR BEAM CRACKED IN FLEXURE (24)

$$V_c^2 = \frac{\pi\ G_c\ c}{\dfrac{\delta F_V}{2} + \left(\dfrac{M}{V}\right)^2 \dfrac{\delta F_B}{2}} \tag{11}$$

where δF_V is the change in shear flexibility and c is calculated as described previously. For the solution of Eq. (11), the critical value of M/V is taken as the shear span length minus d/2. That result is consistent with conditions extant in most tests where concentrated loads have been used. The critical diagonal tension crack is initiated by a flexural crack at a distance d/2 from the load point. That crack then extends towards the load point on a 45 degree line. However, it is equally clear that the same M/V value is not critical for a beam subject to the uniform loading more likely in practice.

Further, it is clear from recent fracture testing that the moment of inertia does not change abruptly from the gross section value to the cracked value as implied by Eq. (10). Thus, Eq. (11) can be expected to underestimate V_c values when the intensity of the flexural loading is low. Fortunately, for those cases Eq. (7) or (9) provides a cut-off.

In Reference (10), Eq. (11) is used to calculate the ultimate capacities for a large number of hypothetical beams with properties as follows: $h < M/V < 7h$; $0.005 < \rho < 0.02$; 6 in. $< b <$ 10 in.; $2b < h < 4b$.

Regression analyses of the results gave:

$$V_c = 8.0 \ (bh)^{0.75} \ \sqrt{f_c'} \ \beta \tag{12}$$

$$\text{where} \quad \beta = \left[1.07 - \frac{0.006}{\sqrt{\rho}} \ \frac{M}{Vh}\right] \leq 1.0 \tag{13}$$

Results predicted by Eqs. (9), (11) and (12) are shown in Fig. 8. Strong effects are predicted of M/Vh and ρ, consistent with test data.

In Reference (10), the predictions of Eq. (12) were compared with the basic shear stress equation, Eq. (11-6) of ACI Code 318-83, and the revision proposed to that equation by ACI-ASCE Committee 426 (13), and to the experimental results obtained in six different investigations in which the effects of concrete strength, f_c', reinforcement ratio, ρ, and the M/Vh ratio were systematically varied. The best overall agreement and the least standard deviation from the mean was for the energy criterion of fracture mechanics described here.

Shown in Fig. 9 are the predicted and measured variations in V_c for data (27) in which the reinforcement ratio ρ was the major variable. The energy approach clearly predicts a variation in V_c with ρ. Shown in Figs. 10 and 11 are variations in V_c with M/Vh and beam depth. Again, there is good agreement between the predictions of the energy version of fracture mechanics and the measured results. The ACI Equation (11-6) contains no size effect. The results of Fig. 11 show clearly a size effect as predicted by Bazant. Fortunately for the public safety issue, the ACI Equation is apparently tuned to a beam depth of about 30 inches.

Diagonal Tension Cracking in Torsion

Torsion causes a state of pure shear stress in a cross-section. In an uncracked section, the shear stress increases from zero at the center of the section to a maximum at the middle of the longer face. Thus, the most probable location for crack initiation is at that

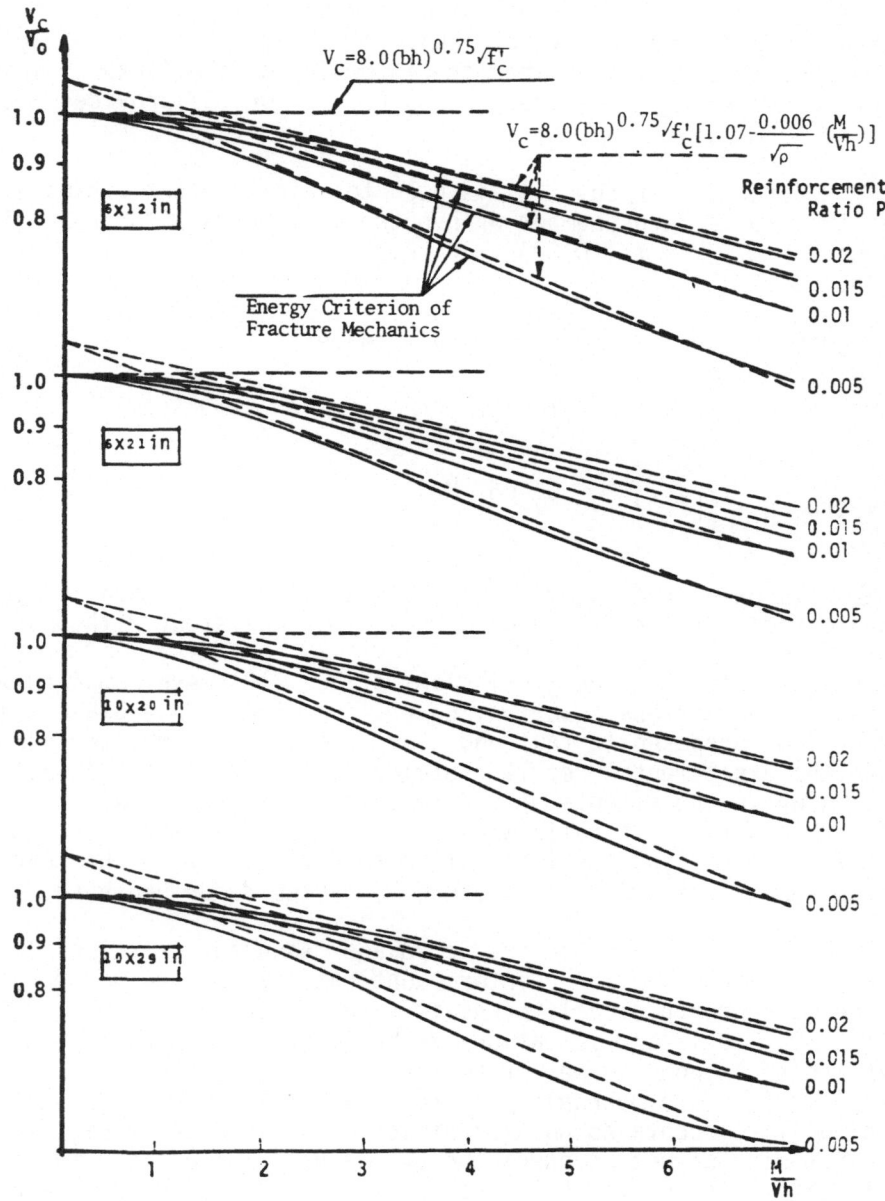

FIG. 8   COMPARISON OF PREDICTIONS OF ENERGY CRITERION
AND REGRESSION EQUATION (10)

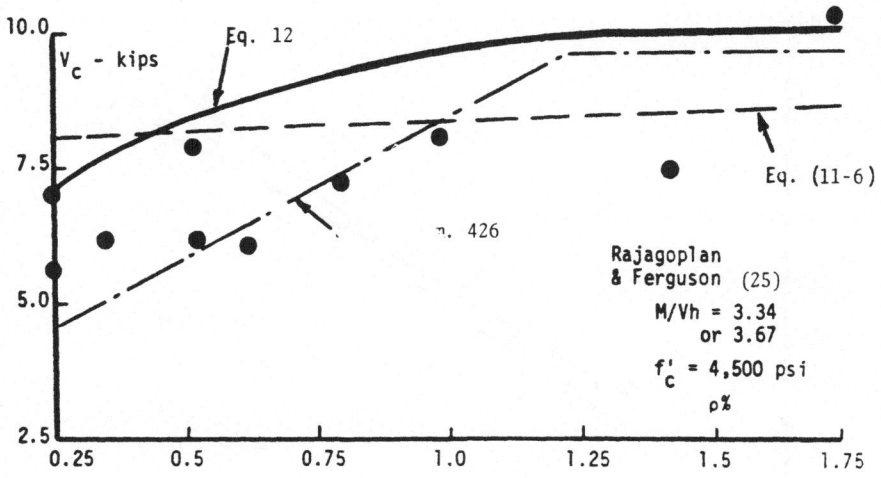

FIG. 9   VARIATION IN SHEAR STRENGTH WITH REINFORCED
                RATIO (10)

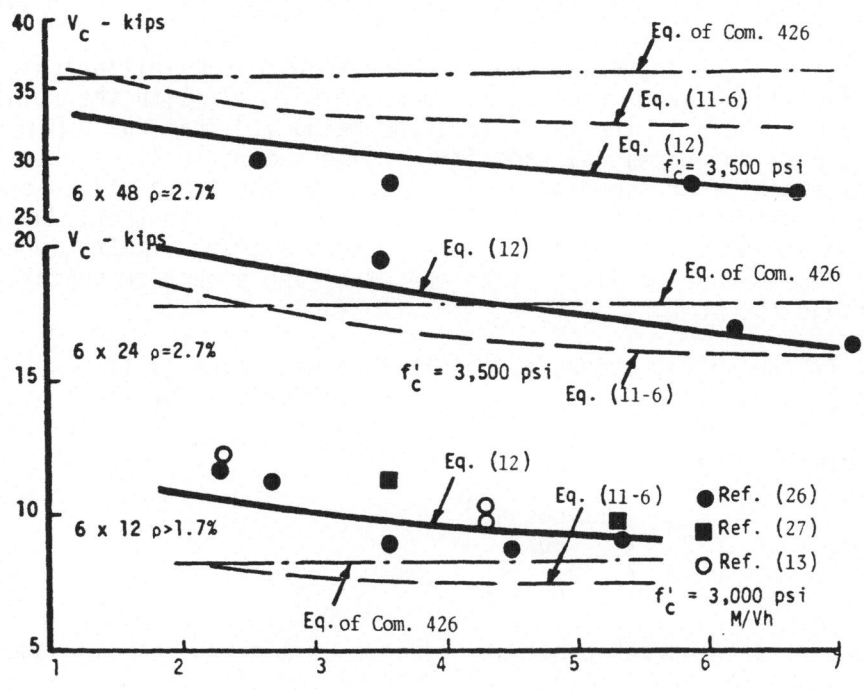

FIG. 10   VARIATION IN SHEAR STRENGTH WITH M/Vh (10)

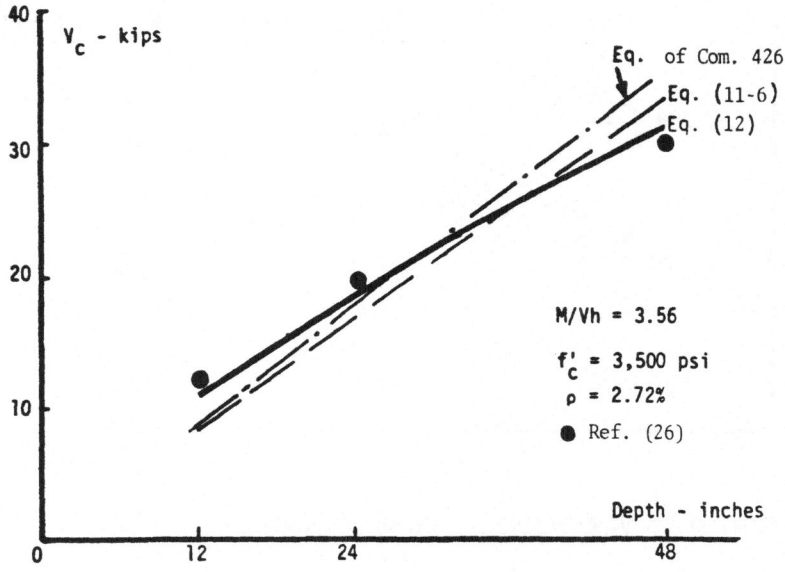

FIG. 11   VARIATION IN SHEAR STRENGTH WITH BEAM DEPTH (10)

location at 45 degrees. The crack develops a semi-elliptical shape during stable crack growth, as shown in Fig. 12, with the relative rate of growth in the two directions being γ. That ratio (21) does not influence the torsional capacity. While theoretically stable crack growth can occur simultaneously on both of the longer faces of the beam, and with those cracks at 90 degrees to one another, due either to an accidental inhomogeneity or lack of a beam symmetry, one flaw will become more critical than the other and propagate unstably through the member.

If the energy criterion of fracture mechanics is to be applied to the torsion problem, expressions must be derived for the change in torsional flexibility, F_T, with crack extension. That flexibility can be expressed as:

$$F_T = \frac{2.32}{E} \int_0^{c\gamma\cos 45^0} \frac{dz}{nb^3h - n_1 bhr^2} \tag{14}$$

where r is the crack penetration depth at a given location and n, n_1 are shape factors for torsional deformation calculations for uncracked and cracked sections, respectively.

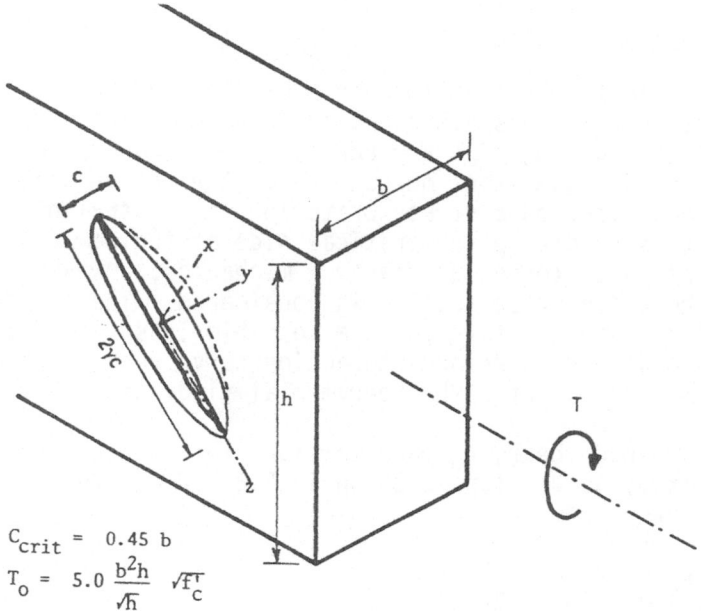

$$C_{crit} = 0.45 \, b$$

$$T_o = 5.0 \frac{b^2 h}{\sqrt{h}} \sqrt{f_c'}$$

FIG. 12   SHAPE AND LOCATION OF CRACK AND STRESS RELIEF
ZONE FOR TORSION LOADING (21)

In Reference (21) it is shown that through integration of Eq. (14) and the derivation of $\frac{\partial F_T}{\partial c}$ it is found that:

$$c_{crit} = 0.393 \sqrt{\frac{\eta}{\eta_1}} \tag{15}$$

and

$$T_o^2 = \frac{G_c E \pi b h}{3.28} \frac{c(\eta b^2 - \eta_1 c^2)^2}{\eta b^2 + \eta_1 c^2} \tag{16}$$

It will be noted that Eq. (16) is independent of the crack shape factor γ.

Values of η, η_1, c_{crit} and T_o were derived for an h/b ratio of 2.0 and it was found that for η equalled 0.229, η_1 equalled 0.175 and c_{crit} equalled 0.45b for:

$$T_o = 5.0 \, (b^2)(\sqrt{h})(\sqrt{f_c'}) \tag{17}$$

Values of T_o calculated from Eq. (17) averaged 86% of those measured in tests reported by Birkeland (28).

Diagonal Tension Cracking in Combined Torsion, Bending and Shear

The combination of torsion and shear loading causes a non-symmetrical shear stress distribution. On one side of the beam the effects of the two loadings are additive, while on the opposite side the two loadings counteract one another. Any shear-torsion inter-action curve, based on a stress criterion, is a straight line, because only the stress state on the critical side of the beam is included. With the energy criterion of fracture mechanics, however, the stress state of the whole cross-section is considered. The beneficial effects of the stress state for the less highly stressed side are included and a shear-torsion interaction curve results which is circular and consistent with experimental findings.

For combined shear, V_U, and torsion, T_U, loading the critical crack is likely to develop as shown in Fig. 12 and the energy equilibrium equation becomes:

$$\frac{T_U^2 \, \delta F_T}{2} + \frac{V_U^2 \, \delta F_V}{2} = \frac{c\gamma\pi E}{4.64} \tag{18}$$

If torsion or shear acts alone, Eq. (18) degenerates to:

$$\frac{T_o^2 \, \delta F_T}{2} = \frac{c\gamma\pi E}{2.32} \tag{19}$$

or

$$\frac{V_c^2 \, \delta F_V}{2} = \frac{c\gamma\pi E}{2.32} \tag{20}$$

For any assumed crack geometry, the changes in flexibility of Eqs. (18), (19) and (20) are the same so that it follows from substitution of the values of δF_T and δF_V from Eqs. (18) and (19) into Eq. (17) that:

$$\left(\frac{T_U}{T_o}\right)^2 + \left(\frac{V_U}{V_c}\right)^2 = 1 \tag{21}$$

The circular interaction curve of Eq. (21) is assumed in ACI Code 318-83 for the development of Eqs. (11-5) and (11-22). The agreement between some typical test data and Eq. (21) is shown in Fig. 13.

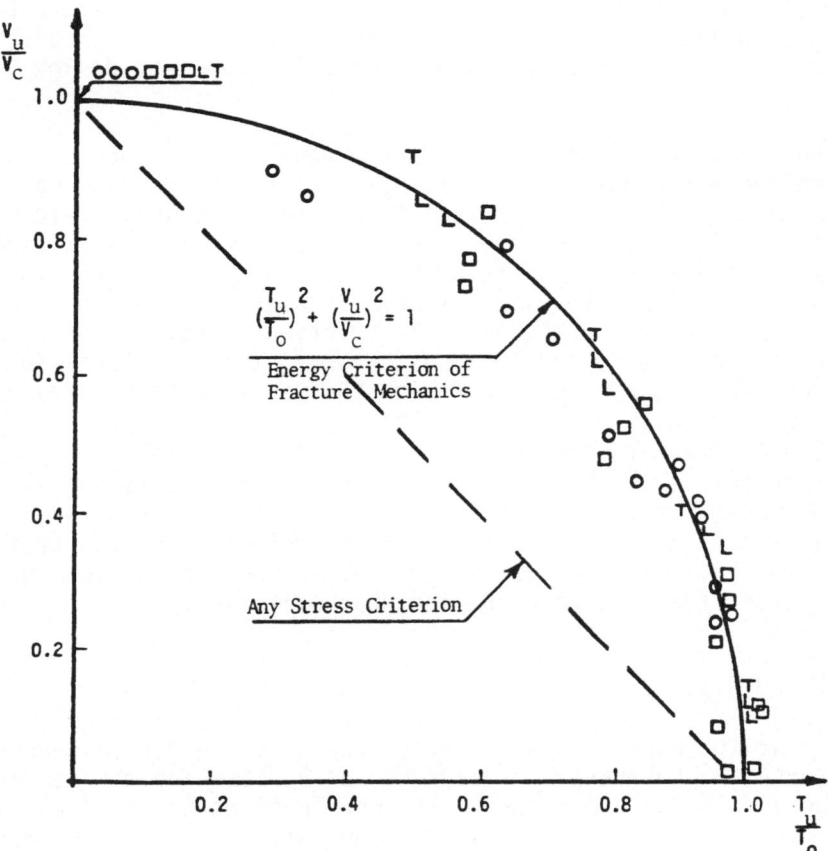

FIG. 13   INTERACTION DIAGRAM FOR TORSION AND SHEAR (21)

CONCLUDING REMARKS

The future role of fracture mechanics in conventional reinforced concrete design will depend primarily on how Code authorities view the public safety issues associated with the use of fracture mechanics. If authorities see that use as a mechanism for reducing public risk by allowing them to prescribe methods for the qualification of new materials or for the reward of designers who utilize more precise methods of analysis, then there is a significant role for fracture mechanics in future reinforced concrete design. If, however, those same authorities see fracture mechanics primarily as a concept that provides a more rigorous basis for accepted provisions of the Code with no associated change in public risk, then the future role for fracture mechanics in conventional reinforced concrete design will be minimal.

The fracture mechanics' fraternity needs to agree on a standardized procedure for determining K_I, K_{II}, etc., for concrete materials, and on the way the results of such tests should be interpreted for adjustments of the implied concrete tensile strength and stiffness values, and the minimum reinforcement ratio and maximum reinforcement spacing values of the Code. Further, fundamental research is needed to reconcile differences between the observed overall deformations for tensile specimens, and the considerably smaller deformations predicted for the same specimens with the use of accepted fracture process zone models. It is clear that application of the energy criterion of fracture mechanics, with its accompanying stress relief zone, can predict crack spacing and crack width effects and diagonal tension cracking effects both in shear and torsion. However, precise guidelines will be needed on how to interpret standardized test results for evaluation of the K_R (or G_C) value, and the form of the stress relief zone, to be used with that criterion.

ACKNOWLEDGMENTS

The writer expresses his deep appreciation to his colleagues Professor A. S. Kobayashi, D. Barker, and B. Liaw and his students F. Jeang and S. Akutagawa for their contributions to the developments of the concepts of this paper and to the National Science Foundation for their support of this work through Grant CEE - 8117029 for which the Program Manager is Dr. M. P. Gauss.

REFERENCES

(1) ACI Committee 318. Building Code Requirements for Reinforced Concrete (ACI 318-83). American Concrete Institute, Detroit, MI, (1983).

(2) American Association of State Highway and Transportation Officials. Standard Specifications for Highway Bridges, 12th Edition, Washington D.C., (1977).

(3) ASME Boiler and Pressure Vessel Code. Concrete Reactor Vessels and Containments, Section III, Division 2, American Society of Mechanical Engineers, New York, NY, (1983).

(4) ACI Committee 344. Design and Construction of Circular Prestressed Concrete Structures, American Concrete Institute, Detroit, MI, (1970).

(5) Hawkins, N.M. Impact of Research on Prestressed Concrete Specifications, Significant Developments in Engineering Practice and Research, American Concrete Institute, Detroit, MI, 1981.

(6) Bazant, A.P. Size Effect in Blunt Fracture: Concrete, Rock, Metal, Engineering Mechanics Journal, ASCE, vol. 110, no. 4, (Apr. 1984).

(7) ACI-ASCE Committee 426. The Shear Strength of Reinforced Concrete Members, Structural Division Journal, ASCE, (June 1973) 1148-1157.

(8) Wecharatana, M and Shah, S.P. Nonlinear Fracture Mechanics Parameters, Fracture Mechanics of Concrete, Elsevier, NY, (F.H. Wittman Editor, 1983) 463-480.

(9) Walsh, P.F. Fracture of Plain Concrete, The Indian Concrete Journal, vol. 46, no. 11, (Nov. 1979) 469-470 and 476.

(10) Hawkins, N.M., Wyss, A.N. and Mattock, A.H. Fracture Analysis of Cracking in Concrete Beams, Structural Division Journal, ASCE, vol. 103, no. STS, (May 1977) 1015-1030.

(11) Hsu, T.T.C. Torsion of Structural Concrete-Plain Concrete Rectangular Sections, S P-18, Torsion of Concrete, American Concrete Institute, Detroit, MI (1968).

(12) Bazant, Z.P. and Cedolin, L. Blunt Crack Band Propagation in Finite Element Analysis, Engineering Mechanics Division Journal, ASCE, vol. 105, no. EM2, (Apr. 1979) 297-315.

(13) ACI-ASCE Committee 426. Suggested Revisions to Shear Provisions for Building Codes, American Concrete Institute, Detroit, MI (1979).

(14) Bazant, Z.P. and Oh, B.H. Spacing of Cracks in Reinforced Concrete, Structural Division Journal, ASCE, vol. 109, no. 9, (Sept. 1983) 2066-2085.

(15) Gergeley, P. and Lutz, L.A. Maximum Crack Width in Reinforced Concrete Flexural Members, Causes, Mechanism, and Control of Cracking in Concrete, SP-20, American Concrete Institute, Detroit, MI (1968) 87-117.

(16) Broms, B.B. Crack Width and Crack Spacing in Reinforced Concrete Members, ACI Journal, vol. 62, no. 10, (Oct. 1965) 1237-1256.

666

(17) Hawkins, N.M., Lin, I.J. and Jeang, F.L. Local Bond Strength of Concrete for Cyclic Reversed Loadings, Proceedings International Conference on Bond in Concrete, College of Technology, Paisley, Scotland (P. Bartos Editor, 1982) 151-161.

(18) Akutagawa, S. Wedge Configuration and Loading History Considerations for Concrete CLWL-DCB Fracture Specimens, MSCE Thesis, University of Washington (August 1984).

(19) Cho, K.Z., Kobayashi, A.S., Hawkins, N.M., Barker, D.B. and Jeang, F.L. Fracture Process Zone of Concrete Cracks, Engineering Mechanics Division Journal, ASCE, vol. 110, no. 8 (August 1984) 1174-1184.

(20) Collins, M.P. and Mitchell, D. Shear and Torsion Design of Pre-stressed and Non-prestressed Concrete Beams, PCI Journal, vol. 25, no. 5, (Sept./Oct. 1980) 32-101.

(21) Wyss, A.N. Application of Fracture Mechanics to Cracking in Concrete Beams, Ph.D. Thesis, University of Washington (1971).

(22) Mattock, A.H. Diagonal Tension Cracking in Concrete Beams with Axial Forces, Structural Division Journal, ASCE, vol. 95, no. ST9, (Sept. 1969).

(23) Kobayashi, A.S., Hawkins, N.M., Barker, D.B. and Liaw, B.M. Fracture Process Zone of Concrete, Proceedings, NATO Advanced Research Workshop on Application of Fracture Mechanics to Cementitious Composites, Northwestern University, Evanston, IL (1984).

(24) Taylor, H.P.J. Investigation of Dowel Shear Forces Carried by the Tensile Steel in Reinforced Concrete Beams, TRA 431, (Nov. 1969); Further Tests to Determine Shear Stresses in Reinforced Concrete Beams, TRA 438, (Feb. 1970); Investigation of the Forces Carried Across Cracks in Reinforced Concrete Beams in Shear by Interlock of Aggregate, TRA 447 (Nov. 1970) Cement and Concrete Association, Slough, Bucks.

(25) Rajagopalan, K.S. and Ferguson, P.M. Exploratory Shear Tests Emphasizing Percentage of Longitudinal Steel, ACI Journal, vol. 65, no. 8 (Aug. 1968) 634-638.

(26) Kani, G.N.J. Basic Facts Concerning Shear Failure, ACI Journal, vol. 63, no. 6 (June 1966) 675-692.

(27) Kani, G.N.J. How Safe are Our Large Reinforced Concrete Beams? ACI Journal, vol. 64, no. 3 (May 1967) 128-141.

(28) Mattock, A.H., Birkeland, C.J. and Hamilton, M.E. Strength of Reinforced Concrete Beams Without Web Reinforcement in Combined Torsion, Shear, and Bending, The Trend in Engineering, University of Washington, vol. 19, no. 4 (Oct. 1967) 8-12 and 29.

(29) Ersoy, U. Combined Torsion in Semi-Continuous Concrete L-Beams Without Stirrups, Ph.D. Thesis, University of Texas (1965).

**APPLICATION OF FRACTURE MECHANICS
TO CEMENTITIOUS COMPOSITES**
NATO-ARW - September 4-7, 1984
Northwestern University, U.S.A.
S. P. Shah, Editor

IMPROVEMENTS IN CONCRETE DESIGN ACHIEVED THROUGH THE APPLICATION
OF FRACTURE MECHANICS

Gustafsson, P J and Hillerborg, A.

Civ. eng. resp. professor, Division of Building Materials, Lund
Institute of Technology, S-220 07 Lund, Sweden.

1. INTRODUCTION

In the "Fictitious Crack Model" (1,2,3) the relation between
deformation and stress is expressed by means of two curves, one
stress-strain-curve and one stress-deformation curve. The stress-
strain curve is valid for all of the material. The stress-deformation curve shows the relation between stress and the additional
deformation of the fracture zone due to strain-localization. In
the general case both these curves can be non-linear and comprise
unloading branches. The curves are determined by means of tensile
tests, and they are assumed to be material properties.

When this model is applied to the analysis of structures, it
is necessary to make numerical calculations, normally by means of
a computer. So far these calculations have been made by means of
the Finite Element Method, although other numerical techniques
could also be used.

Non-linear stress-strain curves make the numerical calculations more complicated and expensive. For this reason all the
results presented below are based on the assumption that the material has a linear elastic stress-strain·curve according to
Fig. 1a.

Non-linear stress-deformation curves for the additional deformation of the fracture zone do not, on the other hand, complicate the numerical calculations very much, as this curve does not
influence the whole body, but the fracture planes only. For the
results presented below two different approximations have been
used according to Fig. 1b and c.

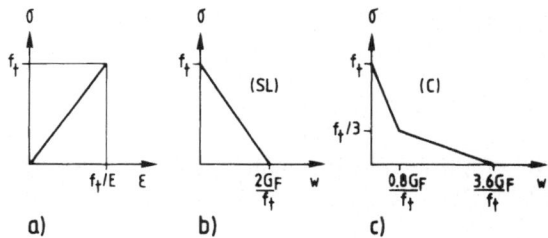

Fig. 1. Simplified stress-strain and stress-deformation curves
in the numerical analyses.

The straight line in Fig. 1b is the simplest possible approximation, whereas the bilinear relationship of Fig. 1c corresponds
closer to the normal shape of the curve for concrete (3). In the
diagrams below those which are based on the straight line in
Fig. 1b will be denoted by (SL), and those based on the more con-
crete-imitating curve of Fig. 1c by (C).

The following material parameters define the properties of the
material:

f_t the tensile strength

E   the modulus of elasticity

G_F the fracture energy = the area below the stress-deformation
       curve.

It is also convenient to use the material parameter

$l_{ch} = EG_F/f_t^2$ characteristic length of the material.

The value of the characteristic length is normally of the
order 200-400 mm for ordinary concrete, but it can take on values
of up to 1 m for concrete with a large aggregate size.

2. FLEXURAL STRENGTH OF RECTANGULAR BEAMS

Fig. 2 shows the result of an analysis of the flexural
strength of a rectangular beam. Curves are given for both the assumptions regarding the shape of the stress-deformation curve. It
can be seen from the Fig. that the shape of the curve is not very
important for the results, at least not in this case.

In Fig. 2. lines are also shown for the flexural strength according to assumptions of elastic brittle failure and perfectly
plastic tensile behaviour. In the latter case the compressive
strength has been assumed to be infinite, which means that the
value is a theoretical upper limit. With a realistic assumption

regarding the ratio between compressive and tensile strength of concrete the value will be about 10 percent lower.

The curves approach the elastic-brittle behaviour when the beam depth increases and they approach the plastic behaviour when the beam depth decreases.

According to elastic-brittle behaviour and to plastic behaviour the strength does not depend on the beam depth. The curves based on fracture mechanics analyses on the other hand show that the flexural strength decreases when the beam depth increases. This is also in agreement with test results. For a normal concrete with a characteristic length of 200-400 mm the curves show that the modulus of rupture, determined on a 100 mm deep beam, can be expected to be about 50 percent higher than the tensile strength. This seems to be in reasonable accordance with practical experience, taking into account that the non-linearity of the stress-strain curve and the strength scatter also have some influence.

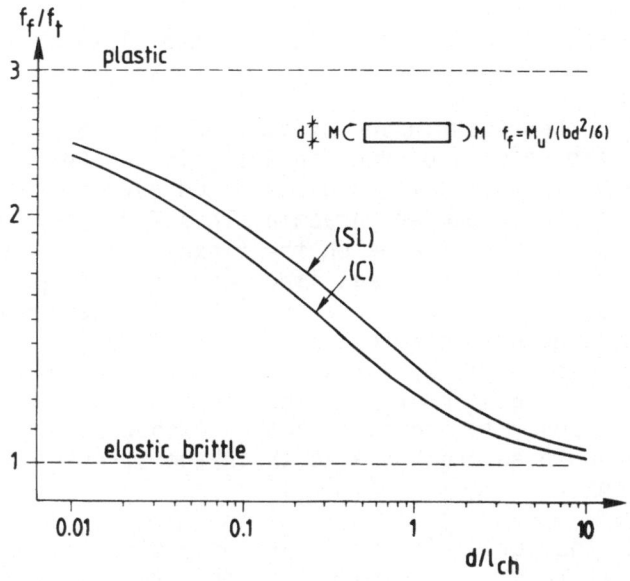

Fig. 2. Theoretical ratio between flexural strength and tensile strength.

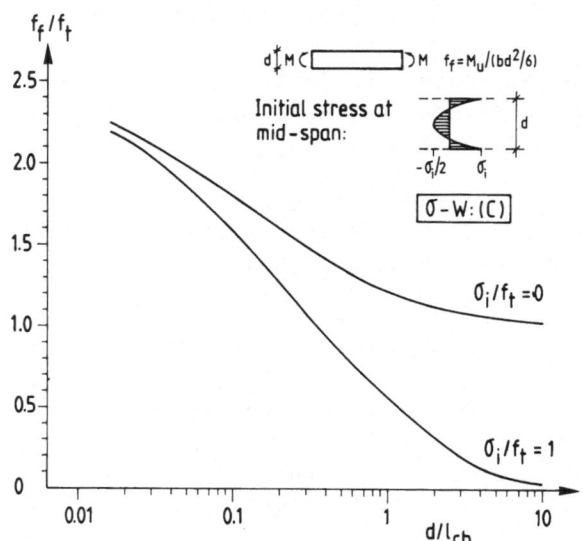

Fig. 3. Influence of shrinkage stress on the flexural strength.

The influence of shrinkage strains has also been analysed. Fig. 3 shows the results of such an analysis, where the shrinkage strains have been assumed to be distributed over the beam depth according to a second degree parabola. This Fig. shows that the influence of shrinkage strains on the flexural strength is very size-dependent. This is of great importance for the possibility of using plain concrete, as concrete is always influenced by shrinkage. The strengths of concrete tiles and thin paving slabs are hardly influenced at all by shrinkage, and they therefore have a high flexural strength under practical conditions. Large concrete beams of plain concrete can be expected to show a very low flexural strength and even crack due to shrinkage stresses, without any exterior moment.

Although the curves in Fig. 3 are only strictly valid for plain concrete, they may indicate that the cracking strength of reinforced beams also depends on the beam depth, especially when shrinkage takes place. The detailed behaviour in this case has not yet been analysed, but this is quite possible.

3. FLEXURAL STRENGTH OF A NOTCHED RECTANGULAR BEAM

The Fictitious Crack Model is general in the sense that it can be applied to uncracked as well as cracked structures, whereas conventional fracture mechanics can only be applied to cracked structures and the ordinary theory of strength of materials can

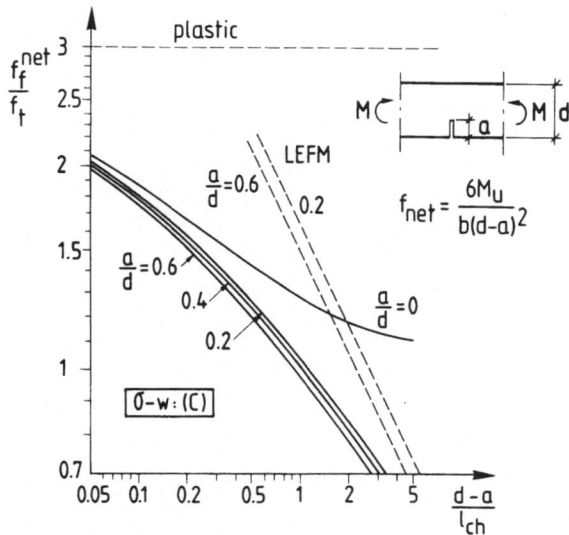

Fig. 4. Theoretical variation in net bending strength of a notched
beam with notch depth and beam depth.

only be applied to uncracked structures.

The model has been applied for analysing different types of
notched specimens used for the determination of conventional fracture mechanics parameters, see (2,3). It has been clearly demonstrated that these tests are unsuitable unless the specimens are
unrealistically large.

Only one example of results from the application to a notched
beam will be demonstrated here. Fig. 4 shows how the ratio between
the net bending strength and the tensile strength of a notched
beam varies with the beam depth. The corresponding relations according to linear elastic fracture mechanics and according to the
theory of plasticity are also shown. As in the previous case the
curve approaches the values according to the theory of plasticity
for small beams and the values according to LEFM for large beams.
Beams of normal size belong to the intermediate part.

4. UNREINFORCED CONCRETE PIPES

Unreinforced concrete pipes fail mainly in two ways, denoted
here as "crushing failure" and "bending failure" respectively,
Fig. 5. Both these types of failures have been analysed by means
of the Fictitious Crack Model. This has been done on request from
one concrete pipe manufacturer, who wished to achieve better design criteria for the pipes.

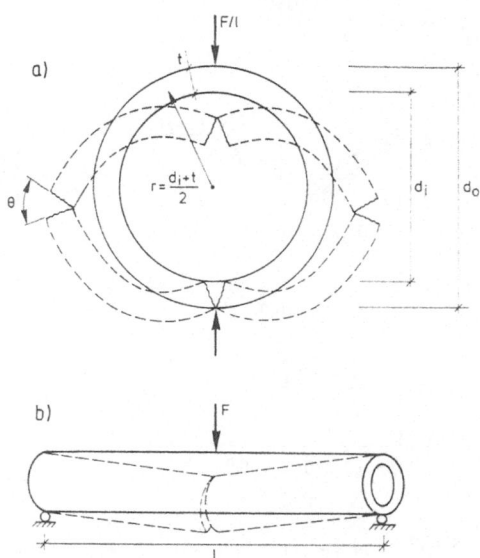

Fig. 5. Crushing failure and bending failure respectively of an unreinforced concrete pipe.

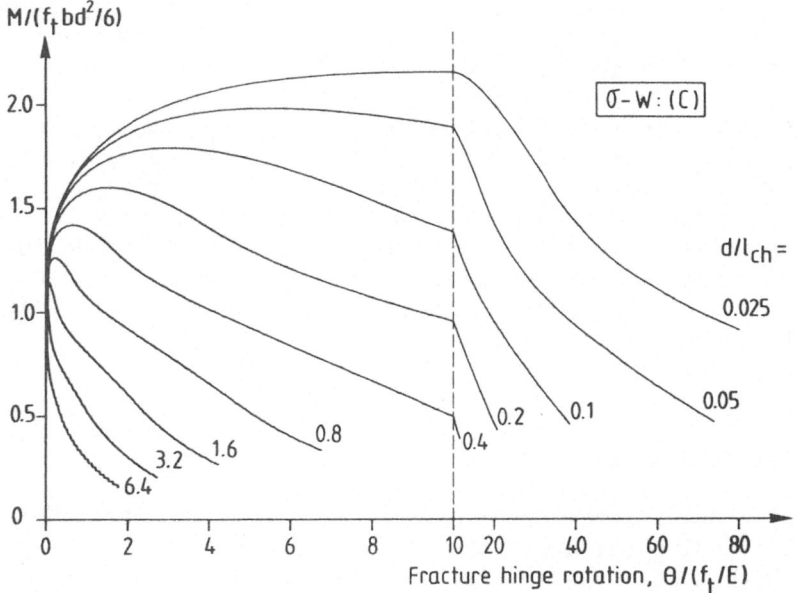

Fig. 6. Relation between hinge rotation and bending moment.

The crushing failure involves the formation of four fracture zones as indicated in Fig. 5. Due to these fracture zones the deformation within these parts of the pipe becomes non-elastic before failure. The fracture zones behave as semi-plastic hinges.

The relation between the hinge rotation and the hinge moment in such a hinge can be calculated by means of the Fictitious Crack Model. Fig. 6 shows the results of such calculations.

After these properties of the hinges have been determined, the analysis of the crushing failure can be made by treating the parts between the hinges as elastic elements and taking the hinge properties into account.

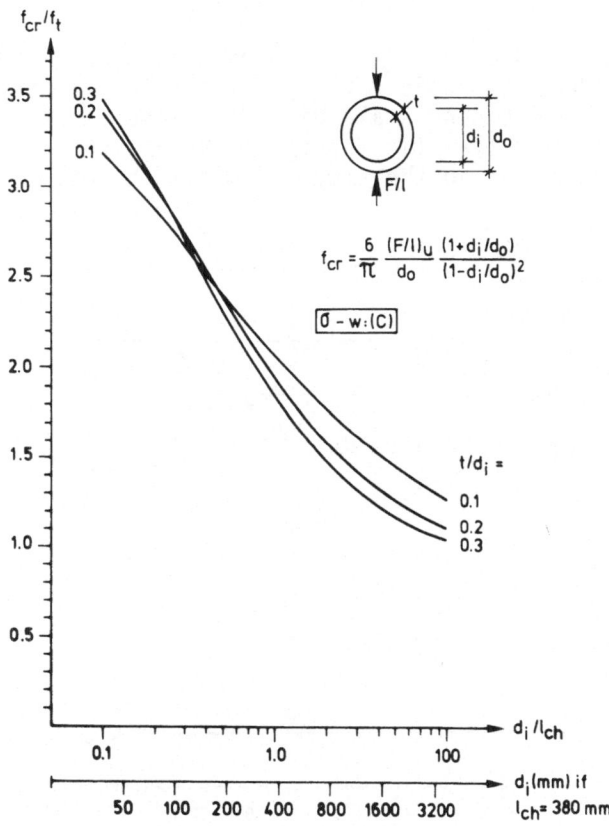

Fig. 7. Theoretical ratio between crushing strength and tensile strength for concrete pipes.

Fig. 7 shows the results of such calculations. It gives the ratio between the formal flexural strength in crushing failure, f_{cr}, and the tensile strength. The formal flexural strength is defined as the maximum tensile stress at failure, calculated according to the ordinary elastic beam theory, which is the normal way of evaluating test results. Thus by definition, with loading according to Fig. 5a

$$M_u = \frac{F_u(d_i + t)}{2\pi}$$

$$f_{cr} = \frac{M_u}{(1t^2/6)}$$

The pipes have also been analysed for bending failure. The result of this analysis is given in Fig. 8.

In Figs. 7 and 8 the ratio between the inner diameter of the pipe and the characteristic length of the material is given on the horisontal axis. A second scale gives the inner diameter corresponding to a characteristic length of 380 mm, which was the value for the concrete used in the studied Swedish pipes.

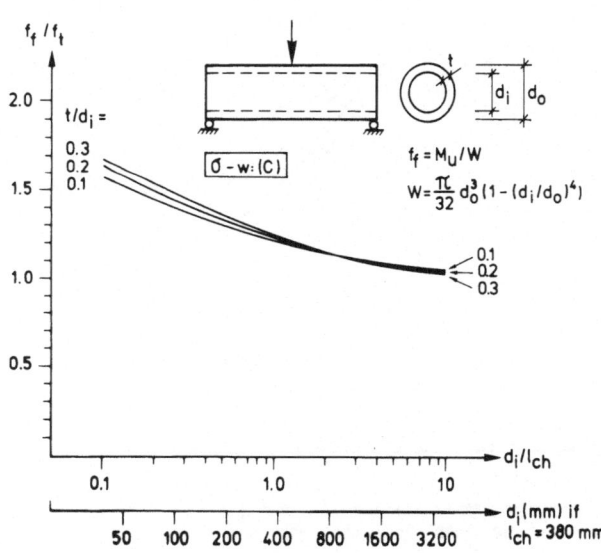

Fig. 8. Theoretical ratio between flexural strength and tensile strength for concrete pipes.

A comparison between Figs. 7 and 8 shows that the ratio between the flexural strength and the tensile strength is much higher for crushing failure than for bending failure. This fact as well as variation of the ratios with the size and the geometrical shape of the pipe shows that the flexural strength cannot be treated as a material property. The difference between the formal flexural strength at bending and at crushing respectively is of such magnitude that the difference is hardly covered by any normal safety factors used in design.

By means of the diagrams the tensile strength of the concrete can be calculated from any test where the formal flexural strength has been determined. If a test has been made for one type of failure it is also possible to calculate the strength of the same pipe for the other type of failure. The strength of the concrete in a pipe can therefore be checked by one type of test or the other.

The sensitivity of the pipes to shrinkage caused by equal drying from the inside and from the outside has also been analysed. It was found that increased wall thickness makes the pipe more sensitive to shrinkage, while alterations in the diameter of the pipe are not predicted to have any substantial influence on the sensitivity to shrinkage. For the studied magnitude of shrinkage, it was found that the formal strength of a pipe with a wall thickness of 50 mm is reduced by only about 5 % in bending and about 10 % in crushing, whereas with a wall thickness of 150 mm the formal strength in bending as well as in crushing is predicted to be reduced by approximatively 20 %.

The theoretical results shown in Figs. 7 and 8 are in agreement with general practical experiences. The diagrams have been applied to a large number of test results concerning crushing, and they have also been applied to all known results where crushing tests and bending tests have been made on pipes of the same kind. The available number of results from the latter type of test is limited. Table 1 shows the only test series which we known about, where a reasonable number of tests have been made using the same concrete quality.

Series I and II represent an evaluation of tests made by Brennan (4). The crushing tests in these series were made in accordance with British Standard. The formal flexural stresses at crushing failure have been calculated with due account to the load distribution. Series III is a Swedish test series. The strength calculated according to the ordinary elastic theory (f_f and f_{cr}) seems to be dependent on the geometry of the pipe and the values differ considerably between crushing failure and bending failure, whereas the values of tensile strength achieved by means of Figs. 7 and 8 (f_t) are almost suprisingly constant for the same concrete quality.

Series	Number of specimens	Type of failure	t mm	d_i mm	f_f MPa	f_{cr} MPa	f_t MPa
I	4	Bend	30	225	5.3		4.1
	4	Bend	42	300	5.4		4.3
	4	Crush	30	225		9.2	4.1
	4	Crush	42	300		8.2	3.9
II	7	Bend	51	300	6.0		4.7
	7	Crush	51	300		9.7	4.7
III	4	Bend	35	100	7.4		4.9
	4	Bend	31	150	6.8		4.9
	1	Bend	34	225	6.6		5.1
	4	Crush	35	225		11.0	4.9
	4	Crush	55	400		9.6	4.9

Table 1. Test results of strength of pipes. The tensile strength f_t has been calculated by means of Figs. 7 and 8, assuming l_{ch} = 380 mm.

The diagrams in Figs. 7 and 8 are now used by manufacturers for the design of pipes and as guidance for establishing suitable control methods.

5. SHEAR STRENGTH OF REINFORCED CONCRETE BEAMS

The shear strength of beams with longitudinal reinforcement but without shear reinforcement has also been analysed by means of the Fictitious Crack Model for beams with geometrical shapes according to Fig. 9. Bond-slip relations for the reinforcement and concrete failure in the compression zone have been taken into account in this analysis, but so far shear forces in a crack (aggregate interlock) and dowel action have been disregarded, and not more than one crack at a time has been studied. These factors can in principle be included, but the numerical work involved will be much greater and the necessary material properties are not yet sufficiently well known.'

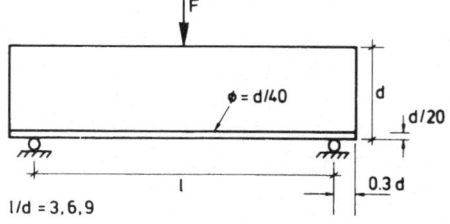

Fig. 9. Reinforced beam analysed for shear failure.

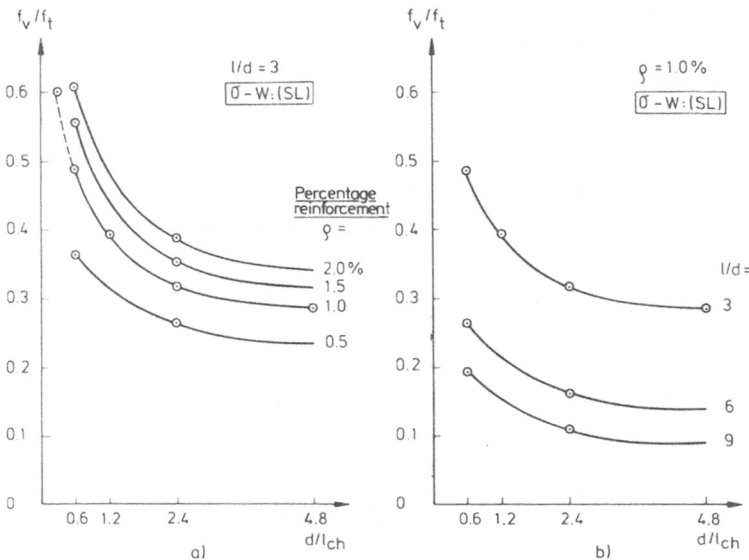

Fig. 10. Theoretical shear strength $f_v = F_u/(2bd)$ of beams according to Fig. 9.

Fig. 10 shows the results reached so far. It illustrates the influence of the amount of longitudinal reinforcement, of the span/depth ratio and of the beam depth.

Some comparisons between these theoretical values and test results are shown in Figs. 11-13. In all the cases the curves show the variation of the shear strength with the factor in question, i e all curves in a diagram have been drawn through one common point. This has been necessary because the tests have been made with varying concrete qualities, and the tensile strength of the tested concrete as a rule has not been reported.

Where the tensile strength of the concrete is known or can be estimated there seems to be a tendency that the diagram of Fig. 10 gives too low values of the shear strength. This is probably due to the fact that the beneficial influence of aggregate interlock and of dowel action have been disregarded.

These results show that it is possible to analyse the shear failure of reinforced concrete in a realistic way by means of suitable fracture mechanics models. This is an appreciable step forwards, as the design formulas so far have been of an empirical nature, where the suitable coefficients have been determined by means of very extensive test series.

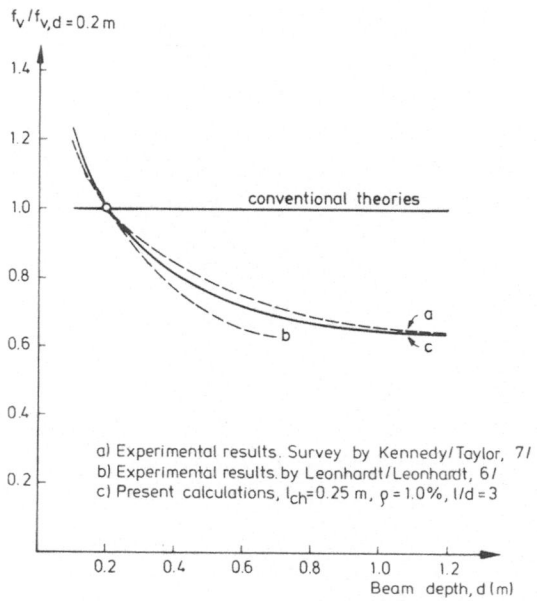

a) Experimental results. Survey by Kennedy/Taylor, 7/
b) Experimental results. by Leonhardt/Leonhardt, 6/
c) Present calculations, l_{ch}=0.25 m, ρ = 1.0%, l/d = 3

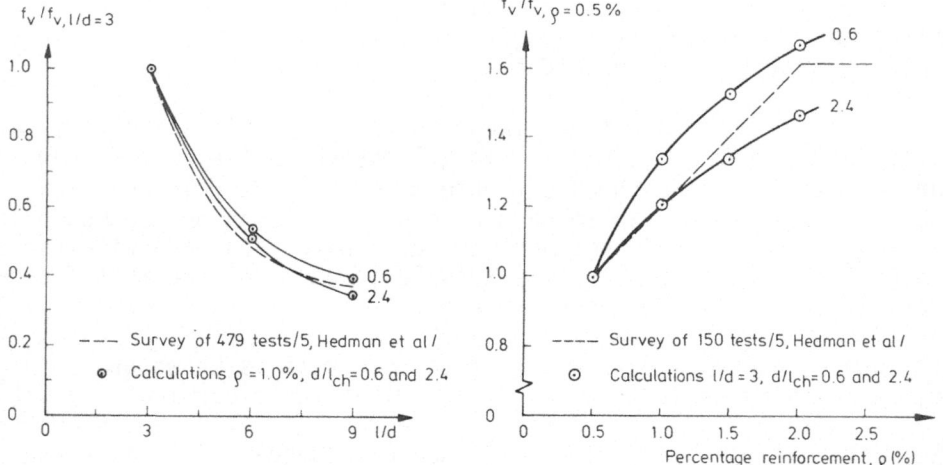

Figs. 11-13. Theoretical shear strength variation with beam depth, span/depth ratio, and amount of reinforcement, compared to test results.

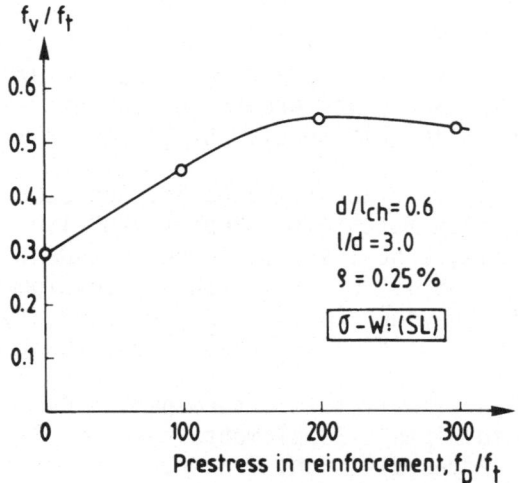

Fig. 14. Theoretical influence of reinforcement prestress on the shear strength. At high prestress bond failure occurs in this case.

Some preliminary analyses have also been made regarding the influence of prestressed reinforcement. The results are presented in Fig. 14.

6. CONCLUDING REMARKS

The possibility of making realistic analyses of the behaviour of structures where tensile fracture or cracking are of importance opens new possibilities of extending the knowledge we need as a basis for our design rules.

One important improvement is that it will now be possible to analyse the influence of one factor at a time on the behaviour of a structure without expensive tests, where the natural scatter of the results makes the conclusions uncertain. By means of rational computer programs it will be possible to simulate thousands of tests at a reasonable cost. Of course the theoretical results have to be checked by means of tests, but for this purpose it will be sufficient to make a very limited number of tests.

This type of approach can be used for the analysis of many other types of structures and also for many other materials, e g ceramics, rock, fiber reinforced materials, wood, particle board etc.

REFERENCES

1. Hillerborg, A., Modeer, M., Petersson, P.E. (1976), "Analysis of Crack Formation and Crack Growth in Concrete by Means of Fracture Mechanics and Finite Elements", Cement and Concrete Research, Vol 6, pp. 773-782.
2. Modeer, M. (1979), "A Fracture Mechanics Approach to Failure Analyses of Concrete Materials", Report TVBM-1001, Div. of Building Materials, Lund Inst. of Techn., Sweden.
3. Petersson, P.E. (1981), "Crack Growth and Development of Fracture Zones in Plain Concrete and Similar Materials", Report TVBM-1006, Div. of Building Materials, Lund Inst. of Techn., Sweden.
4. Brennan, G. (1978), "A Test to Determine the Bending Moment Resistance of Rigid Pipes". Supplementary Report 348, Earthworks and Underground Pipes Division, Transport and Road Research Laboratory Crowthorne, Berkshire, England.
5. Hedman, O., Losberg, A. (1978), "Design of Concrete Structures with Regard to Shear Forces", CEB Bulletin d'Information No 126: Shear and Torsion, pp. 183-210.
6. Leonhardt, F. (1978), "Shear in Concrete Structures", CEB Bulletin d'Information No 126: Shear and Torsion, pp. 66-124.
7. Taylor, H.P.J. (1978), "Basic Behaviour in Shear and the Model Code Provisions for Members without Shear Reinforcement", CEB Bulletin d'Information No 126: Shear and Torsion, pp. 125-140.

**APPLICATION OF FRACTURE MECHANICS
TO CEMENTITIOUS COMPOSITES**
NATO-ARW - September 4-7, 1984
Northwestern Univeristy, U.S.A.
S. P. Shah, Editor

DESIGN CODES AND FRACTURE MECHANICS

T. P. Tassios

Professor, National Technical University, Athens

1.- INTRODUCTION

This is the Reporter's response, mainly to the two lectures of this Session VIII, together with some views regarding post-cracking models of reinforced concrete elements.

The design of concrete structures for safety and economy is intimately connected to r e s i s t a n c e models: Dimensioning of R.C. structural elements is based on appropriate analytical models, depicting the behaviour of critical regions and predicting ultimate situations. And for a long period of time, a simple "estimator" or such an ultimate situation was a maximum stress (or an equivalent stress) acting on one point of one cross-section of the structural element. During the last two decades, however, ultimate action-effects are considered as estimators of ultimate situations. Undoubtedly this is a progress, but it is not enough:

a) Out of a possibly larger (and of unknown extension) critical region, only a cross-section is considered; sometimes its location and orientation might be doubtful.

b) The ultimate situation is expressed in terms of forces (N_u, M_u, V_u), although imposed deformations and degrading environmental actions and rate-effects may occasionally be more decisive.

c) For such a strongly strain-softening material as the concrete, strain gradients along critical a cross-section may mobilise reactions of quite a different nature: Stable resisting pre-pick forces and time-sensitive post-pick resisting forces. It is not unequivocally possible to superimpose these reactions.

Examples of insufficiency of this conventional approach of ultimate limit state have already given some warnings to Structural Engineers. The comments which follow in this Report, are attempting to recall some of these ill-treated cases. This Reporter does not think that the situation is alarming. However, if Fracture Mechanics will prove to be able to remedy this category of problems, Structural Engineering should be impregnated by some new ideas - as it did several times in the past. Thus, the implication of Fracture Mechanics for concrete design will be fruitful.

2.- SCLALE EFFECTS

It is believed that F.M. has already considerably elucidated the old pseudo-dilemma "Elasticity or Plasticity?". In fact, linear elastic fracture mechanics is able to predict that the ultimate average stress τ_u is a function of a scale factor λ between geometrically similar concrete structures (4):

$$\tau_u \sim \lambda^{-1/2} \tag{1}$$

This general phenomenon is better described by the following equation (2) and by Fig. 1:

$$\sigma_u = \beta(1 + \lambda)^{-1/2} . f_{ct} \tag{2}$$

Several authors of this Workshop have referred to this phenomenon too (Hawkins, Sih, Hilsdorf, etc).
Some practical implications are reminded here.
a) Prediction of shear strength of plain concrete cross-sections:
....
Gustafsson and Hillerborg in their Fig. 10a (based on Hillerborg's Fictitious Crack Model) offer numerical solutions related to the size effect on shear strength of R.C. beams. Taking as an approximation of plain concrete their case $\rho_1 = 0,5\%$, we reproduce in Fig. 2 Gustafsson's and Hillerborg's results, normalised over their minimum value, for $l_{ch} = 0,25$ m (characteristic length of the material). On the same Fig. 2 we reproduce the k-values (size effect factor) included in the Model Code of CEB for the design values of shear strength of concrete sections without web reinforcement

$$R_{du_1} = \frac{0,25 . f_{ct,0.05}}{\gamma_m} (1 + 50 . \rho_1) . k . A_c \tag{3}$$

where $k = 1,6 - d^{(m)} \not< 1$
and ρ_1 = ratio of longitudinal reinforcement

Not many other Codes have included such a size effect factor; ACI apparently has tuned its formulae to a section height approximately equal to 0,8 m (according to N. Hawkins).
b) Prediction of flexural strength of plain concrete pipes
....
In the same paper of Gustafsson and Hillerborg, there are convincing numerical examples predicting the crushing strength of concrete pipes, during their transversal and longitudinal test, as a

function of the standard tensile strength of concrete. Fig. 7 and 8
of their paper is a very interesting configuration of the F.M. law
depicted in our Fig. 1. The practical implications for design are
apparent.

c) Design by testing

This Reporter is not much in favour of the principle of de-
signing on the basis only of specific tests made on purpose, even
of full scale models (7). Related to the systematic uncertainties
due to scale effects when testing R.C. building elements, among
others, bond/slip considerations were mainly escaping similitude
laws. The rational approach offered by F.M. on the extreme relati-
vity of the notion of the nominal tensile strength will certainly
reduce one of the many uncertainties generalisations of tests findings are suffering of.

3.- SHEAR STRENGTH OF R.C. CONCRETE

Implications of F.M. on shear strength considerations were more expected because of the relative brittleness of concrete, whose
behaviour is governing many shear phenomena. Some of these implications are recalled here too:

a) The energy criterion as stated by N. Hawkins in his paper
in this Workshop "clearly predicts a variation of shear force resistance with longitudinal reinforcement". Besides, Gustafsson and
Hillerborg, in Fig. 10a of their paper in this Workshop, offer numerical data for such a variation, reproduced in our Fig.3. In order
to compare their results with the expression included in the CEB
formula (our Equ. 3), plain concrete has been approximated with
the case of $\rho_1 = 0,005$.

b) Similarly, the role of the shear ratio $\alpha_s = M:Vh$ on the
nominal shear strength is very clearly depicted by Fig. 10b of
Gustafsson and Hillerborg, whereas Codes have not yet decided to
offer guidance on this subject, except for shear walls.

Taking into account the difficulties encountered in R.C.-Engi-
neering when dealing with unforseen shear failures, these contributions of F.M. have to be welcome. However, further multidirectional
research is needed in this area, since similar results are also predicted by other than F.M. approaches (e.g. aggregate interlock, embrittlement due to biaxial tension-compression).

4.- TIME DEPENDANT PHENOMENA

Concrete, a strain rate sensitive brittle material, needs a
much better knowledge on its behaviour versus fatigue, impact and
long term loading conditions. In this Workshop, Sierakowski, Reinhardt, Wittmann and Mindess have contributed to the elucidation
of these topics.

It is this Reporter's impression that a considerable amount

of research and calibration is needed before directly practical implications of F.M. in this field may be recognised. Nevertheless, we can distinguish already general very positive consequences of this powerful tool:

- Confirmation of some experimental data
- Fundamental understanding of structural behaviour and clear i d e n t i f i c a t i o n of parameters; even empirical investigations may be helped by such achievements.

Some indicative examples may be mentioned here:

- The increased probability of more than one cracks going through the specimen when strain rate increases, may explain the higher strength and c r i t i c a l s t r a i n (4), although a minimum failure strain value is recognised around $\varepsilon \simeq 10^{-2} s^{-1}$.
- The role of disjoining pressure of water in xerogel at high RH values has been recognised, with its increasing effect on creep (Wittman).
- The extreme variability of stress distributions along a notched concrete cross-section during repeated loading (3), unpredictable by conventional means, may prove to be a fundamental finding for fatigue research.
- Improvement of objectivity in finite elements applications.

However, it has to be repeated that it is rather premature to talk in terms of "practical" implications of F.M. in this field.

5.- POST CRACKING MODELS OR R.C. ELEMENTS

Extending the basic notions of Fracture Mechanics into reinforced concrete, further possibilities have been recognised, both for fiber-reinforced and for conventionally reinforced concrete. In fact, instead of the (historically first observed) "surface" forces constituting the source of energy along micro-cracks in Griffith's theory, fibers or steel bars crossing macro-cracks play the role of large crack arrestors.

In real life structural problems, **mixed** modes of cracking are prevailing. Therefore, for the general case of diffused rebars crossing the crack under an arbitrary angle, constitutive laws of several mechanisms of load transfer (or energy absorption) should be known such as pullout/dowel forces and concrete to concrete friction. Several successful approximations have been used to this purpose (see i.a. (1)), in the framework of finite elements' applications. However, in some cases a better and more detailed physical model is needed, in order to build-up more precise analytical models for discrete cracks. An improved smeared crack approach may be derived subsequently.

This Reporter would also like at this point to give a very brief presentation of the work effectuated in the R.C. Laboratory of the National Technical University of Athens, on these constitutive laws of post-cracking load transfer mechanisms.

5.1.- Pre-Cracked and Re-Compressed Concrete

A formalistic model (Fig. 5), both for monotonic and cyclic loading, has been developed, based on experimental results mainly.

5.2.- Friction (or Aggregate Interlock) Along a Crack

Both for monotonic and cyclically imposed shear-displacements, several analytical models have been developed (Fig. 6).

5.3.- Pull-Out Forces

Based on specific local-bond vs. local-slip relationships, axial finite elements computer programmes have been developed for given pullout displacements (Fig. 7). Similar data for cyclic pullout/push-in behaviour are available.

5.4.- Dowel action

Analytical models have also been developed for monotonic and cyclic dowel actions (e.g. Fig. 8).

For a more detailed presentation of these models, the following references may be used: Tassios, 1983 (7) and CEB Bulletin 161 (5).
On the basis of these models, the general problem of post-cracking energy absorption along the crack may be fully computerised.
An output of such a numerical treatment is shown in Fig. 9, based on a hypothesised geometry of the crack.

It is worth to remind here that, for prevailing extensional displacements, geometrical assumptions on crack-shape have led to fruitful results in the case of fibre reinforced concrete (Visalvanich, Naaman, 1983 (8)).

6.- CONCLUSIONS

a) It seems that Fracture Mechanics applications have already offered two first services in the field of concrete structures: First, by extending the knowledge in general, offer a better understanding of failure phenomena which are meant to be avoided by an appropriate design; some improvements of resistance formulae are already envisaged. Second, Fracture Mechanics contributes to a rationalisation and a minimisation of experimental investigations needed.
b) In order to be able to take profit of Fracture Mechanics future achievements, an Umbrella-Code may be needed, containing performance requirements, based on a sound reliability format and including dimensioning principles; separate sub-codes should de-

scribe materials' models and levels of simplification appropriate
for each situation. Thus, more sophisticated approaches may "legi-
timately" be used in specific cases.

c) Structural Engineers may have an interest to create by now
an interactive atmosphere with F.M. - specialists, in order to accelerate realistic developments.

REFERENCES outside this Workshop

1. Bazant Z.P., Cedolin L.: "Fracture mechanics of reinforced concrete", J. Eng. Mech. Div., ASCE, Dec. 1980.
2. Bazant Z.P.: "Size effect in blunt fracture-concrete, rock, metal", J. of Eng. Mechanics, ASCE, 1984, pp. 518.
3. Gylltoft K.: "Fracture mechanics models for fatigue in concrete
   structures", Lulea University of Technology, 1983.
4. Reinhardt H.W.: "Similitude of brittle fracture of stuctural
   concrete", Delft Colloquium Advanced Mechanics of R.C., IABSE,
   1981.
5. "Response of R.C. critical regions under large amplitude reversed actions", CEB Bulletin 161, 1983.
6. Tassios T.P.: "Design based on testing", Scientific papers, Faculty Civil Eng., Nat.Tech.University Athens, April-June 1983.
7. Tassios T.P.: "Physical and mathematical models for re-design
   of damaged structures", Introductory Report, IABSE Symposium,
   Venice, 1983.
8. Visalvanich K., Naaman A.: "Fracture model for fiber reinforced
   concrete", Journal ACI, March-April, 1983.

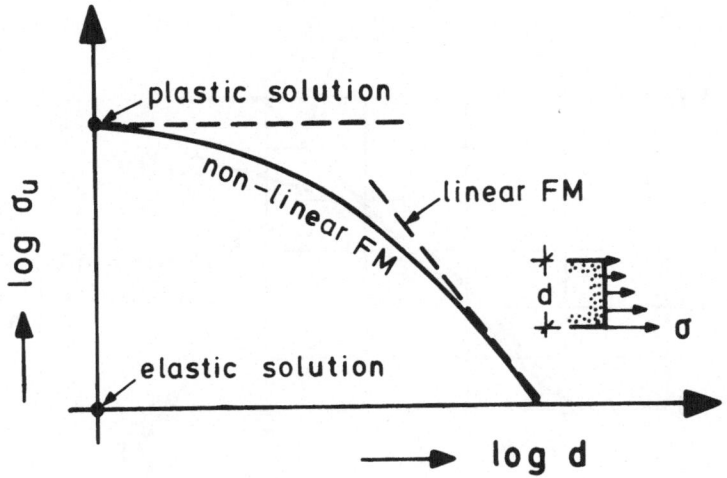

Fig. 1: Size effect ignored by elastic and plastic
solutions; and described by means of Fracture Mechanics

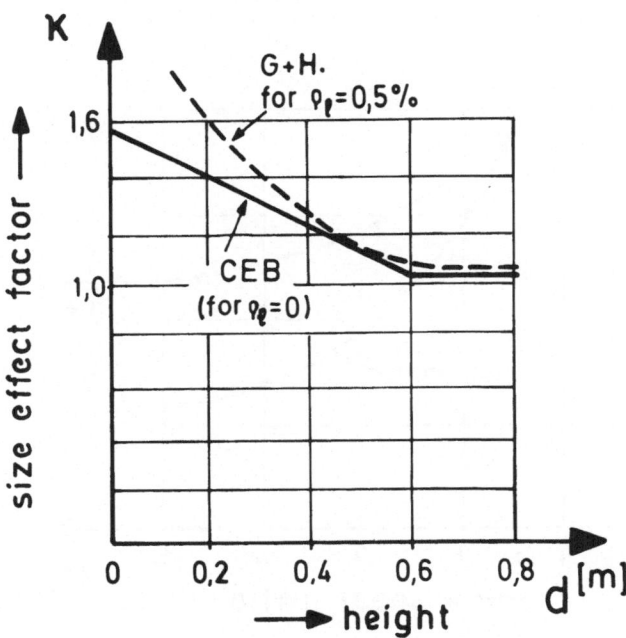

Fig. 2: Size effects for shear strength of
plain concrete by Gustafsson-Hiller-
borg, and by the Model Code of CEB

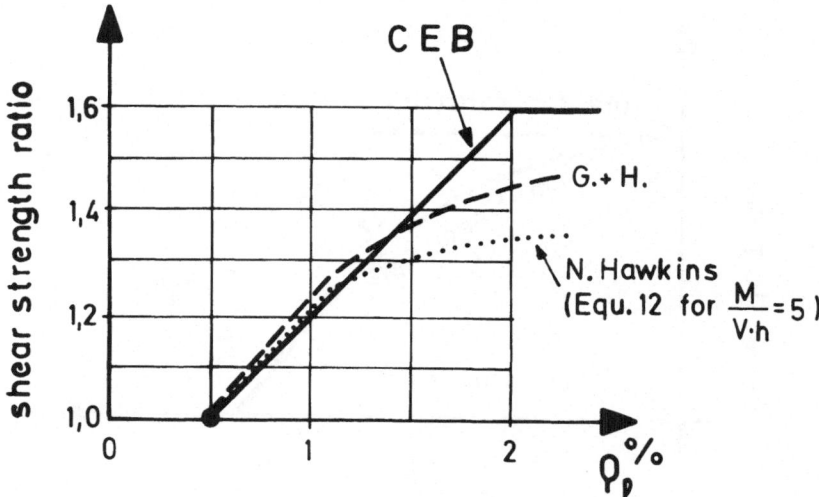

Fig. 3: The role of longitudinal reinforcement ratio ρ_l to the shear strength of concrete beams without web reinforcement

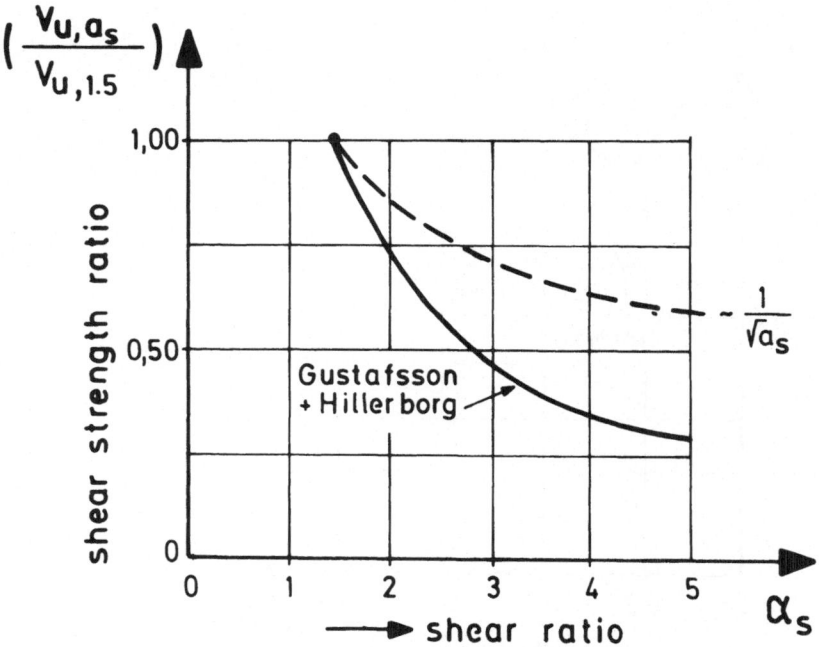

Fig. 4: The role of shear ratio α_s = M:Vh as predicted by F.M. applications

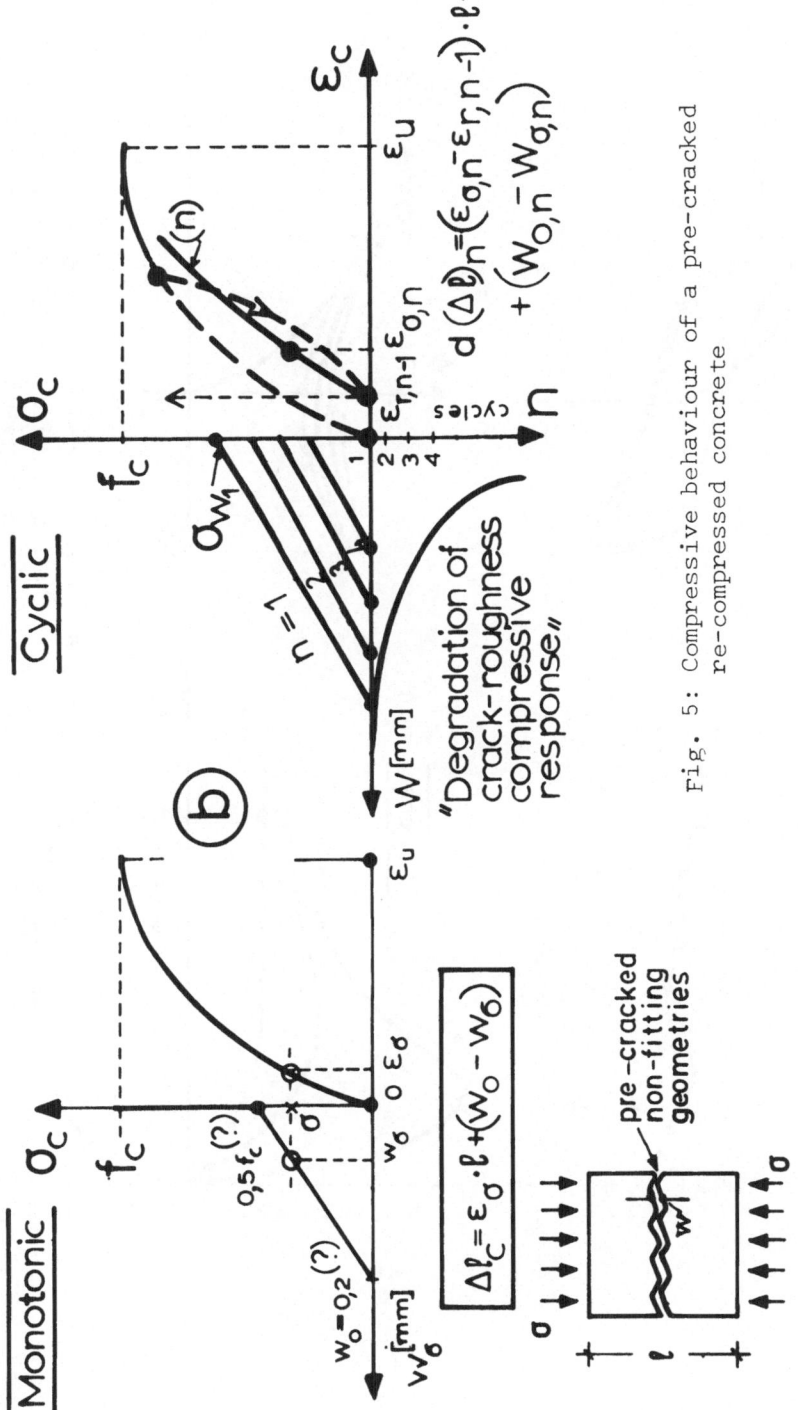

Fig. 5: Compressive behaviour of a pre-cracked re-compressed concrete

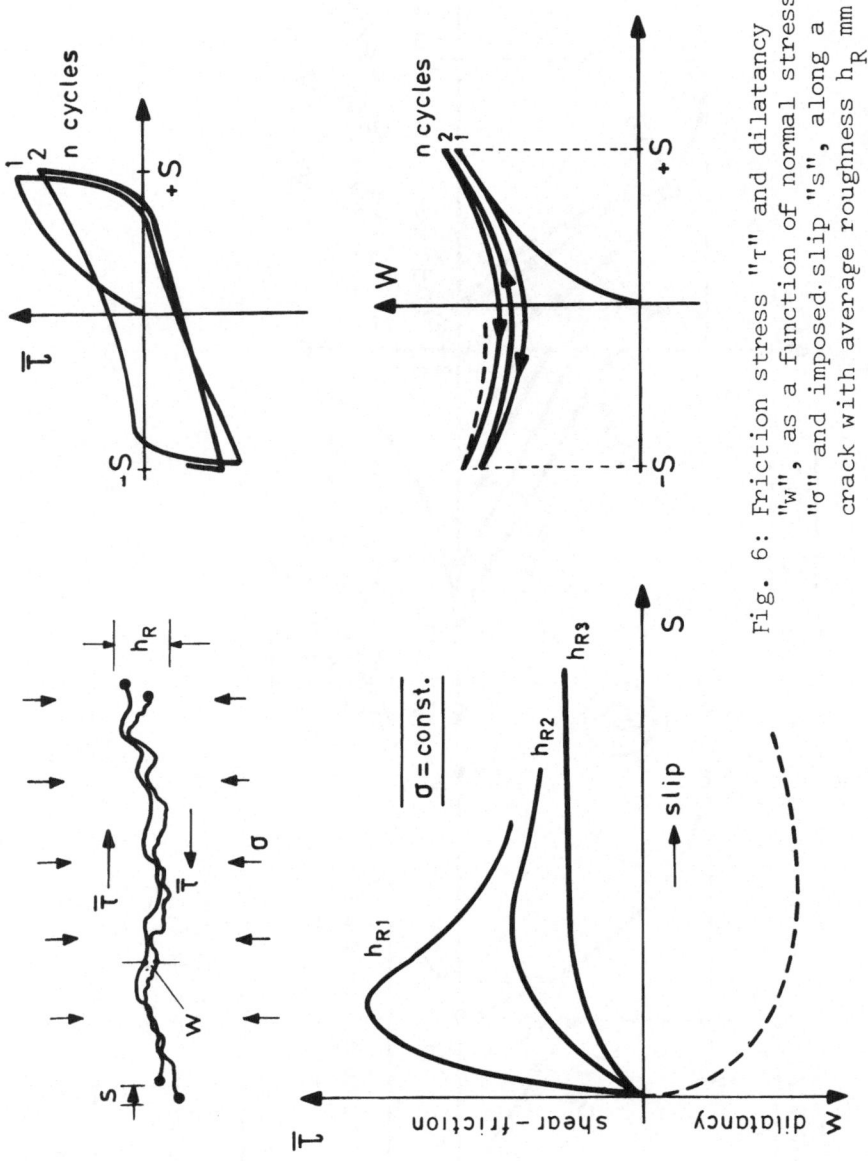

Fig. 6: Friction stress "τ" and dilatancy "w", as a function of normal stress "σ" and imposed slip "s", along a crack with average roughness h_R mm

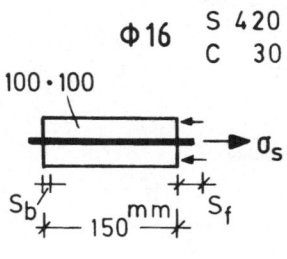

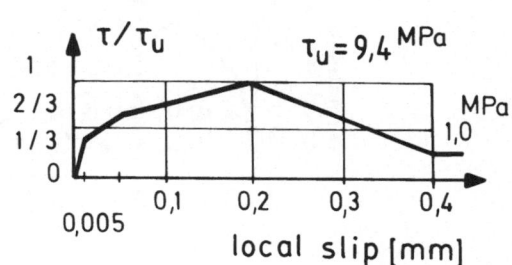

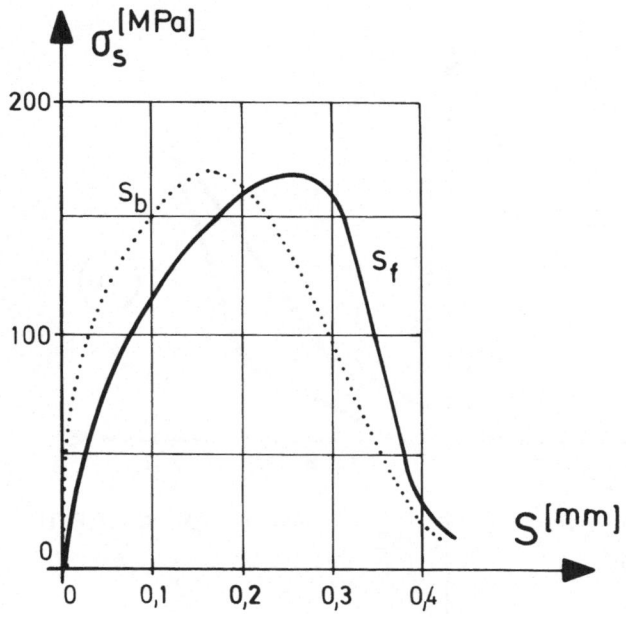

Fig. 7: Computerised predictions of pullout force vs.
pullout displacement relationships

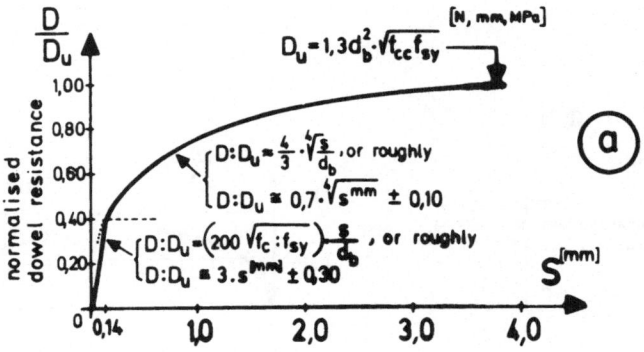

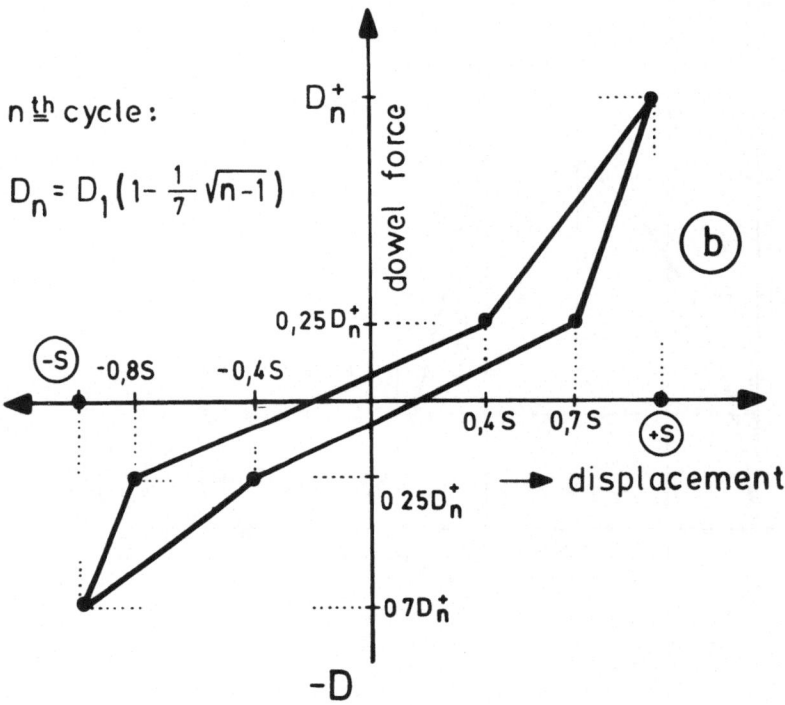

Fig. 8: Analytical model for monotonic and
formalistic model for cyclic dowel
action

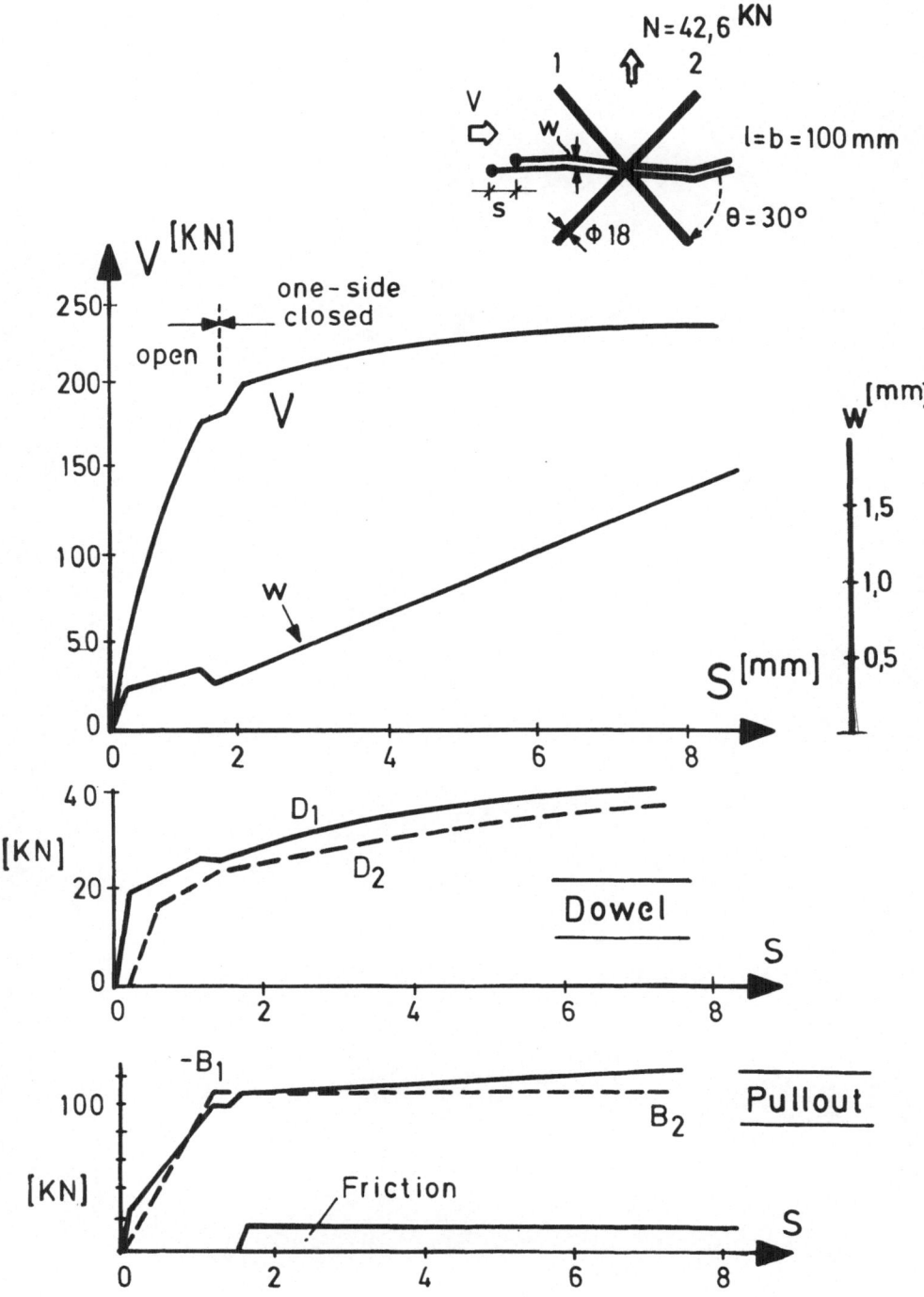

Fig. 9: Post-cracking behaviour: Both shear (s) and dilatancy
(w) displacements are considered taking into account
pullout, dowel and friction forces

LIST OF PARTICIPANTS

Paul Acker, Laboratoire Central des Ponts et Chaussees, Division , Materiaux et Structures, 58 Bd Lefebvre 75732 Paris Cedex 15, France

Gordon Batson, Civil and Environmental Engineering, Clarkson University, Potsdam, New York 13676, U.S.A.

Zdenek P. Bažant, Department of Civil Engineering, Northwestern University, Evanston, Illinois 60201, U.S.A.

James J. Beaudoin, Materials Section, Division of Building Research, National Research Council, Montreal Rd., Ottawa, Canada K1AOR6

Arnon Bentur, Faculty of Civil Engineering, Technion, Israel Institute of Technology, Haifa, Israel

Bruno A. Boley, Dean, The Technological Institute, Northwestern University, Evanston, Illinois 60201, U.S.A.

Boris Bresler, Wiss, Janney, Elstner Associates, Inc., 2200 Powell Street, Suite 925, Emeryville, California 94608, U.S.A.

Oral Buyukozturk, Department of Civil Engineering, Room 1-280, Massachusetts Institute of Technology, Cambridge, Massachusetts 02139, U.S.A.

Robert D. Carnahan, United States Gypsum Company, 101 S. Wacker Drive, Chicago, Illinois 60606, U.S.A.

Alberto Carpinteri, University of Bologna, Istituto di Scienza Delle Costruzioni, Viole Risorgimento 2, 40136 Bologna, Italy

Luigi Cedolin, Dipartimento di Ing. Strutturale, Politecnico di Milano, P. Leonardo da Vinci 32, 20133 Milano, Italy

E. P. Chen, Division 1522, Sandia National Laboratories, Albuquerque, New Mexico 87185, U.S.A.

Ken P. Chong, Department of Civil Engineering, University of Wyoming, Laramie, Wyoming 82071, U.S.A.

John D. Clark, Smith and Nephew, Materials Science, Gilston Park, Harlow, Essex CM20 2RO, U.K.

James Clifton, Leader, Inorganic Materials Group, Building Materials Division, 226/B348, National Bureau of Standards, Wahington, D.C. 20234, U.S.A.

Laurence S. Costin, Div. 1542, Sandia National Laboratories, Albuquerque, New Mexico 87185, U.S.A.

Bruce Cotterman, Metals and Ceramics Lab., AMMRC, Watertown, Massachusetts

David Darwin, Department of Civil Engineering, University of Kansas, Lawrence, Kansas 66045, U.S.A.

Magne Dastol, Department of Civil Engineering, University of British Columbia, Vancouver, Canada

Sidney Diamond, School of Civil Engineering, Purdue University, West Lafayette, Indiana 42906, U.S.A.

Ulrich Diederichs, Institut fur Baustoffe, Massivcbau und Brandschultz, Technische Universitat Braunschweig, Beethovenstr. 52, D 3300 Braunschweig, West Germany

Angelo Di Tommaso, Istituto di Scienza Delle Costruzioni, Fac. Ingegneri, Universita Di Bologna, Viale Risorgimento, 2, I-40134 Bologna, Italy

John W. Dougill, Department of Civil Engineering, Imperial College of Science and Technology, Imperial College Road, London SW7 2BU, U.K.

Lawrence T. Eby, United States Gypsum Company, 700 N.
Highway 45, Libertyville, Illinois 60048, U.S.A.

Manuel Elices, E.T.S. Ingenieros de Caminos, Ciudad
Universitaria s/u, Madrid-3, Spain

Rolf Eligehausen, Institut fur Werkstoffe im Bauwesen,
Universitat Stuttgart, Pfaffenwaldring 4, 7000b
Stuttgart 80, West Germany

Gerardo Ferrara, c/o ENEL-CRIS, Via Ornato 90/14, Milano,
Italy

D. Francois, Directeur du Laboratoire Materiaux, Ecole
Centrale des Arts et Manufactures, Grande Voie des
Vignes, 92290 Chatenay-Malabry, France

Geoffrey Frohnsdorff, National Bureau of Standards, Bldg.
226, Rm. B368, Gaithersburg, Maryland 20899, U.S.A.

Rodney G. Galloway, Civil Engineering Research Division, Air
Force Weapons Laboratory, Kirtland Air Force Base, New
Mexico 87117, U.S.A.

Michael P. Gaus, Director, Structural Mechanics Program,
National Science Foundation, 1800 G Street, N.W.,
Washington, D.C. 20550, U.S.A.

Robert J. Gray, Department of Civil Engineering, University
of British Columbia, 2324 Main Mall, Vancouver, British
Columbia, V6T 1W5, Canada

Per Johan Gustafsson, Division of Building Materials, Lund
Institute of Technology, P. O. Box 725, S-220 07 Lund,
Sweden

Kent Gylltoft, Division of Structural Engineering, Lulea
University of Technology, 951 87 Lulea, Sweden

Grant T. Halvorsen, Department of Civil Engineering, West
Virginia University, P. O. Box 6101, Morgantown, West
Virginia 26506, U.S.A.

Neil M. Hawkins, Department of Civil Engineering, 201 More
    Hall, FX-10, University of Washington, Seattle,
    Washington 98195, U.S.A.

Arne Hillerborg, Division of Building Materials, Lund
    Institute of Technology, Fack 725, S-220 07 Lund, Sweden

Hubert K. Hilsdorf, Institut Massivbau und
    Baustofftchnologie, Universitet Karlsruhe, Kaiserstr.
    12, 75 Karlsruhe, West Germany

Lawrence D. Hokanson, Air Force Office of Scientific
    Research/NA, Bolling Air Force Base, D.C. 20332, U.S.A.

Thomas T. C. Hsu, Department of Civil Engineering,
    University of Houston, University Park, Houston, Texas
    77004, U.S.A.

Anthony R. Ingraffea, Hollister Hall, Cornell University,
    Ithaca, New York 14853, U.S.A.

Mark Kachanov, Mechanical Engineering Department, Tufts
    University, Medford, Massachusetts 02155, U.S.A.

B. L. Karihaloo, Department of Civil Engineering &
    Surveying, The University of Newcastle, N.S.W.,
    Australia

Leon M. Keer, Department of Civil Engineering, Northwestern
    University, Evanston, Illinois 60201, U.S.A.

Lawrence I. Knab, Research Civil Engineer, Inorganic
    Materials Group, Building Materials Division, United
    States Department of Commerce, National Bureau of
    Standards, Washington, D.C. 20234, U.S.A.

Albert S. Kobayashi, Department of Mechanical Engineering,
    FU-10, University of Washington, Seattle, Washington
    98195, U.S.A.

Dusan Krajcinovic, Department of Civil Engineering and
    Mechanics, University of Illinois, Chicago, P. O. Box
    4348, Chicago, Illinois 60680, U.S.A.

J. Le Maitre, Laboratoire de Mecanique de Technologie, ENSET, 61 av de President Wilson, 94230 Cachan, France

Victor C. Li, 1-232 Department of Civil Engineering, Massachusetts Institute of Technology, Cambridge, Massachusetts 02139, U.S.A.

Yiu-Wing Mai, Department of Mechanical Engineering, University of Sydney, Sydney, N.S.W., 2006, Australia

A. J. Majumdar, Department of the Environment, Building Research Establishment, Garston, Waford, Herts, WD2 7JR, U. K.

Lawrence E. Malvern, Engineering Sciences Department, 231 Aero Building, University of Florida, Gainesville, Florida 32611, U.S.A.

Len Margolin, Los Alamos National Laboratory, P. O. Box 1663, Los Alamos, New Mexico 87545, U.S.A.

Jacky Mazars, Laboratoire de Mecanique et Technologie, ENSET/Universite Paris VI/CNRS, 61, av de President Wilson, 94230 Cachan, France

H. Mihashi, Department of Architecture, Faculty of Engineering, Tohoku University, Sendai 980, Japan

Sidney Mindess, Department of Civil Engineering, University of British Columbia, Vancouver, British Columbia, V6T 1W5, Canada

David Mintzer, Vice President of Research and Dean of Science, Northwestern University, Evanston, Illinois 60201, U.S.A.

Antoine E. Naaman, Department of Civil Engineering, The University of Michigan, Ann Arbor, Michigan 48109, U.S.A.

Siavouche Nemat-Nasser, Department of Civil Engineering, Northwestern University, Evanston, Illinois 60201, U.S.A.

C. Dean Norman, Concrete Technology Division, Department of
the Army, Waterways Experiment Station, Corps of
Engineers, P. O. Box 6311, Vicksburg, Mississippi 39180,
U.S.A.

Philip C. Perdikaris, Case Western Reserve University,
Department of Civil Engineering, Cleveland, Ohio 44106,
U.S.A.

Per-Erik Petersson, The National Swedish Institute for
Testing and Metrology, Box 857, S-501 15, Boras, Sweden

Sandor Popovics, Drexel University, Civil Engineering
Department, 32nd & Chestnuts Streets, Philadelphia,
Pennsylvania 19104, U.S.A.

Hans W. Reinhardt, Delft University of Technology,
Department of Civil Engineering, P. O. Box 5048, NL-2600
GA, Delft, The Netherlands

John W. Rudnicki, Department of Civil Engineering,
Northwestern University, Evanston, Illinois 60201,
U.S.A.

Ulrich Schneider, Universitat Kassel, FB14,
Monchebergstrasse 7, D-3500 Kassel, West Germany

Lynn Seaman, 333 Ravenswood Avenue, SRI International, Menlo
Park, California 94025, U.S.A.

Surendra P. Shah, Department of Civil Engineering,
Northwestern University, Evanston, Illinois 60201,
U.S.A.

Robert Sierakowski, Department of Civil Engineering, Ohio
State University, Columbus, Ohio 43210, U.S.A.

George C. Sih, Lehigh University, Institute of Fracture &
Solid Mechanics, Packard Lab. #9, Bethlehem,
Pennsylvania 18015, U.S.A.

Floyd O. Slate, 363 Hollister Hall, Cornell University,
    Ithaca, New York, 14853, U.S.A.

Henrik Stang, Department of Structural Engineering,
    Technical University of Denmark, Bygning 118, DK-2800
    Lyngby, Denmark

Raymond B. Stout, Lawrence Livermore National Laboratory,
    L200, P. O. Box 808, Livermore, California 94550, U.S.A.

Wimal Suaris, Department of Civil Engineering, University of
    Miami, Coral Gables, Florida 33124, U.S.A.

J. M. Summerfield, United States Gypsum Company, 700 N.
    Highway 45, Libertyville, Illinois 60048, U.S.A.

Stuart E. Swartz, Department of Civil Engineering, Kansas
    State University, Manhattan, Kansas 66506, U.S.A.

R. N. Swamy, Department of Civil and Structural Engineering,
    University of Sheffield, Mappin Street, Sheffield S1
    3JD, U.K.

T. P. Tassios, Laboratory of Reinforced Concrete, Athens
    National Technical University, 1 Epirou Street, Athens
    103, Greece

P. S. Theocaris, Department of Theoretical & Applied
    Mechanics, 5 Zographou, Athens National Technical
    University, Athens (625), Greece

Methi Wecharatana, Department of Civil and Environmental
    Engineering New Jersey Institute of Technology, 323 High
    Street, Newark, New Jersey 07102, U.S.A.

Kaspar Willam, Department of Civil Engineering, Box 428,
    University of Colorado, Boulder, Colorado 80309, U.S.A.

Folker H. Wittmann, Ecole Polytechnique Federale de
    Lausanne, Department des Materiaux, Laboratoire des
    Materiaux de Construction, 32, ch. de. Bellerive, CH-
    1007 Lausanne, Switzerland

702

Michael P. Wnuk, College of Engineering and Mathematical
    Sciences, University of Wisconsin-Milwaukee, Milwaukee,
    Wisconsin.

Chien H. Wu, Department of Engineering, Mechanics &
    Metallurgy, University of Illinois, Chicago, Box 4348,
    Chicago, Illinois 60680, U.S.A.

J. Francis Young, Department of Civil Engineering,
    University of Illinois, 208 N. Romine Street, Urbana,
    Illinois 61801, U.S.A.

A. J. Zielinski, Department of Civil Engineering, Delft
    University of Technology, Delft, 1 Stevinweg, P. O. Box
    5048, 1600 GA Delft, The Netherlands

AUTHOR INDEX

Numbers following a name indicate pages where an author is mentioned in this volume. If underlined, they refer to a contribution to this volume.

SUBJECT INDEX

Numbers following a key-word indicate pages where this subject is mentioned in this volume.